PHOTON AND PARTICLE INTERACTIONS WITH SURFACES IN SPACE

ASTROPHYSICS AND
SPACE SCIENCE LIBRARY

A SERIES OF BOOKS ON THE RECENT DEVELOPMENTS

OF SPACE SCIENCE AND OF GENERAL GEOPHYSICS AND ASTROPHYSICS

PUBLISHED IN CONNECTION WITH THE JOURNAL

SPACE SCIENCE REVIEWS

VOLUME 37

PHOTON AND PARTICLE
INTERACTIONS
WITH SURFACES IN SPACE

PROCEEDINGS OF THE 6TH ESLAB SYMPOSIUM,
HELD AT NOORDWIJK, THE NETHERLANDS,
26–29 SEPTEMBER, 1972

Edited by

R. J. L. GRARD

Space Science Department (ESLAB),
European Space Research and Technology Centre,
Noordwijk, The Netherlands

D. REIDEL PUBLISHING COMPANY

DORDRECHT-HOLLAND/BOSTON-U.S.A.

Library of Congress Catalog Card Number 73-83561

ISBN-13:978-94-010-2649-9   e-ISBN-13:978-94-010-2647-5
DOI: 10.1007/978-94-010-2647-5

Published by D. Reidel Publishing Company,
P.O. Box 17, Dordrecht, Holland

Sold and distributed in the U.S.A., Canada, and Mexico
by D. Reidel Publishing Company, Inc.
306 Dartmouth Street, Boston,
Mass. 02116, U.S.A.

TABLE OF CONTENTS*

1. INTRODUCTORY LECTURES

2. INTERACTIONS WITH SPACECRAFT

2.1. *Theoretical Sheath Models*

2.2. *Numerical Analysis and Simulation*

2.3. *Influence of Surface Emission on Experimental Measurements*

* Contributions presented by first author, unless specified otherwise.

PREFACE

The 6th ESLAB Symposium, organised by the Space Science Department (formerly ESLAB) of the European Space Research and Technology Center, was held in Noordwijk from 26–29 September 1972. This year the theme was "Photon and Particle Interactions with Surfaces in Space". More than 60 scientists attended mainly from ESRO Member States and from America.

The first part of the Symposium was devoted to introductory lectures and to papers on interactions with spacecraft. The second half dealt with the photon and particle interactions with celestial objects, and ended with a general discussion and presentations of areas where new developments are required.

The purpose of this Symposium was to throw light on the importance of the problems which are evoked by E. A. Trendelenburg in his introductory remarks, and to sum up our present understanding of these phenomena. It is hoped that this book will prove useful to physicists and engineers who are actually involved in space experiments and are concerned with interactions of these types.

R. J. L. GRARD

OPENING ADDRESS

Gentlemen,

I should like to welcome you to the 6th ESLAB Symposium.

In the past we have always organised this Symposium jointly with our sister institute, ESRIN, in Frascati, but unfortunately reductions in the scientific budget have forced ESRO to terminate the activities of that laboratory. Nevertheless, we have decided to carry on the tradition, and we shall continue on our own organising this series of symposia on specialised subjects.

The reason for choosing this year's particular topic is almost historical. Several years ago, when I was appointed Director of ESLAB by the ESRO Council, I thought that we should devote some effort to Surface Physics. I felt, at that time, that progress in this domain would eventually benefit other fields of space research. Indeed, the importance of Surface Physics in space has outgrown my initial expectations, and today I find it extremely rewarding that the first symposium entirely devoted to this problem is being presented to such a distinguished audience.

Presently, spacecraft are venturing further and further away from the Earth, probing the tenuous medium of the magnetosphere and interplanetary space. More than ever we need to understand the influence of photon and particle fluxes on the electric properties of the surface of a space probe, and to evaluate the extent to which the output of a scientific experiment can be modified by such phenomena. Similar processes also occur at the surface of celestial bodies and may give the clue to phenomena which have been observed, but remain so far unexplained.

It seems to me very promising that physicists coming from different horizons of science – theoreticians, laboratory and space experimenters, as well as lunar specialists – have decided to gather and to compound their knowledge in the very specialised field which deals with the interactions between a surface and its environment in space. It can be expected that the outcome of the discussions between people having such different backgrounds will help to clarify and improve our present understanding of these important problems.

I hope that you will find this meeting useful and profitable, and I sincerely wish you a pleasant stay here in Noordwijk.

I now declare the 6th ESLAB Symposium open.

26 September 1972   E. A. TRENDELENBURG

LIST OF PARTICIPANTS

Anderegg, M., Space Science Department, ESTEC, Domeinweg, Noordwijk, Holland

Andresen, R. D., Space Science Department, ESTEC, Domeinweg, Noordwijk, Holland

Arens, I., Space Science Department, ESTEC, Domeinweg, Noordwijk, Holland

Atzei, A., Energy Conversion Division, ESTEC, Domeinweg, Noordwijk, Holland

Boeckel, J. J. van, Space Science Department, ESTEC, Domeinweg, Noordwijk, Holland

Bujor, M., Groupe de Recherches Ionosphériques, 4, ave. de Neptune, 94, Saint-Maur-des-Fossés, France

Cauffman, D. P., Bldg. 120, Room 1405, The Aerospace Corporation, P.O. Box 92957, Los Angeles, Calif. 90009, U.S.A.

Criswell, D. R., Lunar Science Institute, 3303 NASA Road 1, Houston, Tex. 77058, U.S.A.

Dauphin, J., Reliability Division, ESTEC, Domeinweg, Noordwijk, Holland

DeForest, S. E., Physics Department University of California, San Diego, La Jolla, Calif. 92037, U.S.A.

Domingo, V., Space Science Department, ESTEC, Domeinweg, Noordwijk, Holland

Durney, A. C., Space Science Department, ESTEC, Domeinweg, Noordwijk, Holland

Fahleson, U. V., Department of Plasma Physics, Royal Institute of Technology, S-100 44 Stockholm 70, Sweden

Feuerbacher, B., Space Science Department, ESTEC, Domeinweg, Noordwijk, Holland

Fitton, B., Space Science Department, ESTEC, Domeinweg, Noordwijk, Holland

Fredricks, R. W., TRW Systems Group, Space Sciences Department, One Space Park, Building R-1, Room 1070, Redondo Beach, Calif. 90278, U.S.A.

Freeman, J. W., Jr., Department of Space Science, Rice University, Houston, Tex. 77001, U.S.A.

Gold, T., Space Sciences Building, Cornell University, Ithaca, N.Y. 14850, U.S.A.

Gonfalone, A., Space Science Department, ESTEC, Domeinweg, Noordwijk, Holland

Grard, R., Space Science Department, ESTEC, Domeinweg, Noordwijk, Holland

Haskell, G. P., ESRO, 114, ave. Charles de Gaulle, 92, Neuilly-sur-Seine, France

Isensee, U., Lehrstuhl B für Theoretische Physik, Technische Universität Braunschweig, 33 Braunschweig, Mendelssohnstrasse 1A, Germany

Jones, D., Space Science Department, ESTEC, Domeinweg, Noordwijk, Holland

Kaiser, T. R., Radioastronomy Group, Physics Department, The Hicks Building, The University, Sheffield S3 7RH, England

Kalweit, C. C., GEOS Division, ESTEC, Domeinweg, Noordwijk, Holland

Knott, K., Space Science Department, ESTEC, Domeinweg, Noordwijk, Holland

Köhn, D., Space Science Department, ESTEC, Domeinweg, Noordwijk, Holland

Köneman, B., Lehrstuhl B für Theoretische Physik, Technische Universität Braunschweig, 33 Braunschweig, Mendelssohnstrasse 1A, Germany

Kopp, E., Space Science Department, ESTEC, Domeinweg, Noordwijk, Holland

Köstlin, H., Philips Forschungslaboratorium, 5100 Aachen, Weisshausstrasse, Germany

Laude, L., Space Science Department, ESTEC, Domeinweg, Noordwijk, Holland

Lucas, A., Space Science Department, Domeinweg, Noordwijk, Holland

Manka, R. H., Department of Space Science, Rice University, Houston, Tex. 77001, U.S.A.

Meiner, R. C., Space Science Department, ESTEC, Domeinweg, Noordwijk, Holland

Montgomery, M. D., Max-Planck-Institut für Physik und Astrophysik, Institut für Extraterrestrische Physik, 8046 Garching b. München, Germany

Norman, K., Mullard Space Science Laboratory, Holmbury St. Mary, Dorking, Surrey, England

Page, D. E., Space Science Department, ESTEC, Domeinweg, Noordwijk, Holland

Pedersen, A., Space Science Department, ESTEC, Domeinweg, Noordwijk, Holland

Petit, M., CNET – RSR, 38, ave. du Général Leclerc, 92, Issy-les-Moulineaux, France

Polychronopulos, B., Space Research Department, University of Birmingham, P.O. Box 363, Birmingham B15 2TT, England

Ransome, T., British Aircraft Corporation, Electronic and Space Systems Group, Filton House, Filton, Bristol, England

Reasoner, D. L., Department of Space Science, Rice University, Houston, Texas 77001, U.S.A.

Rosenbauer, H., Max-Planck-Institut für Physik & Astrophysik, Institut für Extraterrestrische Physik, 8046 Garching b. München, Germany

Samir, U., Space Physics Research Building, University of Michigan, Ann Arbor, Mich. 48105, U.S.A., and Department of Environmental Sciences, Tel-Aviv University, Tel-Aviv, Israel.

Sanderson, T. R., Space Science Department, ESTEC, Domeinweg, Noordwijk, Holland

Schröder, H., Lehrstuhl B für Theoretische Physik, Technische Universität Braunschweig, 33 Braunschweig, Mendelssohnstrasse 1A, Germany

Shawhan, S. D., Department of Physics and Astronomy, The University of Iowa, Iowa City, Iowa 52240, U.S.A.

Siscoe, G. L., Department of Meteorology, University of California, Los Angeles, Calif. 90024, U.S.A.

Smith, A. D., GEOS Division, ESTEC, Domeinweg, Noordwijk, Holland

Soop, M., European Space Operations Centre, Robert-Bosch-Strasse 5, 61 Darmstadt, Germany

Srnka, L. J., Culham Laboratory, Room D3-103, UKAEA Research Group, Abingdon, Berkshire, England

Taylor, B., Space Science Department, ESTEC, Domeinweg, Noordwijk, Holland

Thomas, J. O., Imperial College of Science and Technology, Department of Physics, Prince Consort Road, London S.W. 7, England

Trendelenburg, E. A., Space Science Department, ESTEC, Domeinweg, Noordwijk, Holland

Voigt, G.-H., Lehrstuhl B für Theoretische Physik, Technische Universität, 33 Braunschweig, Mendelssohnstrasse 1A, Germany

Walker, E. H., Ballistic Research Laboratories. U.S. Army Aberdeen Research and Development Center, Aberdeen Proving Ground, Md. 21005, U.S.A.

Weil, H., Space Physics Research Laboratory, University of Michigan, Ann Arbor, Mich. 48105, U.S.A.

Wenzel, K. P., Space Science Department, ESTEC Domeinweg, Noordwijk, Holland

Wickramasinghe, N. C., Institute of Theoretical Astronomy, Madingly Road, Cambridge CB3 OEZ, England

Wiesemann, K., Fachbereich Physik der Universität Marburg, Renthof 5, D-355 Marburg/Lahn, Germany

Willis, R. F., Space Science Department, ESTEC, Domeinweg, Noordwijk, Holland

Wills, R. D., Space Science Department, ESTEC, Domeinweg, Noordwijk, Holland

Wrenn, G. L., Mullard Space Science Laboratory, Holmbury St. Mary, Dorking, Surrey, England

Wu, S. T., University of Alabama, P.O. Box 1247, Huntsville, Ala. 35807, U.S.A.

Wynn-Roberts, D., British Aircraft Corporation, Electronic and Space Systems Group, Filton House, Filton, Bristol, England

Young, D. T., Physikalisches Institut, Universität Bern, Sidlerstrasse 5, 3018 Bern, Switzerland

1. INTRODUCTORY LECTURES

FUNDAMENTAL PROCESSES IN PARTICLE AND PHOTON INTERACTIONS WITH SURFACES

A. A. LUCAS*

*Surface Physics Division, European Space Research and Technology Centre,
Noordwijk, The Netherlands*

Abstract. The basic physical principles of several experimental techniques to study photon and particle interaction with surfaces will be reviewed. Photoemission, electron and ion scattering methods are briefly discussed. A few examples of the results of such studies which are directly relevant to space science and technology are also presented.

1. Introduction

The large variety of experimental techniques currently used for studying the interaction of particles with surfaces can conveniently be classified according to the type of excitation source and particle spectrometer used for the experiment. Following such a source/spectrometer classification, a partial list of techniques is given in Table I.

TABLE I

A partial list of currently used experimental techniques to study 'particle' scattering by surfaces, classified according to the source and detected 'particle'

Source spectr.	Photon	Electron	Ion	Fields
Photon	Optical properties	Transition rad. Cerenkov rad. Bremsstrahlung	Radiative neutralisation	
Electron	Photoemission	LEED ILEED RHEED SEE	Auger neutral. Kinetic SEE, INS	Field emiss. Stark ionis.
Ion	Laser heating		Sputtering, channeling rad. damage	Field ion emiss. Field desorpt.

As can be seen from this table the spectroscopic methods fall naturally into two broad classes according to whether photons are primarily involved or not. In the first class involving radiative phenomena, photons are absorbed or emitted as a result of classical or quantum processes, while in the second class the particle emission is essentially a nonradiative process, depending mainly on the quasi-static Coulomb interaction between the incoming particle and bare or screened charges in the solid.

Another broad feature of the classification in Table I is the fact that on going from left to right, i.e. when the mass of the source particle increases, the scattering events occur closer to the surface of the target as a result of the general increase in the atomic

___

* Chercheur Qualifié FNRS. Present address: Institut de Physique, Université de Liège, Sart Tilman, B4000 Liège, Belgium.

R. J. L. Grard (ed.), Photon and Particle Interactions with Surfaces in Space, 3–21. All Rights Reserved

scattering factor. Thus, broadly speaking, techniques of ion scattering (see below) are particularly useful for surface studies while photon scattering methods (optical properties, photoemission...) yield data which are more characteristic of the target bulk properties.

Consider a general scattering experiment, schematized in Figure 1, in which an excitation source provides 'particles' (photons, charged or neutral particles) of known energy and angular distributions. This can be either the measured incoming flux of particles impinging on the surface of a satellite in its real space environment or the controlled source of a simulation experiment in the laboratory. The main objectives of the scattering experiment is to measure the yield Y of outgoing 'particles' and,

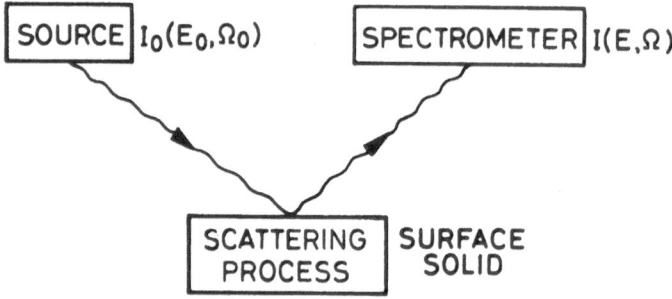

Fig. 1.   Schematic representation of a general scattering experiment.

when the spectrometer is adequate, the energy and angular distributions $I(E, \Omega)$ of scattered particles. These two quantities which are related by a relation of the type

$$Y = \int dE \int d\Omega \, I(E, \Omega) \tag{1}$$

are of primary interest for the determination of such gross properties as the plasma sheath structure, surface potential, surface erosion, etc. of space objects, as will be amply documented in this symposium.

The differential yield $I(E, \Omega)$ contains more fundamental informations concerning the spectrum of elementary excitations of the solid and its surface, information which may eventually be retrieved by using suitable models for the scattering processes and by application of the laws of conservation of energy, momentum, angular momentum, etc.... A few elementary excitations characteristic of the solid state (Pines, 1964) are listed in Table II.

Although photons (i.e. the transverse components of the electromagnetic fields) are always involved to some degree in any time dependent distribution of charges or magnetic moments, the solid state excitations may again be separated according to whether the 'photon content' of the excitation is negligible or not for the experiment under consideration. The nonradiative excitations may further be subdivided into individual or collective types. For individual excitations, large disturbances of the

TABLE II

A few elementary excitations characteristic of the solid state. The word 'polariton' designates any admixture of photons with nonradiative excitations

Nonradiative excitations	Acoustical phonons Optical phonons Plasmons Magnons	Collective
	Intraband trans. Interband trans.	Individual
Radiative excitations	'Polaritons'	

physical parameters (position, momentum, energy ...) of a single real particle (electron, ion, spin, etc. ...) occur while collective modes represent small disturbances of the parameter of a large number of particles. The names given to the excitation are meant to refer to the quantum of energy necessary to excite the system out of its ground state and this quantum itself is often designated as a 'quasiparticle'. For instance a plasmon of wave vector $\mathbf{k}$ in a metal represents a totally delocalized disturbance (involving all the valence electrons of the metal) under which an electron density wave propagates in the medium with a frequency $\omega_k = 10^{16} \, \text{s}^{-1}$ as a result of the Coulomb repulsion between the electrons. The energy quantum (or plasmon quasi-particle) in this case is very large amounting to $\hbar\omega_k = 10 \, \text{eV}$ and is the minimum energy to be spent on the system to destroy the state of average uniform electron density. Ion density waves (or lattice vibrations) and spin waves on the other hand, have rather small elementary quanta $\hbar\omega_k \lesssim 5 \times 10^{-2} \, \text{eV}$ known as phonon and magnon respectively, and in many cases they can be treated as classical oscillations (Kittel, 1971).

As will be illustrated in this and several other talks, the detailed knowledge of the solid state excitation spectrum gained in such scattering experiments, often results in a better understanding and hence a better control of macroscopic material properties of great practical interest. In this introductory paper, we will attempt to describe the very basic physical principles of a few important scattering techniques which are providing us with much of the microscopic data needed for interpretation of large scale behaviour of space objects.

2. Photoelectric Emission

A large number of space science experiments or observations require for their understanding a good knowledge of the differential yield of photo-emission, namely the number of emitted electrons, and their energy and angular distributions, that each photon is capable of ejecting out of the material under illumination. Plasma sheath characteristics around space objects, interstellar grain charging, surface potential of

Moon material are a few examples which will be considered in great detail in this conference.

Going beyond what Einstein (1905) has offered nearly 70 years ago to understand photoemission has proved a very difficult task. Indeed, a detailed account of the process would involve a large fraction of all theoretical solid state knowledge accumulated since the creation of quantum mechanics. In the last ten years however a broad, schematic and relatively simple picture has emerged, mainly under the great experimental incentive of W. Spicer and his school (Berglund and Spicer, 1964). The overall photoemission act is described as a three-step process, each step furnishing to the differential yield a separate convolution factor according to a relation of the type

$$I(E, \Omega) = A [\sigma_v(E - h\nu) \, \sigma_c(E)] \cdot T(E, \Omega) \cdot E_s(E, \Omega). \tag{2}$$

A is the cross section for absorption of the photon of energy $h\nu$ by individual excitation, in the bulk of the material, of an electron initially lying in some valence band state of energy $E-h\nu$, towards an empty conduction state of final energy E. σ_v and σ_c are the density of initial and final states which have the appropriate wave vector to make them candidate to carry an electron in the right direction Ω. The absorption factor A depends on both these quantities in a complicated manner and on optical selection rules. In practically all interpretations of experimental yield curve today, the simplifying assumption has been made that A should be proportional to the so-called joint density of states which is essentially the simple product $\sigma_v(E-h\nu)\sigma_c(E)$ obtained by overlapping the conduction density of state curve with the valence density of states 'lifted' by $h\nu$, as shown in Figure 2. The reason for not performing a detailed calculation of this absorption factor is merely that the required theoretical band structure data are often not available or, when they are, the computational work is prohibitively large. One should notice however that this approximation is basically justified by the circumstance that the absorption is much dominated by the so-called critical points in the band structure, i.e. locations in **k** space where the density of states reaches large values relative to general **k**-points in the Brillouin zone. Thus in a typical photoemission study, one will attempt to correlate the prominent features of the energy distribution of the yield with those **k**-points of a calculated band structure where there is larger joint density of states (Willis *et al.*, 1971).

Next, the factor T represents the so-called transport step of the photoemission act, i.e. the probability that an electron excited somewhere in the material effectively reaches the surface with energy E and appropriate direction Ω. In this convolutional factor are imbedded all scattering phenomena responsible for the finite lifetime of the excited state into which the electron has been lifted in the absorption process. It is often approximated by a function exponentially decaying away from the emission surface with a characteristic decay length given by some mean free path with respect to all sorts of scattering (electron-electron, electron-phonon, electron-impurity, etc. ...). The major influence of the transport step on $I(E, \Omega)$ is to smear out all sharp features which the selection rules and critical points would otherwise produce and to generate a low energy peak contributed by all those electrons which have been multiply scattered inelastically downwards.

Finally, the last factor in (2) represents the third step of the emission act by which the electron travels through the emitting surface into vacuum. Here again, the emission or 'escape' function E_s embodies a process of formidable complexity, namely the detailed quantum mechanical scattering of the electron by the two-dimensional dis-

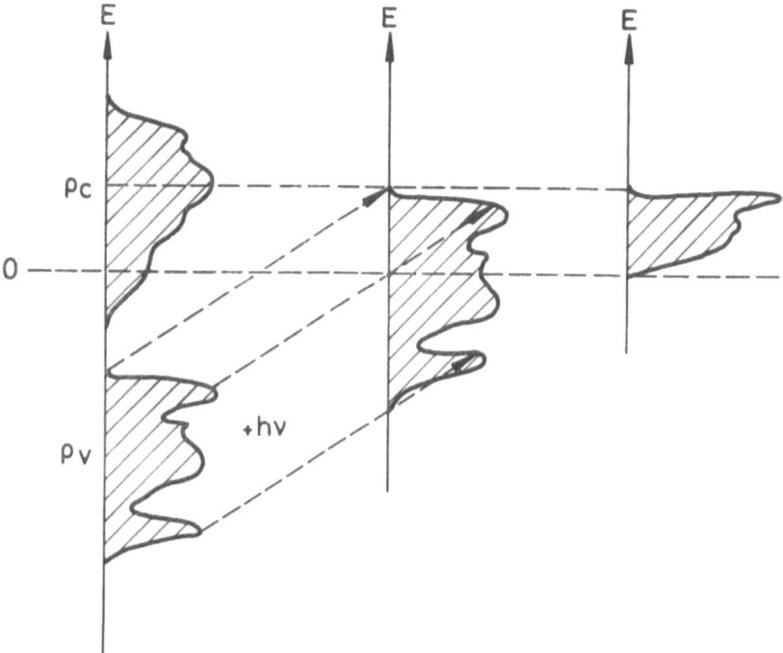

Fig. 2.   Qualitative representation for the generation of the final electron energy distribution in photoelectric emission.

tribution of surface atoms. Until very recently*, this last step has been treated in an extremely crude but pragmatic fashion, simply by allowing all electrons with excited energy E above the Einstein threshold (i.e. above the vacuum zero kinetic energy level) to escape into vacuum. In reality, some of the electrons candidate to escape will be returned to the crystal interior by specular or Bragg reflection or will be diffracted on the surface atomic lattice. The next section will summarize the main characteristics of this scattering process in the inverse but similar situation of LEED where the electron impinges on the surface from the vacuum side.

To close this section, one should remark that the three-step description of photoemission is more a methodological systematisation of a complicated phenomenon than a faithful account of some chronological order in which real processes would

* More rigorous but entirely formal theories of photoemission have recently been advanced (Mahan, 1970).

take place. An obvious case where such a description completely breaks down is photoemission from the so-called 'surface-states', i.e. electronic energy states whose wavefunction is confined to some close neighbourhood of the emitting surface (contrary to ordinary Bloch states which extend over the entire crystal). Considerable attention is currently being focussed on this case in several laboratories (Feuerbacher and Fitton, 1972) owing particularly to the surprisingly large contribution of the surface states to the photoyield in certain materials.

3. Electron Scattering Techniques

In this section we shall state the main physical ideas underlying a few of the large number of electron scattering techniques to investigate surface properties, particularly those which are currently in use in the Surface Physics Division of ESTEC.

First, in this extremely active field, a list of commonly accepted abbreviations may be in order; the references give basic texts or reviews on the subject:

LEED: low energy electron diffraction (Davisson and Germer, 1927; Lander, 1965)
ILEED: inelastic LEED (Duke and Laramore, 1971)
HEED: high energy electron diffraction (Hirsh et al., 1965; Bauer, 1969)
RHEED: reflection HEED (Hirsch et al., 1965)
SEE: secondary electron emission (Dekker, 1967).

The geometries of the LEED, HEED and RHEED experiments are sketched in Figure 3. The main objective of LEED and RHEED is to determine the surface crystallographic arrangement of atoms by exploiting the quantum mechanical scattering of electron waves by the surface atoms. The fundamental principle is Bragg scattering of electron waves by the atomic 'grating' and constitute the two-dimensional equivalent of X-ray scattering by bulk crystals. The scattering power of atoms for low

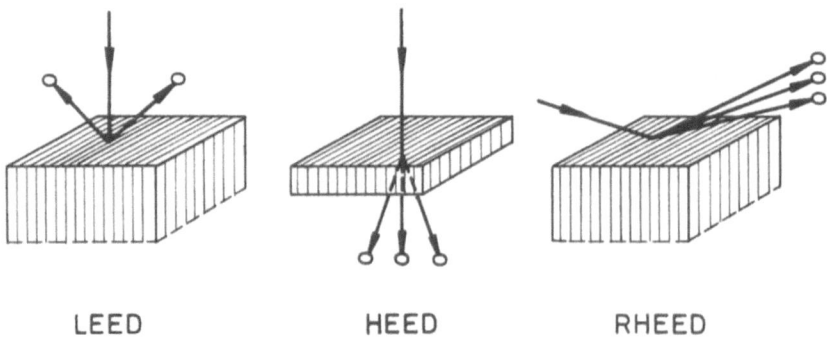

LEED   HEED   RHEED

Fig. 3. Basic geometries of (from left to right) low energy electron diffraction (LEED), high energy electron diffraction (HEED) and reflection HEED. In both HEED and RHEED the scattering angles are very small.

energy electrons $(0 < E < 10^3 \text{ eV})$ is orders of magnitude larger than for X-rays, which allows for substantial backscattering intensities in LEED and RHEED and also implies that the backscattered currents have only probed the outermost atomic layers of the target material.

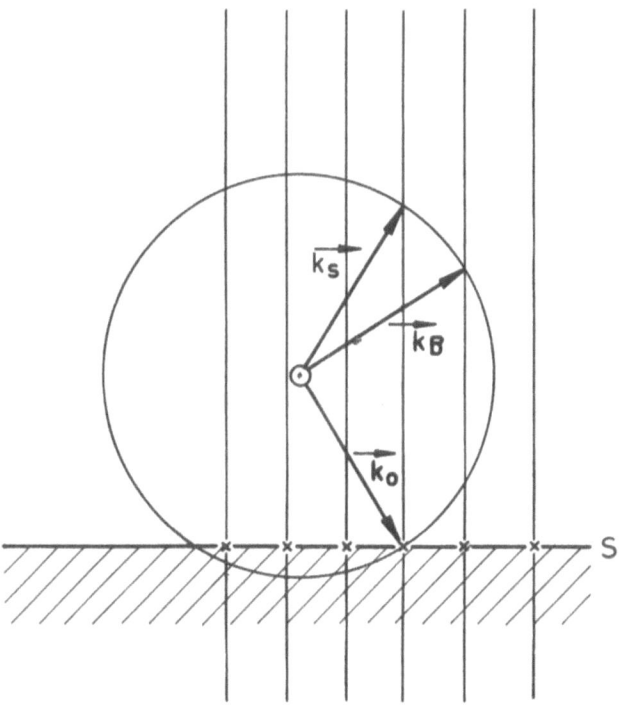

Fig. 4.   Ewald sphere construction in LEED. The primary electron wave-vector is k_0. The vertical lines represent reciprocal lattice 'rods', i.e. rods perpendicular to the scattering surface and passing through the reciprocal points of the two-dimensional surface lattice. The intersections of the rods with the Ewald sphere of radius $|k_0|$ give the direction of reflected and transmitted beams.

There are essentially three basic observations in a LEED (or RHEED) experiment.

(1) *The geometrical pattern* produced e.g. on a fluorescent screen by backscattered electrons. This may be a 'clean' spot pattern, a ring pattern or a diffuse distribution. Since the process of diffraction and interference of waves produces a spatial distribution of diffracted intensities giving the Fourier transform of the spatial arrangement of scattering centers, the above three distributions correspond respectively to a perfectly oriented atomic surface, a mosaïc-like surface made of randomly oriented microcrystal faces and a completely amorphous surface. Of course, in practice, anything intermediate between these idealizations can occur. This may be compared to an X-ray pattern from a monocrystal, a powder of microcrystals or a completely amorphous medium (e.g. a gas). The essential *kinematic* of the formation of a LEED spot pattern is represented in Figure 4 showing a geometrical construction analogous to the Ewald-sphere of the X-ray method.

(2) *The so-called I − V curve* which is the intensity of elastically backscattered electrons in a given spot (either specular or Bragg spot) as a function of primary electron energy (or, equivalently, accelerating gun voltage V). The modulation of backscattered intensity as a function of electron energy, i.e. when the de Broglie wavelength is varied, is produced by interferences between wave amplitudes scattered by successive layers of atoms beneath the surface, in a manner similar to the variation of light reflectivity of a thin plate as a function of wavelength. Thus the measurement of $I − V$ characteristics of surfaces provides information concerning the crystallographic arrangement of atomic layers along the surface normal and also, as can be shown, the arrangement of atoms within the unit cell of the surface Bravais lattice. The observation of the diffraction pattern and the $I − V$ curves are therefore complementary information for the determination of the overall surface structure.

(3) *The energy and angular distributions* of the backscattered currents, i.e. the differential yield $I(E, \Omega)$ defined before. Figure 5 shows a schematic energy distribution of the specular spot with its conventionally separated regions of 'elastic' primaries, inelastic primaries and 'true' secondaries. The ILEED technique essentially concentrates on a study of the first two regions made of electrons which, in addition to having been elastically reflected by the atomic cores, may have lost a small or large amount of energy by creation of one or several of the elementary excitations listed in Table II. Since the inelastic portion of the spectrum is substracted out of the elastic peak (for particle number conservation), it is clear that this effect can also modulate the $I − V$ curve and this interferes with pure crystallographic information. It is therefore crucial that inelastic effects be understood if the LEED technique is to be used as a reliable tool of surface analysis. Moreover, ILEED measurements bring direct information on the excitation spectrum of the solid, particularly at its surface (Powell, 1968; Ibach, 1970; Lucas and Sunjic, 1972) and constitute a complementary method to other tech-

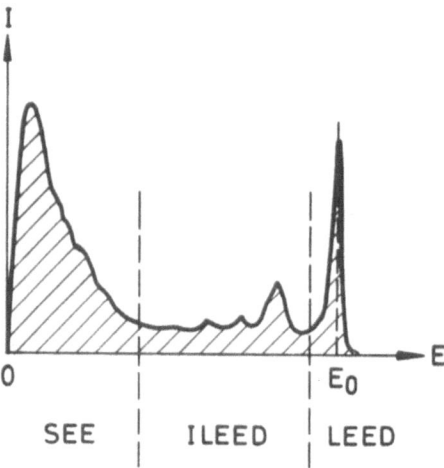

Fig. 5. The three conventional parts of the energy distribution of backscattered electrons in a LEED experiment.

niques such as optical absorption, Raman scattering, neutron scattering, etc. ... in understanding the interaction of space objects with their environmental radiation. The third portion of the energy spectrum (*SEE* in Figure 5) is made of those primary electrons which have been scattered so many times in the solid (both elastically and inelastically) that they have 'lost memory' of their internal origin and also those electrons which, initially in the valence band of the sample, have been knocked out in some conduction state by the primary electrons and emitted into vacuum. as in the photoemission process. The 'true' secondaries can be separated from the other parts of the backscattering current by the property that their spectrum is positionally independent of the primary energy E_0. The analysis of the true structure lying on top of the continuous secondary emission background has recently been developed,

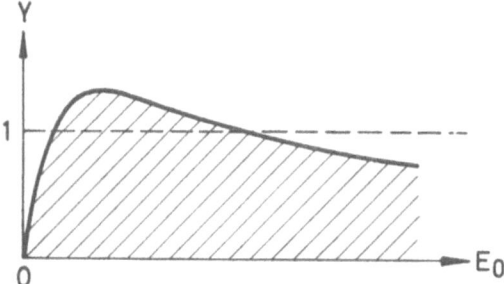

Fig. 6.   Dependence of backscattered current on primary electron energy E_0 in a SEE experiment. The yield is measured in number of secondary electrons per incident electron.

especially in this laboratory, into a powerfull technique of studying the band structure and collective excitation spectrum of materials (Anderson, 1972). An important property for the macroscopic electrical behaviour of objects subjected to electron bombardment, is the total secondary yield Y, defined in Equation (1), as a function of primary electron energy. Schematically, the yield may look like the curve of Figure 6 and is very sensitive to material and surface conditions (Willis *et al.*, this volume). Such yield curve will be considered in detail in this conference, particularly by Professor Gold (this volume) in his studies of Moon dust dynamics.

4. Ion Scattering Techniques

In answer to the question 'what happens when an energetic ion hits a surface' one may produce the star diagram of Figure 7 showing the flurry of events generated by such a violent collision. Several 'rays' or channels of this diagram will be the object of particular attention in several papers of this conference. For instance, electron emission will have to enter the current balance equations in the study of plasma sheath; sputtering, i.e. ejection of particles in neutral, ionized or metastable excited states, will be important for erosion processes of the surface of the Moon and other celestial objects by the solar wind, etc.....

Detailed, specialized information concerning ion scattering by solids can be found in recent books (Carter and Colligon, 1968; Kaminsky, 1965). Here we shall have to limit ourselves to a few remarks which can only give a taste of the richness of the overall phenomenon.

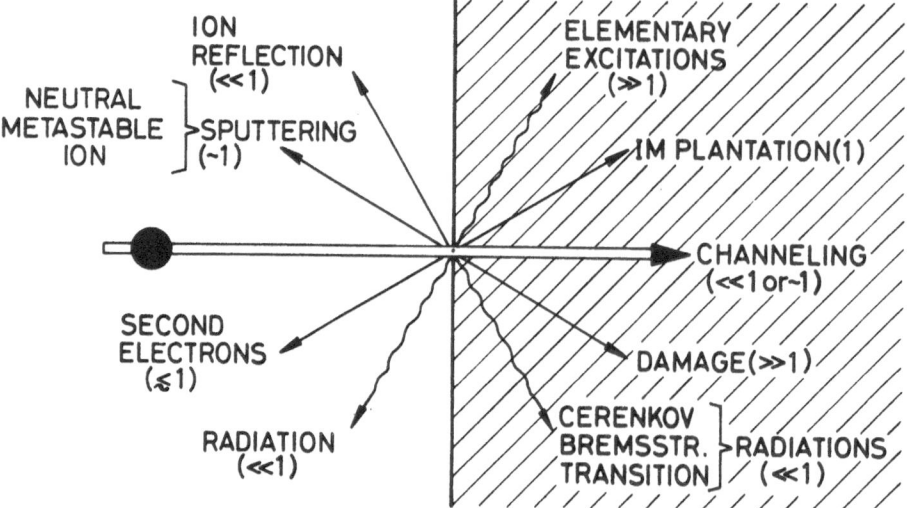

Fig. 7.   A 'star' of possible events when a fast ion hits a solid surface. The numbers in parentheses give the order of magnitude of the number of events per incident ion.

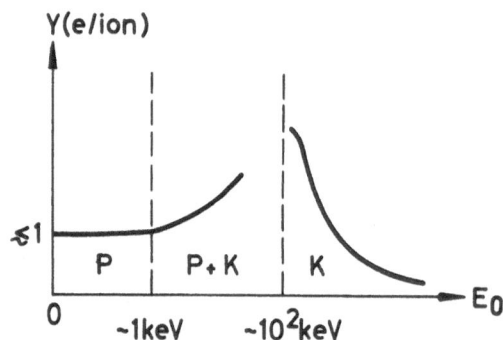

Fig. 8.   The yield of secondary electron emission from ion-surface scattering, as a function of primary ion energy.

In the ejection of secondary electrons, it has been possible to separate two regimes in the electron yield – primary ion energy, as shown in Figure 8. Below say a few keV (for protons), one has the so-called *potential emission regime*. This consists in a transfer of the potential energy available in the deep empty level of the incoming ion (the ionization potential) to some valence electron of the bombarded material which may then be capable of overcoming the work function and escape into vacuum. The actual process is rather complicated, as shown in Figure 9 and involves an Auger-like neutralization of the primary ion by tunneling of a first valence electron through the surface potential barrier and simulataneous absorption of the liberated energy by a second valence electron which may then be emitted. This premature neutralization by tunneling turns out to be more efficient than ordinary Auger or radiative neutralization when the ion has penetrated into the solid. The measurement of the energy distribution of the secondary yield provides indirect information on the structure of the valence band. This important technique, known as Ion Neutralization Spectroscopy (INS, Table I), has been pioneered by Hagstrum (Hagstrum, 1953).

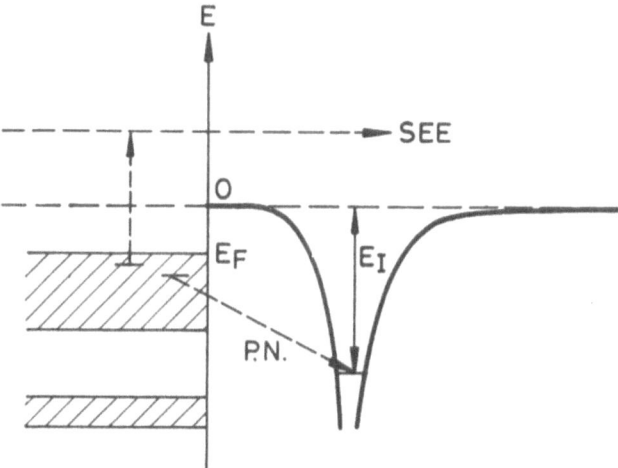

Fig. 9.   Energy diagram for an Auger-like neutralisation process of an ion outside a metal surface. The energy released by potential, tunneling neutralisation (P. N.) is picked up by another valence electron which may be emitted as a secondary electron (SEE).

The second clear regime of secondary electron emission occurs at high energies, say above 100 keV (for protons), and is referred to as kinetic emission (Figure 8). Here the exact mechanism of electron ejection is rather poorly understood but would likely involve the direct Coulomb scattering of the electrons or nucleus of the primary ion with the electrons of the target material. Between these two regimes, an intermediate energy region exists where the particularly high yield results from the combined contributions of both potential and kinetic mechanisms. The overall energy dependence of the phenomenon appears to be consistent with a direct application of the fundamental adiabatic (Born-Oppenheimer) theorem (Messiah, 1959) and the sudden approximation theorem. Indeed for low ion energies, the valence electrons of the target are swift enough to adjust adiabatically to the Coulomb field of the slowly incoming charge (by collectively creating an average 'image charge'), so that neutralization only can supply the ejection energy. On the other hand, for high ion energies, i.e. when the incoming velocity becomes comparable to or greater than the Fermi velocity of the material (say a metal), the time-dependent perturbation created by the ion is fast enough to produce 'direct hits' with individual target electrons. Finally, when the energy is very high ($\gtrsim 1$ MeV for protons) the secondary yield decreases rapidly in agreement with the sudden theorem (or Born approximation) which states that if a perturbation is too fast, the system has no time to respond (Messiah, 1959).

Ion reflection and sputtering are relatively simpler processes, following essentially a 'billiard balls' kind of dynamics.

Techniques of radiation damage, chanelling and ion implantation (Mayer *et al.*, 1970) have developed rapidly in recent years and have been extremely beneficial in both their scientific and technological applications. For instance, chanelling is currently being exploited as a very sensitive technique for surface chemical analysis. The essential idea is that, as a result of the large ion-ion or ion-atom scattering cross sections, even small amounts of foreign material adsorbed on the crystal surface will spoil the chanelling by obstruction of the propagation channels and hence produce a measurable increase in the backscattering yield.

Ion bombardment effects are not always beneficial, however, as examplified by the rather catastrophic phenomenon of material swelling under high doses of particle irradiation (see next section).

We may close this section by mentioning the very spectacular and fruitful technique of Field Ion Emission (FIE, Table I). Ion emission by a surface may be viewed as the reverse process of ion neutralisation described above. However, a slow neutral atom approaching the surface of a neutral body will never be backscattered as an ion unless some external agent lifts its occupied electronic ground state level above the Fermi level of the target material (a metal, say), thus allowing the tunneling ionization otherwise forbidden by the Pauli exclusion principle. An ingenious way to operate this level promotion is to lift it by a very strong electrostatic field, as shown in Figure 10. The fields needed are so high $(\gtrsim 1 \text{ V Å}^{-1}$ or $10^8 \text{ V cm}^{-1})$ that they can only be

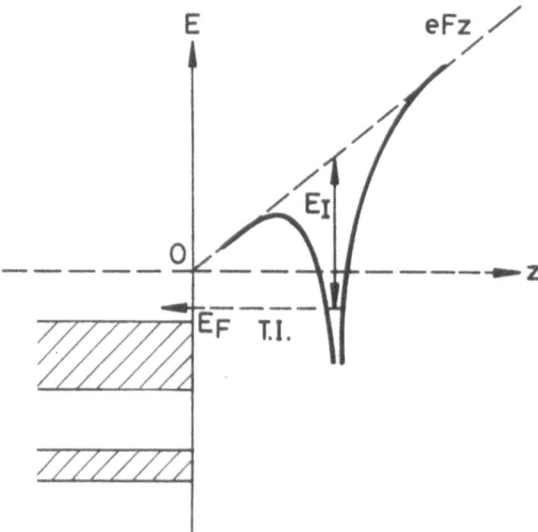

Fig. 10.   Energy diagram for Field Ion Emission (FIE). The occupied ground state level of a neutral atom is lifted by an external field F above the metal Fermi level E_F to allow tunneling ionization (T.I.).

produced (excluding dynamic fields of powerful lasers) around extremely sharp, positively charged metal points (tip radii less than 1000 Å, tip voltages $\simeq$ 10 kV). The ions produced by electron tunneling into the tip are then strongly repelled by the positive tip voltage, fly radially away from the tip surface and create a visible impact on a fluorescent screen. This radial projection principle has been exploited by Müller (1969) in his invention of the Field Ion Microscope, an instrument of atomic resolution which, for the first time, has allowed direct imaging of individual surface atoms. As a visualisation technique, field ion microscopy has vastly contributed to our understanding of physical and chemical phenomena taking place at metal surfaces.

5. Model Problems

In this section we describe in some details a few recent advances in surface physical problems which find direct applications in space science and technology. These examples are chosen to illustrate how accumulated studies of fundamental processes in photon and particle scattering by surfaces in laboratory experiments often lead to a vastly improved understanding of the behaviour of celestial objects in their space radiation environment.

(1) Our first example is related to the properties of *infrared optical absorption of interstellar dust* (van de Hulst, 1957). In the laboratory, the problem consists in producing a fine powder of an appropriate model solid material whose bulk optical properties are assumed to be understood and to measure and interpret those of small particles in various states of dispersion. By small particles we mean here micron- or submicron-size objects, much smaller than the wavelength of the incoming radiation $(+10\,\mu)$. A general discussion may be found in a recent review by Ruppin and Englman (1970) who have emphasized the role of surface modes vs ordinary bulk modes for the absorption behaviour of such systems. But the clear experimental identification of surface collective modes either plasmons (Raether, 1965) or optical phonons (Boersch *et al.*, 1968), is precisely one of the great achievements of electron energy loss spectroscopy in the last 20 yr. Initially investigated on systems with planar interface, pure surface modes have also been identified in the spherical geometry (Genzel and Martin, 1972), which, after all, should be more favourable (although theoretically less simple) than the planar case since the surface to volume ratio may be increased more easily by decreasing the particle size. On the theoretical side, it may be amusing to notice, as pointed out by Ruppin and Englman (1970), that the predicted absorption behaviour of small spheres including surface modes was already fully contained in calculations due to Mie (1908) at the beginning of this century, at a time when lattice dynamics had not been developed (and the word phonon did not yet exist!).

An elementary account of this problem may be given as follows. Suppose we have a material optically active in the infrared, i.e. with a dielectric function of the form

$$\varepsilon(\omega) = \varepsilon_\infty \frac{\omega_{LO}^2 - \omega^2}{\omega_{TO}^2 - \omega^2}, \tag{3}$$

where ε_∞ is the high-frequency ($\omega \gg \omega_{TO}$) dielectric constant and ω_{LO}, ω_{TO} are the longitudinal and transverse optical bulk phonon frequencies, respectively (see Figure 11). If the particle size is much smaller than the wavelength $\lambda \simeq 2\pi c/\omega_{TO}$ of the incoming radiation, then retardation effects can be ignored. This means that over the diameter of the small object, one can neglect the variation of the radiation amplitude, so that the particle is essentially bathed in a uniform, time-dependent electric field.

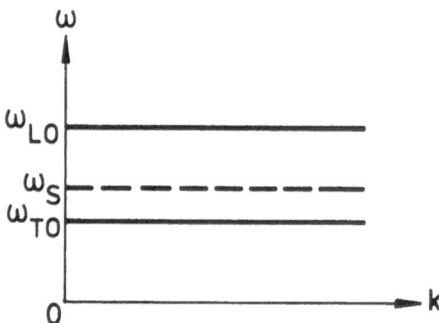

Fig. 11.   Dispersion relation for a polar insulator possessing one longitudinal optical phonon branch (ω_{LO}) and two degenerate transverse optical branches (ω_{TO}). The dipolar surface mode of a small sphere of this material would have a frequency ω_s lying in the gap.

Now the polarization of a spherical dielectric in a uniform field **E** is given, in elementary electrostatics (Jackson, 1962), by

$$\mathbf{P} = \frac{3}{4\pi} \frac{\varepsilon - 1}{\varepsilon + 2} \mathbf{E}. \tag{4}$$

One immediately sees that the polarization and hence the absorption is resonant when the relation

$$\varepsilon(\omega) + 2 = 0 \tag{5}$$

is satisfied. This is precisely the definition of the nonretarded surface phonon frequency in spherical geometry. The frequency ω_s satisfying (5) lies in the TO-LO band gap (see Figure 11), a general feature for true surface states, and this is where the peak in the absorption spectrum should be observed.

The sphere simply behaves as an elementary point electric dipole of eigenfrequency ω_s and, in the external radiation, its polarization vector goes 'up and down' in a resonant manner if the radiation frequency coincides with ω_s (Figure 12). This result is beautifully illustrated by the observed infrared transmission spectra of small particle specimens of NiO (Hunt *et al.*, 1973). The peak absorption indeed occurs at the surface frequency $\omega_s \simeq 500$ cm^{-1} and not at the bulk restrahlen frequency $\omega_{TO} \simeq 400$ cm^{-1}. However, an apparent disagreement between theory and experiment exists as the calculated peak is much narrower than the measured one (Hunt *et al.*, 1973). We believe that the observed broadening can be interpreted as a direct manifestation of the adhesion (Lucas, 1973) between identical powder particles as well as an indication of the irregularity of particle shapes and sizes (Hunt *et al.*, 1973). Needless to say, in the interpretation of infra-red spectra of interstellar dust, this dependence on parameters of shape, size and state of aggregation may considerably complicate the chemical identification on the simple basis of assumed bulk dielectric properties.

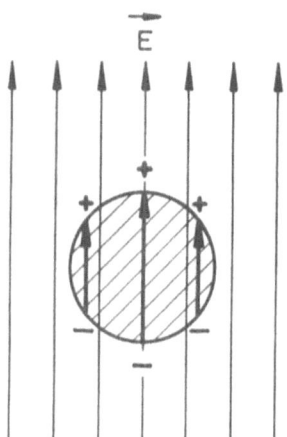

Fig. 12. Dipolar oscillation leading to resonant absorption of a small sphere in a long wavelength external field **E**.

(2) Our second example has to do with a very severe and unexpected secondary form of *radiation damage*, namely the formation of microporosity in metals bombarded at high temperatures by large doses of fast particle irradiation. This phenomenon, which was observed (Pugh *et al.*, 1971) in prototype fast reactors as well as in laboratory simulation experiments, may also be expected to develop in the structural components of nuclear energy sources on board of future space vehicles and also in some components of ion propulsion rockets. Radiation damage at room temperature generally takes the form of single atomic vacancies, small vacancy clusters and self interstitials, resulting from atoms being knocked out of their regular lattice positions by the primary or secondary particles along the radiation cascade. However, at higher temperatures, the above elementary defects become mobile and agglomerate into large defect clusters. If the dose is sufficient, the vacancy clusters themselves develop into small cavities or voids of submacroscopic size, leading to an overall swelling or volume increase of the irradiated sample. Under typical conditions of operations, the swelling may reach exceedingly large values (15% of relative volume increase) creating impossible problems of stability for the mechanical structure of the system. Thus voids in irradiated metals have posed scientific and technological problems of great magnitude and several international conferences (Pugh *et al.*, 1971) have already been devoted to their study. What we should like to mention here is the fact that some new understanding of the void phenomenon has been gained from considerations of the

basic nature of elementary excitations (Lucas, 1972), especially plasmons, in porous media. A small spherical void in a metal behaves in many respects like its anti-system, i.e. a small sphere in vacuum. Both systems possess characteristic surface plasmon modes which can respond to excitation by light or charged particles as described above. In fact, an elegant technique of directly imaging voids through excitation of these modes by inelastic electron microscopy has been used by French workers (Natta, 1969). A fundamental difference between a void and a solid sphere, however, is that outside a void there can occur radial plasmon fluctuations with uniform accumulation of charges on the void surface. This monopole 'breathing' mode (Lucas, 1973) does not exist on a solid sphere whose lowest order multipole mode is dipolar. As a result one can show that between neighbouring voids, there exists an interaction of the van der Waals type which decreases only as the inverse square distance, in contrast to the equivalent adhesive interaction between solid spheres which behaves like the inverse sixth power of their separation. Such a strong void-void interaction is likely to play an important role in the early stages of void growth and is probably also partly responsible for the extremely spectacular phenomenon observed in some samples where millions of voids are seen to organize themselves into a single superlattice of high perfection (Evans, 1971).

(3) In our third example we will discuss a new method of detection of ultrarelativistic charged particles, whose principle relies entirely on fundamental radiation and particle scattering processes at solid-vacuum and solid-solid interfaces. The method, initially proposed by Frank (1966), consists in detecting the so-called *Transition Radiation* i.e. emission of electromagnetic radiation by fast charged particles when they 'transit' through an interface between two media of different dielectric constants (see Figure 7). Although the detector is still at a stage of development (Wang *et al.*, 1972; Alikhanian *et al.*, 1973), it holds great promises, possibly for space applications, as it can advantageously take over the Cherenkov and other counters which become inefficient at high energies. If $E = m_0 c^2 \gamma$ represents the particle energy where $\gamma = (1 - \beta^2)^{-1/2}$ and $\beta = v/c$, one sees that a detector sensitive to γ would be much more efficient for mass discrimination of high energy particles than a detector sensitive to β, since $\delta\gamma/\delta\beta = \beta\gamma^2 \gg 1$ at high values of $\gamma (\simeq 10^3$ for GeV electrons). The transition radiation detector is such a device as it has been shown both theoretically and experimentally (Yuan *et al.*, 1970; Ter Mikalian, 1972) that the transition radiation yield Y in the X-ray frequency range is proportional to γ.

The phenomenology of Transition Radiation is contained in classical Maxwell's equations according to which electromagnetic radiation is emitted whenever the velocity v of a charged particle varies relatively to the local phase velocity of light. This relative variation can be obtained either by changing v and keeping c fixed (cyclotron radiation, Bremsstrahlung, radiative β-decay, etc...) (Jackson, 1962) or by keeping v fixed and changing c/n, i.e. by interposing material inhomogeneities with variable refraction index n. Thus, at an interface, light must be emitted by the passage of a uniformly moving charge. A simple physical picture of Transition Radiation is to describe it as due to the anihilation of the incoming charge with its image at the interface and immediate recreation of an opposite pair when the charge penetrates the solid. Another, more technical way to describe the process is to decompose the phenomena in two steps. First the charge excites plasmons, both non-radiative and radiative ('polaritons' in Table II). Then, while the non-radiative plasmons remain trapped in the medium, the radiative ones decay by photon emission into vacuum. This more fundamental point of view has been taken by Ferrel (1958), Stern (1967) and more recently criticized by Economou (1969).

The spectrum of emitted radiation, contrary to the Cherenkov effect, has no energy threshold features so that nonrelativistic as well as relativistic particles can emit transition radiation. At optical frequencies, the shape of the spectrum is dictated by the optical properties $\varepsilon(\omega)$ of the bombarded material and this has led to a very active branch of solid state research (Boersh, 1965; Tomas et al., 1972). Similar to Cherenkov light, there is a high frequency cutoff defined as follows.

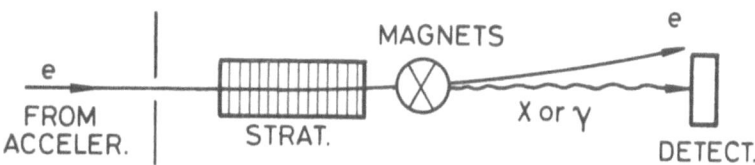

Fig. 13.   Emission and detection of X-ray or γ radiation from a stratified radiator bombarded by ultra-relativistic charges.

The minimum radiation period (maximum frequency) is of the order of the time required by the charge to move through the spatial inhomogeneity of the interface. For non relativistic particles, this gives $\omega \lesssim v/\Delta Z$ where ΔZ is of the order of 1 Å. For relativistic particles, time or length dilatation effects must be taken into account and a more general concept is then the so-called *formation zone* or *coherent length* giving the path length of the particle trajectory required to generate the radiation of frequency ω:

$$Z = \frac{c}{\omega} \beta [1 - \beta(\varepsilon - \varepsilon \sin^2 \theta^{1/2}], \qquad (6)$$

where θ is the emission angle measured from the particle trajectory. The usefulness of this concept is manifest when several interfaces are separated by distances less than Z. When this occurs, the light intensity is not given by the incoherent addition of the intensities delivered by each interface but depends on the coherent addition of the amplitudes which may interfere destructively or constructively. Thus in artificially produced periodic media, the emitted light may be qualitatively different from ordinary Transition Radiation and is called *Resonance Radiation* in the russian literature. It is this coherent resonant emission effect which has been at the origin of the hopes to construct a practical detector device in which the number of interfaces is multiplied by using a stratified medium as the radiator (Harutunian, 1965) (see Figure 13). It appears that this hope is about to materialize and the new detector may find important applications in the coming generation of high energy accelerators.

6. Conclusions

In this introductory paper, we have attempted to discuss some fundamentals of the interaction between photon and particles with solid surfaces. Although we have covered but a few of the existing techniques, one may feel bewildered by the enormous variety of phenomena, and one could wonder whether there exists a unifying principle underlying this complexity. In this respect, it seems appropriate to remember that the processes described in this superficial review are all manifestations of the 'electromagnetic force' which dominates solid state physics. Thus, all photon and particle interactions with surfaces may be viewed as processes of absorption and emission of photons, real or virtual. Fermi (1924) was the first to suggest an approximation method where charged particles are represented by 'a bunch' of virtual or pseudo-photons.

TABLE III

Comparison between charged particle and photon interactions.
(see Jackson, 1962)

Charged particle interactions	Photon interactions
Electron-solid scattering (LEED, ILEED, RHEED)	Photoelectron emission (real photons)
Bremsstrahlung (electron-nucleus collision)	Photon scattering (virtual-real photons)
Collisional atom ionization	Photoionization (virtual photons)
Transition radiation	Photon refraction, reflection (virtual-real photons)

To calculate the scattering of the particle, one determines the cross-section $\sigma(\omega)$ for scattering by the target of a photon of frequency ω and integrate over the set of particle pseudo-photons. This is called the virtual or pseudo-photon method in electrodynamics. The similarity between photon and particle scattering processes which is the basis of this method is illustrated in Table III where several radiative and nonradiative collisions are compared.

Acknowledgements

The author is grateful to Dr E. A. Trendelenburg, Dr B. Fitton and the staff of the Surface Physics Division for hospitality and encouragements.

References

Alikhanian, A. I., Kankanian, S. A., Oganeshan, A. G., and Tamanian, A. G.: 1973, *Phys. Rev. Letters* **30**, 109.
Anderson, S.: 1972, *Solid State Comm.* **11**, 1401.

Bauer, E.: 1969, in *Techniques of Metals Research*, Vol. II, part 2, Chap. 15, Wiley, New York.
Berglund, C. N. and Spicer, W. E.: 1964, *Phys. Rev.* **136**, 1030.
Boersch, H.: 1965, *Z. Physik* **187**, 97.
Boersch, H., Geiger, J., and Stickel, W.: 1968, *Z. Physik* **212**, 130.
Carter, G. and Colligon, J. S.: 1968, *Ion Bombardment of Solids*, Heinemann Educational Books Ltd.
Davisson, C. J. and Germer, L. H.: 1927, *Phys. Rev.* **30**, 705.
Dekker, A. J.: 1967, *Solid State Physics*, Prentice-Hall.
Duke, C. B. and Laramore, G. E.: 1971, *Phys. Rev.* **B3**, 3183, 3198.
Economou, E. N.: 1969, *Phys. Rev.* **182**, 539.
Einstein, A.: 1905, *Ann. Phys.* **17**, 132.
Evans, J. H.: 1971, *Radiation Effects* **10**, 55.
Fermi, E.: 1924, *Z. Phys.* **29**, 315.
Ferrell, R. A.: 1958, *Phys. Rev.* **111**, 1214.
Feuerbacher, B. and Fitton, B.: 1972, *Phys. Rev. Letters* **29**, 786.
Frank, I. M.: 1966, *Soviet Phys. USPEKHI* **8**, 729.
Genzel, L. and Martin, T. P.: 1972, *Phys. Stat. Sol.* (b)**51**, 91.
Hagstrum, H. D.: 1953, *Phys. Rev.* **91**, 543.
Harutunian, F. R., Ispirian, K. A., and Oganesian, A. G.: 1965, *Nucl. Phys.* **1**, 842.
Hirsh, P. B., Howie, A., Nicholson, R. B., Pashley, D. W., and Whelan, M. J.: 1965, *Electron Microscopy of Thin Crystals*, Butterworths.
Hunt, A. J., Steyer, T. R., and Huffman, D. R.: 1973, *Conf. on Surface Properties and Surface States of Electronic Materials*, to be published.
Ibach, H.: 1970, *Phys. Rev. Letters* **24**, 1416.
Jackson, J. D.: 1962, *Classical Electrodynamics*, Wiley, New York.
Kaminsky, M.: 1965, *Atomic and Ionic Impact Phenomena on Metal Surfaces*, Springer-Verlag.
Kittel, C.: 1971, *Introduction to Solid State Physics*, Wiley, New York.
Lander, J. J.: 1965, in *Progress in Solid State Chemistry*, Vol. 2, Pergamon, New York.
Lucas, A. A. and Sunjic, M.: 1972, *Progr. Surface Sci.* **2**, 75.
Lucas, A. A.: 1972, *Phys. Letters* **41A**, 375.
Lucas, A. A.: 1973, *Phys. Rev.* **4**, 2939.
Lucas, A. A.: 1973, to be published.
Mahan, G. D.: 1970, *Phys. Rev. Letters* **24**, 1068.
Mayer, J. W., Eriksson, L., and Davies, J. A.: 1970, *Ion Implantation in Semiconductors*, Academic Press, New York.
Messiah, A.: 1959, *Mécanique quantique*, Vol. II, Dunod, Paris.
Mie, G.: 1908, *Ann. Phys.* **25**, 377.
Müller, E. W. and Tsong, Tien Tsou: 1969, *Field Ion Microscopy*, Elsevier, New York.
Natta, N.: 1969, *Solid State Comm.* **7**, 823.
Pines, D.: 1964, *Elementary Excitations in Solids*, Benjamin, New York.
Powell, C. J.: 1968, *Phys. Rev.* **175**, 972.
Pugh, S. F., Loretto, M. H., and Norris, D. I. R., (eds.): 1971, *Proc. 1971 Int. Conf. on Radiation-Induced Voids in Metals*, Albany, New York, in press.
Raether, H.: 1965, *Springer Tracts in Mod. Phys.* **38**, 84.
Ruppin, R. and Englman, R.: 1970, *Rep. Prog. Phys.* **33**, 149.
Stern, E. A.: 1967, *Phys. Rev. Letters* **19**, 1321.
Ter Mikaelian: 1972, *Interscience Tracts on Physics and Astronomy*, No. 29.
Tomas, M. S., Lucas, A. A., and Sunjic, M.: 1972, *Solid State Comm.* **10**, 1181.
Van de Hulst, H. C.: 1957, *Light Scattering by Small Particles*, Wiley, New York.
Wang, C. L., Dell, G. F., Jr., Uto, H., and Yuan, Luke C.: 1972, *Phys. Rev. Letters* **29**, 814.
Willis, R. F., Feuerbacher, B., and Fitton, B.: 1971, *Phys. Rev.* **B4**, 2441.
Yuan, Luke C., Wang, C. L., Uto, H., and Prünster, S.,: 1970, *Phys. Rev. Letters* **25**, 1513.

THE PARTICLE ENVIRONMENT IN SPACE

G. L. SISCOE

Dept. of Meteorology, University of California, Los Angeles, Calif. 90024, U.S.A.

Abstract. This article reviews the charged and neutral particles in the solar system and particularly in the vicinity of Earth. The average values and ranges of the parameters characterizing the charged particle flux from the sun (the solar wind), mainly protons and electrons, are given and also the correlations between the parameters for large scale fluctuations. Aerodynamics calculations give the physical conditions of the shocked solar wind plasma in the region between the Earth's bow shock and the boundary with the geomagnetic cavity. The charged particle population inside the geomagnetic cavity and the geomagnetic tail is highly structured both in its spatial distribution and in energy. The average properties and dynamical behavior of the polar cusp, plasma sheet, ring current, and plasmasphere populations are described. Models and dynamical behavior of the neutral particles in the Earth's exosphere are given.

This review concerns the particle environment of Earth and interplanetary space primarily as revealed by spacecraft measurements. Direct measurements presently extend from the orbit of Venus to the orbit of Mars; and missions to Jupiter are now collecting data at greater distances. Future missions are planned to go inside the orbit of Mercury and beyond Saturn. There may also be an out-of-the-ecliptic flight that would reach high heliographic latitudes at a radial distance of approximately one astronomical unit (1 AU). Indirect methods such as the scattering and scintillation of cosmic radio waves and the behavior of comet tails give information on the interplanetary medium over a considerable spatial range (see the review by Axford, 1968), but they are much less quantitative and complete as direct measurements.

Numerous data-collecting Earth satellites have provided a fairly complete picture of the distribution and properties of the near-Earth particle population out to approximately the lunar orbit. In this region which includes the magnetosheath, the magnetosphere, and the geomagnetic tail, the particle distribution is highly structured and exhibits large fluctuations. The interplanetary region between Venus and Mars is less well sampled, and although the particle population – the solar wind – is not so sharply structured, our picture of it is less precise and detailed as that of the near-Earth population.

We begin with the solar wind particles. From space probe measurements extending over approximately one decade and from indirect indicators as mentioned above, it is believed that the solar wind blows continuously at all heliographic longitudes and latitudes and radially out to at least the orbit of Jupiter (5 AU). The last statement is an inference based on the correlation of Jupiter radio emissions with solar activity. The properties given here primarily refer to data from a heliocentric distance close to 1 AU, and undisturbed by the Earth's presence.

1. Solar Wind Particles

Measurements over sufficiently long time intervals to provide reasonable statistics of

R. J. L. Grard (ed.), Photon and Particle Interactions with Surfaces in Space, 23–45. All Rights Reserved
Copyright © 1973 by D. Reidel Publishing Company, Dordrecht-Holland

solar wind parameters began with the 1962 Mariner II mission to Venus. The radial speed, density, and ion temperature were measured. Subsequently, spaceprobe plasma instruments have provided data also on the other two components of the flow speed, electron temperatures, ion and electron temperature anisotropies and heat fluxes and ionic composition. (See Vasyliunas, 1971, for a discussion of space plasma instruments.)

2. Averages and Ranges

Several reviews and data summaries giving averages and ranges of these parameters and the correlations between them are now available (Neugebauer and Snyder, 1966; Axford, 1968; Olbert, 1968; Hundhausen, 1968, 1970a; Hundhausen et al., 1970a; Ogilvie et al., 1968; Kavanagh et al., 1970; Howe et al., 1971; Wolfe, 1972; Goldstein and Siscoe, 1972; Mihalov and Wolfe, 1971). Table I gives a summary of averages, standard deviations, and ranges of solar wind proton properties. The data come from many spacecraft and cover the time period September, 1962, to January, 1970, a little less than one complete solar cycle. Throughout this period one finds typical speeds near 400 km s^{-1}, densities near 6 cm^{-3}, temperatures near 1×10^5 K, and fluxes near 3×10^8 cm^{-2} s^{-1}. The flow speed is typically 10 times the proton thermal speed. Thus, the solar wind is hypersonic with respect to the protons.

TABLE I

Averages, standard deviations, and ranges of solar wind proton properties from different spacecraft. Letters (a) and (b) in the first column indicate different plasma instruments on the same spacecraft. (Modified from Wolfe, 1972)

SPACECRAFT	VELOCITY km sec^{-1}			DENSITY cm^{-3}			PROTON TEMP $\cdot 10^5$ °K			PROTON FLUX 10^8 cm^{-2} sec^{-1}			~DATE
	AVE	S.D.	RANGE	AVE	S.D.	RANGE	AVE	S.D.	RANGE	AVE	S.D.	RANGE	
MARINER-2	504		319-771	5.4		.44-54	1.5-1.8		0.3-8	2.4			9/62-12/62
IMP-1(a)	360		190-610	7		1-28							12/63-2/64
IMP-1(b)	378												12/63-2/64
VELA-2	420						1.4						7/64-7/65
VELA-3	400	80	290-550	7.7	4.6	2.8-16.2	0.91	0.74	0.1-2.5	3.0	1.8	1.2-6.5	7/65-7/67
PIONEER-6(a)	430			6			0.38						12/65-2/66
PIONEER-6(b)	422	79	280-640	5.7	3.5	<1-20	1.0	0.72	0.1-4.8	2.3	1.2	.4-9	12/65-2/66
PIONEER-7(a)	460			6									8/66-10/66
PIONEER-7(b)	455	80	300-750	4.4	3.4		1.6	1.25	0.1-9.8	2.0	1.5	2-10	8/66-10/66
EXPLORER-34	438					<1-20	0.46						6/67-12/67
MARINER-5	410	81	290-690	6.2	4.4	<1-22	1.1	.25	0.1-5.0	2.3	1.5	<1-8	6/67-11/67
HEOS-1	409			4.3			0.66						12/68-1/70

Examples of data which are typical of those of all spacecraft are listed in Table I. Figure 1 shows histograms of proton flow speeds, densities, temperatures, and fluxes from Pioneers 6 and 7. The histograms show a broad, nearly symmetric distribution of flow speeds, and highly skewed density, temperature, and flux distributions.

The major variations in these parameters are associated with large scale coronal inhomogeneities which as the solar rotation moves them past an essentially fixed spacecraft give time scales of the order of several days to ten days for the largest fluctuations. There is little dependence of the averages on the 11 yr cycle of solar

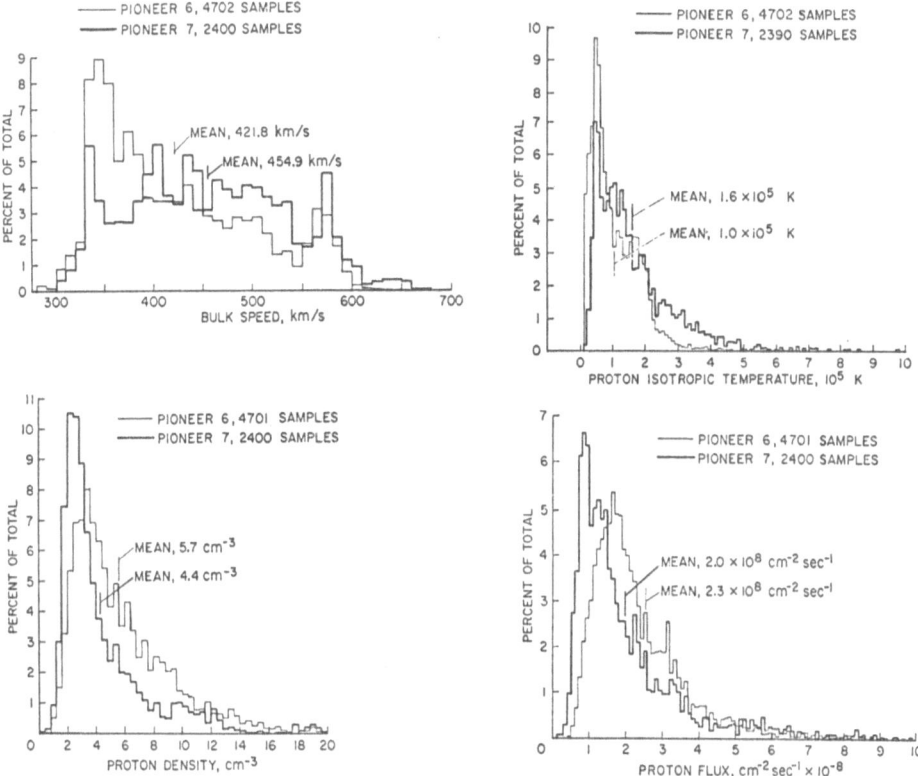

Fig. 1.   Histograms of solar wind proton bulk speed, temperature, density and flux from the Pioneer 6 and 7 spacecraft. (From Mihalov and Wolfe, 1971.)

activity (Gosling *et al.*, 1971). Using the 7° heliographic latitude range available from the inclination of the Earth's orbit to the solar equator, Hundhausen *et al.* (1971) infer that flow speeds are bigger and densities smaller at non-equatorial latitudes than at the equator.

Solar wind electrons are more difficult to measure than the ions, and fewer data exist. However fairly firm upper limits on electric space charge and electric current densities permit the conclusion that electron densities and flow speeds are very nearly the same as those of the ion component; and this has been confirmed by direct measurements (Montgomery *et al.*, 1968). Figure 2 shows a solar wind electron velocity spectrum. One sees that the mean electron velocity is much higher than typical solar wind flow speeds. The solar wind is subsonic with respect to the electrons. The separation of the two curves, toward-Sun fluxes and away-from-Sun fluxes, is due to the electron bulk flow speed. The macroscopic, bulk flow parameters derived from the data in Figure 2 are flow speed $= 360$ km sec^{-1}, density $= 9$ cm^{-3}, and temperature 1.5×10^5 K.

Electron flow speeds and densities can be assumed to be the same as those for the

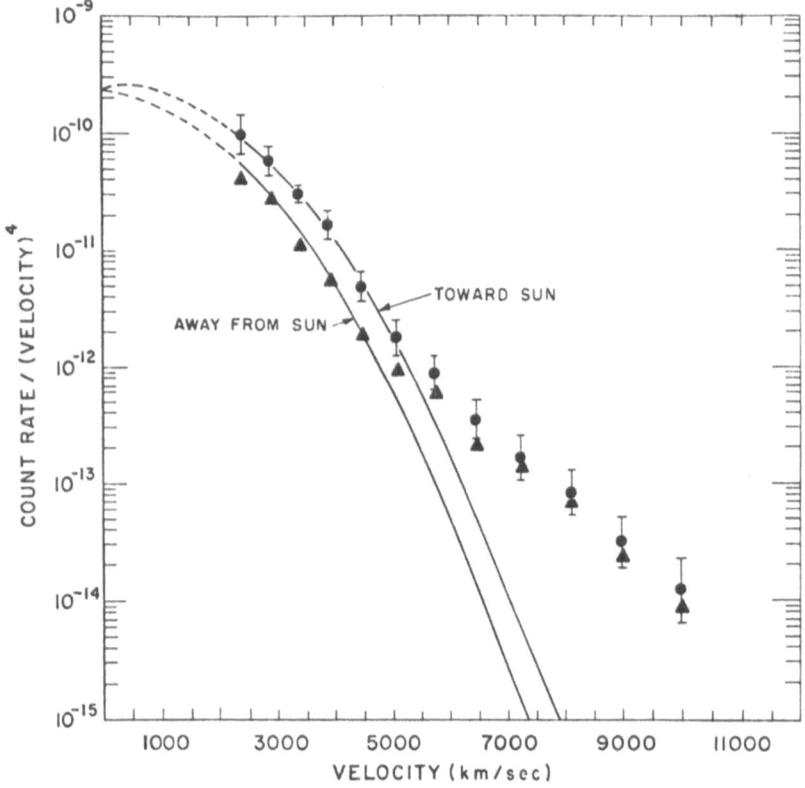

Fig. 2.   A solar wind electron spectrum derived from Vela 4B electron measurements.
(From Montgomery *et al.*, 1968.)

ions, but the electron temperature has a remarkable tendency to remain nearly constant at approximately 1.5×10^5 K, with variations from 1 to 2×10^5 K. Since the proton temperature is highly variable, the ratio of electron to proton temperatures is also highly variable, as is seen in Figure 3. When the solar wind flow speed is less than 370 km s^{-1} (labeled quiet), T_E/T_p is generally greater than 1 and is typically 4. The curve labeled disturbed (solar wind speed greater than 370 km s^{-1}) peaks at a ratio near unity and ranges on either side by approximately a factor of 3. The difference in the two curves is due to a correlation between T_p and the solar wind flow speed, as is discussed later.

Since the flow is hypersonic with respect to the protons, the flux of protons on a surface is to a good approximation the same as that associated with the bulk flow of the solar wind (given as proton flux in Table I and Figure 1) times the cosine of the angle between the normal to the surface and the flow direction. However, for the electrons both the thermal flux and directed flux must be considered. At 1.5×10^5 K and a flow speed of 400 km s^{-1} and a number density n (cm^{-3}) the flux on a surface

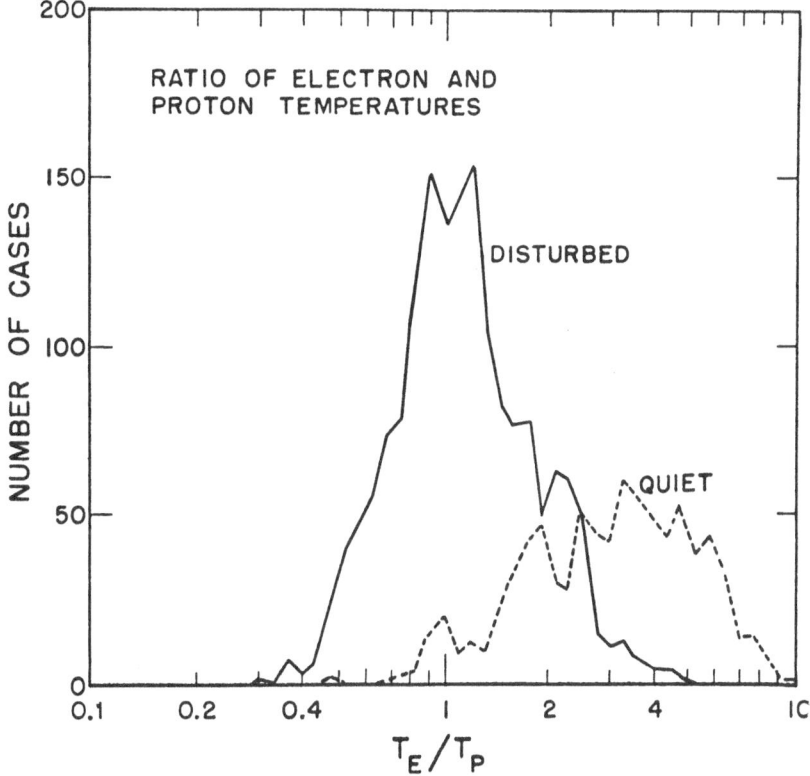

Fig. 3. Distributions of the ratios of electron to proton temperatures in the solar wind from Vela 4 measurements between 12 May 1967 and 5 July, 1967. The dashed curved (labeled 'quiet') is for solar wind speeds less than 370 km s^{-1}. The solid curve ('disturbed') is for speeds greater than 370 km s^{-1}. (From Bame et al., 1969.)

facing into the wind is 8×10^7 n cm^{-2} s^{-1} and away from the wind it is 3.7×10^7 n cm^{-2} s^{-1}.

The proton and electron temperatures have been given above as scalar quantities corresponding to isotropic temperatures. In fact the temperatures are generally slightly anisotropic. The anisotropy is aligned with the magnetic field such that the temperature based on thermal motion parallel to the magnetic field is greater than that based on thermal motion perpendicular to the field $(T_{\parallel} > T_{\perp})$. The temperature based on thermal motion in the plane perpendicular to the field is isotropic, and therefore, it is only necessary to specify the two temperatures, $T_{\parallel}$ and $T_{\perp}$. Statistics on temperature anisotropies compiled from Vela 4 spacecraft measurements are given in Table II.

The dominant ion in the solar wind is H$^+$. Doubly ionized helium, He^{++}, is also detected in variable percentage relative to H$^+$. The ratio of He^{++} density to H$^+$ density is found to be typically 0.05 with a range from 0.01 to 0.24 (Ogilvie and Wilkerson, 1969). Large ratios tend to occur with disturbed, high flow speed conditions; and the infrequent very large ratios are associated with the solar wind dis-

TABLE II

Proton and electron thermal anisotropy ratios. Vela 4 measurements
from May 1967 to May 1968 (Montgomery, 1972)

	Average	Median	Standard deviation	Range
Proton T_{max}/T_{min}	1.48	0.4	1.36	1–3.5
Electron T_{max}/T_{min}	1.1	0.08	1.08	1–1.5

turbances caused by solar flares. The flow speed of the He^{++} ions are very nearly the same as those of the H^{+} ions; but the helium to hydrogen temperature ratio has a mean value of approximately 4.0 (Robbins *et al.*, 1970). Under unusual conditions of very high fluxes and very low ion temperatures, other ions have been detected. The flux of the next most abundant after He^{++} is O^{6+} for which a flux 1/500 of the H^{+} flux has been given (Bame *et al.*, 1968).

3. Correlations between Solar Wind Parameters

Solar wind parameters, flow speed, density, and temperature, show large variations as discussed in the previous section. These variations exhibit considerable correlations between the parameters. Figure 4 shows proton flow speeds, densities, and temperatures (thermal speeds) for a one month interval measured with the Mariner 5 spaceprobe. The flow speed shows variations between intervals of high and low values with a characteristic time of 5 to 8 days, sometimes described as alternating fast and slow

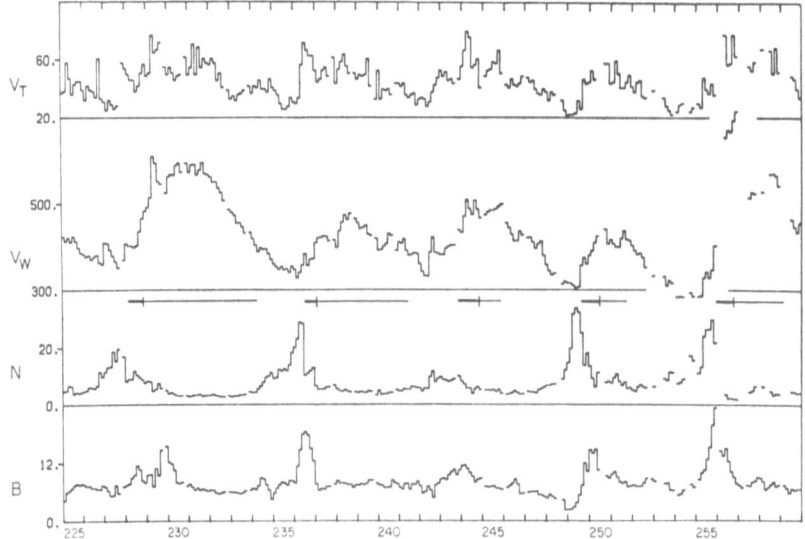

Fig. 4. Three-hour averages of solar wind parameters (V_T = most probable proton thermal speed, V_w = radial proton bulk speed, N = proton number density, B = magnetic field strength) for 35 days from Mariner 5. (From Belcher and Davis, 1971.)

streams. The proton temperature in the fast streams is seen to be consistently greater than in the slow streams; that is, the temperature and flow speed are positively correlated. The density also shows regular variations consisting of positive density spikes which precede the sharp rise of the high speed streams. Such density spikes are likened to a 'snow-plow' effect in which density piles up as a result of the compression as a fast stream pushes into a preceding slow stream. Density variations and slow speed variations thus tend to be anti-correlated.

The correlations evident in Figure 4 are shown explicitly in Figure 5 where proton density and temperature (both averaged over 25 km s^{-1} flow speed intervals) are given as functions of the flow speed. Also shown is the electron temperature, which, as discussed earlier has little variation in the solar wind.

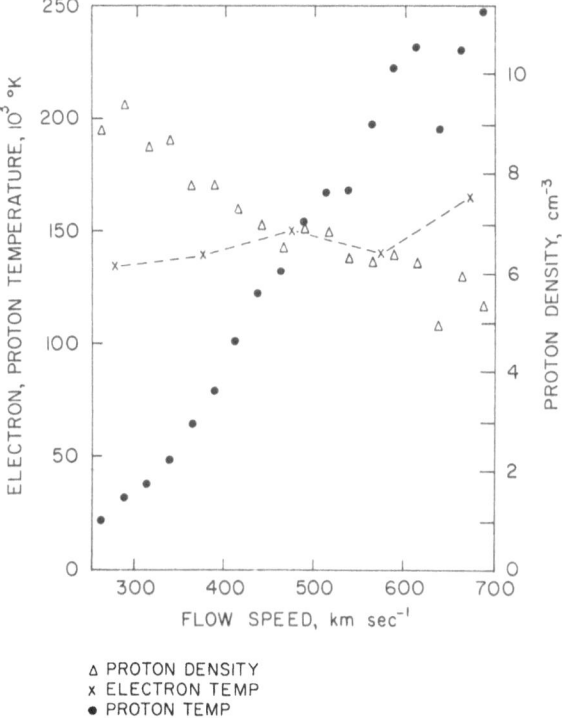

Fig. 5. Statistical variation of solar wind proton density and temperature and electron temperature with the flow speed from Vela 3 and 4. Proton densities and temperatures have been averaged in 25 km s^{-1} flow speed intervals. Electron temperatures have been averaged in 100 km s^{-1} flow speed intervals. (From Hundhausen *et al.*, 1970a; Montgomery, 1972.)

The correlations described above refer to variations with periods of several days and greater. These are presumed to be related to large scale inhomogeneities in the solar corona which rotate with the sun past a fixed observer causing temporal variations in solar wind parameters. Power spectra of solar wind variations reveal that the greatest power resides at long periods, up to approximately 10 days. The correlated variations shown in Figures 4 and 5 are therefore associated with the greatest amplitude fluctuations in the solar wind.

At shorter periods (less than approximately one day) the character of the fluctuations and the correlations change (Goldstein and Siscoe, 1972). These shorter period, smaller amplitude variations appear to be due to propagating hydromagnetic waves and discontinuities in the solar wind.

4. Special Events

Solar flares often cause interplanetary events of unusual severity. Flare generated shock waves are commonly experienced at earth orbit and have recently been reported from beyond the orbit of Mars by the Pioneer-Jupiter spaceprobe. A passing shock wave produces sudden increases in the flow speed, density, and temperature of the

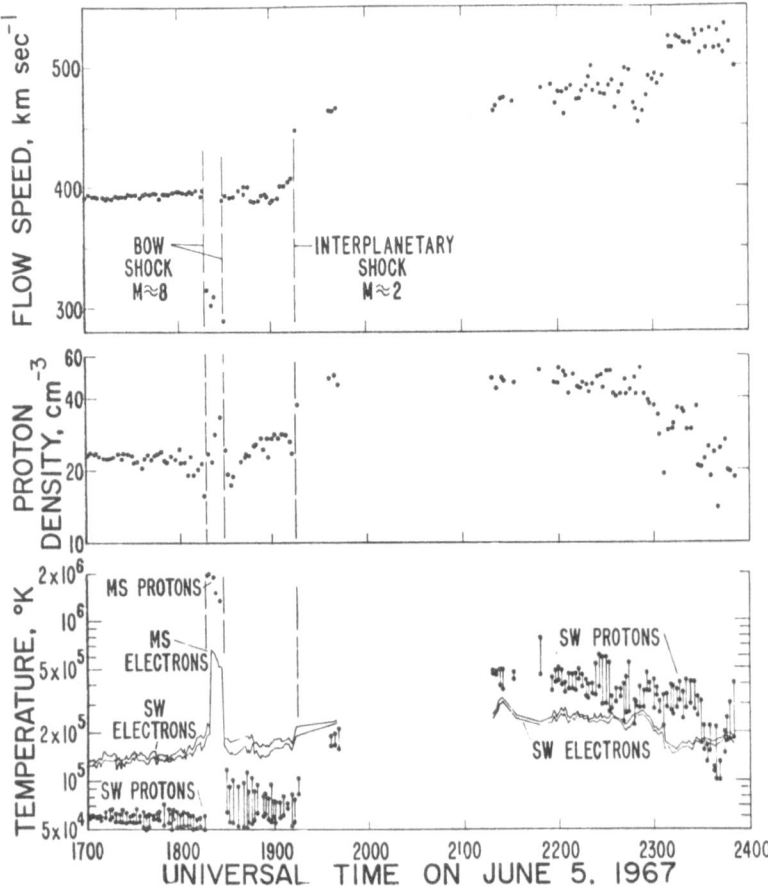

Fig. 6.   Vela 4B measurements of solar wind parameters before and after the passage of an interplanetary shock wave on 5 June, 1967. A brief penetration into the Earth's magnetosheath occurred between 1820 and 1830 UT. (From Hundhausen, 1970b.)

plasma. Typical values are 100 km s^{-1} increase in the flow speed, density increase of a factor of 2 to 3, and proton temperature increases of a factor up to an order of magnitude (Gosling *et al.*, 1968; Hundhausen *et al.*, 1970b). Figure 6 shows solar wind parameters at the time of an interplanetary shock on June 4, 1967 as observed by Vela 4B. The shock passed at 1915 UT producing abrupt increases in the three proton parameters as mentioned above. This event also illustrates another important feature of interplanetary shocks: the electron temperature increases only slightly across the shock compared to the proton temperature increase. Although there is evidence that very energetic shock waves can produce significant increases in electron temperature, Figure 6 is typical of the majority of interplanetary events. The figure also shows data from a brief encounter with the magnetosheath plasma behind the Earth's bow shock; and this is the subject of the next section.

5. Magnetosheath Particles

The interaction between the solar wind and the magnetic field of Earth (and probably also with the atmospheres of Venus and Mars) produces a bow shock upstream from the planet which slows and heats the plasma and deflects the solar wind around the planet. The situation is analogous to a hypersonic flow interaction with a blunt body, and numerical calculations based on the analogy have been performed (see reviews by Spreiter and Alksne, 1969; Dryer, 1970). Given the solar wind parameters upstream from the bow shock, flow parameters in the magnetosheath can be predicted (Figure 7). The figure, based on assumed Mach 8 flow, shows increased densitites and temperatures and decreased flow speeds, especially in the forward-most region of

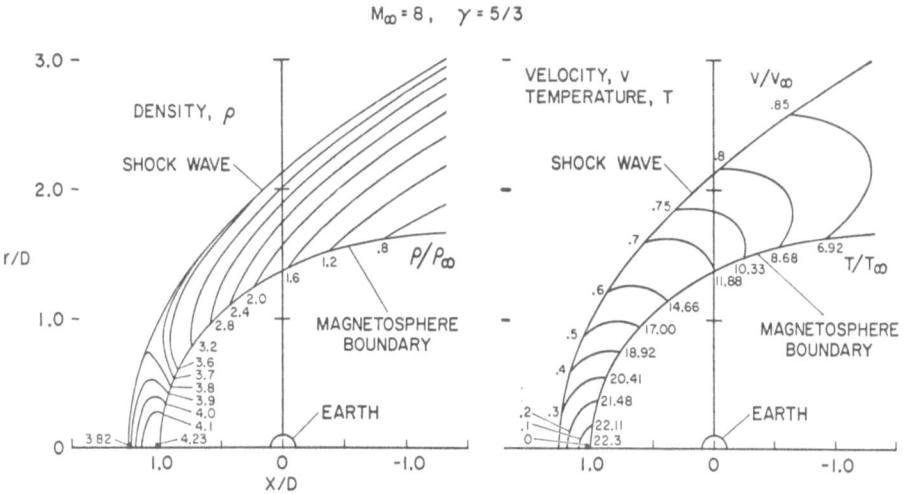

Fig. 7. Isopleths of density, velocity, and temperature between the Earth's bow shock and the magnetopause (i.e. in the magnetosheath) calculated with supersonic, hydromagnetic flow theory. The upstream Mach number is $M_\infty = 8$, typically assumed for the solar wind, and the polytropic index is $\gamma = 5/3$, appropriate to an ideal gas. (Spreiter and Alksne, 1966.)

the magnetosheath. Here the flow is subsonic, and particle fluxes are nearly isotropic.

Comparisons between predicted values and observations of solar wind ions for the magnetosheath passage of Pioneer 6 show good agreement, except that magnetosheath densities appear to be somewhat too low (Spreiter and Alksne, 1968; Howe, 1970). The calculations, based on single-fluid hydro-dynamics, give reasonable values for the proton temperatures. It is evident from the brief magnetosheath encounter displayed in Figure 6 that the electrons are also heated by the bow shock, but their temperature is not well predicted by single-fluid calculations. The relative amounts of electron and proton heating depends on which dissipation mechanism dominate in the shock structure, and this is not completely understood, except that apparently protons are heated to a greater extent, which explains why the single fluid model gives a good approximation to their temperature. Analysis of Vela 4 magnetosheath data shows that the proton temperature is typically 2 to 4 times greater than the electron temperature (Montgomery *et al.*, 1970).

6. Magnetosphere Particles

The charged particle populations in the magnetosphere and geomagnetic tail extend in energy from the thermal plasma in the plasmasphere with a temperature of a few $\times 10^3$ K to the relativistic electrons in the trapped radiation belts. Their properties as they were known up to 1967 are reviewed by Frank (1968) and up to 1968 by Van Allen (1969) and Gringauz (1969), and the trapped radiation is fully discussed in a recent book by Roederer (1970). Subsequent information, especially on the low-energy (< 100 keV) populations, has added considerably to completing the picture of the spatial distribution and dynamic behavior. We review here just the low-energy populations; they have the largest fluxes in the magnetosphere and geomagnetic tail.

Figure 8, a noon-midnight cross-section through the magnetosphere including the geomagnetic tail, shows the main low-energy populations: the plasma sheet, the polar cusp and the plasmasphere. The region labeled the particle cusp contains the inner edge of the plasma sheet and quasi-trapped energetic electrons. The plasmasphere is also the region of space where the radiation belts occur. The neutral sheet is part of the plasma sheet as far as the particle population is concerned, and merely labels the region where the magnetic field reverses direction in the tail from away-from-Earth in the southern lobe to toward-Earth in the northern lobe.

6.1. The plasma sheet

The plasma sheet occupies the central portion of the tail in the north-south direction but extends completely across the tail in the east-west direction. A cross-section through the tail giving the location of the plasma sheet is shown in Figure 9. The cross-section is intended to be representative of 20 Earth radii (R_E) geocentric distance where the tail radius is approximately 20 R_E. The plasma sheet is thinner in the middle of the tail (4–6 R_E) than near the boundary (10–12) (Bame *et al.*, 1967; Hones, 1968). The plasma sheet maintains essentially uniform shape and properties out at least to the orbit of

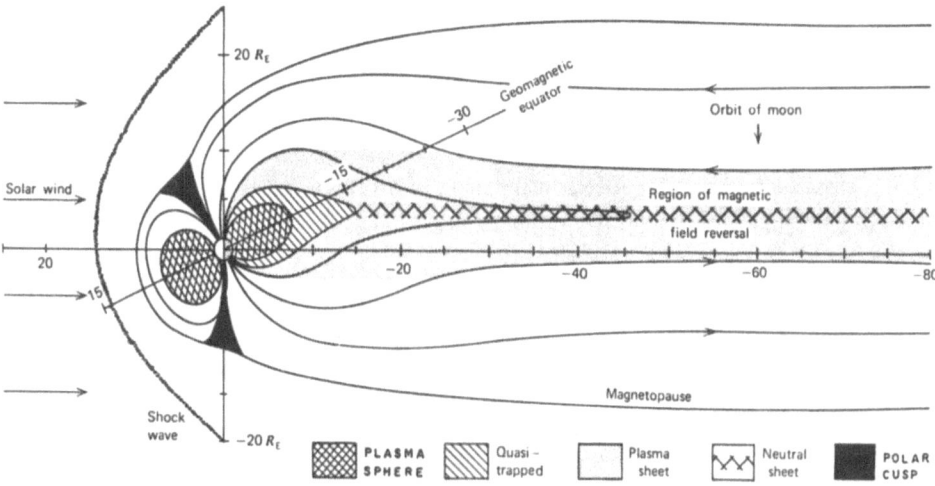

Fig. 8. A noon-midnight meridian plane cross-section through the magnetosphere and geomagnetic tail showing several components of the charged particle populations. (Modified from Ness, 1969.)

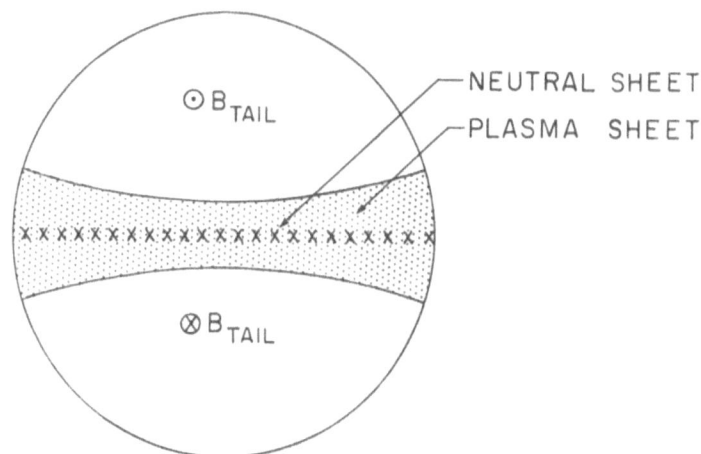

Fig. 9. A cross-section of the geomagnetic tail at approximately 20 R_E geocentric distance, general location and shape of the plasma sheet and neutral sheet.

Moon (60 R_E) (Meng and Mihalov, 1972; Nishida and Lyon, 1972). As is indicated in Figure 8, the near-Earth portion of the plasma sheet extends along magnetic field lines down to the Earth's atmosphere at auroral latitudes (roughly 60° to 75°). The inner edge of the plasma sheet in the equatorial plane is different for electrons and protons. The electron inner edge lies typically between 6 and 8 R_E in the midnight sector. It maintains this distance in the midnight sector at least to the dawn terminator. In the pre-midnight sector it goes from the midnight location out to greater distances, swinging in an approximate 10 R_E circle past the dusk terminator to touch the magnetopause near the subsolar point (Vasyliunas, 1968a, b; Frank, 1971a). The proton component of the plasma sheet emerges continuously with the ring current protons which circle the Earth in closed drift shells. The inner edge of the ring current tends to lie within one Earth radius inside the plasmapause (the outer edge of the plasmasphere).

The absolute and relative locations of these various features are quite variable as is indicated in Figure 10, which shows the position in the local midnight, near equatorial sector of the magnetosphere, of the inner edge of the ring current, the plasma sheet electrons (labeled plasma sheet), the region of decreasing plasma sheet electron fluxes (labeled earthward edge of plasma sheet), and the trapping boundary for >40 keV electrons from six consecutive series of observations from the OGO-3 satellite. Although the ring current and plasma sheet are shown as having sharp outer edges for graphical convenience, these populations merge smoothly with the plasma sheet protons and electrons in the tail. Considerable variations in the locations are evident from one series of observations to the next; however certain correlations can be seen. The protons extend closer to the Earth than the electrons, and the ring current inner edge is closely associated with the plasmapause.

The location of the geostationary orbit (6.6 R_E), as shown by a dashed line in the

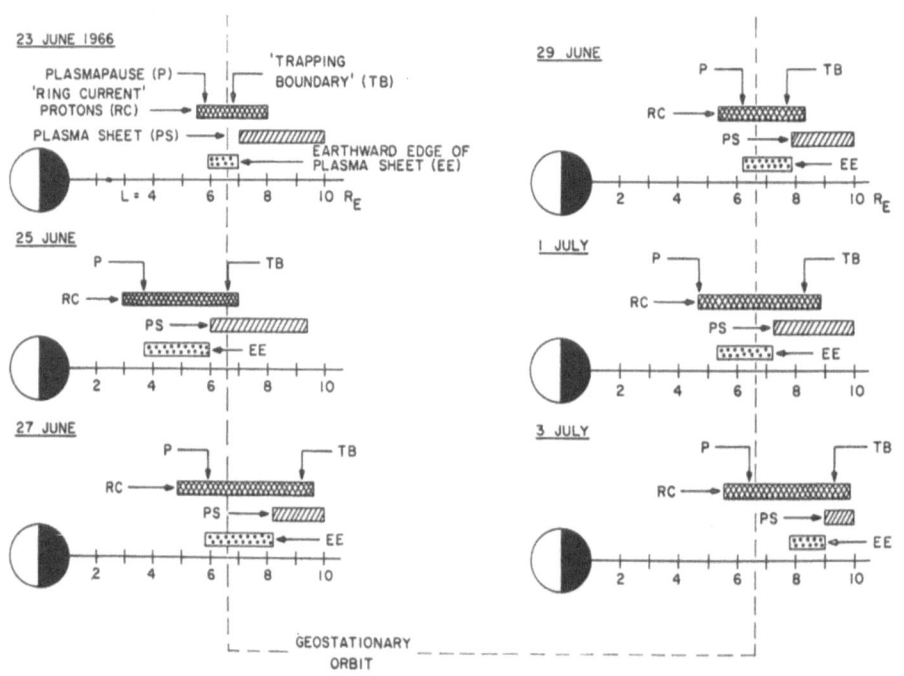

Fig. 10.   Relative locations of the plasmapause, ring current, plasma sheet, and trapping boundary for > 40 keV electrons as determined from OGO 3 data in the near-equatorial, local midnight sector on six consecutive series of observations. (From Frank, 1971a.)

figure is subject to a highly variable particle environment. It is sometimes inside and sometimes outside the region of plasma sheet electrons. Although not indicated in this series, other measurements show that it is also sometimes within the plasmasphere.

Figure 11 shows three electron energy spectra measured at different times in the plasma sheet by Vela satellites and also a spectrum from the magnetosheath for comparison. Whereas the magnetosheath electrons have a mean energy of approximately 0.1 keV, the mean energy of plasma sheet electrons ranges from several hundred eV to 10 keV, but it is typically 1 keV. Further statistics of the electron population are given in Figure 12. The number density is typically 1 cm^{-3} but ranges

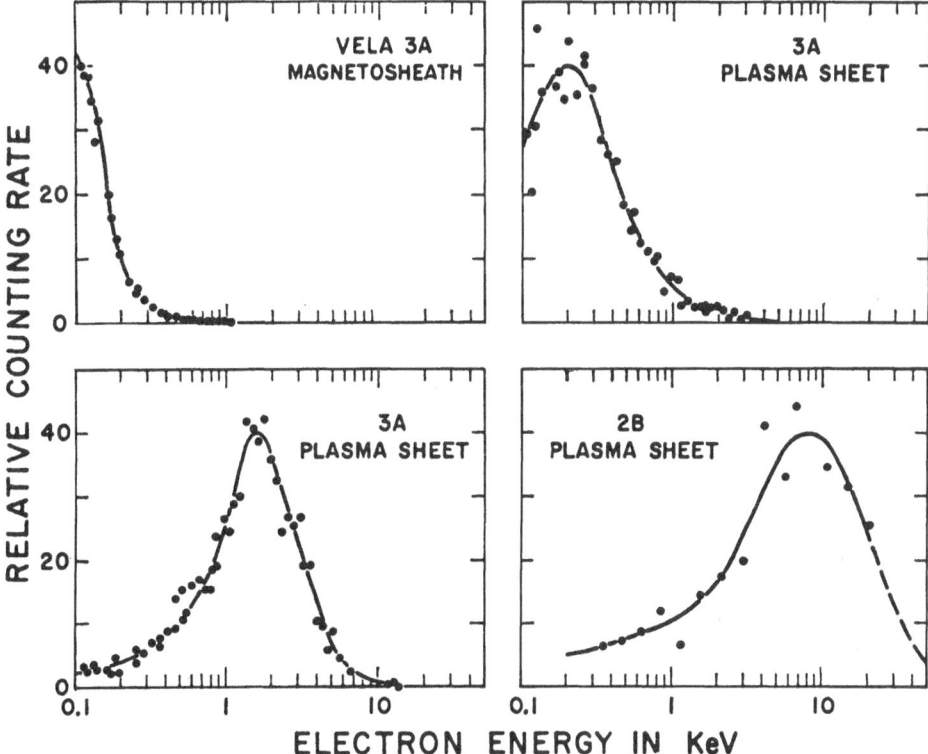

Fig. 11. Vela determined spectra of electrons in the magnetosheath and three samples from the plasma sheet showing range of thermal properties. (From Bame *et al.*, 1967.)

roughly from 0.1 cm^{-3} to 10 cm^{-3}. The omnidirectional flux generally lies in the range 10^9 to 10^{10} cm^{-2} s^{-1}. There are order of magnitude variations in these parameters, but there is a tendency for the mean energy per particle and number density to be anticorrelated such that the energy density remains roughly constant with a value 1 keV cm^{-3} (Vasyliunas, 1968a). It is found that high mean energies and low densities are associated with quiet magnetospheric conditions, and low mean energies with high densities characterize magnetic disturbances.

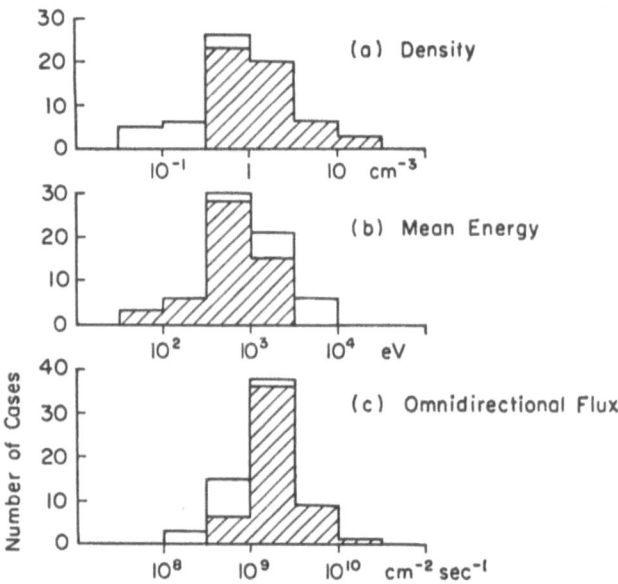

Fig. 12. Distributions of plasma sheet electron parameters determined from
OGO 1 and 3 observations. (From Vasyliunas, 1968a.)

The proton number density is very nearly the same as that of the electrons, as is required for charge neutrality. In the near-Earth region where there are protons but no plasma sheet electrons, charge neutrality is maintained by low energy electrons. The inner edge of the electron plasma sheet is not a density edge, but rather a temperature edge (Vasyliunas, 1968a; Schield and Frank, 1970). In the plasma sheet average proton energy is approximately 6 times that of the electrons (6 keV for protons, 1 keV for electrons) (Hones, 1970). The proton and electron fluxes are isotropic to within experimental uncertainties. In the case of the protons, the uncertainties are fairly large and substantial anisotropy could be present.

During magnetically quiet times, no flows are observed in the plasma sheet, again within fairly large experimental uncertainties. But, flows with speeds of the order of several hundred km s^{-1} have been reported to occur in association with magnetic substorms at Vela satellite distance ($\sim 18\ R_E$) (Hones et al., 1972) and substorm associated ion bursts or flows have been observed at lunar distance (Garret et al., 1971; Prakash, 1972). The plasma sheet is observed to become thin prior to and during the expansion phase of magnetic substorms ($< 1\ R_E$ thickness) and to expand during the recovery phase of the storm (Hones, 1970). As the plasma sheet thins, less energetic particles are observed. More energetic particles populate the post-storm expanded plasma sheet.

The near-Earth portion of the plasma sheet responds to magnetic disturbance by Earthward motion of the inner edge of the electron plasma sheet (Vasyliunas, 1968a) and injection of plasma sheet protons into the proton ring current in the evening sector (Frank, 1970). Ring current proton intensities vary with geomagnetic activity. An

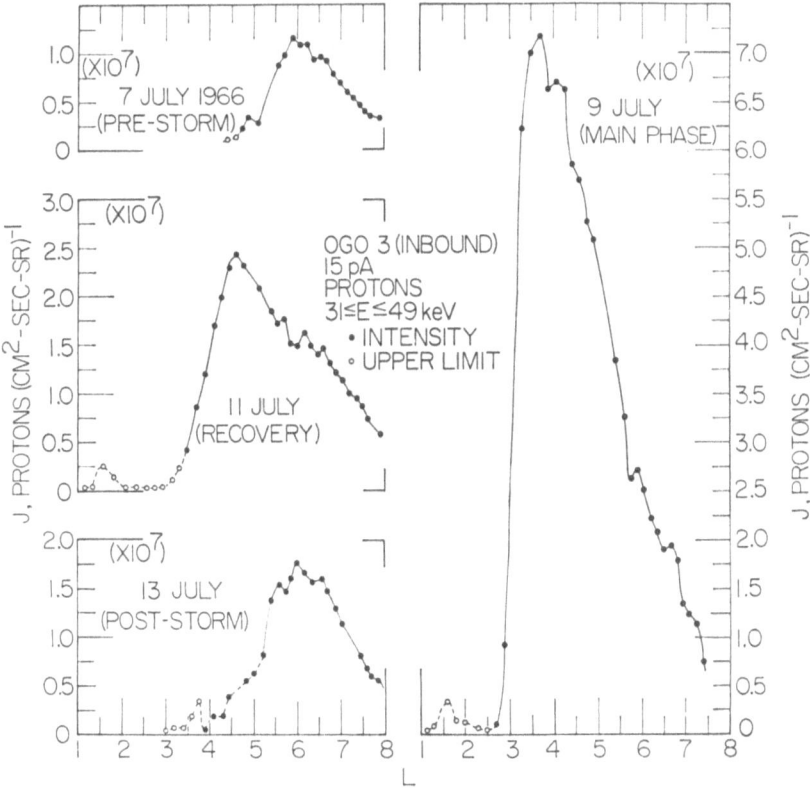

Fig. 13. Directional intensities of ring current protons in the energy range 31 keV to 49 keV as functions of geocentric distance (measured by the L parameter identifying the equatorial crossing distance in units of R_E of the field line through the observation point) at low magnetic latitudes measured by OGO 3 during the pre-storm, main phase, recovery phase, and post storm periods of a magnetic storm in July, 1966. (From Frank, 1967.)

example of pre-storm, storm, and post-storm proton directional intensities as a function of distance from Earth (L in Earth radii) near the equatorial plane for a moderate magnetic storm in July 1966 is shown in Figure 13. The intensity varies from a pre-storm value of 1×10^7 cm^{-2} s^{-1} sr^{-1} to a main phase value of 7×10^7 cm^{-2} s^{-1} sr^{-1}. The location of the peak intensity moves Earth-ward as magnetic activity increases (from $L=6$ pre-storm to $L=3.5$ main phase). The energy density of ring current electrons is, for this example, a factor of 3 to 10 less than that of the protons, the ratio varying with L; thus, electron fluxes are an order of magnitude greater than the proton fluxes.

6.2. THE POLAR CUSP

The regions labeled polar cusp in Figure 8 extend in the noon-midnight meridian plane from the north and south dayside auroral ionosphere in funnel-shaped regions into the polar magnetosphere, intersecting the magnetopause in the vicinity of the magnetic

neutral points predicted by the early models of the solar wind-magnetosphere interaction (for example, Mead and Beard, 1964). Apparently the weak magnetic field near
the magnetopause at these two locations allows solar wind plasma from the magnetosheath to enter the magnetosphere and extend down to the ionosphere along
magnetic field lines (Gringauz, 1969; Heikkila and Winningham, 1971; Frank, 1971b,
c). Space-craft observations show electron and proton properties in the polar cusps to
be the same as in the magnetosheath; namely electron energies peaked at about 100 to
200 eV with a flux of the order of 10^9 cm^{-2} s^{-1} sr^{-1}, proton energies peak near
300 eV and a flux typically $> 10^7$ cm^{-2} s^{-1} sr^{-1}.

During magnetically quiet times, the north polar cusp intersects the dayside
auroral zone at a latitude near 79° and extends north and south a distance between 20
to 400 km. During magnetic disturbances the ionospheric intersection of the polar
cusps moves equatorward by several degrees and the north-south width increases by
approximately a factor of two or less.

The shape of the polar cusps away from the noon-midnight meridian plane is not
yet determined from spacecraft observations. A popular notion is that polar cusp
plasma merges continuously with the plasma sheet. In this model (Frank, 1971b) there
is one intricately contoured region of plasma comprising both the polar cusps and
plasma sheet. This view is not necessarily inconsistent with the observed differences in
the thermal properties of the plasma in polar cusps and plasma sheet since the model
considers only the spatial distribution of plasma and not the history of the plasma as
a function of location. An attempt to depict this one-plasma-region model is shown in
Figure 14, which is a 'quartered' magnetosphere formed by cutting through the noon-
midnight meridian plane and through the equatorial plane. In this figure only the
electron component is represented. The quarter selected as viewed from the sun is
the south-east quadrant, and the view given in the figure is as seen approximately
from the moon at a little past third quarter phase and looking down from above
the equatorial plane. The outer-most boundary (furthest into the page in this three
dimensional representation) is the magnetopause. The areas where the meridian plane
and equatorial plane cut through the plasma sheet and polar cusp are indicated by
stripes. The remaining two curved boundaries are the upper and lower boundaries to
the plasma region, the upper boundary is its earthward edge and the lower one is its
boundary with the southern polar cap region and the southern plasma-free tail lobe.
The plasma sheet and polar cusp are seen to be parts of a single circumpolar plasma
curtain. The lines drawn on the magnetopause and plasma curtain boundaries are
magnetic field lines, that is, the magnetic field lines lie in (do not cross) these surfaces;
although this is not an essential requirement for the shown plasma configuration.
Although the model is probably inaccurate in detail, especially at high altitudes and
near the magnetopause, it represents a minimum-boundary configuration consistent
with the known regions of plasma. It is known that dayside auroras exist and that
auroras are associated with the plasma sheet electrons. The plasma sheet must therefore circumscribe the Earth. If the polar cusp is distinct from the plasma sheet, there
must be more plasma boundaries on the dayside than shown in Figure 14.

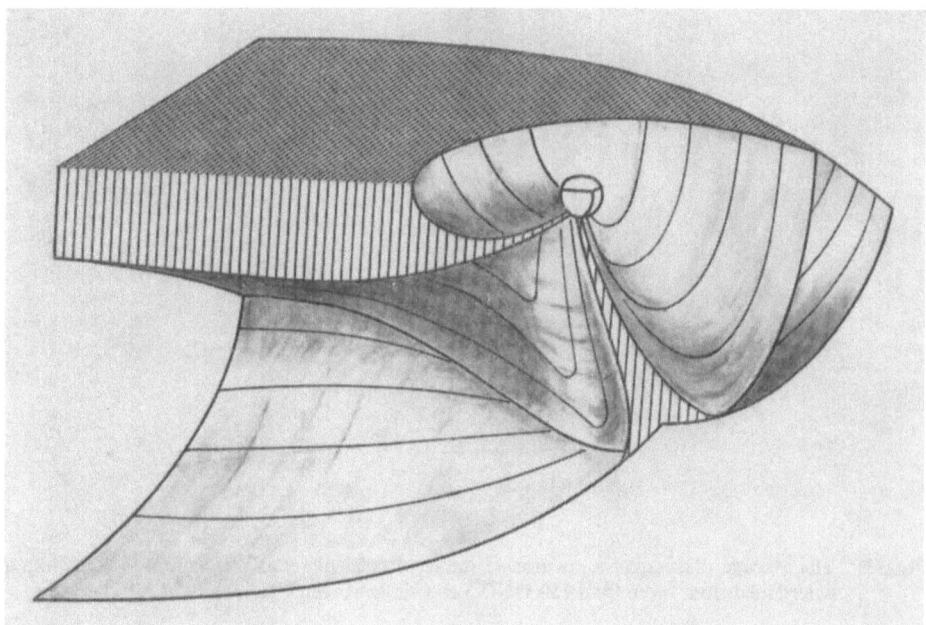

Fig. 14. An attempt at a three-dimensional representation of the merged plasma sheet-polar cusp model. The figure shows a 'quartered' magnetosphere formed by cutting through the noon-midnight meridian plane and through the equatorial plane. The plasma sheet and polar cusp plasma are shown as being different parts of a single volume of plasma, a plasma curtain, almost completely circling the Earth and intersecting the ionosphere in a circumpolar circle in the auroral zone. Boundaries marked with straight, parallel lines are cuts through plasma curtain. Curved lines are magnetic field lines lying in the plasma curtain boundary and magnetopause (the outermost boundary).

Inside the Earthward edge of the plasma curtain lies the plasmasphere, a doughnut-shaped region of several thousand degree temperature plasma derived from the upward extension of the Earth's ionosphere.

6.3. THE PLASMASPHERE

The region of space bounded below by the ionosphere and above by the shell formed by geomagnetic field lines which intersect the Earth at latitudes near $60°$ is occupied by the plasmasphere, essentially the upward extension of the ionosphere. At the outer boundary of the plasmasphere, the plasmapause, the H^+ density drops from several hundred cm^{-3} to between 1 and 10 cm^{-3} in a radial distance small compared to 1 R_E. The equatorial plasmapause lies on average between 4 and 5 R_E, but it deviates from circular symmetry in being further from the Earth in the evening sector than in the morning sector, as is shown in Figure 15. Number densities of H^+, He^+, and O^+ ions on a typical pass through the plasmasphere are shown in Figure 16. The plasmapause is evident by drops in both the H^+ and He^+ densities. Significant densities and structure beyond the plasmapause can be seen in the H^+ profile. The figure also shows the transition from the O^+ dominated ionosphere below $L \sim 2$ to the H^+ dominated protonsphere for larger values of L.

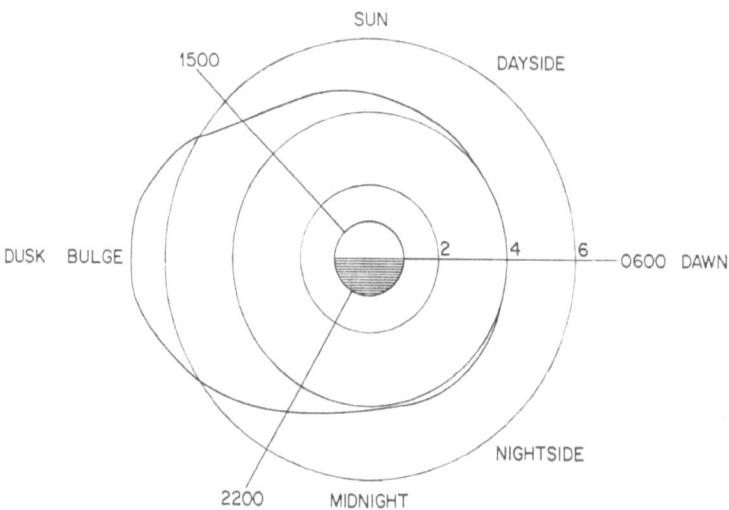

Fig. 15 The average plasmapause position in the equatorial plane as a function of local time as determined from more than 150 OGO 5 crossings. (From Chappell *et al.*, 1971.)

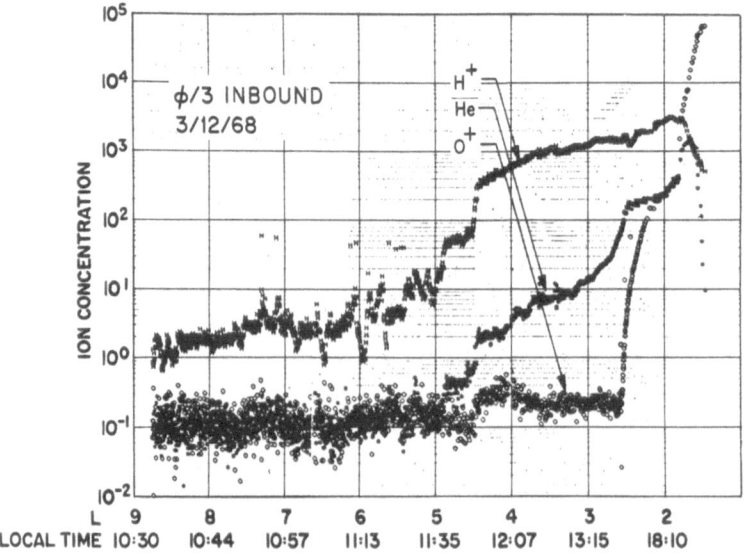

Fig. 16. A typical OGO 5 pass through plasmasphere showing profiles of H^+, He^+, and O^+ ion concentrations as a function of L (see caption of Figure 13). (From Chappell *et al.*, 1970.)

The location of the plasmapause shows considerable variation as a function of magnetic activity. The equatorial location can vary over a radial range of 3 R_E or more, coming closer to the Earth at times of magnetic disturbance and moving out during magnetically calm conditions. Also associated with magnetic disturbances, portions of the outer plasmasphere are observed to separate from the inner plasmasphere (to peel off) and move further out into the magnetosphere, primarily toward the dayside. The peeling off is observed most frequently in the evening sector (Chappell *et al.*, 1971).

The formation of the plasmasphere, or equivalently the reason for a sharp outer boundary to the ionosphere, appears to be due to magnetospheric convection, or motion, which sweeps plasma away beyond the boundary. The distance from the Earth where the plasma sweeping action is effective moves Earthward with increasing magnetic activity, thus accounting for the variation in the plasmapause location. The time scale for sweeping away as magnetic activity increases and for filling up, as magnetic activity declines is on the order of several hours to a day. The filling up of the plasmasphere after magnetic disturbances occurs from upward fluxes of ions (of the order of 3×10^8 cm^{-2} s^{-1}) from the dayside ionosphere (Chappell *et al.*, 1971).

The motion of the plasma in the plasmasphere according to theoretical models, is characterized by corotation with the Earth at low altitudes but with substantial deviations from corotation occurring at greater altitudes, and these deviations depend on local time position. Near the plasmapause in the morning sector ($\sim$0600 local time), the theory predicts motion to be super-rotation, approximately 17% greater than corotation. At the plasmapause near the evening terminator (1800 local time), there occurs a stagnation point between corotating motion and magnetospheric convection; the plasma there is very nearly at rest and remains at that local time.

6.4. THE NEUTRAL EXOSPHERE

Thus far we have considered only components of the charged particle populations in space, the sources of which are the Sun (solar wind and solar energetic particles), the ionosphere, and galactic cosmic rays. The sources of neutral particles in space are planetary atmospheres, comets, and the interstellar medium. Although there has been some discussion on the point, it seems likely that the high temperature of the solar corona prevents a significant flux of neutral particles from the Sun.

In a review of the interaction of the solar wind with the interstellar medium, Axford (1972) gives a range for the interstellar atomic hydrogen density in the neighborhood of the solar system of $n(\text{H}) \approx 0.03$–0.12 from interpretations of backscattered solar Lα measurements. At 1 AU, this number is reduced by ionization to a value on the order of 1% of the interstellar density. The local interstellar neutral population is usually assumed to be a cold gas moving at a speed of the order of 20 km s^{-1} relative to the Sun. Fluxes of fast neutrals can result from charge exchange between the solar wind and atmospheric neutrals or interstellar neutrals. In the charge exchange process, the fast solar wind H$^+$ ion is changed into a fast H neutral, and the cold (atmospheric or interstellar) H neutral is converted to a cold H$^+$ ion. The interaction with the interstellar medium produces fluxes of solar-wind-speed neutrals moving away from the Sun. The flux as a function of heliocentric distance has a very broad peak near 15 AU with a value of approximately 10^4 cm^{-2} s^{-1}, the actual value depends on the interstellar neutral density and velocity relative to the Sun (Siscoe and Vasyliunas, 1972).

The interaction of the solar wind with the atmospheres of Mercury, Venus, and

Mars will also produce fluxes of fast neutrals, but numerical values have not been calculated. The solar wind will interact with the small densities of neutrals in the Earth's exosphere beyond the magnetopause (geocentric distance$\gtrsim 10\,R_E$ except during unusual solar wind disturbances that can push it closer to the Earth) to produce small neutral fluxes. Fast neutrals will result from charge exchange with the magnetospheric charged populations (the plasma sheet and ring current protons) especially during magnetic storms when the ring current is strong and close to the Earth. A major contributor to the decay of the ring current in the recovery phase of a magnetic storm, which has a time scale on the order of 1 day, is thought to be charge exchange with exospheric neutrals.

Turning now to the atmospheric neutrals, above 200 to 300 km altitude the atmosphere is essentially isothermal and the composition varies as each species decreases exponentially with 'geopotential height' (which allows for the decrease of gravity with height) according to its own scale height. However, the exosphere temperature varies with the level of solar activity and the local time. Over a solar cycle, the temperature ranges from a nighttime, solar minimum value of 700 K to a daytime, solar maximum value of 1800 K. An average value of 1200 K is sometimes assumed. The daytime maximum occurs near 1400 local time and the nighttime minimum near 0300 local time. Exospheric composition and densities are determined by the temperature and therefore also vary with time of day and solar cycle. Figure 17 shows derived density-height profiles for H, He, A, O, O_2, and N_2 up to a geometric altitude of 2500 km for three different temperatures covering the normal range of temperatures. At a given altitude the density of a given species increases with temperature, except for H which shows the reverse behavior. The lightest species dominate at high altitudes, with H dominating at the greatest heights (the region referred to as the hydrogen geocorona). The altitude where H begins to dominate is near 500 for the lowest temperature and increases to more than 2500 km at the highest temperature. The total mass density varies in positive correlation with the temperature. At 350 km the density varies an order of magnitude (from $\sim 2 \times 10^{-15}$ g cm^{-3} to 2×10^{-14} g cm^{-3}) over a solar cycle, while the temperature ranges from 700 K to 1800 K (less than a factor of 3).

In addition to the regular solar cycle and diurnal variations, there are a number of other variations which have been reviewed by Jacchia (1971). Solar activity associated with individual active regions on the Sun and solar flares at any time during the solar cycle produce increases in the exospheric temperature, and related density and composition changes. These changes are well correlated with the emission of solar 10.7 cm wavelength radiation. The exospheric temperature also increases with geomagnetic activity as much as 500 K during magnetic storms. The storm-time energy input to the atmosphere from the magnetosphere is probably localized in latitude to the auroral zone region (60° to 70°), giving geographic variations of temperature and composition structure during storms. However, it is believed that the deposited energy is translated to other latitudes by gravity waves and winds in a time scale of the order of hours. Finally, there is an approximately 20% semiannual variation in mass density with

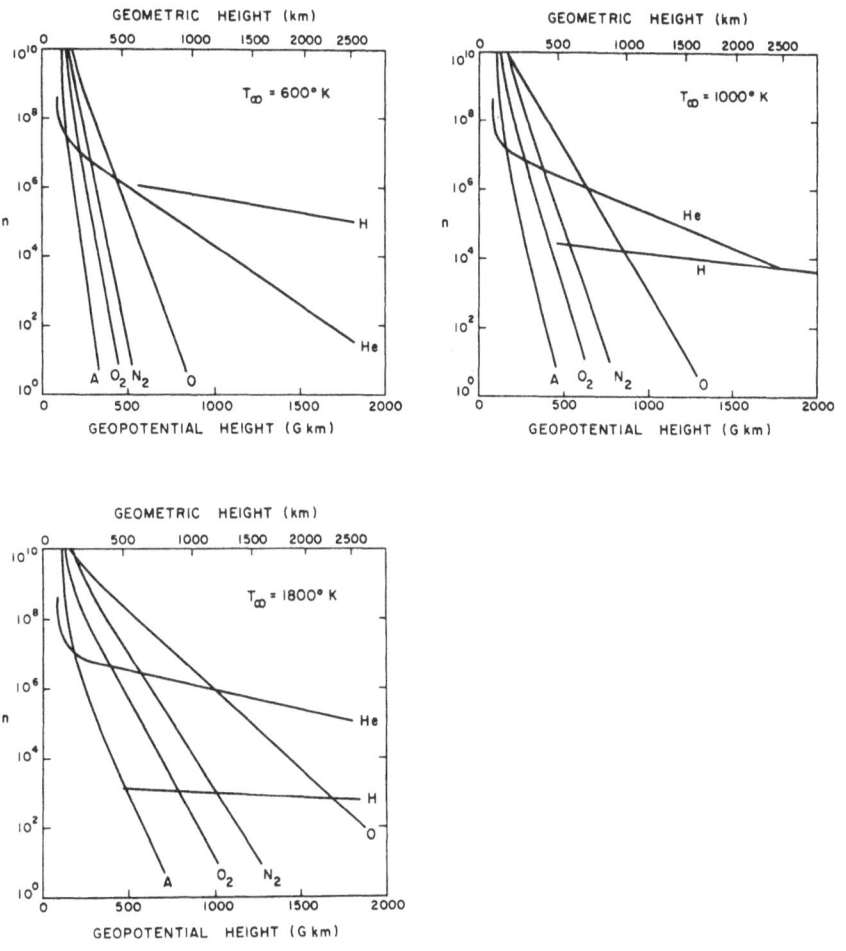

Fig. 17. Calculated atmospheric compositions for three values of the exospheric temperatures. Number densities for atomic and molecular species H, He, A, O, O_2, and N_2 are given as a function of geopotential height (bottom of figures) and geometric height (top of figures). (From Jacchia, 1971.)

maxima near the equinoxes and minima near the solstices, for which there is as yet no satisfactory explanation.

Acknowledgement

This work was supported by the Atmospheric Sciences Section of the National Science Foundation under Grant NSF-GA 31842.

References

Axford, W. I.: 1968, *Space Sci. Rev.* **8**, 331.
Axford, W. I.: 1972, in C. P. Sonnet, P. J. Coleman, and J. M. Wilcox (eds.), *Solar Wind* (Proceedings of the Asilomar Solar Wind Conference, March 1971) U.S. Government Printing Office, p. 609.

Bame, S. J., Asbridge, J. R., Felthauser, H. E., Hones, E. W., and Strong, I. B.: 1967, *J. Geophys. Res.* **72**, 113.

Bame, S. J., Hundhausen, A. J., Asbridge, J. R., and Strong, I. B.: 1968, *Phys. Rev. Letters* **20**, 393.

Bame, S. J., Asbridge, J. R., Hundhausen, A. J., and Montgomery, M. D.: 1969, *Trans. Am. Geophys. Union* **50**, 301.

Belcher, J. W. and Davis, L., Jr.: 1971, *J. Geophys. Res.* **76**, 3534.

Chappell, C. R., Harris, K. K., and Sharp, G. W.: 1970, *J. Geophys. Res.* **75**, 50.

Chappell, C. R., Harris, K. K., and Sharp, G. W.: 1971, *J. Geophys. Res.* **76**, 7623.

Dryer, M.:1970, *Cosmic Electrodyn.* **1**, 115.

Frank, L. A.: 1967, *J. Geophys. Res.* **72**, 3753.

Frank, L. A.: 1968, in B. M. McCormac (ed.), *Earth's Particles and Fields*, Reinhold, New York, p. 67.

Frank, L. A.: 1970, *J. Geophys. Res.* **75**, 1263.

Frank, L. A.: 1971a, *J. Geophys. Res.* **76**, 2265.

Frank, L. A.: 1971b, *J. Geophys. Res.* **76**, 2512.

Frank, L. A.: 1971c, *J. Geophys. Res.* **76**, 5202.

Garret, H. B., Hill, T. W., and Fenner, M. A.: 1971, *EOS* **52**, 326.

Goldstein, B. and Siscoe, G. L.: 1972, in C. P. Sonett, P. J. Coleman and J. M. Wilcox *Solar Wind* (Proceedings of the Asilomar Solar Wind Conference, March, 1971), U. S. Government Printing Office, p. 506.

Gosling, J. T., Asbridge, J. R., Bame, S. J., Hundhausen, A. J., and Strong, I. B.: 1968, *J. Geophys. Res.* **73**, 43.

Gosling, J. T., Hansen, R. T., and Bame, S. J.: 1971, *J. Geophys. Res.* **76**, 1811–1815.

Gringauz, K. I.: 1969, *Rev. Geophys.* **7**, 339.

Heikkila, W. J. and Winningham, J. D.: 1971, *J. Geophys. Res.* **76**, 883.

Hones, E. W., Jr.: 1968, in Carovillano, McClay, and Radoski (eds.), *Physics of the Magnetosphere*, Springer-Verlag New York Inc., New York, p. 392.

Hones, E. W., Jr.: 1970, in B. M. McCormac (ed.), *Particles and Fields in the Magnetosphere*, D. Reidel Publishing Co., Dordrecht, Holland, p. 24-34.

Hones, E. W., Jr., Asbridge, J. R., Bame, S. J., Montgomery, M. D., Singer, S., and Akasofu, S. I.: 1972, *J. Geophys. Res.*, in press.

Howe, H. C., Jr.: 1970, *J. Geophys. Res.* **75**, 2429.

Howe, H., Binsack, J., Wang, C. G., and Clapp, E.: 1971, MIT-Center for Space Research, Technical Report-4.

Hundhausen, A. J.: 1968, *Space Sci. Rev.* **8**, 690.

Hundhausen, A. J.: 1970a, *Rev. Geophys. Space Phys.* **8**, 729.

Hundhausen, A. J.: 1970b, in B. M. McCormac (ed.), *Particles and Fields in the Magnetosphere*, Springer-Verlag, New York, p. 79.

Hundhausen, A. J., Bame, S. J., Asbridge, J. R., and Sydoriak, S. J.: 1970a, *J. Geophys. Res.* **75**, 4643.

Hundhausen, A. J., Bame, S. J., and Montgomery, M. D.: 1970b, *J. Geophys. Res.* **75**, 4631–4642.

Hundhausen, A. J., Bame, S. J., and Montgomery, M. D.: 1971, *J. Geophys. Res.* **76**, 5145–5154.

Jacchia, L. G.: 1971, Smithsonian Astrophysical Observatory Special Report 332.

Kavanagh, L. D., Jr., Schardt, A. W., and Roelof, E. C.: 1970, *Rev. Geophys. Space Phys.* **8**, 389.

Mead, G. D. and Beard, D. B.: 1964, *J. Geophys. Res.* **69**, 1169.

Meng, C. I. and Mihalov, J. D.: 1972, *J. Geophys. Res.* **77**, 1739.

Mihalov, J. D. and Wolfe, J. H.: 1971, *Cosmic Electrodyn.* **2**, 326–339.

Montgomery, M. D., Bame, S. J., and Hundhausen, A. J.: 1968, *J. Geophys. Res.* **73**, 4999–5003.

Montgomery, M. D., Asbridge, J. R., and Bame, S. J.: 1970, *J. Geophys. Res.* **75**, 1217.

Montgomery, M. D., in C. P. Sonett, P. J. Coleman and J. M. Wilcox (eds.), *Solar Wind* (Proceedings of the Asilomar Solar Wind Conference, March, 1961), U.S. Government Printing Office, p. 208.

Neugebauer, M. and Snyder, C. W.: 1966, *J. Geophys. Res.* **71**, 4469.

Ness, N. F.: 1969, *Rev. Geophys.* **7**, 97.

Nishida, A. and Lyon, E. F.: 1972, *J. Geophys. Res.* **77**, 4086.

Ogilvie, K. W., Burlaga, L. F., and Wilkerson, T. D.: 1968, *J. Geophys. Res.* **73**, 6809.

Ogilvie, K. W. and Wilkerson, T. D.: 1969, *Solar Phys.* **8**, 435.

Olbert, S.: 1968, in Carovillano, McClay, and Radoski (eds.), *Physics of the Magnetosphere*, D. Reidel Publishing Co, Dordrecht-Holland, p. 641.

Prakash, A.: 1972, *J. Geophys. Res.* **77**, 5633.

Robbins, D. E., Hundhausen, A. J., and Bame, S. J.: 1970, *J. Geophys. Res.* **75**, 1178.

Roederer, J. G.:1970, *Dynamics of Geomagnetically Trapped Radiation*, Springer-Verlag, New York.

Schield, M. A. and Frank, L. A.: 1970, *J. Geophys. Res.* **75**, 5401.

Siscoe, G. L. and Vasyliunas, V. M.: 1972, *EOS* **53**, 1109.

Spreiter, J. R., Summers, A. L., and Alksne, A. Y.: 1966, *Planetary Space Sci.* **14**, 223–254.

Spreiter, J. R. and Alksne, A. Y.: 1968, *Planetary Space Sci.* **16**, 971.

Spreiter, J. R. and Alksne, A. Y.: 1969, *Rev. Geophys.* **7**, 11–50.

Van Allen, J. A.: 1969, *Rev. Geophys.* **7**, 233.

Vasyliunas, V. M.: 1968a, *J. Geophys. Res.* **73**, 2839.

Vasyliunas, V. M.: 1968b, *J. Geophys. Res.* **73**, 7519.

Vasyliunas, V. M.: 1971, in R. H. Looberg (ed.), *Methods of Experimental Physics*, Vol. 9B of *Plasma Physics*, Academic Press, pp. 49–88.

Wolfe, J. H.: 1972, in C. P. Sonett, P. J. Coleman, and J. M. Wilcox (eds.), *Solar Wind* (Proceedings of the Asilomar Solar Wind Conference, March, 1971), U.S. Government Printing Office, p. 170.

DISCUSSION

Gold: Do we have any information concerning the mean motion of protons and electrons in the magnetotail? This mean motion will define the difference between the bombardment on the front and the back of the Moon.

Siscoe: The available data are not very definitive and give a rough upper limit on a steady flow of approximately 100 km s^{-1}. Flows of short duration, up to approx. 10 min, with speeds of several hundred km s^{-1} have been seen in connection with magnetic substorms. These sporadic flows have been seen by the Vela satellite at 18 R_E in the tail, by the lunar orbiting Explorer 35, and by a detector on the lunar surface. Except for the event observed by the detector on the lunar surface, these flows were directed toward the Earth.

Freeman: Regarding the question by Dr Gold about shadowing effects by the moon in the plasma-sheet, with the ALSEP SIDE we have seen streaming ion enhancements from the earth direction during a magnetic substorm. Indeed it seems that ion flow enhancement occurs principally when there is a magnetic disturbance. The SIDE experiment being located on the lunar surface is not in a position to see the ion fluxes on the far side of the Moon, however, to really look for lunar shadowing, a lunar orbiting vehicle is required. Recently Nishida and Lyons in *J. Geophys. Res.* have reported such a shadowing effect seen by a plasmaprobe on Explorer 35.

Kaiser: The plasmasphere boundary (plasmapause) does not respond in a simple way to magnetic activity. Recurrent substorms generate a longitudinal modulation of the boundary which co-rotates with the earth and isolated plasma clouds, which travel from the evening bulge towards the Sun.

2. INTERACTIONS WITH SPACECRAFT

2.1. THEORETICAL SHEATH MODELS

SPHERICALLY SYMMETRIC MODEL OF THE PHOTOELECTRON SHEATH FOR MODERATELY LARGE PLASMA DEBYE LENGTHS

HEINZ SCHRÖDER

Lehrstuhl B für Theoretische Physik, Technische Universität Braunschweig,
33 Braunschweig, Mendelssohnstraße 1 A, Deutschland

Abstract. The potential distribution is investigated in the neighbourhood of a conducting, electron emitting sphere surrounded by a collisionless plasma. The energy dependence of the emitted electrons is adopted in accordance with data on photoemission, but their spatial distribution is treated spherically symmetric. A Maxwellian velocity distribution with zero average velocity is assumed for the undisturbed plasma electrons. The ions are taken as uniform background. A self consistent solution of the collisionless Boltzmann equation for the electrons and Poisson's equation has been obtained by an iterative scheme. Results are presented for plasma parameters corresponding to solar wind conditions at 1 AU. The role of the plasma- and photoelectron Debye lengths is discussed.

Photoemission is the source of strong variations of the density and potential near a space probe in the interplanetary medium. As the photoelectron flux is normally an order of magnitude larger than the thermal plasma electron flux, a positive potential will develop on a sunlit surface. Some work has been done to determine its value from the current balance (Whipple, 1965; Wyatt, 1969; Middendorf *et al.*, 1970). As the mean energy of the photoelectrons is of the order of one to three volts, the photoelectron flux reacts most sensitively to increasing surface potential and a positive potential of some volts will arise.

The non-rationalized CGS electrostatic system of units is used throughout this paper. For the photoelectron sheath a rapid decrease of potential and density is expected due to the small photoelectron Debye-length $H_{Ph} = (kT_{Ph}/4\pi e^2 N_{Pho})^{1/2}$, where k is Boltzmann's constant (Grard and Tunaley, 1971).

The low effective temperature T_{Ph} of the photoelectrons and their high density N_{Pho} at the surface may cause nonmonotonic potential distributions. The case of small plasma Debye length and spherical symmetry has been investigated by Chang and Bienkowski (1970), a plane sheath has been treated by Guernsey and Fu (1970).

One of the aims of this note is to give an estimate of the magnitude of the potential minimum. The following basic *assumptions* are made:

(1) existence of a time independent solution;

(2) spherical symmetry (this includes uniform potential and photoemission over the whole surface);

(3) collisionless plasma;

(4) absence of surface effects other than absorption and photoemission;

(5) absence of magnetic field (gyro radius much larger then probe radius R_0).

Simplifying – but not necessary – assumptions are, that the ion density N_i is uniform, that a Maxwellian velocity distribution is assumed for the plasma electrons, and that trapped particles are not considered.

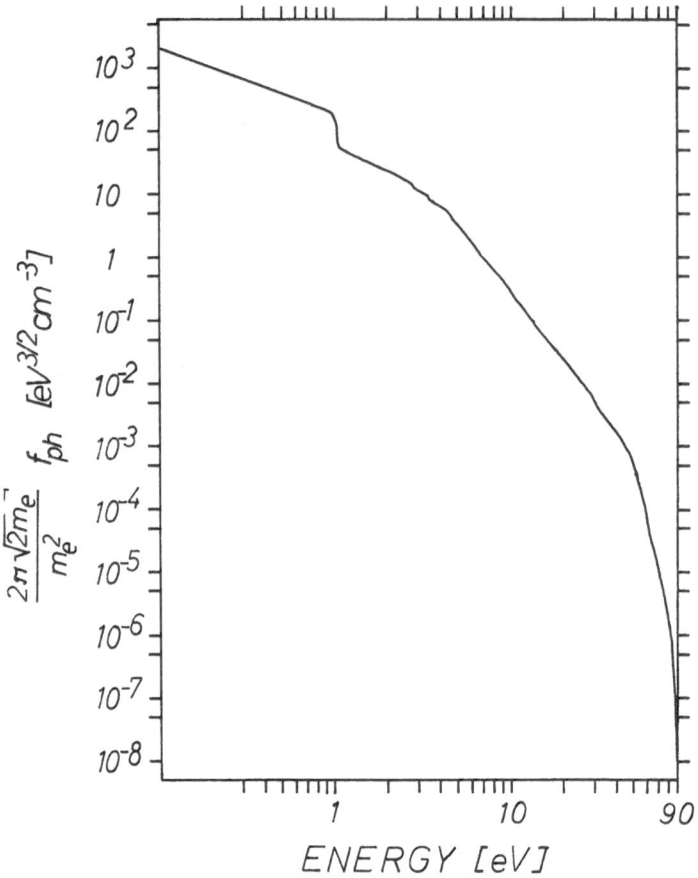

Fig. 1. Photoelectron energy distribution.

The *velocity distribution* $f_{Ph}(\tfrac{1}{2}m_e w^2)$ used in this study is presented in Figure 1; f_{Ph} gives the density of photoelectrons per velocity and space interval $d^3w\,d^3x$ and depends only on the energy of the photoelectrons. This curve has been derived (Middendorf *et al.*, 1970) from an estimation of the yield function and photoelectron energy distribution of quartz (material used for second surface mirrors and covers of solar cells) and measured values of the solar spectrum (Hinteregger, 1965; Detwiler *et al.*, 1961). The step in the curve at the energy of 1 eV is due to the Lα line and the adopted threshold energy for photoemission of 9.2 eV. The surface density of photo-electrons with positive normal velocity N_{Pho+}, the mean energy $\langle E \rangle$, and the maximum flux j are

$$\begin{bmatrix} N_{Pho+} \\ \langle E \rangle \\ j \end{bmatrix} = \int\limits_0^\infty dE\, \frac{2\pi}{m_e^2}\sqrt{2m_e}\,f_{Ph}(E) \begin{bmatrix} E^{1/2} \\ E^{3/2}/N_{Pho+} \\ E/\sqrt{2m_e} \end{bmatrix} = \begin{bmatrix} 460\ (\mathrm{cm}^{-3}) \\ 1.2\ (\mathrm{eV}) \\ 1.2\ 10^{10}\ (\mathrm{s}^{-1}\ \mathrm{cm}^{-2}) \end{bmatrix}.$$

These average values fit well the recent measurements of Feuerbacher and Fitton

(1972) and Grard (1972); only at very low energies <1 eV strong discrepancies exist between the photoelectron distribution functions.

The set of *equations* that governs the potential V and the density distribution is the collisionless Boltzmann equation for the electrons

$$\mathbf{w} \cdot \nabla f + \frac{e}{m_e} \nabla V \cdot \nabla_w f = 0$$

and Poisson's equation

$$\Delta V = - 4\pi e \left(N_i - N_e \right),$$

where the electron density has to be determined from the velocity distribution function of the electrons $N_e = \int d^3 w f$.

As *boundary conditions* for Poisson's equation we prescribe $V(R_0) = V_0$ (the case of floating potential is not considered here but could be included later) and $V(\infty) = 0$.

For the Boltzmann equation we have

$$f(\mathbf{w}, R_0) = f_{\text{Ph}}(\tfrac{1}{2} m_e w^2) \quad \text{for} \quad \mathbf{w} \cdot \nabla R \geqslant 0 \qquad (R: \text{radial variable})$$
$$f(\mathbf{w}, R \to \infty) = f_{\text{Maxw}}(w) \quad \text{for} \quad \mathbf{w} \cdot \nabla R \leqslant 0$$
$$f(\mathbf{w}, R) = 0 \qquad \text{if the orbit is bounded but does not strike the probe.}$$

The *solution* of the system follows closely the formalism of Bernstein and Rabinowitz (1959), which has been developed for Maxwellian particles in the field of an attracting probe by Laframboise (1966).

For a given spherically symmetric potential the solution of the Boltzmann equation can be represented as a (multivalued) function of the integrals of motion energy E and angular momentum $J: f = f(E, J)$. This solution is sufficiently general because of the spherical symmetry: considering an electron at a point $\mathbf{r}$ with velocity $\mathbf{w}$, the orbit can be traced back using the integrals of motion until the proper boundary value can be assigned to f. Keeping the point $\mathbf{r}$ fixed, and varying the velocity $\mathbf{w}$, it is seen, that the whole velocity space can be divided into different regions, where $f(\mathbf{r}, \mathbf{w})$ equals $f_{\text{Ph}}(\tfrac{1}{2} m_e w^2)$, $f_{\text{Maxw}}(w)$, or zero respectively. For the determination of f at the point $\mathbf{r}$ it is therefore necessary, to determine the separation curves between these regions only. For the calculation of the density the velocity integral is then transformed to an integral over E, J.

For a given electron density Poisson's equation is solved numerically after a transformation of the independent variale $R \to 1/R$.

These two operations are performed in one step of the iteration procedure. Using this direct iteration scheme results in a sequence of computed densities which in general oscillates with increasing amplitude about the solution. This overstable behaviour can be suppressed by using a weighted average of the last two calculated densities in the following iteration step (Laframboise, 1966).

The iterative process can be started with the following trial function for the potential:

$$V = (A/R) \exp(- R/H_{\text{pl}}) + B/(R^2 + H_{\text{pl}}^2),$$

where H_{pl} is the plasma Debye length.

The coefficient B was determined from a rough asymptotic analysis neglecting the nonmonotonic behaviour of the electric potential and the effective radial potential (sum of electric and centrifugal potential). The coefficient A was then defined to give the chosen surface potential V_0.

Considering the above assumptions on the ion and the photoelectron energy distributions and taking the probe radius $R_0 = 1$ m, three independent parameters are left, namely the surface potential V_0, the plasma density at infinity N_∞ and the plasma electron temperature T_e. This investigation is not yet complete and only a few preliminary *results* are given.

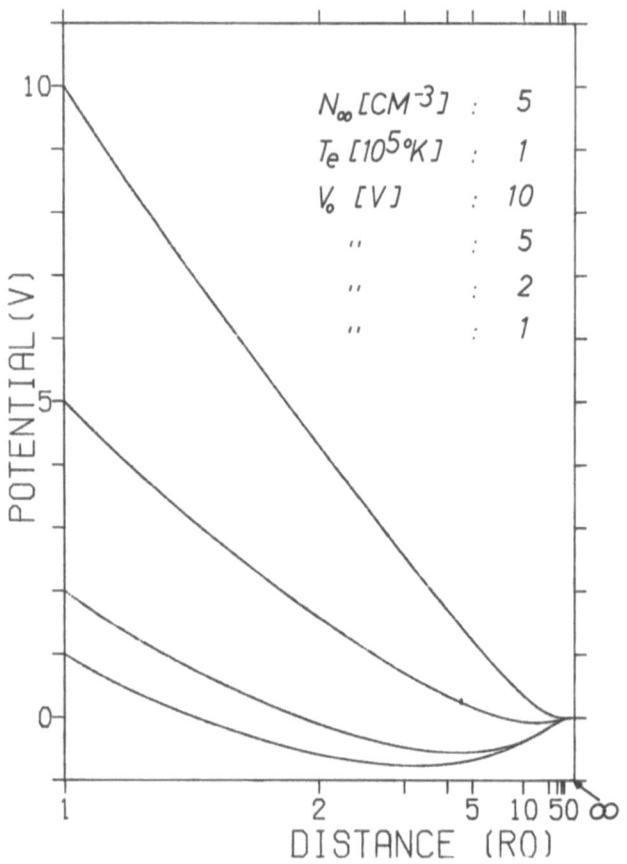

Fig. 2.   Potential distribution around a probe for various surface potentials.

Figure 2 shows the potential profiles for various surface potentials. As R_0/R is chosen as abscissa on a linear scale a vacuum potential gives a straight line. It is clearly seen, that the higher surface potentials are mainly shielded by the surrounding plasma ($H_{pl} = 9.8$ m). If we consider that the photoelectron density decreases like $1/R^2$ only, the drop in potential over one radius due to the photoelectron space charge would be significant if

$$V_0 < 4\pi e N_{\text{Pho}} R_0^2.$$

The r.h. side of the inequality is about 20 V for $N_{\text{Pho}} = 1000$ cm^{-3}. Thus a surface potential up to 20 V should be effectively shielded by the photoelectrons. However, because of the strong dependence of the electron density on the potential, as shown in Figure 3, the shielding is effective only for much smaller surface potentials.

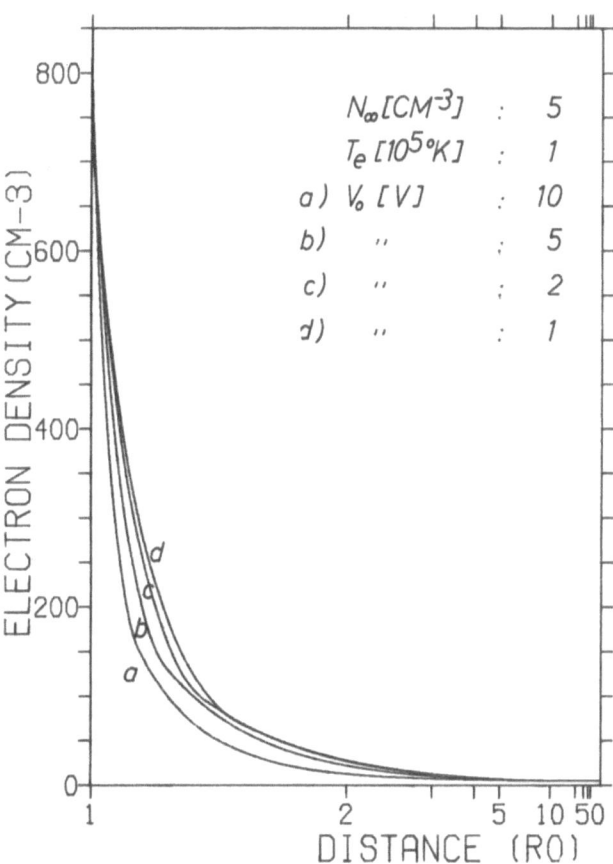

Fig. 3.   Electron density distribution around a probe for various surface potentials.

The irregularities in the slope of the density curves is caused by the discontinuity due to the Lα line in the photoelectron velocity distribution, as it can be seen by examining the corresponding value of the potential.

The dependence of the potential distribution on the plasma density for a fixed surface potential of 5 V and a plasma temperature of 10^5 K is seen in Figure 4. The position of the potential minimum does not vary much in this case. As the potential at a distance of a few probe radii is much less than the characteristic e-folding energy

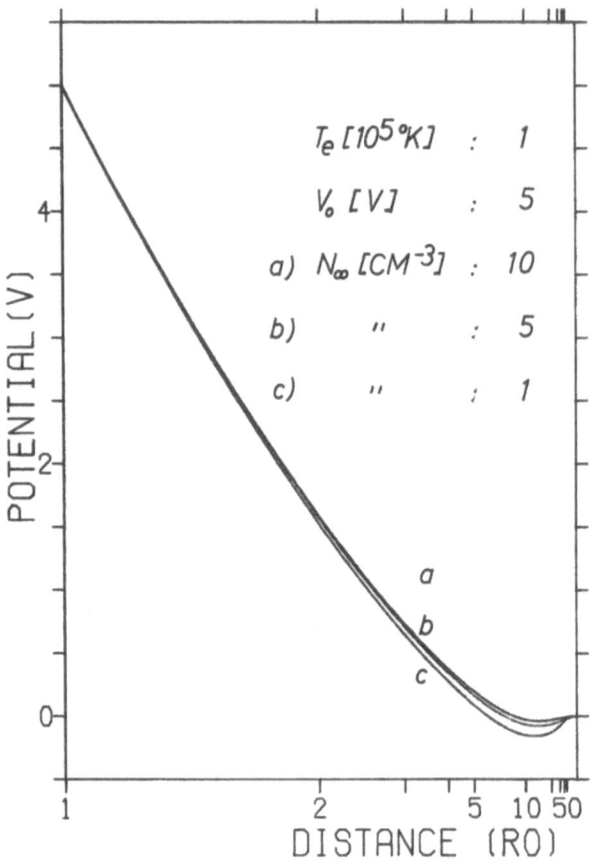

Fig. 4.   Potential distribution for various densities of the ambient plasma.

of photoelectrons, we find for the asymptotic photoelectron density $N_{Ph} =$ $= \text{const.}\ (R_0/R)^2$, where the constant is determined primarily by the surface potential and the vacuum-like near zone. The distance over which the surface charge is compensated (equal to the distance, at which the potential minimum occurs) depends therefore primarily on the surface potential. The uncompensated photoelectron density at larger distances is the cause of the potential minimum, which is shielded with the plasma Debye length. As a result the magnitude of the potential minimum increases with the plasma Debye length.

The potential variations due to a change in plasma electron temperature are for a given plasma Debye length nearly identical to those shown in Figure 4, and are therefore not presented.

The dominant influence of the surface potential on the photoelectron density can be seen in Figure 5.

These preliminary results seem to indicate, that using the photoelectron Debye length as a shielding distance, must be done with caution. The magnitude of the

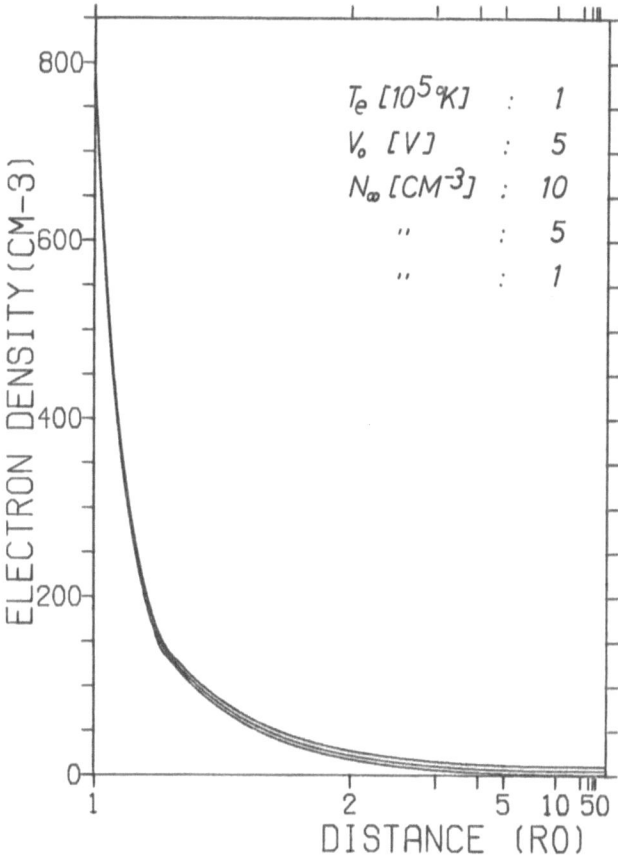

Fig. 5. Electron density distribution for various densities of the ambient plasma.

potential minima is certainly overestimated, because spherical symmetry was assumed, but it is hoped that these results represent the qualitative behaviour of a photoemissive probe in a diluted plasma.

References

Bernstein, I. B. and Rabinowitz, I. N.: 1959, *Phys. Fluids* **2**, 112.
Chang, K. W. and Bienkoski, G. K.: 1970, *Phys. Fluids* **13**, 902.
Detwiler, C. R., Garret, D. L., Purcell, J. P., and Tousey, R.: 1961, *Ann. Geophys.* **17**, 263.
Feuerbacher, B. and Fitton, B.: 1972, *J. Appl. Phys.* **43**, 1563.
Grard, R. J. L. and Tunaley, J. K. E.: 1971, *J. Geoph. Res.* **76**, 2498.
Grard, R. J. L.: 1972, Europ. Sp. Res. Techn. Centre Noordwijk, ESTEC Internal. Work Paper No. 663.
Guernsey, R. L. and Fu, J. H. M.: 1970, *J. Geophys. Res.* **75**, 3193.
Hinteregger, H. E.: 1965, *Space Sci. Rev.* **4**, 461.
Laframboise, J.: 1966, in J. H. de Leeuw (ed.), *Rarefied Gas Dynamics*, Vol. 2, Academic Press, New York-London, p. 22.
Middendorf, H. D., Schröder, H., Seehusen, J., and Abromeit, C.: 1970, Final Report 1969/1970 Physikzentrum Technische Universität Braunschweig, unpublished.

Whipple, E. C., Jr.: 1965, NASA X-615-65-296.
Wyatt, S. P.: 1969, *Planetary Space Sci.* **17**, 155.

DISCUSSION

Wiesemann: Did you check the consistency of your model by considering the possible ion trapping in the calculated potential well? Trapped ions would influence your results severely.

Schröder: No, ion trapping is not probable in the solar wind.

Grard: How did you define the potential of the surface; which is one of your boundary conditions?

Schröder: The potential has been chosen from rough considerations, but it could eventually be computed from the current balance.

Grard: How would this affect your results?

Schröder: The potential minimum would also occur if the current balance were established, because the slower photoelectrons contribute more to the density than the plasma electrons striking the probe. As a consequence the potential becomes negative at some distance from the probe.

THE PHOTOELECTRON SHEATH
AROUND A SPHERICAL BODY

J. K. E. TUNALEY and J. JONES

The University of Western Ontario, Dept. of Physics, London, Canada

Abstract. An approximate method based on a variational principle is used to calculate the surface electric field on a spherical body which is photo-emitting electrons. The treatment also allows an estimate of the sheath dimensions to be made. It is shown that, unlike the planar surface, the sheath parameters are strongly dependent on the velocity distributions of the emitted electrons. As the sphere radius is reduced the sheath shrinks for a monoenergetic, isotropic, velocity distribution of electrons at the surface. However for monoenergetic electrons directed radially it expands. The technique is developed because of the possibility of employing it for more complex systems where other methods would prove intractable.

The same problems will be treated by solving the differential equations using a series expansion; this will indicate the accuracy of the method for small spheres.

1. Introduction

A knowledge of the photoelectron sheath around a body in a tenuous plasma, such as that existing in interplanetary space or in the magnetosphere, is important for the interpretation of satellite borne probe system. Also it allows the interaction of charged dust grains, illuminated by the sun, to be calculated. Previous treatments of the photoelectron sheath have been concerned largely with planar geometry where solutions to the equations describing the problem can be readily obtained (for example Singer and Walker, 1962; Grard and Tunaley, 1971). In the latter publication the sheath parameters are evaluated for various velocity distributions of the electrons emitted from the surface. It is shown that the sheath thickness will generally be of the order of the photoelectron screening length, λ, (defined below). We may expect that for a spherical body of radius much greater than λ the planar approximation should be valid. Unfortunately the dimensions of typical satellites are of the same order as λ namely of the order of 1 m so that we cannot rely on the planar approximations. For dust particles these approximations are clearly invalid.

The spherical problem has not received so much attention owing to its intractibility. It may be divided into two parts. The first part consists of finding the charge density as a function of potential for a given velocity distribution of emitted electrons. This may be achieved by employing the method of Bernstein and Rabinowitz (1959). The resulting double integral can be performed without approximation for simple velocity distributions and the charge density obtained especially when the presence of an ambient plasma is ignored.

The second part involves the solution of Poisson's equation using the charge density already calculated. Even with spherical symmetry the exact solution is difficult to obtain analytically and in the present study we made use of a technique based on a variational principle. This gives approximate solutions which should be particularly

R. J. L. Grard (ed.), Photon and Particle Interactions with Surfaces in Space, 59–71. All Rights Reserved
Copyright © 1973 by D. Reidel Publishing Company, Dordrecht-Holland

useful for estimating the surface charge density on the body and the sheath dimensions. A comparison will be made between solutions obtained with the variational technique and the exact solutions of Grard and Tunaley to estimate the accuracy of the method in the planar case. However, for small spheres the accuracy will be discussed more fully in the appendix.

To simplify the problem we assume that the body is illuminated uniformly and that the work function and photo-yield are constant over the surface; this results in spherical symmetry. Singer and Walker have treated the photosheath under these conditions using a numerical method for a very small sphere. However, unlike in their treatment no approximations will be required in finding the charge density and approximate analytic solutions of Poisson's equation will be found valid both for very small and very large sphere radii.

Since the ambient plasma is neglected, all photoelectrons must eventually return to the sphere. Thus the floating potential of the sphere must be such that the most energetic particle is pulled back (as described by Grard and Tunaley).

2. Theory

2.1. NOMENCLATURE

The following nomenclature will be employed

- e electronic charge
- m electronic mass
- N number density
- ϕ potential
- v velocity
- r radial space variable
- ε_0 permittivity of free space
- F velocity distribution function
- E energy
- J angular momentum
- u axial velocity component
- w radial velocity component
- α aximuthal angle
- λ photoelectric screening length equal to $(\varepsilon_0\phi_0/N_0 e)^{1/2}$; analogous to Debye length
- ϱ charge density

Unless otherwise indicated the subscript 'zero' indicates quantities at the surface of the body.

2.2. THE CHARGE DENSITY

We require the charge density as a function of potential for a given velocity distribution of emitted photoelectrons. Since spherical symmetry is assumed we are dealing with a central field problem and we can express the solution of the equation of motion for each electron as a function of quantities that are conserved. There are the energy and angular momentum where

$$E = \tfrac{1}{2}m(u^2 + w^2) - e(\phi - \phi_0) \tag{1}$$

$$J = mwr. \tag{2}$$

As in the treatment of Bernstein and Rabinowitz (1959), u and w are the components of velocity expressed in a cylindrical coordinate system with axis passing through the centre of the sphere; we define α to be the azimuthal angle in this system. An element of volume in velocity space can be expressed in terms of E and J by finding the Jacobean of the transformation which results in the equation

$$du\, w\, dw\, d\alpha = \frac{J\, dJ\, dE\, d\alpha}{m^3 r^2 u}. \tag{3}$$

Thus, from (1), (2) and (3) the volume element becomes:

$$du\, w\, dw\, d\alpha = \frac{J\, dJ\, dE\, d\alpha}{(mr)^2 \left\{ 2m[E - e(\phi_0 - \phi)] - \dfrac{J^2}{r^2} \right\}^{1/2}}. \tag{4}$$

At a point r, those electrons with small initial kinetic energy and low angular momentum will have already reached their maximum distance from the sphere. Only those with $u^2 > 0$ will be present. From (1) and (2), the electrons which are present satisfy the inequality:

$$E \geqslant e(\phi_0 - \phi) + \frac{J^2}{2mr^2}. \tag{5}$$

Hence the electron density at a given point is given by

$$N = N_0 \int\!\!\!\int\!\!\!\int_0^{2\pi} \frac{F(E, J)\, J\, dJ\, dE\, d\alpha}{(mr)^2 \left\{ 2m[E - e(\phi_0 - \phi)] - \dfrac{J^2}{r^2} \right\}^{1/2}}, \tag{6}$$

where $F(E, J)$ is the normalized velocity distribution function (having dimensions of velocity^{-3}) expressed in terms of E and J and the integration is to be carried out over the domain indicated in Figure 1. We may remark that the situation under consideration, where the ambient plasma is neglected is somewhat simpler than that covered by Bernstein and Rabinowitz in that only one class of particle is treated; all electrons are in bound orbits.

We will consider two types of velocity distribution namely the case of monoenergetic electrons with isotropic distribution and monoenergetic electrons with velocity directed radially. For isotropic emission, it is clear that the velocity distribution function is a function only of the energy; the contours of constant speed are spheres in velocity space. Hence after integrating with respect to α and J over the range given in Figure 1 (6) becomes

$$N = 2\pi N_0 \int_{e(\phi_0 - \phi)}^{\infty} \frac{F(E)}{m^2} \{2m[E - e(\phi_0 - \phi)]^{1/2}\} \, dE. \tag{7}$$

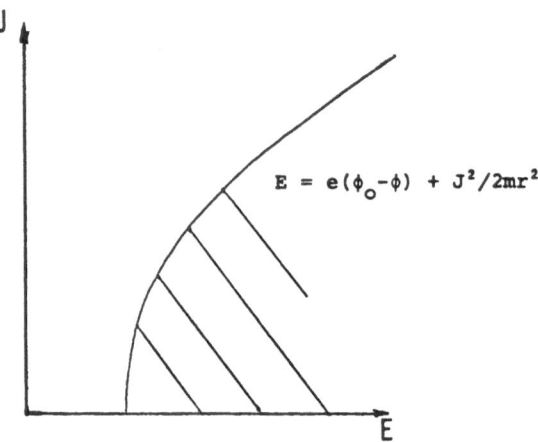

Fig. 1.   Domain of integration.

The distribution function $F(E)$ is best evaluated at the sphere surface where $E = = m(u^2 + w^2)/2$ and for monoenergetic electrons of energy E_0:

$$F(E) = \frac{m\delta(E - E_0)}{2\pi v_0}, \tag{8}$$

where v_0 is the electron speed at the sphere. Inserting this into (7) yields

$$N = \frac{N_0}{v_0} \sqrt{\frac{2e\phi}{m}}, \tag{9}$$

where we have used the fact that the floating potential ϕ_0 is given by

$$e\phi_0 = E_0. \tag{10}$$

It should be noted that (9) differs from the equivalent equation of Singer and Walker (1959) by a factor of four, taking into account that our N_0 includes both incoming as well as outgoing electrons and the factor two customarily employed in other publications has been omitted.

On the other hand the normalised velocity distribution function for monoenergetic particles moving radially is

$$F(u, w, \alpha) = \frac{1}{\pi} \delta(u - u_0) \delta(w^2) \tag{11}$$

or in terms of E and J at the sphere surface

$$F(E, J) = \frac{2}{\pi} m^3 u_0 r_0^2 \, \delta \, (E - E_0) \, \delta \, (J^2).$$ (12)

Performing the integrations in (6) and using (10) gives

$$N = N_0 \frac{r_0^2}{r^2} \sqrt{\frac{m}{2e\phi}} \cdot u_0.$$ (13)

2.3. SOLUTION OF POISSON'S EQUATION

The potential within the photosheath can be derived by using Poisson's equation with the charge density calculated in the previous section. With spherical symmetry we have:

$$\frac{1}{r^2} \frac{\partial}{\partial r} \left(r^2 \frac{\partial \phi}{\partial r} \right) = \frac{- \varrho \, (\phi, r)}{\varepsilon_0}.$$ (14)

A simplification may be made by writing $\psi = r\phi$ when (14) becomes

$$\frac{\partial^2 \psi}{\partial r^2} = \frac{- r\varrho \, (\psi, r)}{\varepsilon_0}.$$ (15)

Inserting the electron densities given either by (9) or (13) causes (15) to become non-linear and its exact solution is difficult in terms of elementary functions. However, an approximate solution may be obtained using a variational principle.

We require a functional, $U[\psi(r)]$ which takes a minimum value when $\psi(r)$ is chosen to satisfy (15). In general the functional given by

$$U = \int_a^b F(\psi, \dot{\psi}, r) \, dr$$ (16)

takes a minimum (or maximum) value for a path running from $r=a$ to $r=b$ when

$$\frac{\partial F}{\partial \psi} - \frac{d}{dr} \left(\frac{\partial F}{\partial \dot{\psi}} \right) = 0,$$ (17)

where $\dot{\psi} = d\psi/dr$ and the value of ψ at $r=a, b$ is known (Matthews and Walker, 1965). Comparing (15) and (17) it may be deduced that a suitable functional for the present problem* is

$$U = \int_{r_0}^{\infty} \left\{ \frac{\varepsilon_0}{2} \left(\frac{d\psi}{dr} \right)^2 - r \int_0^{\psi} \varrho \, (\psi, r) \, d\psi \right\} dr,$$ (18)

where ψ and r are treated as independent variables.

---

* It is possible to deduce a functional directly applicable to (14) by noting that (14) is a special case of the Sturm-Liouville equation as described by Matthews and Walker (1965). However, although this yields the same results, the calculations are more complex.

To obtain an approximate solution of (15) we can choose a trial function for ψ which satisfies the boundary conditions at r_0 and for example ∞ and minimise U with respect to some parameter (say s) in the trial function. The value of s which gives the minimum U corresponds to the form of ψ closest to the exact solution of (15). The method is capable of considerable accuracy. However, the choice of trial function should in general bear some resemblance to the real one as well as satisfying the boundary conditions. Furthermore the method becomes unattractive if the manipulations become complicated owing to a complex choice of the trial function.

The choice of trial function ψ also depends to some extent upon the parameter which we are trying to calculate. This is because it is difficult to find a reasonably simple form of function to represent the potential applicable to a wide range of sphere radii. For small sphere radii compared with the screening length λ, it would be expected that the potential would become close to the Laplacian solution, at least close to the sphere. On the other hand, for large radii spheres the fields are almost wholly produced by space charge effects. Thus for small spheres ϕ should contain r^{-1} term whereas for large spheres the solutions of the type presented by Grard and Tunaley (1971) can be employed. It is of course possible in principle to combine the two types of function and insert a r^{-1} factor in the planar solution. However whether or not this is desirable depends upon the parameters to be optimised and the complexity of the resulting manipulations.

To calculate the approximate surface charge density the trial function need not be particularly exact in form and may be chosen to simplify the manipulations. An examination of the curves of Grard and Tunaley show that the potential falls smoothly as the distance from the surface increases and, for the velocity distributions discussed here, the potential falls to zero at a finite distance from the surface. However for the radially directed monoenergetic electrons the second derivative of the potential (namely the charge density) exhibits an infinity at the sheath edge as can be seen from (13). However, if we were to consider the presence of an ambient plasma these irregularities would probably be smoothed out. Hence for this present application we will choose a potential of the Debye form to anticipate the inclusion of this important case, i.e.,

$$\phi = \phi_0 \frac{r_0}{r} \exp\left[-s(r - r_0)\right], \tag{19}$$

bearing in mind that the choice is likely to be less satisfactory for the radially directed monoenergetic distribution of particles. In terms of ψ (19) becomes

$$\psi = \psi_0 \exp\left[-s(r - r_0)\right]. \tag{20}$$

An important parameter in practice is the thickness of the photosheath. This could be particularly useful for estimating the rate of collection of ambient particles in a tenuous plasma. In this case it is necessary to use as a trial function a form which is closer to the exact solution especially near to the sheath edge. It will be shown in the appendix that, for isotropic monoenergetically emitted photoelectrons, the potential varies as the fourth power of the distance from the sheath boundary. Hence a form

$$
\phi = \begin{cases} 0 & r > r_0 + a \\ \dfrac{\phi_0}{a^4}(a + r_0 - r)^4 & r_0 < r < r_0 + a \end{cases} \tag{21}
$$

will be employed where the parameter a is equal to the sheath thickness. This is similar to the appropriate exact solution of Grard and Tunaley for the planar case but it has not been generalised by including a r^{-1} term since this renders the integrations too complex. Thus it is only a good trial function for large sphere radii. In terms of ψ we have

$$
\psi = \frac{\psi_0}{a^4} \frac{r}{r_0} (a + r_0 - r)^4. \tag{22}
$$

2.3.1. *Approximate Surface Charge Density for the Isotropic Monoenergetic Distribution*

In terms of ψ the charge density ϱ is given from (9), which after some manipulation becomes

$$
\varrho = \frac{\varepsilon_0}{\lambda^2} \left(\frac{\psi_0}{r_0}\right)^{1/2} \left(\frac{\psi}{r}\right)^{1/2}. \tag{23}
$$

Thus the functional U given by (18) is

$$
U = \int_{r_0}^{\infty} \left\{ \frac{\varepsilon_0}{2} \left(\frac{d\psi}{dr}\right)^2 + \frac{2}{3} \frac{\varepsilon_0 r^{1/2}}{\lambda^2} \left(\frac{\psi_0}{r_0}\right)^{1/2} \psi^{3/2} \right\} dr. \tag{24A}
$$

After substituting the trial ψ of (20) and integrating with respect to r we have

$$
U = \varepsilon_0 \psi_0^2 \left\{ \frac{s}{4} + \tfrac{4}{3} \exp\left(\tfrac{3}{2} s r_0\right) - \frac{1}{\lambda^2 r_0^{1/2}} \left(\frac{3s}{2}\right)^{-3/2} \int_{3s r_0/2}^{\infty} x^2 e^{-r^2} dx \right\}. \tag{25}
$$

This functional is made complicated by virtue of the integral over x. However, the integral can be evaluated to a fair approximation in terms of elementary functions for the cases of either large or small sphere radius. From the considerations of Grard and Tunaley (1971) or from simple dimensional arguments it might be expected that the order of magnitude of s will be λ^{-1}. Therefore we will treat the cases $r_0 \ll \lambda$ and $r_0 \gg \lambda$ so that lower limit in the integral of (25) is either much smaller or much larger than unity.

First we will consider the use of a large sphere. It may be shown that the integral in (25) may be evaluated in terms of a convergent series for large values of the lower limit (cf. Matthews and Walker, 1965). It is easily verified that the first two terms in this series are

$$
\tfrac{1}{2}\left(\tfrac{3}{2} s r_0\right)^{1/2} \exp\left(-\tfrac{3}{2} s r_0\right) + \tfrac{1}{4}\left(\tfrac{3}{2} s r_0\right)^{-1/2} \exp^2\left(-\tfrac{3}{2} s r_0\right). \tag{26}
$$

On substituting (24) and (23) we have:

$$U = \varepsilon_0 \psi_0^2 \left\{ \frac{s}{4} + \frac{4}{9} \frac{1}{\lambda^2 s} + \frac{4}{27} \frac{1}{\lambda^2 r_0 s^2} \right\}. \tag{27}$$

The minimum value of U may be found by differentiating with respect to s and, if r_0 is large compared to λ as in the planar case, we can neglect the last term and show that U is a minimum when

$$s = \frac{4}{3\lambda}. \tag{28}$$

Reinserting this value in $\partial U / \partial s = 0$, we can correct the value of s for finite radii which yields:

$$s \simeq \frac{4}{3\lambda} \left(1 - \frac{\lambda}{2r_0} \right)^{-1/2}. \tag{29}$$

Thus for large sphere radii the potential near to the sphere surface is given by

$$\phi \simeq \phi_0 \frac{r_0}{r} \exp\left\{ \frac{4}{3\lambda} \left(1 - \frac{\lambda}{2r_0} \right)^{-1/2} (r - r_0) \right\} \tag{30}$$

and the electric field at the surface is

$$\mathscr{E}_0 \simeq \mathscr{E}_{OL} \left[1 + \frac{4}{3} \frac{r_0}{\lambda} \left(1 - \frac{\lambda}{2r_0} \right)^{-1/2} \right], \tag{31}$$

where $\mathscr{E}_{OL}$ is the field that would be present in the absence of photoemission i.e. the Laplacian solution given by ϕ_0 / r_0.

An examination of (30) shows that, as the radius of the sphere decreases, the argument of exponential term becomes greater for a given distance from the surface. Thus the sheath tends to shrink. From (31) the surface field, and hence the surface charge density are controlled by space charge for large sphere radii, the Laplacian term becoming more important as the radius is reduced.

We may now compare the surface fields given by this approximate method with those of Grard and Tunaley (1971). Taking into account differences in notation the exact treatment of the planar case gives

$$\mathscr{E}_0 = \sqrt{\frac{4}{3}} \frac{\phi_0}{\lambda}. \tag{32}$$

The present method results in an error only of about 15% in spite of the crudity of the trial ψ.

For small spheres the procedure must be repeated but in this case the integral in (15) must be expanded for small values of the lower limit. It may be shown that the minimum U occurs for small spheres when

$$s = \frac{2}{3} \left(\frac{16\pi}{9} \right)^{1/5} \frac{1}{\lambda^{4/5} r_0^{1/5}}. \tag{33}$$

Thus the field at the surface is given by

$$\mathscr{E}_0 \simeq \mathscr{E}_{OL} \left[1 + \frac{2}{3} \left(\frac{16\pi}{9} \right)^{1/5} \left(\frac{r_0}{\lambda} \right)^{4/5} \right] \tag{34}$$

which is of the same order as the Laplace field. As r_0 decreases the sheath radius continues to shrink as may be seen from (33) where we may interpret s^{-1} as some length of the same order of magnitude as the sheath thickness. However the sheath thickness is only relatively weakly influenced by the sphere radius and even for dust grains of radius one micron the sheath thickness will be typically 10 cm.

It should be noted that the values of s which have been calculated are consistent with the lower limit of the integral in (25) being either large or small.

2.3.2. *Approximate Surface Charge Density for the Directed Monoenergetic Distribution*

After inserting the charge density given by (13) the appropriate functional becomes

$$U = \int_{r_0}^{\infty} \left\{ \frac{\varepsilon_0}{2} \left(\frac{\partial \psi}{\partial r} \right)^2 + \frac{2}{\lambda^2} \frac{r_0^{1/2}}{r^{1/2}} \psi_0^{3/2} \psi^{1/2} \right\} dr \tag{35}$$

and by performing similar manipulations as before with a trial ψ given by (20), it may be shown that for large sphere radii the value of s which yields the minimum U is given to the same approximation as before by

$$s \simeq \frac{4}{\lambda} (1 + \lambda/2r_0)^{-1/2}. \tag{36}$$

On the other hand for small sphere radii we have

$$s \simeq (2\pi r_0)^{1/3} \left(\frac{2}{\lambda} \right)^{4/3}. \tag{37}$$

To compare these results with those of Grard and Tunaley the electric field at the surface of the sphere may be calculated in the planar situation. This is given by

$$\mathscr{E}_0 = \frac{4\phi_0}{\lambda} \tag{38}$$

and is a factor of two too large. As expected the trial function is not a good choice for this velocity distribution for large spheres. Nevertheless for small spheres it may be better and (37) should yield a fair approximation for the surface field, namely

$$\mathscr{E}_0 = \mathscr{E}_{OL} \left(1 + (2\pi)^{1/3} \left(\frac{2r_0}{\lambda} \right)^{4/3} \right). \tag{39}$$

If we take s^{-1} as an order of magnitude for the sheath radius we see that, unlike for the case of isotropic monoenergetic electron emission, the sheath tends to expand as the sphere size is reduced.

2.3.3. *The Sheath Thickness for Large Spheres with Isotropic Emission*

The trial function given by (20) is not suitable for calculating the sheath thickness since it does not represent the potential near the sheath edge. Inserting the trial function given by (22) into (24) gives after integration

$$U = \frac{\varepsilon_0 \, \psi_0^2}{2 \, a r_0^2} \left[\frac{(a - 4r_0)^2}{7} - \frac{10}{7 \times 8} a \, (a - 4r_0) + \frac{50}{7 \times 8 \times 9} a^2 \right] +$$
$$+ \frac{2}{3} \frac{\varepsilon_0 \, \psi_0^2}{\lambda^2 \, r_0^2} \left[\frac{r_0^2 a}{7} + \frac{2r_0 a^2}{7 \times 8} + \frac{2a^3}{7 \times 8 \times 9} \right]. \tag{40}$$

When $r_0 \to \infty$ we have

$$U = \varepsilon_0 \psi_0^2 \left(\frac{8}{7a} + \frac{2a}{21\lambda^2} \right) \tag{41}$$

and this takes a minimum value when

$$a = 2\sqrt{3}\,\lambda \tag{42}$$

(this result agrees with that of Grard and Tunaley (1971)). Inserting this value in $\partial U/\partial a = 0$ gives for $r_0 \gg \lambda$:

$$a = 2\sqrt{3}\,\lambda \left(1 + \frac{\sqrt{3}\,\lambda}{r_0} \right)^{-1/2}. \tag{43}$$

Thus the sheath thickness, a, falls as the sphere radius is reduced.

Since the trial function does not become Laplacian for small spheres this case will not be treated.

3. Discussion

Approximate surface electric fields have been calculated for both very large and very small spheres for two velocity distributions of photo-emitted electrons. In the case of very large sphere the variational method with a Debye type of potential is clearly better for the isotropic monoenergetic electrons than for the radially directed ones; this was expected. Nevertheless, the treatment indicates that for an isotropic mono-energetic distribution the sheath tends to shrink as the sphere radius is reduced, whilst for the radially directed electrons it tends to expand. For the first velocity distribution this is confirmed using a better choice of potential distribution more suited to sheath thickness calculations. Although we have only solved this case for the sheath radius when $r_0 \gg \lambda$, we could in principle use this trial function for $r \sim \lambda$ and expect it to yield reasonably accurate results.

On the other hand, for both velocity distributions, it is clear that the electric field becomes closely Laplacian near to the sphere for spheres of small radius. However to estimate the suitability of the Debye type of potential and the accuracy of the surface fields is not so straight forward because it is the small deviations from the Laplace solutions that are of interest. To estimate the accuracies it is convenient to use another technique for solving the differential equation and herein lies the disadvantage of the variational method applied to this problem. However it is shown in the appendix that an error of only 10% is introduced in the surface field for small spheres emitting monoenergetic electrons isotropically; this is sufficiently small for many problems. Furthermore the method still results in an acceptable order of magnitude estimate for the sheath radius.

Thus, if we carefully consider the form of the charge density for a given velocity distribution of electrons, we may be able to choose a trial potential which is physically reasonable. In this case the variational technique may be employed with some confidence. For example of the velocity distribution is isotropic and varies smoothly with respect to speed the Debye potential will be a sufficiently accurate choice. In general the velocity distribution for a given material is likely to be more complicated than those treated have and to make further progress the variational method may be the only one to yield analytic results easily. The series expansion technique dealt with in the appendix may become unattractive.

Finally two points of practical interest should be noted. First we have illustrated that the surface fields and sheath dimensions will be much more affected by the velocity distribution than in the planar case (see Grard and Tunaley, 1971). Second in this treatment the effect of an ambient plasma has been neglected. To include this the charge density can be modified but the integrals involved in the variational principle become complex for the velocity distributions chosen here. However the technique can still be used in principle. This is also the case when we consider the possible existence of minima in the potential (Guernsey and Fu, 1970).

Appendix

In the case of monoenergetic electrons with an isotropic velocity distribution we have to solve the differential equation

$$\frac{\partial^2 \psi}{\partial r^2} = \frac{r^{1/2} N_0}{\varepsilon_0 v_0} \sqrt{\frac{2e\psi}{m}} \tag{44}$$

which, on writing $V = \psi/\psi_o$, gives

$$\frac{d^2 V}{dr^2} = \frac{1}{\lambda^2} \left(\frac{Vr}{r_0}\right)^{1/2}. \tag{45}$$

We will consider the non-dimensional form of (45), i.e.

$$\frac{d^2 V}{dx^2} = (xV)^{1/2} \tag{46}$$

subject to the boundary conditions $V=0$, $\partial V/\partial x=0$ at the sheath edge; these correspond to the potential and the electric field being zero. Thus provided that the sheath has a finite thickness we can expand V about the sheath edge in a power series:

$$V = \text{const. } h^n \sum_i a_i h^i, \tag{47}$$

where n is a positive number and h is the distance from the sheath edge. On substituting (47) into (46) and equating coefficients of powers of h, we find

$$V = \frac{x_s h^4}{144} \left\{ 1 - \frac{3h}{7x_s} - \frac{h^2}{49x_s^2} - \cdots \right\}, \tag{48}$$

where x_s is the value of x at the sheath edge, so that

$$x = x_s - h. \tag{49}$$

It is apparent that, if $h < x_s$, accurate values of V are obtained by neglecting all but the first few terms even in the case of very small spheres.

The value of x_0, which corresponds to the sphere surface, can be determined by noting that at this point $V = 1$. The sheath thickness, h_0, satisfies the equation

$$1 = \frac{x_s h^4}{144} \left\{ 1 - \frac{3h_0}{7x_s} - \frac{1}{49} \frac{h_0^2}{x_s^2} + \cdots \right\}. \tag{50}$$

Using successive approximations we find that h_0 is given by

$$h_0 \simeq \frac{3.46}{x_s^{1/4}} \left\{ 1 + \frac{0.334}{x_s^{5/4}} \right\}. \tag{51}$$

To scale these results so that they are appropriate to (45) we can let

$$r = \lambda^{4/5} r_0^{1/5} x, \tag{52}$$

so that the sheath thickness becomes approximately equal to

$$\lambda^{4/5} r_0^{1/5} h_0.$$

To remove the parameter corresponding to x_s in (51), (49) can be employed.

Of special interest is the case of small spheres for which $h_0 \sim x_s$. Using (50) directly it can be seen that $h_0 \simeq 3.05$.

Thus the sheath thickness is given by (c.f. Equation (33))

$$3.05 \, \lambda^{4/5} r_0^{1/5}.$$

Using (52) the electric field at the surface of the sphere is given by

$$\mathscr{E}_0 = \frac{\phi_0}{r_0} \left\{ 1 + \frac{r_0^{4/5}}{\lambda} \left(\frac{\partial V}{\partial h} \right)_{h_0} \right\} \tag{53}$$

and by differentiating (48) and putting $h_0 = 3.05$ we find that for small spheres

$$\mathscr{E}_0 = \frac{\phi_0}{r_0} \left\{ 1 + 1.04 \left(\frac{r_0}{\lambda} \right)^{4/5} \right\} \tag{54}$$

which is close to the result given by (34).

The procedure for the radially directed monoenergetic velocity distribution is similar. For spheres of very small radius we find a sheath thickness given by

$$0.531 \, r_0^{-1/3} \lambda^{4/3} \tag{55}$$

and a surface electric field given by

$$\mathscr{E}_0 = \frac{\phi_0}{r_0} \left\{ 1 + 6.19 \left(\frac{r_0}{\lambda} \right)^{4/3} \right\}. \tag{56}$$

The constant 6.19 must be compared with that given by the variational approach, namely 4.67.

References

Bernstein, I. B. and Rabinowitz, I. N.: 1959, *Phys. Fluids* **2** (2), 122.
Grard, R. J. L. and Tunaley, J. K. E.: 1971, *J. Geophys. Res.* **76** (10), 2498.
Guernsey, R. L. and Fu, J. H. M.: 1970, *J. Geophys. Res.* **75** (16), 3193.
Matthews, J. and Walker, R. L.: 1965, *Mathematical Methods of Physics*, Benjamin, New York.
Singer, S. F. and Walker, E. H.: 1962, *Icarus* **1** (1), 7.

PLASMA SHEATH AND SCREENING OF CHARGED BODIES

EVAN HARRIS WALKER

U.S. Army Ballistic Research Laboratories, Aberdeen Proving Ground, Md., U.S.A.

Abstract. The potential and charge density distributions are derived quite generally for both a stationary charged sphere and a charged body moving rapidly through a plasma. The work of previous authors on the screening of stationary charged bodies has been limited by a failure to recognize the importance of spiral orbits on the space charge density and a failure to cope with the problem of bound orbits. Most of the previous work on the screening of rapidly moving bodies has been limited to cases where the body, the potential, or the Debye length are assumed small. In the few cases where these assumptions have not been made, iterative solutions have been calculated, but here the uniqueness of the solution is seriously in question.

We have calculated the potential and charge density as a function of position about a stationary charged sphere, using both monoenergetic and Maxwellian velocity distributions for the ions and electrons of the ambient plasma. The potential decreases with distance more slowly than in the case of local thermodynamic equilibrium; the density of the ions (if the body is negative, electrons if positive) is generally much smaller than given by the barometric formula and varies in a complicated way. We also calculate the ion and electron voltage-current probe characteristics and the equilibrium potential as a function of the radius of the body. We find the Mott-Smith and Langmuir equations for the ion current (if the body is negative, electrons if positive) are unsatisfactory unless the sheath thickness is expressed as a function of the potential and radius of the body. For a spherical body, the appropriate expression for the sheath thickness σ is found to be $\sigma = 1.17\psi_s^{1/2}\varrho_s^{1/3} - 0.34\psi_c^{1/2}\varrho_c^{1/3}$ inside the pericritical surface and $0.83\psi_s^{1/2}\varrho_s^{1/3}$ outside the pericritical surface, where ψ_s and ϱ_s are the nondimensional potential and radius for the body, and ψ_c, ϱ_c are the values of these quantities on the pericritical surface.

Eigenvalue solutions are obtained if the charged body neutralizes most of the ions and electrons that strike its surface; i.e., if the reflection coefficients for the surface of the body are small. Under these conditions, the potential is found to vary more slowly than r^{-2} for small values of the potential.

For a rapidly moving body, we have developed a self-consistent method for solving the screening problem which does not require iterative calculations. Equations for the solution of the screening of axially symmetric bodies are derived for plasmas in which the thermal motion of the ions can be neglected and for plasmas with a Maxwellian velocity distribution. We have calculated the potential and density variation in the wake, the probe characteristics, and the impact and electric drag characteristic curves for various bodies. These calculations show that there is a trough in the ion density surrounding a highly charged body. The drag calculations show that under certain conditions a negative drag is obtained if the potential on the body is large and if the ions are neutralized and elastically reflected at the surface of the body.

The difficulties encountered in considering the interaction of a charged body with a plasma, including magnetic fields, particle interactions with the surface, time-dependent fluctuations, semi-collisionless plasmas, and plasmas with realistic velocity distributions at infinity, suggest a hierarchy of hopeless problems. Even when one greatly simplifies the problem to the case of the interaction of a stationary sphere with a collisionless plasma and excludes magnetic fields, the problem remains quite complicated.

If we neglect the possibility of time-dependent fluctuations, the general formulation of the problem is quite simple. One uses the Poisson equation, together with the appropriate expression giving the space charge distribution as a function of the

R. J. L. Grard (ed.), Photon and Particle Interactions with Surfaces in Space, 73–89. All Rights Reserved
Copyright © 1973 by D. Reidel Publishing Company, Dordrecht-Holland

potential distribution; i.e., by means of the Boltzman equation. One can consider problems in which the mean free path is much greater than the Debye length or, conversely, problems in which the mean free path is of the same order of magnitude or less than the Debye length. It can be shown, however, that the use of the Poisson equation is *not consistent* with this latter assumption, since individual charge effects become important (i.e., the sheath cannot be treated as homogeneous and time-independent). The use of the Poisson equation restricts us to problems in which the mean free path is much larger than the Debye length; e.g., to problems in which the plasma is to be treated as collisionless. If the plasma is mixed with a neutral gas, this restriction is not necessarily binding. For most physically important problems, however, the mean free path is large compared with the Debye length.

The derivation of suitable expressions for the density of ions and electrons for the problem of the screening of a charged sphere embedded in a plasma for which the mean free path is large compared with the Debye length requires a detailed treatment of the trajectories of the particles involved. In particular, it requires an understanding of two important types of orbits and how they contribute to the screening of the charged body. These are pericritical orbits and trapped orbits.

If the potential field in the vicinity of the charged body varies more slowly than $1/r^2$, then the orbits of the particles moving in this field will be hyperbolic-like; however, if the field varies more rapidly than $1/r^2$ the orbits will be spirals or, as we refer to them, pericritical orbits. In a potential field in which pericritical orbits become important, an *isotropic* velocity distribution at one point in the field will become *non-isotropic* as we approach the body, since for spiral orbits, the pitch angle is always decreasing. The fact that the velocity distribution is not isotropic will have a very important effect upon the expression for the density of the spiraling particles (see Figure 1).

The equations for the solution of the spherical and cylindrical probe characteristics in the case of a monoenergetic plasma were first given by Bernstein and Rabinowitz (1958). Walker (1965) gave the first treatment of the screening problem for spherical probes, including a Maxwellian distribution for the ambient plasma. This work pointed out a difficulty that exists for the treatment of probes that absorb all or most of the particles reaching the probe surface. Using the exact equations for this problem, it was shown that, in general, time-independent solutions do not exist for these problems. The problem is particularly severe for small values of the potential. The neglect of this fact, while allowing one, nevertheless, to obtain numerical solutions, is unlikely to yield valid results since the numerical results are strongly dependent on the choice of boundary condition. As a consequence, Walker gives numerical calculations only for specularly reflecting spheres while indicating the modification required for their approximate application to charge-collecting probes.

This restriction on the application of these equations was not recognized by either Bernstein and Rabinowitz (1958) or Laframboise (1966). This latter author carried out detailed numerical calculations for both spherical and cylindrical *collecting* probes. The validity of these results must be held in question until a resolution of the time

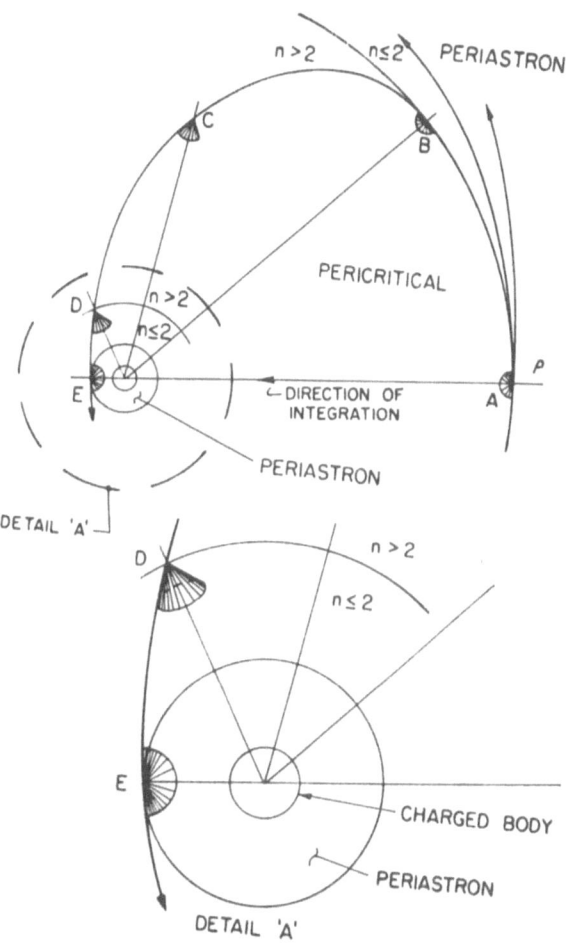

Fig. 1. An illustration of the dependence of the pericritical region on the value of the exponent n. Moving from outside in there are the four regions: Periastron with $n \leqslant 2$, pericritical with $n > 2$, pericritical with $n \leqslant 2$, and again periastron with $n \leqslant 2$. Along a typical trajectory A B C D E we have: A: the initial situation in which $n \leqslant 2$ and all pitch angles $\alpha = 0$ to $\pi/2$ are allowed, the periastron case. B: there is a transition from $n \leqslant 2$ to $n > 2$. Here the allowed cone is still from 0 to $\pi/2$ but is now beginning to become smaller. C: we have the pericritical case with the allowed pitch angles from 0 to α_{max}, $\alpha_{max} < \pi/2$. D: the transition from $n > 2$ to $n \leqslant 2$ occurs. The allowed cone now begins to increase but α_{max} is still less than $\pi/2$. E: the trajectory reaches its point of closest approach to the origin. Here its pitch angle is $\pi/2$ and the allowed cone of pitch angles will be $\alpha = 0$ to $\pi/2$. Point E marks a boundary between the pericritical region on the outside and periastron region inside.

dependence on the probe characteristics can be determined. Further, Laframboise (1966) did not recognize the equivalence of his equations to those of Walker, nor did he acknowledge the validity of the specularly reflecting problem (albeit an idealized case).

These authors (Bernstein and Rabinowitz, 1958; Laframboise, 1966) have also been concerned about the buildup of trapped orbits in the potential field of the charged body. An accumulation of such orbits would greatly enhance the space charge. Bernstein and Rabinowitz (1958) proposed the possibility of the buildup of such orbits but suggested no way for dealing with bound orbits and as a result have neglected all problems in which such orbits might contribute.

The exclusion of the region in which bound orbits can exist is totally unnecessary because it is easily shown that bound orbits cannot contribute significantly to the space charge. The reason for this is simple. There is an asymmetry between the injection cross section and the removal cross section, arising from the well-known (Spitzer, 1956) dominance of long range, small-angle Coulomb scattering over short-range, large-angle Coulomb scattering. Consider, for example, a particle moving along a hyperbolic-like trajectory in the potential field. In order for this particle to enter a bound orbit, it must experience a large angle scatter (or a large transfer of energy). Let us assume that for this particle to enter into a certain class of bound orbits it must experience a deflection of 60 deg. The chance that the particle would undergo such a collision for a semi-collisionless plasma might be of the order of, say,* one chance in 10^6. If the removal cross section were exactly the same as the injection cross section, this particle would remain bound for about 10^6 orbits before experiencing a similar collision that would remove it. However, for Coulomb scattering, it is well known that the long-range particle interactions are much more important than the short-range, large-deflection interactions; as a result, after about 10^4 orbits, the particle would have undergone a sufficient deflection to have been removed from the set of bound orbits. Thus we see that the cumulative effects of long-range scatters, which for Coulomb scattering is much more important than the short-range, large-deflection event, result in a more rapid removal of trapped orbits than would otherwise be expected. It should be kept in mind, however, that in most cases bound orbits are not allowed in any event, since only the metastable circular orbits are allowed in fields that vary more rapidly than r^{-2}. These facts are not recognized in the work of Laframboise (1966) nor by Wasserstrom et al. (1964).

Particles reaching the surface of a body are almost always neutralized. A proper formulation of the problem should take into account this neutralization of the particles when they reach the surface of the body. Although this offers no additional difficulty to the derivation of appropriate equations, it does affect the problem of obtaining solutions to the equations as mentioned above. It raises the question of whether or not *time-independent* solutions do exist. For these reasons, let us consider a problem in which one assumes either that all the particles reaching the surface are reflected from the surface or that virtually all the particles entering into the screening do eventually reach the surface and are neutralized. That is to say, we will introduce a return factor Q which has a value of 1 where all the particles reach the surface and are neutralized and which has a value of 2 when all the particles that do reach the surface are specularly reflected. Under these assumptions, we obtain the following set of equations for the density of the ions and electrons in the screened spherically symmetric potential field

---

* The actual value is not important to the present argument. The important point is that the ratio of the injection cross section to removal cross section is small.

of a body taken to be negatively charged with a collisionless plasma having a Maxwellian velocity distribution at infinity (Walker, 1965).

For the electrons (or, in general, for the repelled particles)

$$N_- = \tfrac{1}{2}QN_0 e^{\psi} \qquad (\psi < 0, \text{ repulsive field}).$$

For the ions (or, in general, the attracted particles) in the periastron region

$$N_a = \tfrac{1}{2}QN_0 \left[2\sqrt{-\psi/\pi} + \left(\frac{2}{\sqrt{\pi}} \int_{\sqrt{-\psi}}^{\infty} e^{-y^2} dy \right) e^{-\psi} \right]$$

$$(\psi < 0, \text{ attractive field}, n < 2).$$

For the ions (or attracted particles) in the pericritical region

$$N_c = \tfrac{1}{2}QN_0 \left\{ 2\sqrt{-\psi/\pi} + \left(\frac{2}{\sqrt{\pi}} \int_{\sqrt{-\psi}}^{\infty} e^{-y^2} dy \right) e^{-\psi} + \right.$$

$$\left. - \frac{2}{\sqrt{\pi}} \int_0^{Z_c} \left[Z - \psi - \frac{\varrho_c^2}{\varrho^2}(Z - \psi_c) \right]^{1/2} e^{-Z} dZ \right\}$$

$$(\psi < 0, \text{ attractive field}, n > 2),$$

where

$$Z_c = \frac{2 - n}{2} \psi$$

$$n = - \varrho\psi'/\psi.$$

ψ is the potential energy in terms of kT, N_0 the ambient density, Q the return factor, ϱ the radius in Debye lengths, and ϱ_c and ψ_c are functions of Z giving the radius and potential where Z became equal to $(2-n)/2\psi_c$.

The numerical solution of Poisson's equation, using the above expressions for the space charge, yield the potential vs radius curve shown in Figure 2. Here the non-dimensional potential $(-\psi)$ is plotted against the non-dimensional radius ϱ for screening by a plasma with a Maxwellian velocity distribution at infinity. The labels $\varrho_0 = 1, 2$, etc., give the value of ϱ where the potential has a value of -0.001, the initial point used in the numerical calculation of the curves. These curves are used as follows: for a body of a given radius and potential, one finds the corresponding point in the figure. The potential as one moves away from the body will vary in the manner indicated by the family of potential vs radius curves (the part of the curve corresponding to the interior of the charged body will have no significance for that particular example).

Since the ambient plasma is Maxwellian, there are different pericritical surfaces for particles of different initial energies; thus, typical pericritical surfaces have been drawn in for particles having an energy U/kT of 0.25, 0.5, 1, 2, 4, and 8.

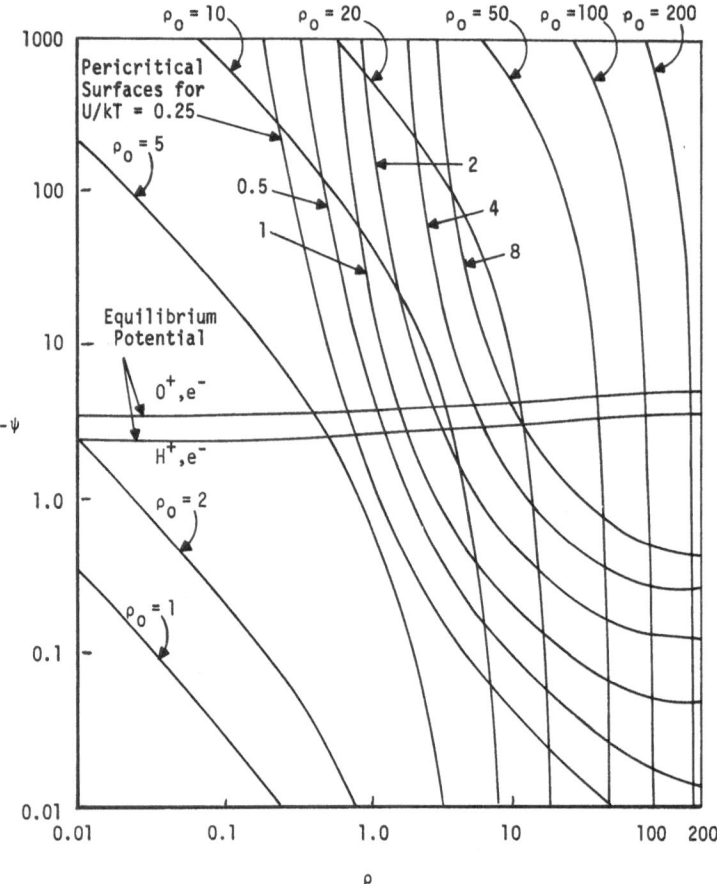

Fig. 2. The non-dimensional potential $(-\psi)$ plotted against ϱ, for screening by a plasma with a Maxwellian velocity distribution at infinity. The labels $\varrho_0 = 1$, $\varrho_0 = 2$, etc., give the value of ϱ where $\psi = -0.001$, the initial point used in the numerical calculation of the curve. Since the ambient plasma is Maxwellian, there are different pericritical surfaces for particles of different initial energy. Thus, typical pericritical surfaces have been drawn in. We also show the equilibrium potential curve for bodies in a hydrogen plasma and in a singly ionized oxygen plasma.

Ignoring the problem of the return factor Q, the equilibrium potential of the body that yields a net zero accretion of ions and electrons to the surface of the body is indicated in Figure 2 for the case of an oxygen plasma and for the case of a hydrogen plasma.

The density of the ions and the total space charge density have also been calculated. Figures 3 and 4 display some typical results for curves corresponding to the potential vs radius curves shown in Figure 2. Note here that the abscissa used in this plot is not the radius, but is rather the distance from the initial point in the integration moving toward the center of the charged body. For small bodies, there is a simple increase in both the density of ions and in the net space charge density. As we go to larger bodies,

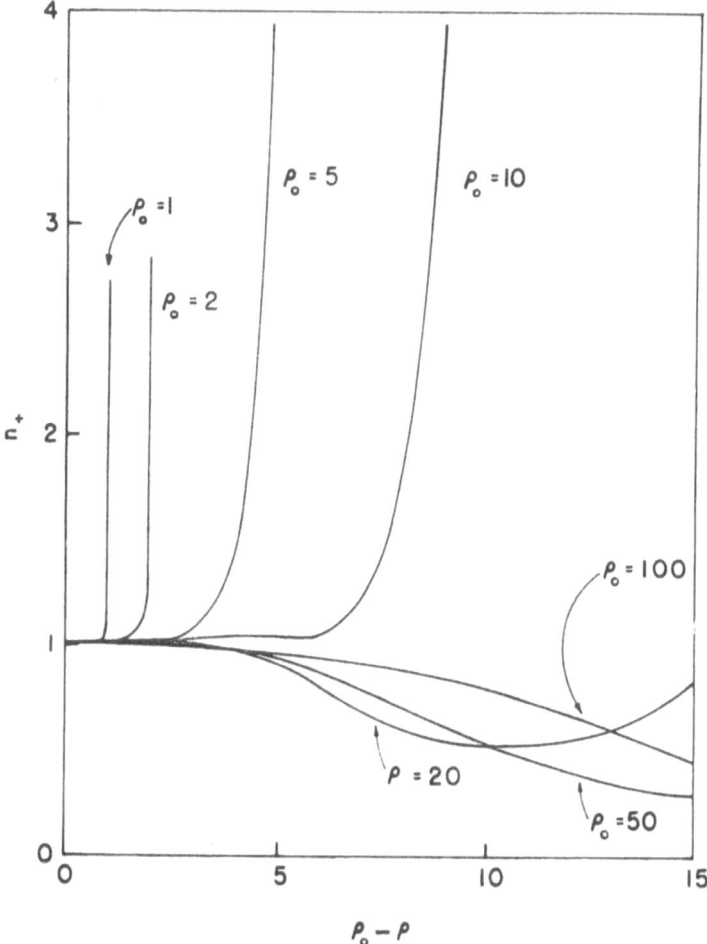

Fig. 3. n_+ plotted against $\varrho_0 - \varrho$. These curves show the variation in the ion density, starting outside the body at a distance ϱ_0 from the center where the potential is $\psi = -0.001$ and moving towards the surface of the body. The curves shown here correspond to the ψ vs ϱ plots of Figure 2 as indicated by the value of ϱ_0. To use these curves when ψ_s and ϱ_s (the values of ψ and ϱ at the surface of the body) are given, find the appropriate curve in Figure 2 and then find the corresponding curve (from the family of curves) here.

there is a slight initial increase in the density of the ions, followed subsequently by a decrease in the density and, finally, with a rise in the density. The total space charge also reflects this behavior, there being an initial rise as the body is approached, then somewhat of a leveling off or even a drop in density, and finally a rise in density. The drop in the density reflects the existence of a pericritical region in which all the particles move along spiraling orbits.

The accretion of particles to the surface of the body is also obtained as a result of these calculations. Assuming the return factor $Q = 2$, the accretion of particles ex-

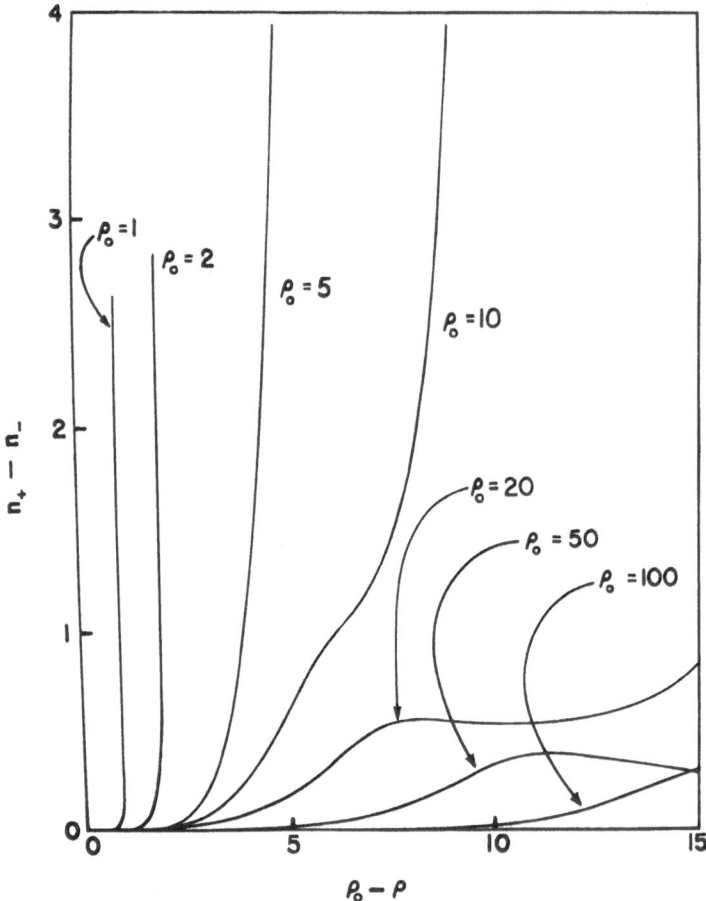

Fig. 4. $n_+ - n_-$ plotted against $\varrho_0 - \varrho$. These curves show the variation in the total space charge density starting outside the body at a distance ϱ_0 from the center where the potential is $\psi = -0.001$, and moving towards the surface of the body. The use of the curves for specific examples is the same as for Figure 3. These curves shown here correspond to the ψ vs ϱ plots of Figure 2 as indicated by the value of ϱ_0

pressed in a non-dimensional form is shown in Figure 5. Also shown in this figure are several comparitive examples, giving the accretion of ions to a negatively charged body as computed using the Mott-Smith and Langmuir (1926) probe equations for spherical symmetry. In their equation, the sheath thickness σ ($\alpha = \varrho + \sigma$) is an adjustable parameter. It will be noticed in Figure 5 that the use of the Mott-Smith and Langmuir equation can be quite misleading unless one knows rather accurately the variation in the sheath thickness as functions of the potential and radius of the body. Since the Mott-Smith and Langmuir equation is in a closed form, it would be very useful if this expression could be used together with an appropriate expression for the sheath thickness to give the results of our calculations in a closed form. Using an expression

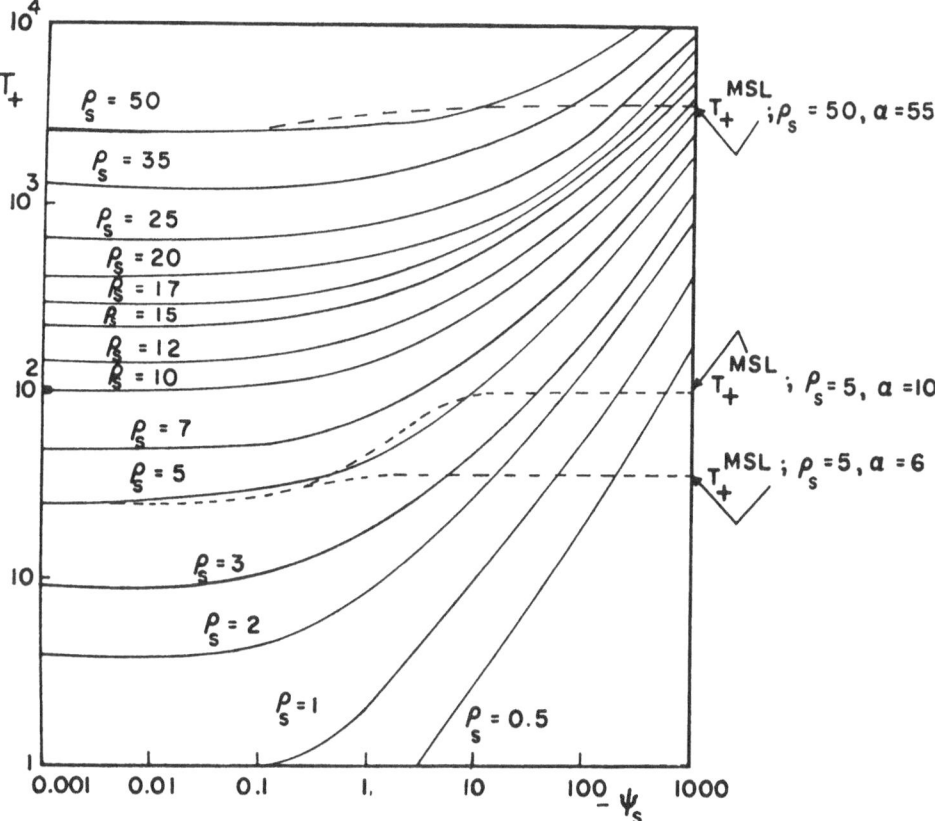

Fig. 5. A plot of the ion current-voltage characteristics obtained from the screening calculations. Here ϱ_s, the non-dimensional radius of the charged sphere, appears as the parameter in the plot of the non-dimensional ion current I_+ against the non-dimensional potential. In addition I_+^{MSL} (Mott-Smith, Langmuir ion current to a negative body, in non-dimensional form) is plotted for three cases: $\varrho_0 = 50$, $\alpha = 55$; $\varrho_0 = 5$, $\alpha = 6$; $\varrho_0 = 10$, $\alpha = 10$.

for the sheath thickness

$$\sigma = 0.83 \, \psi_s^{1/2} \varrho_s^{1/3},$$

where the subscript s refers to the surface values of the quantities, it is found (Bettinger and Walker, 1965) that the Mott-Smith and Langmuir equation will reproduce our curves for the accretion to the body quite accurately. If the return factor is set equal to 1, then the appropriate expression for σ is

$$\sigma = 1.17 \, \psi_s^{1/2} \varrho_s^{1/3}.$$

A much more satisfactory expression than this can be obtained. In the pericritical region, all the particles moving along spiral trajectories eventually intersect the surface of the body so that the expression for the sheath thickness corresponds to that appropriate to $Q=1$ (i.e., a coefficient of 1.17). Outside of the pericritical surface, however, only a portion of the particles are moving along trajectories that will eventually intersect the surface of the body and, for this reason, $Q=2$ is more appropriate. Using Figure 2 to give us the pericritical radius ϱ_c and the corresponding potential on that surface ψ_c one can write for the sheath thickness

$$\sigma = \begin{cases} 1.17\,\psi_s^{1/2}\varrho_s^{1/3} - 0.34\,\psi_c^{1/2}\varrho_c^{1/3} & \varrho_c > \varrho_s \\ 0.83\,\psi_s^{1/2}\varrho_s^{1/3} & \varrho_c \leqslant \varrho_s. \end{cases}$$

We have already mentioned the shortcomings of equations derived using the assumption that the return factor $Q=2$. Of course, the correct expressions would allow for a return factor 2 for those particles that do not reach the surface of the charged body and a return factor of 1 for those particles that strike the surface of the body and are neutralized (Hagstrum, 1954; Fowler and Farnsworth, 1958). The difficulty that arises here is not in deriving the appropriate equations. The derivation of the appropriate equations is actually no more complicated than the case where we assume that all the particles are reflected at the surface of the body. The difficulty arises with the resulting equations. If one linearizes these equations (Walker, 1965) so as to consider the nature of the potential field at a large distance from the body, one finds that the solution to the linearized equations are eigen functions and the eigen values of these functions impose certain limitations on the problem which cannot always be satisfied. As a result, there will not be, in general, a solution for a screening problem that satisfies these equations. The conclusion is that, generally speaking, *the screening problem for real bodies does not result in a time-independent configuration for the sheath*. One might argue that the linearized equations exclude solutions which might be allowed in the complete non-linear equation. This, however, is not the case. The linearized equations do hold in the boundary region of large radius with an accuracy that can be made as great as desired. As we go to infinity, these equations approach being exactly correct and, therefore, the eigenvalue conditions become exact.

There is an additional interesting feature to be discerned from these exact equations which include the neutralization of particles reaching the surface of the body. The equations show that the field must vary quite slowly. At large distances, one can show that the potential must vary more slowly than $1/r^2$ but more rapidly than $1/r$, not as would be expected, e^{-ar}/r.

Let us now consider the problem of the screening of a rapidly moving charged body embedded in a collisionless plasma in which there are no magnetic fields. A great many authors (Jastrow and Pearse, 1957; Beard and Johnson, 1960, 1961; Öpik, 1964; Kraus and Watson, 1958; Pitaevskii, 1963; Rand, 1959; Lundgren and Chang, 1963; Davis and Harris, 1961) have dealt with the problem of the screening of a rapidly moving body, but there are a number of objections that can be brought against much of that work. First, most of the papers essentially eliminate the interesting aspects of the physics of screening before they begin the problem. These papers limit their attention to those physical situations in which the potential is very small or the sheath is so thin that it can essentially be neglected. In other approaches in which the potential and sheath are taken to be large, iterative approaches are used to obtain solutions. Unfortunately, when iterative appraoches are used, the results are open to the serious objection that the results may very well not be unique. The present approach to this problem avoids these two types of difficulties. The approach allows us to calculate the screened field and the trajectories of particles in a self-consistent manner and, furthermore, to treat problems in which the potential is large and the ions are deflected significantly.

Figure 6 shows a diagram outlining the approach to the problem. One begins at an assumed boundary surface. On this boundary surface, the electric field and corresponding potential can be established using linearized equations. Since this boundary surface is taken at a large distance, the potential will be small and, as a result, the screening of the ions will not have been disturbed as yet. A step-by-step calculation is used to obtain the trajectories of the ions as they depend upon the potential field in the neighborhood of the particles. The extension of the trajectories thus made allows one to calculate the potential field in the adjacent region. Thus, step by step, the trajectories are extended and then the field is extended until the entire space has been mapped out. The equations for this approach have been reported previously (Walker, 1965).

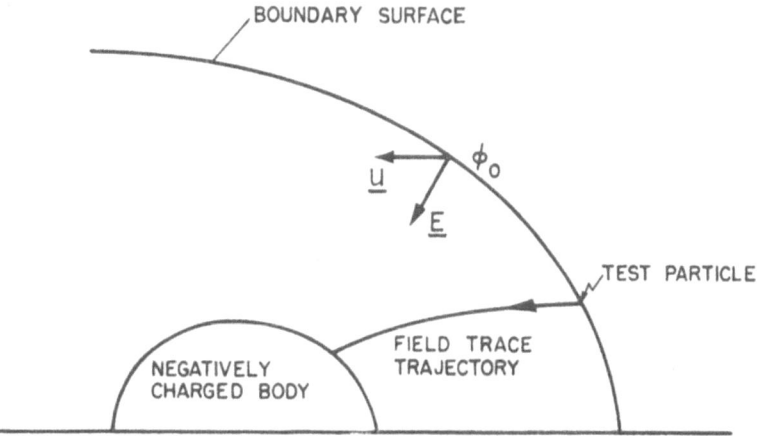

Fig. 6. The initial parameters involved in the self-consistent calculation of screening for a moving sphere. The figure shows an arbitrary boundary surface on which the potential, electric field and initial velocity of the test particles are given. A typical test particle trajectory is shown along with the surface of the negative body.

The initial boundary surface is selected arbitrarily. This boundary leads to a *real* solution, although, in general, a solution that only approximates the shape of the charged body for which a solution is sought. As a result of this, to obtain the screening of a body of a given shape, one must calculate the solution for a number of boundary surfaces successively approximating to the required boundary surface that will result in a charged body of the appropriate configuration. This approach differs from most iterative procedures in that each solution obtained is a real solution. There is no problem of convergence to spurious solutions.

Let us now consider the results of the calculations of the screening of a charged body as obtained using these equations. Figure 7 shows a set of equal potential surfaces and particle trajectories appropriate to the problem of a body having a diameter of about 15 Debye lengths, moving at a velocity of 7 km s^{-1} through an oxygen plasma at a temperature of 1500 K, or in the solar wind (hydrogen) at a relative velocity of 320 km s^{-1}, $T = 2 \times 10^5$ K. The body (actually any equipotential surface for which $\psi \gg 1$ can be selected as the surface of the body) is assumed to be charged to a high potential of 40 times the energy of the ions measured relative to the moving charged body. Figure 8 shows the corresponding space charge density for these ions. It will be noticed that

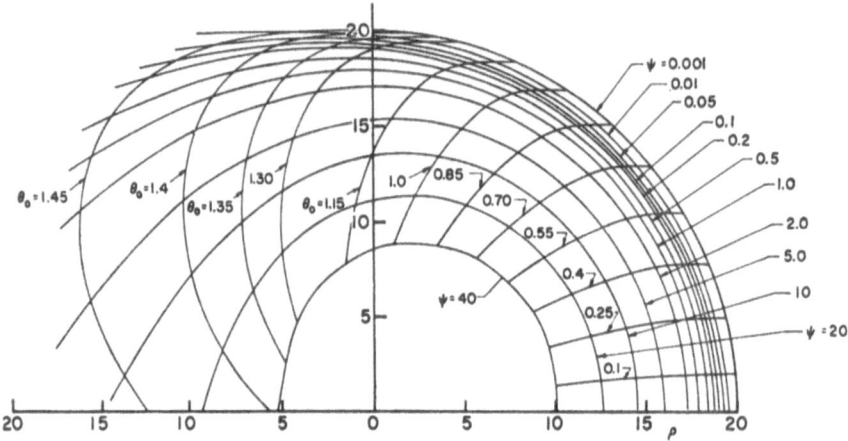

Fig. 7. A plot of the trajectories and equipotential surfaces for a rapidly moving body. Here $\varrho_0 = 20$ and $\psi_0 = -0.001$; the ions have a Maxwellian velocity distribution with $kT_i/U = 0.031812$. The initial angle θ_0 of each trajectory is given in radians.

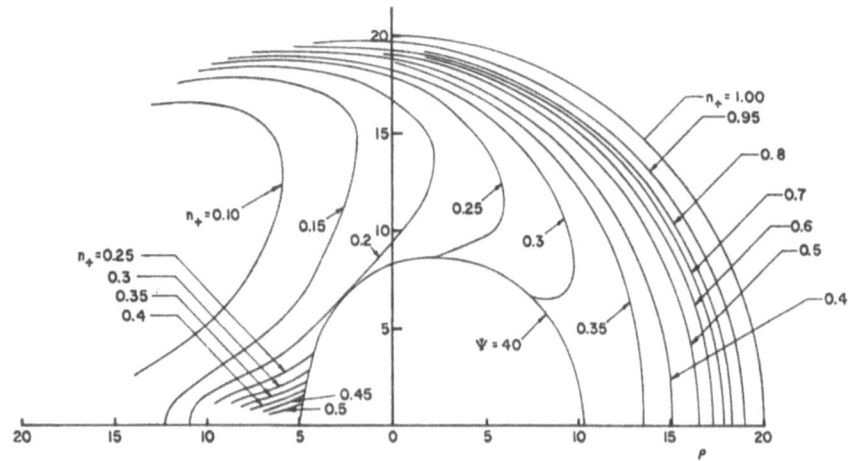

Fig. 8. A plot of the equidensity surfaces for a rapidly moving body. Here $\varrho_0 = 20$ and $\psi_0 = -0.001$; the ions have a Maxwellian velocity distribution with $kT_i/U = 0.031812$. An equipotential surface from Figure 7 is also shown.

there is a drop in the density of the particles not only behind the body but also in front of the body. There is a slight increase in the density at the rear of the body. In the case of small bodies, this increase in density can become larger than the ambient density but for large bodies there will never be more than a slight increase over the general low value of the density behind the body.

We find that the results of these calculations are not in agreement with the iterative calculations performed by Davis and Harris (1961); this discrepancy would indicate difficulties with one of the two approaches. Since there are, however, some obvious difficulties in the results obtained by Davis and Harris, namely, singularities in the space charge density along the axis in front of the body and additional, somewhat peculiar highs and lows in the space charge density scattered throughout the screened field, we feel that the inconsistency is to be ascribed to shortcomings of the iterative calculations. There is a confirmation of this work to be obtained from a calculation of the ion currents to the charged body shown in Figure 9. These curves are quite consistent with the curves obtained for the case of a stationary sphere embedded in a

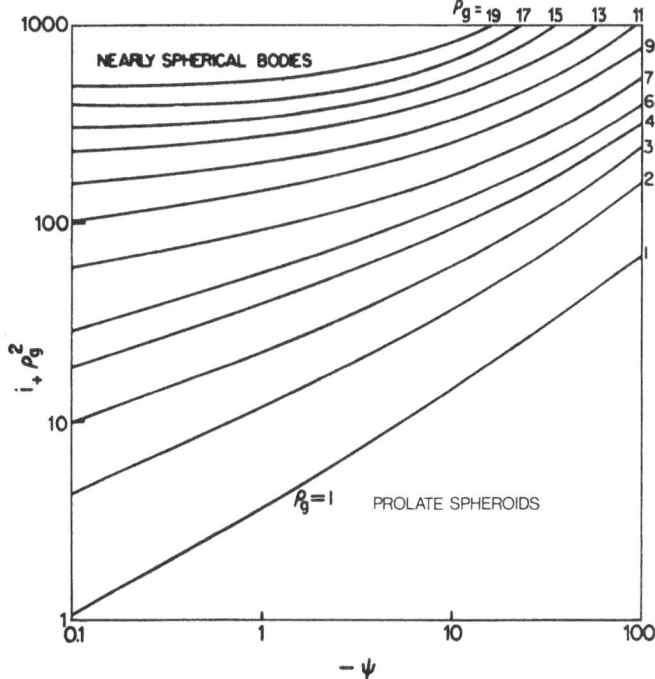

Fig. 9. A plot of the current-voltage characteristics for bodies moving rapidly through a plasma. We have plotted here the non-dimensional quantities $i_+ \varrho_g^2$ vs $-\psi$. The geometric radius ϱ_g appears as a parameter. The data used is drawn from a limited number of calculations in which the shape of the charged body varied so that the curves in the lower part of the graph apply best to somewhat prolate spheroids. The first order correction for a Maxwellian velocity distribution of the ambient ions has been included here; we have taken $kT_e/U = kT_i/U = 0.031812$ (corresponding to a relative velocity $U = 7$ km s^{-1} for a singly ionized oxygen plasma with $T = 1500$k). The curves should be only weakly dependent upon the value of T_e and T_i if kT_i or $kT_e \ll U$.

plasma. In fact, one can use the same expression for the sheath thickness together with the Mott-Smith and Langmuir equation in order to reproduce the curve shown in Figure 9, which indicates that the sheath thickness is at least of the correct order of magnitude.

Let us finally consider the drag experienced by the rapidly moving charged body as a result of the ionic impacts and the electric fields acting on the body. We assume that all the ions reaching the surface of the body are neutralized. There are two assumptions concerning the momentum transfer on impact. One extreme is to assume that the ions impact completely inelastically; the other is to assume that the collision is completely elastic. The drag that results for these two cases is shown in Figure 10. The curves labeled 'A' are for the assumption of completley elastic collision and the curves labeled 'B' correspond to a completely inelastic collision with a body. It was at first quite annoying to discover that the assumption of completely elastic collision gave

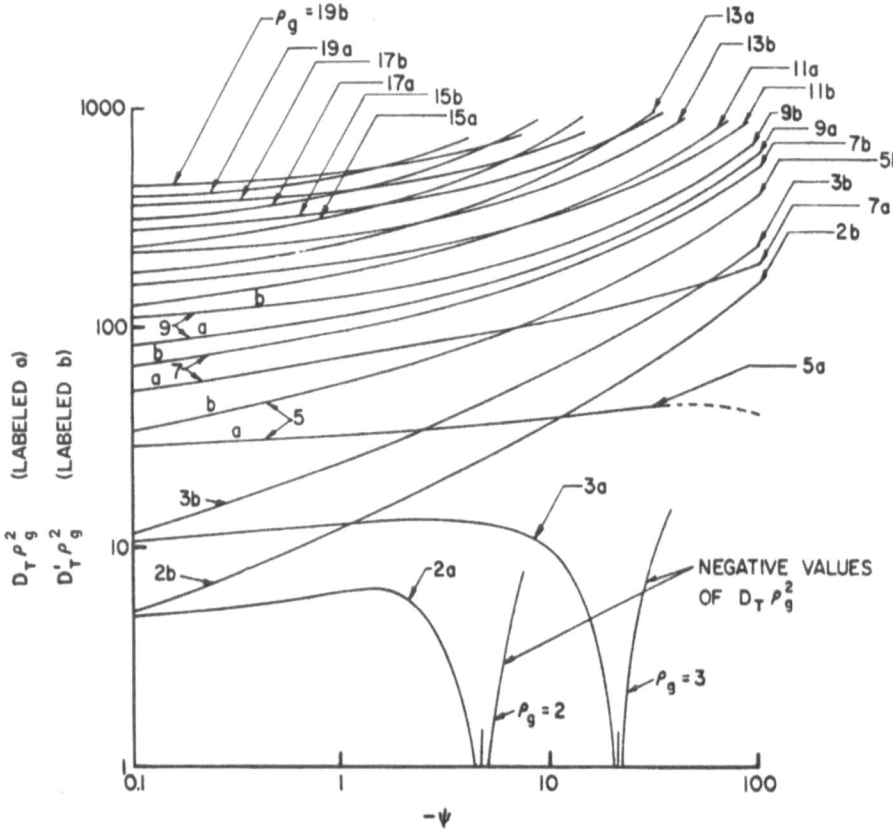

Fig. 10.   Plots of $D_T \varrho_g^2$ and $D'_T \varrho_g^2$ vs $(-\psi)$ for bodies moving rapidly through a plasma. The geometric radius ϱ_g appears as a parameter in both sets of curves. The curves near the bottom of the graph apply best to somewhat prolate spheroids. These curves include the correction for a Maxwellian velocity distribution of the ambient ions, here $kT_e/U = kT_i/U = 0.031812$. The occurrence of negative values of D_T (for $\varrho_g = 2$, $\psi \gtrsim 4.6$ and $\varrho_g = 3$, $\psi \gtrsim 22$) is discussed in the text.

negative values for the drag in some cases. It was thought that this was due to some error in the numerical calculations. After some reflection on this question, however, it became obvious that not only are negative values of the drag possible but quite reasonable.

Consider a test particle of mass m moving with a velocity u of 7 km s^{-1} toward a body with a potential $\psi = 24$ (see Figure 11). Upon reaching the surface of the body, it has a velocity

$$v = u\sqrt{1 + \psi} = 35 \text{ km s}^{-1}.$$

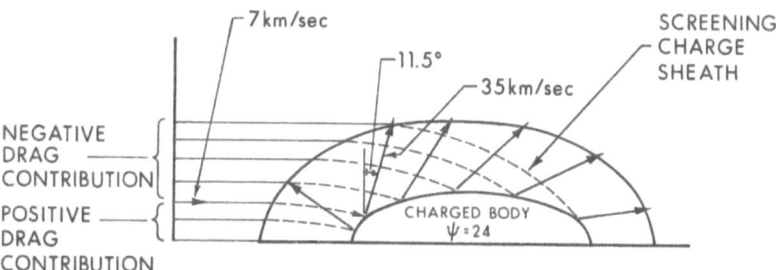

Fig. 11.   Individual plasma ion trajectories and velocities contributing to the negative drag of a charged prolate body. Velocity vectors are given for a coordinate system at rest relative to the body.

Since the particle will be Auger neutralized and, according to our assumption, specularly reflected at the surface, it is not decelerated as it moves away from the surface of the charged body. If it strikes the body in such a way that it moves away in the rearward direction, then the net momentum transfer will be

$$\Delta P = m(7 - 35) \text{ km s}^{-1}$$
$$= -m\,28 \text{ km s}^{-1},$$

thereby increasing the momentum of the body. If more particles leave in the rearward direction than in the forward direction as occurs for end-on, highly prolate spheroids having a sheath thickness large compared with their radius of curvature (perpendicular to the axis aligned with the velocity vector), then there will be a net force accelerating the body, i.e., a negative drag. For the case of small bodies or bodies that are prolate and charged to a high potential, it is most likely that there will be more particles leaving in the rearward direction than in the forward direction and, thus, that there will be a negative drag for the case of specular reflection of the ions at the surface. Such a principle may have application as an externally fueled ion engine.

A by-product of the above approach to the problem of the screening of rapidly moving bodies is that the technique allows a single-calculation design of an ion engine. If one begins with a streaming plasma that is well columnated and already removed from the region of high electric fields and calculates back along the trajectories using these equations, one will obtain the appropriate configuration for the electrodes necessary to produce such a beam. Sample calculations for this problem have been performed. Figure 12 is a schematic of a typical solution.

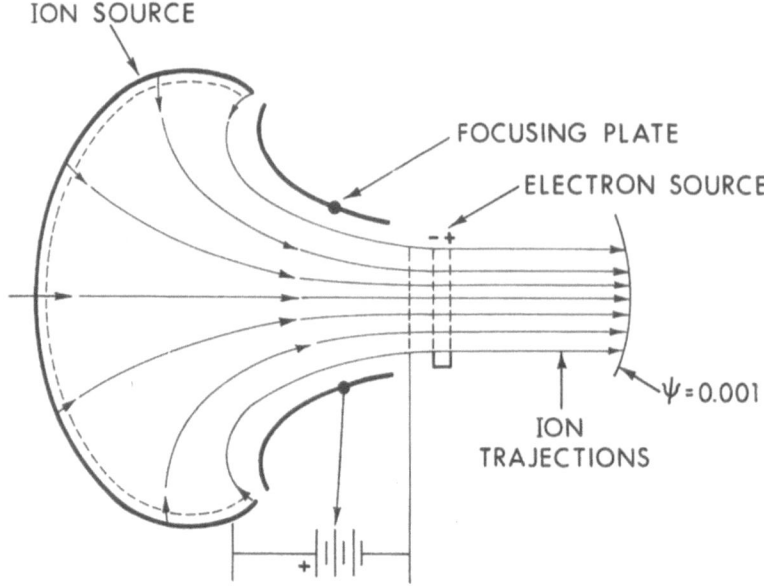

Fig. 12.   Schematic of a typical application of the trajectory trace method of designing ion engines.

References

Beard, D. B. and Johnson, F. S.: 1960, *J. Geophys. Res.* **65**, 1.
Beard, D. B. and Johnson, F. S.: 1961, *J. Geophys. Res.* **66**, 12.
Bernstein, I. B. and Rabinowitz, I. N.: 1958, Project Matterhorn Report PM-S-38.
Bettinger, R. T. and Walker, E. H.: 1965, *Phys. Fluids* **8**, 748–751.
Davis, A. H. and Harris, I.: 1961 in L. Talbot (ed.), *Rarefield Gas Dynamics*, Academic Press, New York, pp. 691–699.
Fowler, H. A. and Farnsworth, L. E.: 1958, *Phys. Rev.* **111**, 103.
Hagstrum, H.: 1954, *Phys. Rev.* **96**, 325.
Jastrow, R. and Pearse, C. A.: 1957, *J. Geophys. Res.* **62**, 3.
Kraus, L. and Watson, K. M.: 1958, *Phys. Fluids* **1**, 480.
Laframboise, J. G.: 1966, Institute for Aerospace Studies, University of Toronto, UTIAS Report No. 100.
Lundgren, T. S. and Chang, C. C.: 1963, Aerospace Corporation Report No. ATN-63 (9226)-2.
Mott-Smith, H. M. and Langmuir, I.: 1926, *Phys. Rev.* **28**, 727.
Öpik, E. J.: 1964, in S. F. Singer (ed.), *Interactions of Space Vehicles with an Ionized Atmosphere* (*Am. Astronaut. Soc. Symp.*, March 1961), Pergamon Press, New York, pp. 3–60.
Pitaevskii, L. P.: 1963, *AIAA* **1**, 994.
Rand, S.: 1959, *Phys. Fluids* **2**, 265.
Spitzer, L.: 1956, *Physics of Fully Ionized Gases*, Interscience, New York.
Walker, E. H.: 1965, in S. F. Singer (ed.), *Interaction of Space Vehicles with an Ionized Atmosphere*, Pergamon Press, London, pp. 61–162.
Wasserstrom, E., Su, C. H., and Probstein, R. F.: 1964, Fluid Mechanics Laboratory, Dept. of Mechanical Engineering, MIT.

DISCUSSION

Wrenn: You indicated that in some circumstances there could be a build-up of plasma density in the front of a body; can you elaborate on the conditions for this to occur?

Walker: The build-up referred to is that found in these calculations well into the field after a sizeable density drop has occurred. This build-up occurs for very high body potential energies (like 10–20 times the kinetic energy of the ions) as the ions are focussed onto the body, i.e. by convergence of the trajectories. For stationary bodies, under certain circumstances an initial build-up does occur. Such a build-up might occur for bodies moving relative to the plasma, but the effect become less and less as the kinetic energy of the ions becomes greater. For the cases computed, I believe the original data showed a build-up of something like 0.001 over ambient at potentials less than kT. Such an effect probably would be difficult to notice experimentally.

Thomas: Downstream contours of electron density were not shown for values of $\varrho > 20$. Is this because the calculation is inapplicable when $\varrho > 20$ or thereabouts and if so, why?

Walker: The calculation can be carried further. The question that should be asked first is to what distance will a time independent solution be meaningful. In actual fact the calculations were terminated arbitrarily at this point.

Polychronopulos: Can the net drag on a body moving in a plasma be negative?

Walker: Yes, it can be under certain conditions and for certain body geometries (e.g. prolate).

THE DISTORTION OF AN ELECTROMAGNETIC
WAVE FIELD BY A CAVITY IN A COLD MAGNETO-PLASMA

T. R. KAISER and P. C. KENDALL

Depts. of Physics and Applied Mathematics, The University, Sheffield 10, England

Abstract. The distortion of an em wave field by a satellite such as GEOS and its attendant sheaths creates problems in interpreting AC electric field measurements, especially with electric sensors which may not extend beyond the region of perturbed plasma. In order to gain insight into this, a simple model of a spherical cavity in a cold magneto plasma is considered. Equations are derived which relate the amplitudes, phases and directions of the perturbed and unperturbed fields.

1. Introduction

The geostationary satellite (GEOS), to be launched in the ESRO programme during the period of the International Magnetospheric Survey, aims to measure natural AC electric fields over a wide range of frequencies. Probes on long booms will measure a single component of the **E**-field while a shorter boom system carries probes to measure three orthogonal components.

A serious problem may arise from the distortion of the wave field due to the presence of the satellite vehicle. Even in vacuo, the vehicle causes distortion; in a plasma it also produces a local plasma perturbation which may be in the form of electron depletion in the Debye sheath and electron enhancement in the photo-electron sheath. The short boom probes will generally be within the region of distorted field and the purpose of this contribution is to gain some insight into the nature of the problem.

The model considered is that of a spherical vacuum cavity in the magneto-plasma, the unperturbed wave field being assumed uniform with components parallel and transverse to the geomagnetic field, **B**. A solution for the perturbed field is found for the limiting case of a cold plasma in the quasi-static approximation (wavelengths of characteristic waves are assumed large compared with the cavity dimensions). This might be expected to have some relevance to the case of the Debye sheath, however some limitations should be emphasised; these are:

(i) The plasma in the Debye sheath is not quasi-neutral; only the electron density is depressed.

(ii) The tensor permittivity is assumed to change abruptly at the cavity boundary; even in the cold plasma case a finite electron density gradient at this boundary can have a significant effect (see Section 2).

(iii) The cold plasma approximation assumes that the particle gyro-radii are sufficiently small compared with the cavity dimensions and that a particle makes many oscillations in the wave field in travelling a distance comparable with the cavity dimensions. The former requirement may be reasonably satisfied for the electrons (except perhaps near resonances) but not for the ions. The latter condition will fail at

R. J. L. Grard (ed.), Photon and Particle Interactions with Surfaces in Space, 91–97. All Rights Reserved

frequencies which are small compared with the electron plasma frequency (assuming the dimensions to be only a few Debye lengths).

In the isotropic case ($B=0$), the field in the cavity differs only in magnitude from the unperturbed field while the external perturbation is dipolar; in general ($B \neq 0$) the field in the cavity will differ also in its polarization. We will consider, first, the simpler isotropic case for comparison with the general one. As representative of conditions at the geostationary orbit, the following parameters have been assumed for numerical computation:

Electron and ion densities, $N_e = N_i = 10$ cm^{-3}
Magnetic field, $B = 150 \ \gamma$
yielding the characteristic frequencies:
ion gyro-frequency $f_i = 2.29$ Hz
lower hybrid frequency $f_{LH} = 98.0$ Hz
electron gyro-frequency $f_e = 4.20 \times 10^3$ Hz
plasma frequency $f_p = 2.84 \times 10^4$ Hz
upper hybrid frequency $f_{UH} = 2.87 \times 10^4$ Hz
Note that $0 < f_i < f_{LH} < f_e < f_p < f_{UH}$.

The justification for considering the effects of the cavity only on the electric wave field (E) is that, if the incident wave has wavelength λ, then it may be shown that the cavity is magnetically transparent to order $(a/\lambda)^2$ where a is the cavity radius.

2. Perturbation in an Isotropic Plasma ($B = 0$)

We will consider here the somewhat more general case where the dielectric constant in the unperturbed plasma is $\kappa = 1 - f_p/f^2$ where f is the wave frequency and inside a spherical cavity of radius a is κ_i. To make the model a little more realistic we will assume that κ varies linearly between these limits in a thin boundary layer (thickness $\ll a$) with gradient κ'.

Kaiser and Closs (1952), in considering the scattering of an electromagnetic wave by a spherical plasma cloud, have solved the problem for the case $\kappa = 1$, $\kappa_i \neq 1$. Using the method of these authors we obtain the following results: – The potential is

$$
\begin{aligned}
V &= -E_i r \cos\theta \quad \text{for} \quad r < a \\
V &= -\left(E + \frac{A}{r^3}\right) r \cos\theta \quad \text{for} \quad r > a,
\end{aligned} \tag{1}
$$

where r is radius measured from the centre of the sphere and θ is the polar angle measured from the direction of the unperturbed field, E.

The internal field E_i, which is uniform and parallel to E, is given by

$$
\frac{E_i}{E} = \frac{3\kappa}{2\kappa + \kappa_i + \dfrac{2\kappa\kappa_i}{\kappa' a}\left[\ln\left|\dfrac{\kappa}{\kappa_i}\right| + i\delta\right]}, \tag{2}
$$

where $\delta = 0$ for κ and κ_i of the same sign

$$\delta = +\pi \quad \text{for} \quad \kappa > 0, \kappa_i < 0$$
$$\delta = -\pi \quad \text{for} \quad \kappa < 0, \kappa_i > 0.$$

The external perturbation is dipolar with

$$\frac{A}{a^3 E} = \frac{\kappa - \kappa_i + \dfrac{\kappa \kappa_i}{\kappa' a} \left[\ln \left| \dfrac{\kappa}{\kappa_i} \right| + i\delta \right]}{2\kappa + \kappa_i + \dfrac{2\kappa \kappa_i}{\kappa' a} \left[\ln \left| \dfrac{\kappa}{\kappa_i} \right| + i\delta \right]}. \tag{3}$$

When $\kappa' \to \infty$ (sharp boundary) we see, from (2) and (3), that a resonance occurs ($E_i \to \infty$, $A \to \infty$) when $\kappa_i = -2\kappa$. If κ' is finite but the second term in the denominator of (2) is sufficiently small, the resonance still occurs but the resonant perturbation has finite amplitude with

$$\left(\frac{E_i}{E} \right)_{res} = \left(\frac{A}{a^3 E} \right)_{res} = -i \frac{3\kappa' a}{2\kappa_i \delta} = i \frac{3\kappa' a}{4\kappa \delta}. \tag{4}$$

When $\kappa = 0$, i.e. $f = f_p$, the internal field vanishes.

If $\kappa_i = 1$ (vacuum cavity) the resonance occurs for $\kappa = -\frac{1}{2}$, i.e. at frequency f_0 where $f_0/f_p = (\frac{2}{3})^{1/2}$ and, when $\kappa' \to \infty$,

$$\frac{E_i}{E} = \frac{3(f^2 - f_p^2)}{3f^2 - 2f_p^2}. \tag{5}$$

For the above numerical parameters, $f_0 = 2.31 \times 10^4$ Hz.

3. Cavity in a Magneto-Plasma

We restrict ourselves here to a spherical vacuum cavity of radius a in a magneto-plasma. In a later work it is proposed to extend the analysis to the case where the internal plasma density is non-zero.

Taking the z-axis of cartesian coordinates to be parallel to the steady magnetic field, B, the dielectric tensor for the unperturbed plasma is

$$\tilde{\kappa} = \begin{pmatrix} \kappa_1 & -\kappa_2 & 0 \\ \kappa_2 & \kappa_1 & 0 \\ 0 & 0 & \kappa_3 \end{pmatrix}, \tag{6}$$

where

$$\kappa_1 = \frac{(f^2 - f_{LH}^2)(f^2 - f_{UH}^2)}{(f^2 - f_e^2)(f^2 - f_i^2)},$$

$$\kappa_2 = i \frac{f f_p^2 (f_e - f_i)}{(f^2 - f_e^2)(f^2 - f_i^2)},$$

$$\kappa_3 = (f^2 - f_p^2)/f^2.$$

Now $\nabla \cdot (\tilde{\kappa} \mathbf{E}) = 0$ and, in the quasi-static approximation, $\mathbf{E} = -\nabla V$. Thus we get, external to the cavity,

$$\alpha \left(\frac{\partial^2 V}{\partial x^2} + \frac{\partial^2 V}{\partial y^2} \right) + \frac{\partial^2 V}{\partial z^2} = 0, \tag{7}$$

where

$$\alpha = \kappa_1/\kappa_3 = \frac{f^2 (f^2 - f_{LH}^2)(f^2 - f_{UH}^2)}{(f^2 - f_e^2)(f^2 - f_i^2)(f^2 - f_p^2)}.$$

Note that (7) may be elliptic or hyperbolic depending upon the sign of α.

Inside the cavity $\nabla^2 V = 0$ and the following boundary conditions apply at $r = a$:

(i) $\hat{n} \wedge \mathbf{E}$ is continuous

(ii) $\hat{n} \cdot \mathbf{D}$ is continuous

where the electric displacement, $\mathbf{D}$, is given by

$$\mathbf{D}/\varepsilon_0 = \kappa_1 \mathbf{E}^\perp + \kappa_3 \mathbf{E}^\| - \kappa_2 (\mathbf{E} \wedge \hat{z}).$$

$\hat{n}$ is a unit radial vector at the cavity boundary, $\hat{z}$ is a unit vector and the symbols $\perp$ and $\|$ refer to the directions transverse and parallel to $\mathbf{B}$.

Thus condition (ii) requires

$$(f^2 - f_p^2)(\mathbf{E}^\| + \alpha \mathbf{E}^\perp - i\beta \mathbf{E}^\perp \wedge \hat{z})$$

to be continuous, where

$$i\beta = \kappa_2/\kappa_3, \qquad \beta = \frac{f^3 f_p^2 (f_e - f_i)}{(f^2 - f_e^2)(f^2 - f_i^2)(f^2 - f_p^2)}.$$

To solve (7) we follow Kaiser (1962) and Kaiser and Tunaley (1968) who dealt have with the behaviour of a conducting AC probe in a magneto-plasma and define coordinates such that (7) becomes $\nabla_1^2 V = 0$ where $x_1 = x\alpha^{-1/2}$, $y_1 = y\alpha^{-1/2}$, $z_1 = z$. The cavity surface is thus defined by

$$\alpha (x_1^2 + y_1^2) + z_1^2 = a^2, \quad \text{or} \quad \alpha R_1^2 + z_1^2 = a^2.$$

3.1. SOLUTION FOR $\alpha > 1$

Following Lamb (1932), we define prolate spheroidal coordinates μ_1, ζ_1, ϕ whence (7) has solutions of the form $P_n^m (\mu_1) Q_n^m (\zeta_1) e^{jm\phi}$ where P_n^m and Q_n^m are associated Legendre functions of the first and second kinds, respectively. At the cavity boundary $\zeta_1^2 = = \zeta^2 = \alpha/(\alpha - 1)$ and $\mu_1 = \mu = \cos\theta$ and the appropriate potential, external to the cavity, becomes

$$V = AP_1 (\mu_1) Q_1 (\zeta_1) + BP_1^1 (\mu_1) Q_1^1 (\zeta_1) e^{j\phi}. \tag{8}$$

Inside the cavity, $\nabla^2 V = 0$, and we have

$$V = CrP_1 (\mu) + DrP_1^1 (\mu) e^{j\phi}. \tag{9}$$

The constants A, B, C, D are obtained by applying the boundary conditions.

If E_x, E_y and E_z are the (x, y, z) components within the cavity and E_1, E_2, E_3 are those external, we finally obtain

$$E_z = E_3 (\lambda_3 - 1)/[\lambda_3 - (1 - f_p^2/f^2)^{-1}] \tag{10}$$

and

$$(E_x - jE_y) = (E_1 - jE_2)(\lambda_0 - \alpha)/[\lambda_0 - (1 - f_p^2/f^2)^{-1} - ij\beta], \tag{11}$$

where

$$\lambda_3 = \zeta Q_1'(\zeta)/Q_1(\zeta),$$
$$\lambda_0 = \zeta Q_1^{1'}(\zeta)/Q_1^1(\zeta),$$

both evaluated for $\zeta^2 = \alpha/(\alpha - 1)$

It should be noted that there is a double complex notation in (9), both i and j being $\sqrt{-1}$, the former referring to time variation as $\exp(i\omega t)$ and the latter to the spatial variation $\exp(j\phi)$. Thus, to obtain E_x in real terms, we take the real part of the right hand side with respect to j, then multiply by $\exp(i\omega t)$ and take the real part with respect to i. For E_y, we take the imaginary part with respect to j, multiply by $\exp(i\omega t)$ and take the real part with respect to i.

We have also obtained solutions for the perturbed field external to the cavity; these will not be discussed here but will be presented in a further work.

3.2. SOLUTION FOR $0 \leqslant \alpha < 1$

Oblate spheroidal coordinates, in the transformed system are now appropriate, but their use nevertheless leads to the same analytic solutions (10) and (11), as for $\alpha > 1$. However, the Q's are functions involving $\ln[(\zeta + 1)/(\zeta - 1)]$ which is real and straightforward only for the preceding case, $\alpha > 1$, making ζ real and greater than unity. In the present case we can write $\zeta = i\zeta'$ where $\zeta' = [\alpha/(1 - \alpha)]^{1/2}$ whence

$$\ln[(\zeta + 1)/(\zeta - 1)] = -2i \tan^{-1}(1/\zeta') \tag{12}$$

3.3. SOLUTION FOR $\alpha < 0$

In this case both the prolate and oblate coordinate systems become complex and appear to have no geometrical meaning. Nevertheless the solutions (8) and (9) and the cavity fields (10), (11) will be physically meaningful provided appropriate conditions of continuity are imposed when making the coordinate transformations. These require us to define, on the cavity boundary

$$\ln[(\zeta + 1)/(\zeta - 1)] = \ln|(\zeta + 1)/(\zeta - 1)| - i\pi \tag{13}$$

(note that in this case ζ is positive and less than unity).

3.4. A SIMPLIFICATION FOR $(f/f_e)^2 \gg 1$

In this case we find

$$\alpha \simeq 1 - \frac{f_e^2 f_p^2}{f^2(f^2 - f_p^2)} = 1 - \varepsilon, \qquad \beta \simeq \frac{f_e f_p^2}{f(f^2 - f_p^2)}. \tag{14}$$

Except near f_p, ε is small compared with unity $(\varepsilon \sim n^{-1}$ for $f = f_p \pm n(f_{UH} - f_p)$ and we obtain:

$$\zeta^2 \simeq \varepsilon^{-1}, \qquad Q_1(\zeta) \simeq \varepsilon/3, \qquad Q_1'(\zeta) \simeq - 2\varepsilon^{3/2}/3, \qquad Q_1^1(\zeta) = - 2\varepsilon/3,$$

$$Q_1^{1'}(\zeta) = 4\varepsilon^{3/2}/3 \quad \text{and hence} \quad \lambda_0 \simeq \lambda_3 \simeq - 2.$$

Equations (10) and (11) now simplify to

$$\frac{E_z}{E_3} = \frac{3(f - f_p^2)}{3f^2 - 2f_p^2} \tag{15}$$

$$\frac{E_x}{E_1} = \frac{3(f^2 - f_p^2)(3f^2 - 2f_p^2)}{(3f^2 - 2f_p^2)^2 - (f_e f_p^2/f)^2} \tag{16}$$

$$\frac{E_y}{E_1} = \frac{3i(f^2 - f_p^2) f_e f_p^2/f}{(3f^2 - 2f_p^2)^2 - (f_e f_p^2/f)^2}. \tag{17}$$

We see that the expression for the longitudinal component is identical with the isotropic case, exhibiting the plasma resonance at $f_0 = (\frac{2}{3})^{1/2} f_p$. Equations (16) and (17) give the transverse fields in the cavity for the transverse unperturbed field in the direction of the x-axis $(E_2 = 0)$. We note that the plasma resonance is now split with E_x zero and E_y minimum at $f = f_0$; both E_x and E_y become infinite at $f = f_0 \pm f_e/4$, i.e. the splitting is equal to one half of the electron gyrofrequency.

It should be noted that (15), (16) and (17) are valid when f is sufficiently large compared with f_e, except for a region around the plasma and upper hybrid frequencies. Also, the transverse field reduces to the isotropic case when $(f_e f_p^2/f)^2 \ll |3f^2 - 2f_p|^2$, i.e. sufficiently far from the resonance. This inequality is of order $1/n$ when $|f - f_0| = = nf_e/4$.

4. Summary

While the cavity model cannot be said to simulate, in any close way, the conditions to be expected in the GEOS environment, some general conclusions can be reached. Clearly the distortion of the wave field will create problems of interpretation, especially for those probes mounted on the short booms but also for those on the long ones (since although they may be external to the region of plasma perturbation they will still be in a region of distorted field). Some of the limitations of the model are discussed in Section 1; we can say, however, that we will expect the measured fields within the perturbed plasma region to differ in magnitude, phase and polarization from the unperturbed wave field. An encouraging feature is that, over a considerable frequency range, the magneto-plasma results are similar to the isotropic case which is itself amenable to a more general treatment (including a finite density gradient, warm plasma etc.). It should be noted that, in the isotropic cold plasma case when κ_1 and κ_2 are opposite in sign, the perturbed field becomes infinite at the boundary where $\kappa = 0$. This infinity will be removed, due to Landau damping, for a warm plasma.

The ratios of internal to perturbed field have been computed, using the numerical parameters in Section 1, for a frequency range which embraces all the characteristic frequencies. These indicate, as expected, strong field perturbations in the vicinity of f_i, f_{LH} and f_e as well as those already discussed. Plans are presently in hand to extend the work to include a more realistic density profile in the isotropic approximation and to the warm plasma case; we will also hope to extend the magneto-plasma calculations to a cavity with a finite plasma density, differing from that of the ambient plasma. It may also be possible to consider a different geometry, e.g. a cylindrical plasma perturbation. It would therefore be of interest to obtain, from other workers, a more realistic model of the plasma perturbation to be expected in the neighbourhood of the GEOS satellite.

Acknowledgement

We are indebted to Mrs L. Wilkinson and Mr J. Ashworth for the not inconsiderable task of computing numerical values for the field perturbations; these will be presented in a further work which will contain a more comprehensive account of the mathematical treatment than is given here. In referring to the paper by Kaiser and Closs (1952) we should mention the indebtedness of these authors to the work of Herlofson (1951).

References

Herlofson, N.: 1951, *Arkiv Fysik* **3**, 247.
Kaiser, T. R.: 1962, *Planetary Space Sci.* **9**, 639–657.
Kaiser, T. R. and Closs, R. L.: 1952, *Phil. Mag.* **43**, 1–32.
Kaiser, T. R. and Tunaley, J. K. E.: 1968, *Space Sci. Rev.* **8**, 32–73.
Lamb, H.: 1932, *Hydrodynamics* (6th ed.), Dover, p. 139.

DISCUSSION

Thomas: Would you comment on the result which you mentioned of a 'splitting' of the resonance. Does this occur at one half the electron gyrofrequency or is the situation such that there are two resonances separated by $\omega_e/2$?

Kaiser: The resonance, for an isotropic plasma, occurs at a frequency equal to $(\frac{2}{3})^{1/2}$ times the plasma frequency internal to the cavity. In the presence of a magnetic field, **B**, we find still a single resonance for the component of electric field parallel to **B**, but the resonance for the transverse component of the electric field is split into two, the separation being in the case investigated, numerically equal to one-half of the electron gyrofrequency.

Grard: Would it be feasible to also consider a cylindrical cavity, since this problem has some relevance to the plasma inhomogeneity existing around a long electric aerial.

Kaiser: The case of a cylindrical inhomogeneity in an isotropic plasma has already been treated in connection with the scattering of radio waves from meteor trails. In the case of a magnetoplasma the problem is still simple when the axis of the cylinder is parallel to the direction of the magnetic field; the complete solution, however, may not be as straightforward for an arbitrary orientation.

2.2. NUMERICAL ANALYSIS AND SIMULATION

STRUCTURE OF THE IONOSPHERIC DISTURBANCES
ABOUT PLANETARY ENTRY PROBES

HERSCHEL WEIL and HOWARD JEW

Space Physics Research Laboratory, University of Michigan, Ann Arbor, Mich., 48105, U.S.A.

and

URI SAMIR

Space Physics Research Laboratory, University of Michigan, Ann Arbor, Mich., 48105, U.S.A.
and Dept. of Environmental Sciences, Tel-Aviv University, Ramat-Aviv, Israel

Abstract. Local ionospheric disturbances which would be created by a planetary entry probe are investigated. Competing theories of spacecraft-ambient plasma interactions are used to estimate computationally the perturbations of the plasma, particularly the structure of the near wake behind planetary entry vehicles.

The results have bearing on the location and operation of plasma diagnostic instrumentation aboard planetary entry vehicles. It is therefore appropriate to draw attention to this far-from-negligible phenomenon.

Recent estimates of Mars ionospheric properties plus vehicle dimensions and speeds similar to those of the Viking Mars Lander are used to define the parameters essential to the theory. Smaller entry bodies are also considered. Comparisons are made of the results based on the different theories for a given assumed planetary atmosphere, and also with the perturbations a similar vehicle would generate in the Earth's ionosphere.

1. Introduction

The purpose of this paper is to obtain, by application of theoretical-numerical methods, models of the wake of a planetary entry vehicle penetrating the planet's ionosphere. For concrete examples we consider the proposed Mars Viking lander aeroshell as described by Soffen and Young (1972) in the Martian ionosphere and also a much smaller object of the size of an Explorer Satellite. We compute the perturbations in electron and ion density and in the potential which constitute the wake region.

A number of different formulations of the space vehicle-ambient plasma interaction problem are available in the literature. Since it would require excessively large scale computations to obtain numerical solutions to the general equations, each formulation embodies simplifying physical and mathematical assumptions which are not necessarily valid for many practical situations. The assumptions differ in the various formulations (see review papers by Gurevich *et al.* (1969), Liu (1969), and Samir (1972)) so that, for a given set of planetary plasma parameters, only qualitative agreement on the wake structure can be obtained. Also, because of computational difficulty, the only three dimensional problems which have been fully formulated and investigated in numerical examples assume spherical, spheroidal or flat disc bodies. The Mars lander aeroshell is, however, a 1.75 m radius, 140° cone. It is fortunate, however, for a fixed ambient temperature, body potential and velocity the geometrical blocking effect of the body on the ion flow is primarily influenced by the maximum body cross sectional size and shape in a plane perpendicular to the vehicle velocity direction. The blocking is generally less sensitive to body shape and size in the flow direction.

R. J. L. Grard (ed.), Photon and Particle Interactions with Surfaces in Space, 101–126. All Rights Reserved
Copyright © 1973 by D. Reidel Publishing Company, Dordrecht-Holland

Experimental support for these remarks on body shape and size effects are given by
Troy et al. (1970) and Samir and Weil (1973). Hence it is reasonable to consider
results for spheres and discs in place of actual body shapes.

Geometric blocking of the flow, although an important factor in the wake formation is not necessarily the dominant factor in cases when the magnitude of the body
potential and/or the ratio, S of vehicle speed v_s to most probable ion speed $v_i =$
$= \sqrt{2KT/m_i}$ vary over wide intervals. (Here T is ion temperature in K, m_i is the ion
mass and K is Boltzmann's constant).

Because of the limitations on the accuracy or applicability of the available theoretical-numerical procedures we have carried out the computations using three separate
methods. Two of these are for spheres, namely the formulations of Liu (1969), Jew
(1968), and Kiel et al. (1968) while the third formulation due to Call (1969), is for
a flat circular disc moving broadside. We shall refer to these respectively as the LJ,
KGG and Call methods in the remainder of the paper. To simulate the Mars Viking
lander shell on entry we choose the sphere or disc radius R_0 to be 1.75 m, v_s to be 5 to
7 km s^{-1} and assume the unperturbed ambient plasma properties to be typical values
taken from Fjeldbro et al. (1970) for the Martian ionosphere at 120 km altitude where
the peak electron number density, $n_e \sim 7 \times 10^{10}$ m^{-3}, occurs. The temperature there is
$T \sim 400$ K and the principal ion is CO_2^+. Electron and ion temperatures are taken to be
equal. The magnetic field is < 50 γ. These values correspond to $v_i \sim 0.39$ km s^{-1}, and
Debye length $\lambda_D \sim 6.9 \sqrt{T/n_e} \sim 0.5$ cm. The satellite is thus intermediate in speed to
the most probable thermal ion and electron speeds, $(v_e = 43 v_i)$ with speed ratio
$S = v_s/v_i \sim 13$ to 19 and a normalized size $R_0/\lambda_D \sim 350$. The electron ion mean free path
is of the order of 500 meters and the electron-electron ($=$ion-ion) mean free path is
of the order of 700 m. The Larmor radius of ions is well above 1 km and the Larmor
radius of electrons exceeds 100 m. The parameters are in the realm for which the
theories apply, namely body speed intermediate between ion and electron speeds,
collisionless plasma, and negligible magnetic field effects. The normalized size ratio is
however significantly larger than has been used in prior calculations.

The calculations we report for the Martian atmosphere are for the above S and
R_0/λ_D values as well as for a smaller body $R_0/\lambda_D \sim 100$ to illustrate body size influence
while remaining in the realm of very large R_0/λ_D ratios. We do not however consider
the variations in the structure of the wake due to pointwise variation of the surface
potential. The S range chosen is also quite limited. In that sense the computations
presented here are limited and require further study.

A comparison with the wake which similar bodies would create in the Earth's
ionosphere in the altitude range of about the F2 region is also presented. We made
computations using $n_e \sim 7 \times 10^{10}$ m^{-3}, $T_e = 2000$ K $\sim T_i$ and principal ion, O$^+$. We
assume a satellite velocity of ~ 8 km s^{-1} which yields $S = 5.6$. Also, $\lambda_D \approx 1$ cm so the
Viking Mars lander size body would have $R_0/\lambda_D \sim 188$. We present results for $S = 5.6$
and a series of R_0/λ_D values from 188 to 19, again to illustrate size effects and going
down to a size corresponding to 19 cm radius probe such as might perhaps be mounted
on a boom.

The two methods used for the spherical body namely LJ and KGG, were chosen because they embody quite different simplifying assumptions, yet there is nothing definitive in the literature to indicate which of these sets of assumptions does the least violence to the true physics of the problem.

The KGG method ignores electric field effects on the ion trajectories, hence it considers only geometric blocking of the ion flow just as a flow of neutral particles would be blocked. For the electrons an approximate distribution is assumed which takes into account the fact that, for bodies very large compared to the Debye length (as measured in the unperturbed ionosphere), a potential well develops behind the body. In particular KGG assumed a Boltzmann distribution $n_e = n_0 \exp(-e\phi/KT)$ for radial distances $r > r_0$ where r_0 is radial distance to the bottom of the potential well, n_0 is the unperturbed electron density, ϕ the local potential (which is taken to be zero far from the body), K is Boltzmann's constant and T is local temperature in degrees Kelvin. For $r < r_0$ they used the constant value $n_e = n_0 \exp(-e\phi(r_0)/KT)$ as an approximation to a slowly varying analytic function which they have derived. By contrast the LJ formulation and also Call's work do not neglect electric field effects on the ion trajectories but use an unmodified Boltzmann electron distribution which is not valid, when there is an appreciable potential well. The LJ method in turn differs from Call's method in invoking an approximately invariant integral to avoid the necessity of detailed computation of individual particle trajectories.

All the methods used assume no photoionization takes place and that all ions which strike the surface are neutralized.

2. Results

The figures presented in this paper are results of our computations of

$$
\begin{array}{ll}
\text{normalized electron density} & (n_e/n_0) \\
\text{normalized ion density} & (n_i/n_0) \\
\text{negative normalized potential} & \left[\dfrac{-|e|\,\phi}{KT_e} \right]
\end{array}
$$

along radial lines from the center of the body. Here n_0 is the unperturbed electron ($=$ion) density and $|e|$ is the magnitude of the electronic charge. All the figures use normalised distance from the center (R/R_0) as abscissa and each contains a set of curves for different angles θ between the wake's symmetry axis and the direction of the radial line. $\theta = 180°$ is the ray directed downstream along the wake axis. Along these curves (n_e/n_0), (n_i/n_0) and $[-|e|\phi/KT_e]$ vary with R/R_0 for constant angles θ. The computations represented in Figures 1–15 are for spherical bodies while Figure 16 is for a flat disc moving broadside. Figures 1–3 represent (n_e/n_0), (n_i/n_0) and $[-|e|\phi/|KT_e] = f(R/R_0)$ computed by the KGG and the LJ methods for $S = 19$ and $[R_0/\lambda_D] = 350$. Figures 4–6 represent the results for $S = 13$ and $[R_0/\lambda_D] = 350$. Therefore these figures provide the changes in the structure of the wake of a Viking type vehicle for two different S values given that the rest of the main plasma parameters are constant.

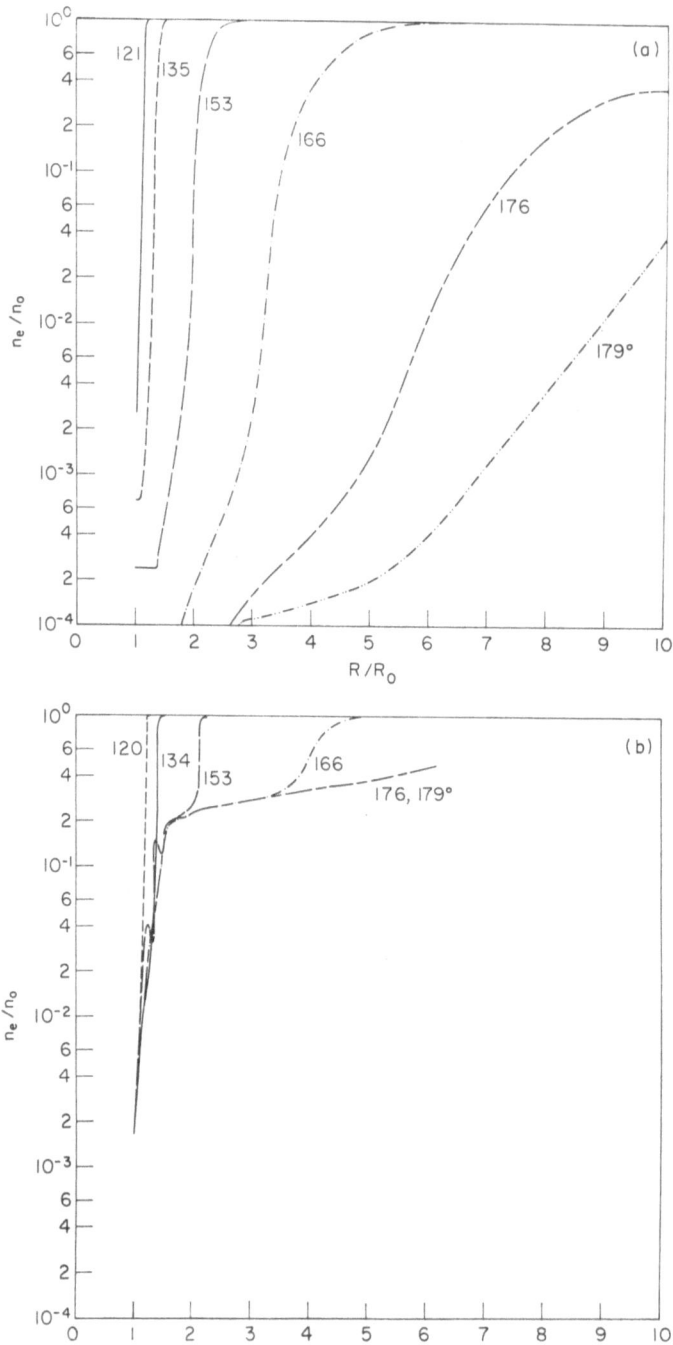

Fig. 1.   Normalized electron density vs normalized radial distance from the center of a sphere
$S = 19$, $R_0/\lambda_D = 350$ (Martian parameters): (a) KGG; (b) LJ.

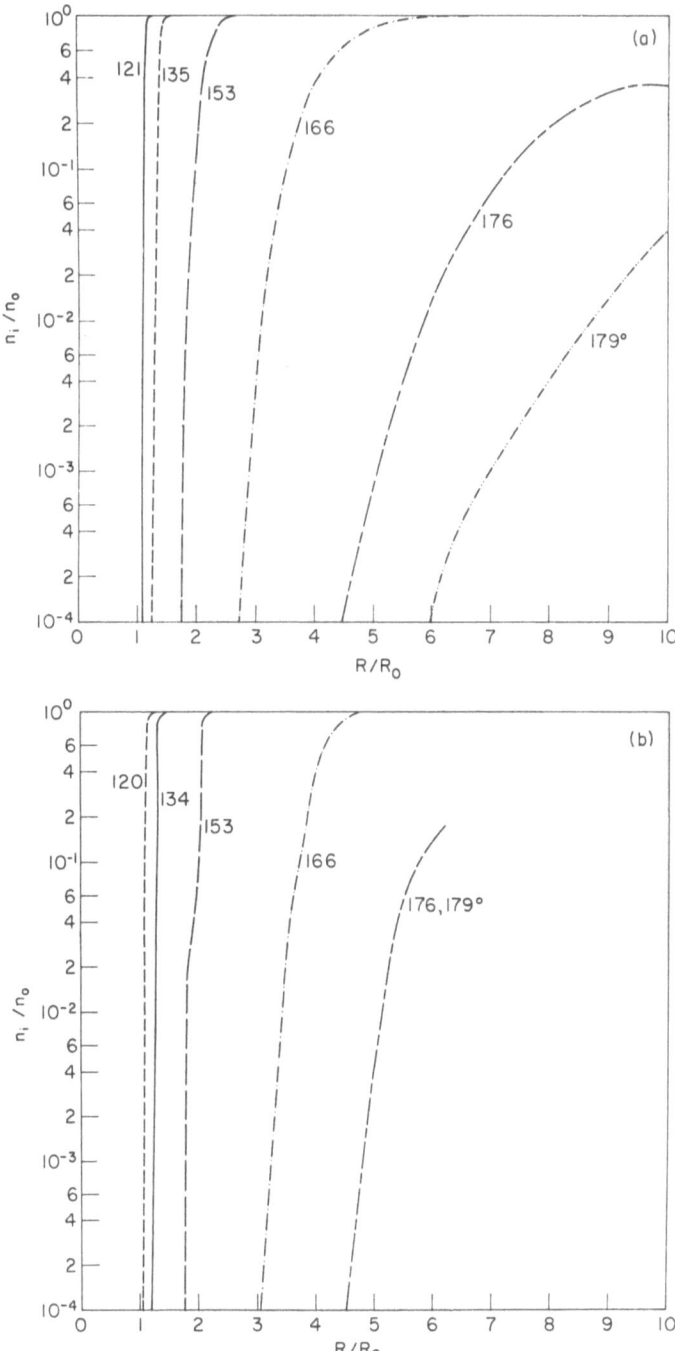

Fig. 2. Normalized ion density vs normalized radial distance from the center of a sphere $S = 19$ $R_0/\lambda_D = 350$ (Martian parameters): (a) KGG; (b) LJ.

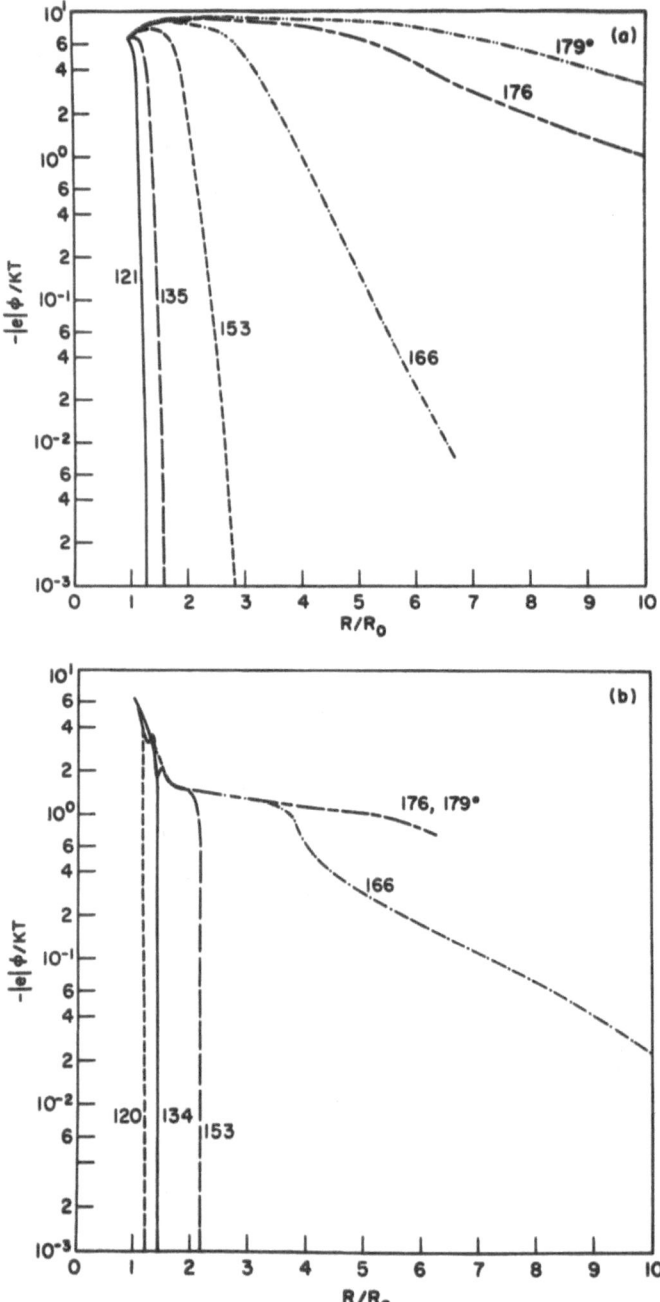

Fig. 3. Normalized potential vs normalized radial distance from the center of a sphere $S = 19$,
$R_0/\lambda_D = 350$ (Martian parameters): (a) KGG; (b) LJ.

Figure 4–6 and Figures 7–9 yield the comparison between (n_e/n_0), (n_i/n_0) and $[-|e|\phi/KT]$ for the same value of S ($=13$) but for different $[R_0/\lambda_D]$ values, namely for $[R_0/\lambda_D]=350$ and $[R_0/\lambda_D]=100$.

Figures 10–12 present curves of constant (n_e/n_0), (n_i/n_0), and $[-|e|\phi/KT_e]$ based on the KGG method in the wake of a sphere for the case $S=13$, $[R_0/\lambda_D]=350$ which is in line with Martian parameters.

While Figures 1–12 are appropriate for Martian parameters, Figures 13–16 refer to situations relevant for Earth parameters and are presented here to demonstrate the importance of the parameter $[R_0/\lambda_D]$ on the structure of the wake for a fixed speed ratio $(S=5.6)$. For interpreting Figure 14 note that in the KGG approximation the ion distribution is independent of the potential. Hence the KGG density curves plotted as function of distance measured in body radii are the same for each value of R_0/λ_D. Hence only a single set of ion density curves is needed. Note also in the KGG results that, as discussed previously, $n_e=$ constant between potential well minima and the body. Similar comments hold for the other KGG results and Figure 5(a) is identical with 8(a).

In all cases the computations were carried out using a lower bound approximation to the floating surface potential in a steady state where no net current is drawn from the plasma. The approximate bound which was computed by neglecting electro-static forces on the ions, particle emissions from the surface, photoemission and magnetic field effects is a fraction of a volt negative.

A numerical exploration of the effect of surface potential variation was carried out by both the LJ and KGG methods. The quantitative effects of a decrease of surface potential by a factor of two are almost entirely confined to the region $1<R/R_0 \lesssim 2$. Even in this region the surface potential change on the ion distribution is negligible, but the curves of $e\phi/KT$ while maintaining their general form are 'pulled' over to meet the specified wall potential, while the electron density at the surface increases with decreasing $|e\phi/KT|$. Thus for example, the surface value of $n_e/n_0 \sim 10^3$ shown for LJ in Figure 1(b) increases to 4×10^{-1}, but the detailed structure (including the humps on the 153° curve) are preserved; the increase by the KGG theory is much less marked.

It is evident that the two methods give results which differ considerably in detail, particularly in the complex near wake structure of the potential field for the very large radius spheres such as those corresponding to the Viking lander aeroshell radius. In addition the neglect of electrostatic pull by the body on the ions in all stages of the Kiel *et al.* formulation shows up in the much longer wake which is computed from this model than from the Liu-Jew model.

The figures presented enable a detailed quantitative comparison of the results by the two methods in the near wake region. In the near wake, because of the rapid variation of particle density with distance, particle densities computed at fixed points often differ considerably by the two methods in these regions even though relatively small horizontal shifts of the curves would often permit agreement. It should be noted that, while in the LJ wake model (Figure 2(b)) the (n_i/n_0) for $\theta=176$, 179° is identical, this is not the case for the GG model under the same conditions (see Figure 2(a)).

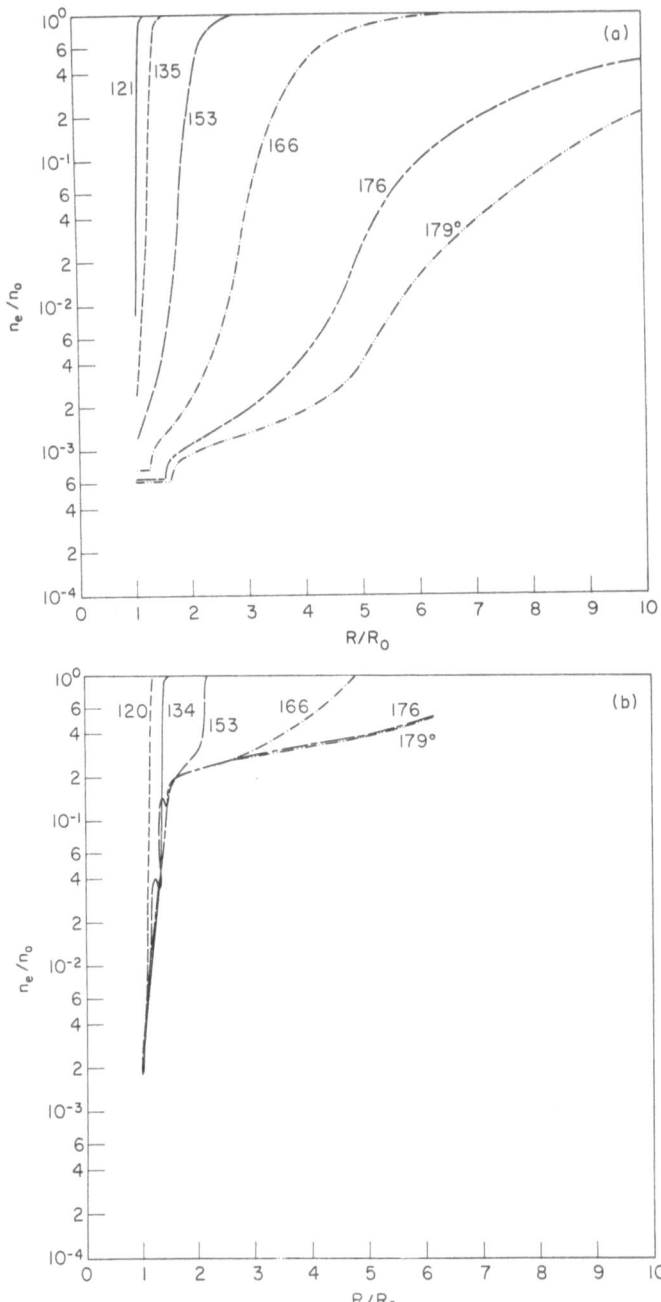

Fig. 4.   Normalized electron density vs normalized radial distance from the center of a sphere $S = 13$, $R_0/\lambda_D = 350$ (Martian parameters): (a) KGG; (b) LJ.

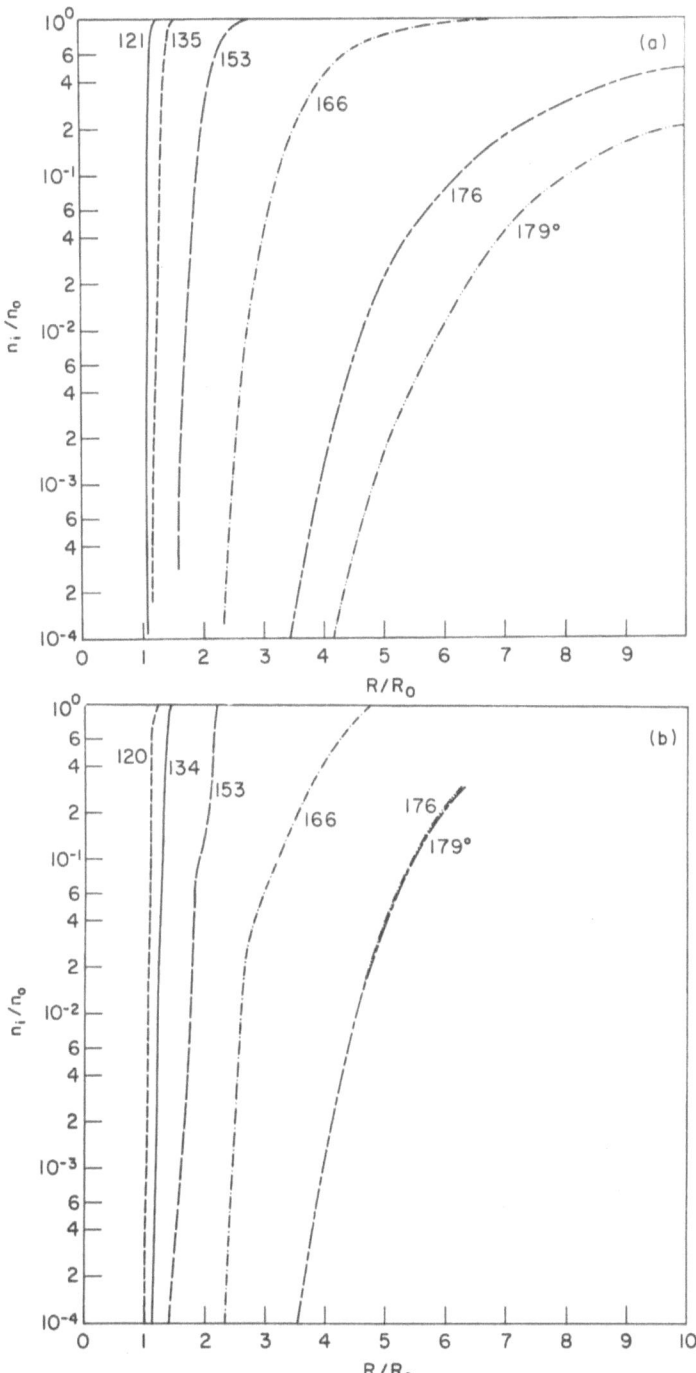

Fig. 5. Normalized ion density vs normalized radial distance from the center of a sphere $S = 13$, $R_0/\lambda_D = 350$ (Martian parameters): (a) KGG; (b) LJ. Note that Figure 5(a) is identical to Figure 8(a). (See Text.)

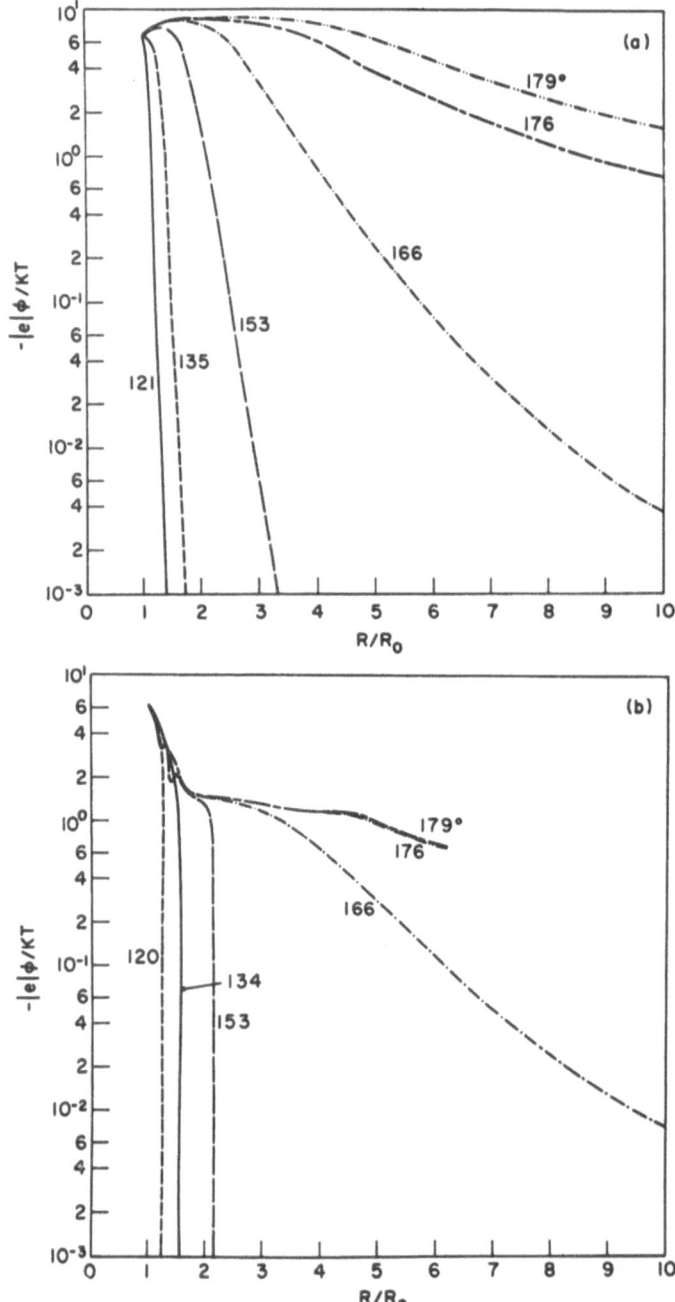

Fig. 6. Normalized potential vs normalized radial distance from the center of a sphere $S = 13$, $R_0/\lambda_D = 350$ (Martian parameters): (a) KGG; (b) LJ.

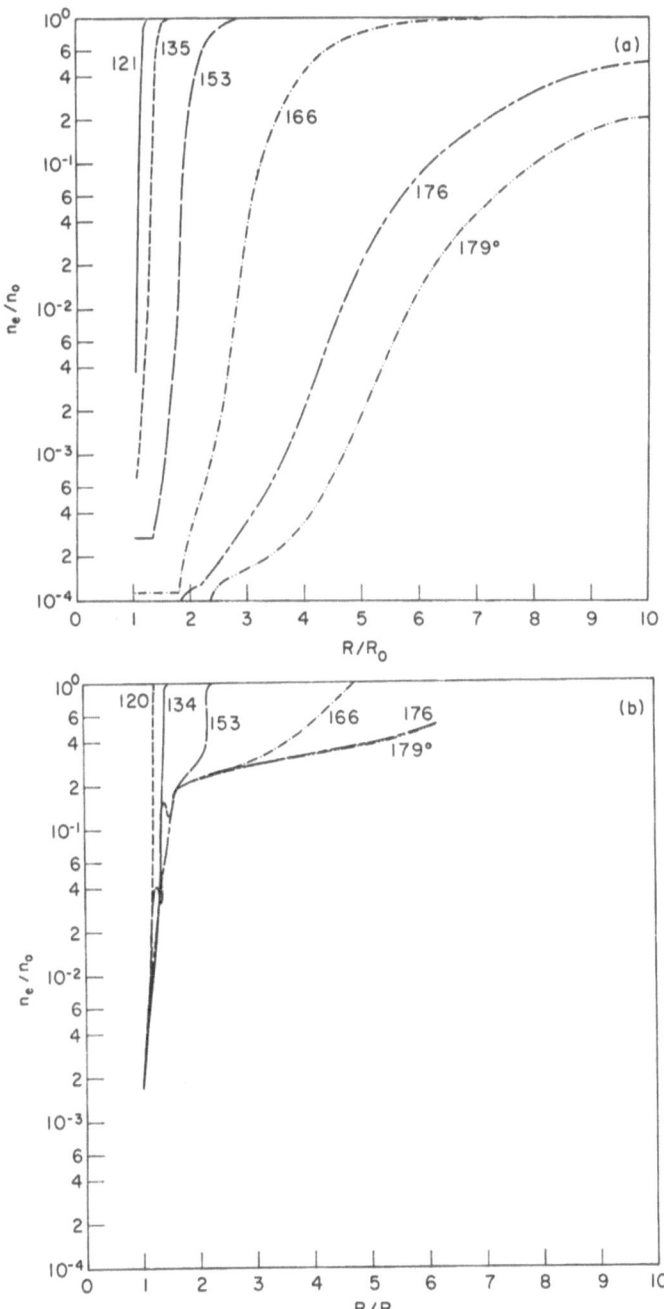

Fig. 7. Normalized electron density vs normalized radial distance from the center of a sphere $S = 13$, $R_0/\lambda_D = 100$ (Martian parameters): (a) KGG; (b) LJ.

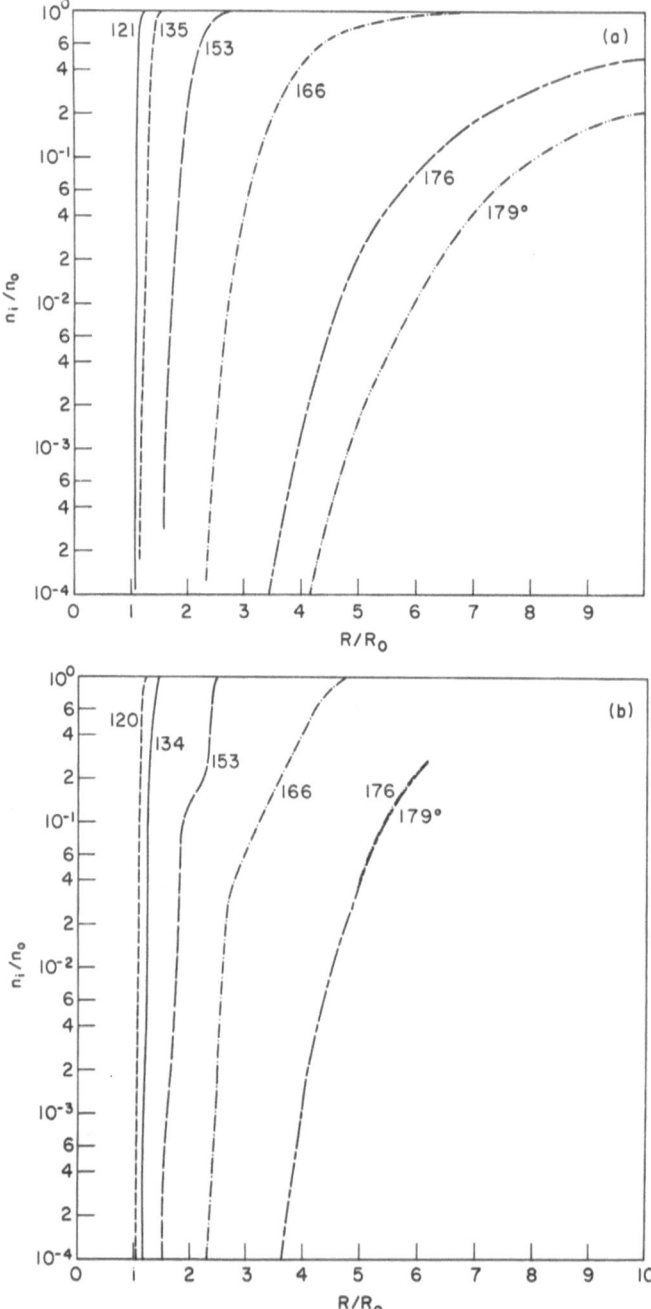

Fig. 8. Normalized ion density vs normalized radial distance from the center of a sphere $S = 13$, $R_0/\lambda_D = 100$ (Martian parameters): (a) KGG; (b) LJ. Note that Figure 8(a) is identical to Figure 5(a), (see text).

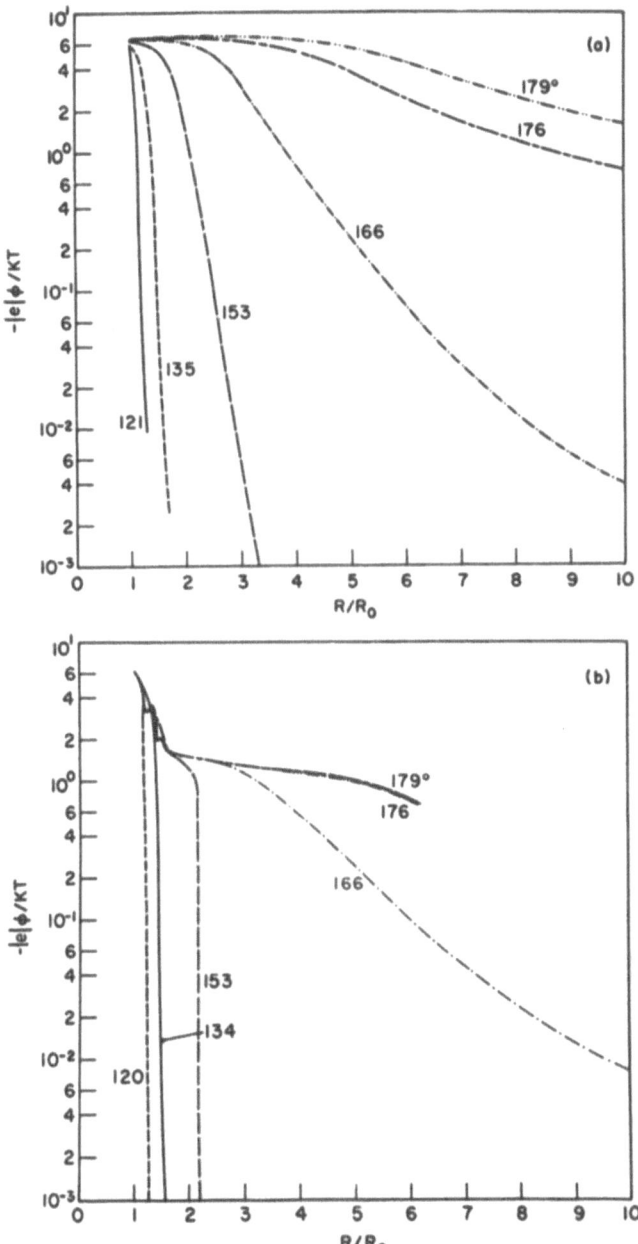

Fig. 9. Normalized potential vs normalized radial distance from the center of a sphere $S = 13$, $R_0/\lambda_D = 100$ (Martian parameters): (a) KGG; (b) LJ.

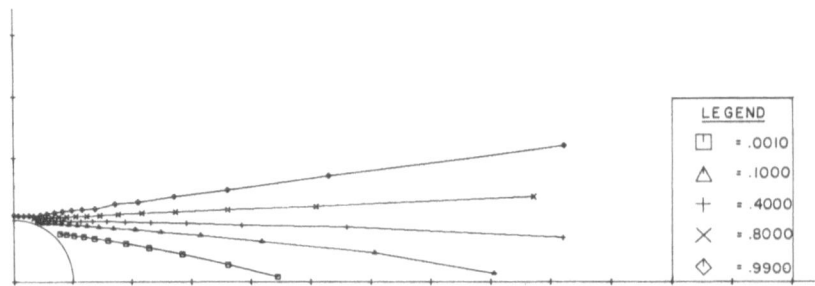

Fig. 10.   Curves of constant electron density in the wake of a sphere for the case $S = 13$, $R_0/\lambda_D = 350$
(Martian parameters) based on the KGG formulation.

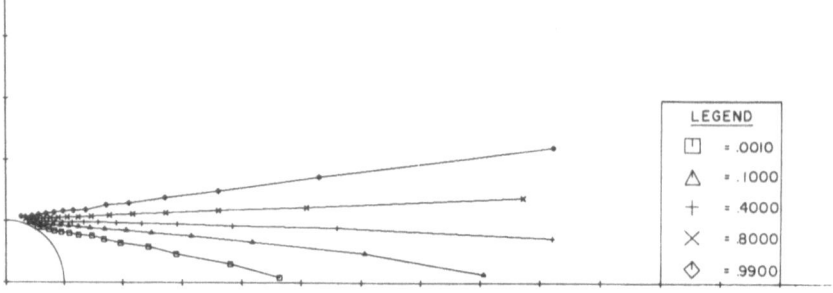

Fig. 11.   Curves of constant ion density for the case $S = 13$, $R_0/\lambda_D = 350$ (Martian parameters)
based on the KGG formulation.

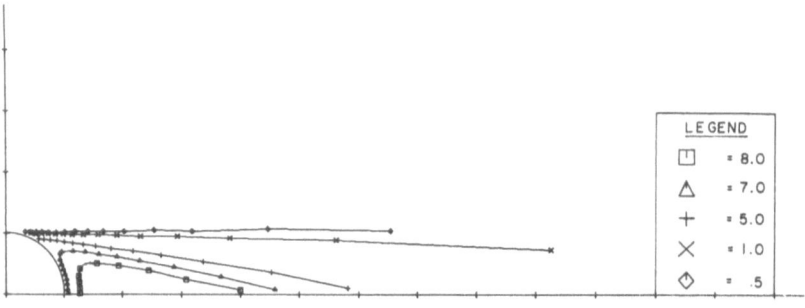

Fig. 12.   Curves of constant potential for the case $S = 13$, $R_0/\lambda_D = 350$ (Martian parameters) based
on the KGG formulation.

A similar comment applies to the (n_e/n_0) as seen in Figure 1. Also, the potential
$[-|e|\phi/KT_e]$ drop with (R/R_0) is much steeper in the LJ model compared to that of
KGG. (See Figure 3). This is at least partly due to the fact that in the LJ model the
potential field influence on the ion motion is considered. Similar comments apply to
Figures 4–6 and 7–9. In spite of the differences in structural detail the general trend
in the wake structure with varying body radius and with varying satellite velocity
were found to be similar by either method.

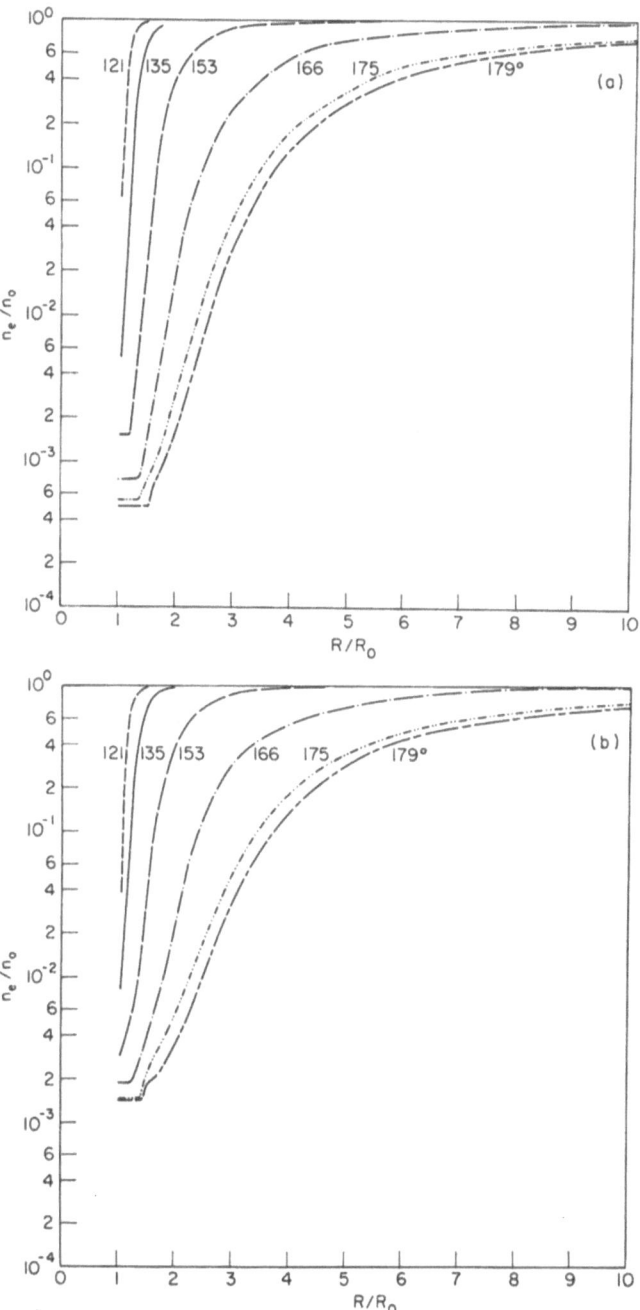

Fig. 13a–b.

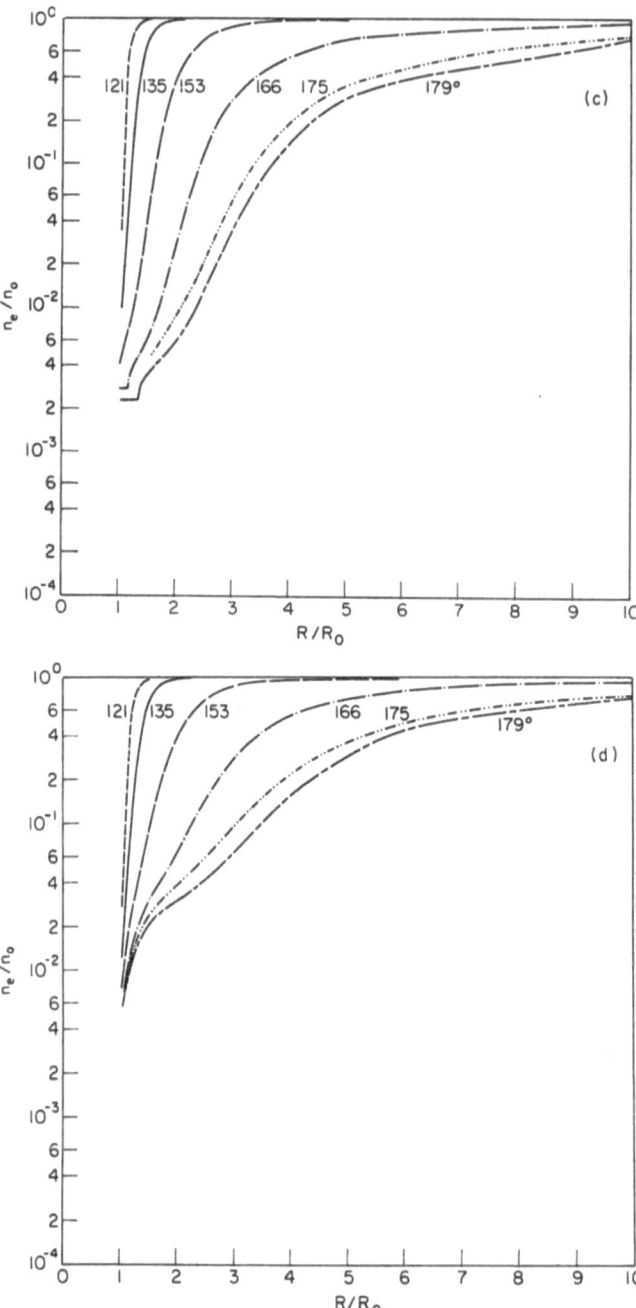

Figs. 13c–d.

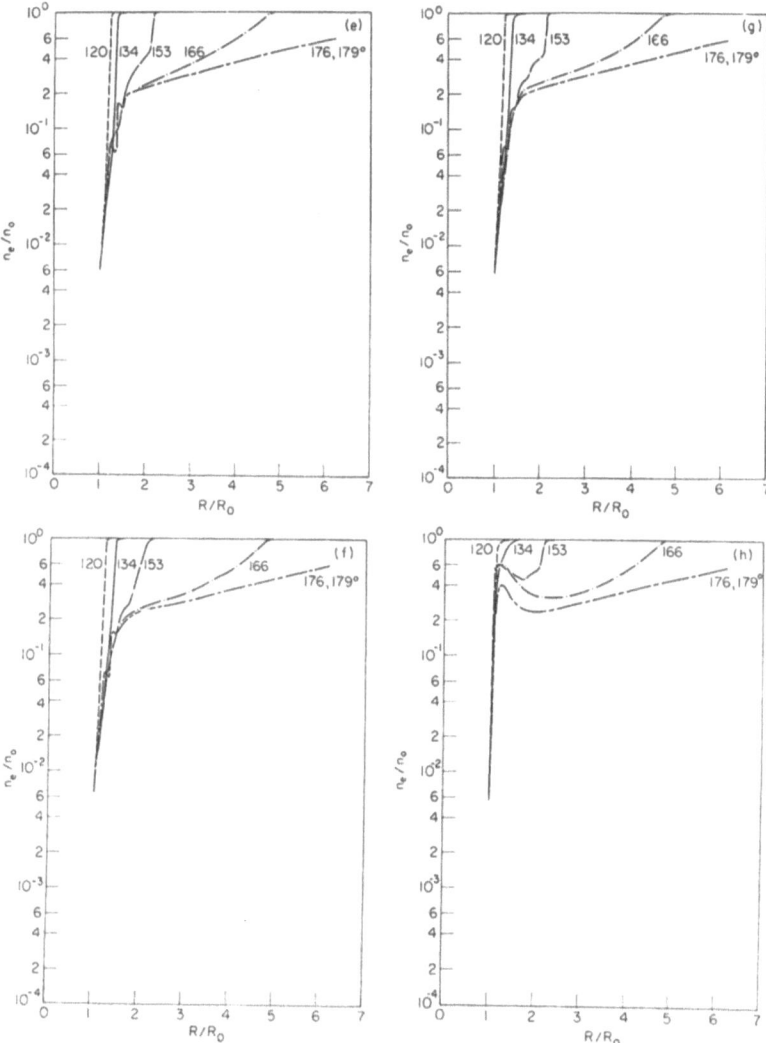

Fig. 13.   Normalized electron density vs normalized radial distance from the center of a sphere, $S = 5.6$ (Earth Parameters) for $R_0/\lambda_D = 188$ – (a) KGG and (e) LJ; 94 – (b) KGG and (f) LJ; 63 – (c) KGG and (g) LJ; 19 – (d) KGG and (h) LJ.

In all the cases considered it is evident from the figures that the downstream region of the ambient plasma is subject to considerable perturbation. The strongly perturbed regions extend at least several body radii downstream in the central (axial) region of the wake. Although in this region there is the greatest reduction of ion and electron density the relative deviation from charge neutrality is greatest in the 'shoulder' of the wake. This fact is reflected in the potential field where the most rapid potential variations take place in directions through the shoulders ($\theta = 120°$–$167°$ in our plots) and not in directions near the wake axis ($\theta = 176°$, $179°$).

Potential wells develop in the near wake ($R/R_0 < S/4$) for the large radius but are not

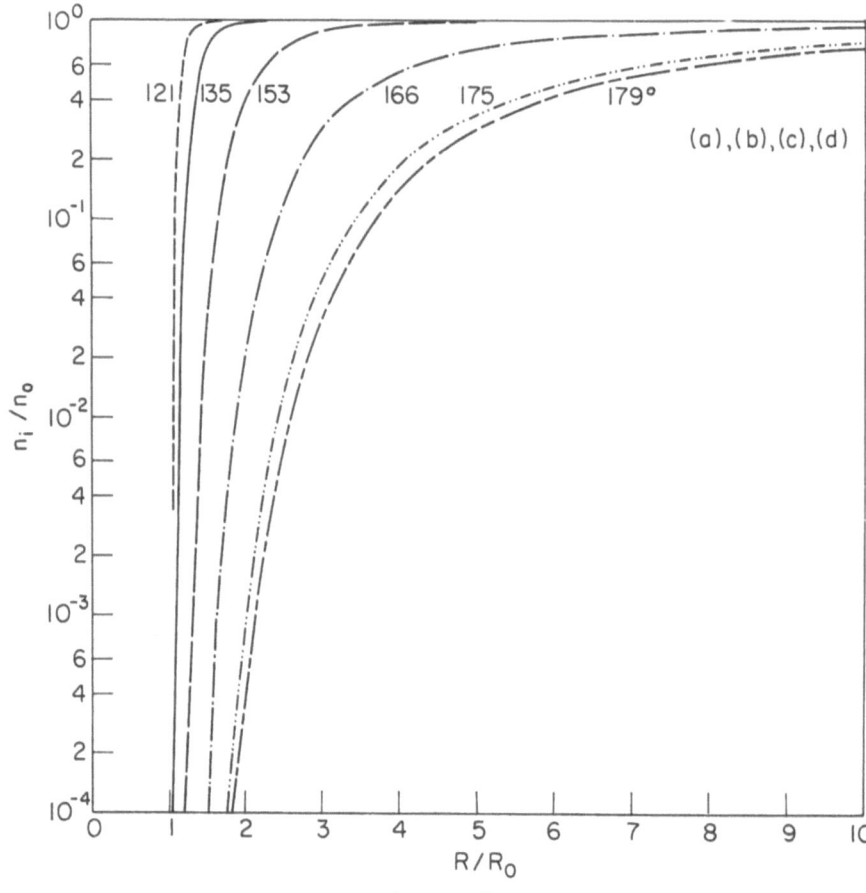

Figs. 14a–d.

evident for the small radius body in the KGG results (see Figures 3, 6, 9, 12, and 15). In the LJ model the wells appear for all R_0/λ_D but interestingly, the wells become toroidal in shape for $R_0/\lambda_D > 63$. This is evident from the structure of the potential curves which exhibit extrema along $\theta = 153°$ but not near the axis at say $\theta = 176$ and $179°$ nor in the 'shoulders' on the $\theta = 134$ and $120°$ curves. These differences in potential field structure are probably tied up with the different assumed analytic forms for electron density since, in the near wake $(R/R_0 < 3)$ where the potential wells occur, there is not much difference in the ion distributions computed by the two methods (see Figures 2, 5, 8, and 14).

Further effects of body size may be judged from the sequences in Figure 4–9 and 13–15 as well as from the corresponding Tables I and II. For the size ranges considered; namely $R/\lambda_D = 19$–188 for $S = 5.6$ and 100–350 for $S = 13$ the only influence of body size on wake structure at a downstream distance of $R = SR_0$ is to directly scale the wake size proportional to R_0. In fact, except very near the axis, direct scaling is the only body size influence at $R = 0.5(SR_0)$. On the other hand the near wake at $R < 0.25(SR_0)$ is quite definitely modified with body size in a more complex fashion.

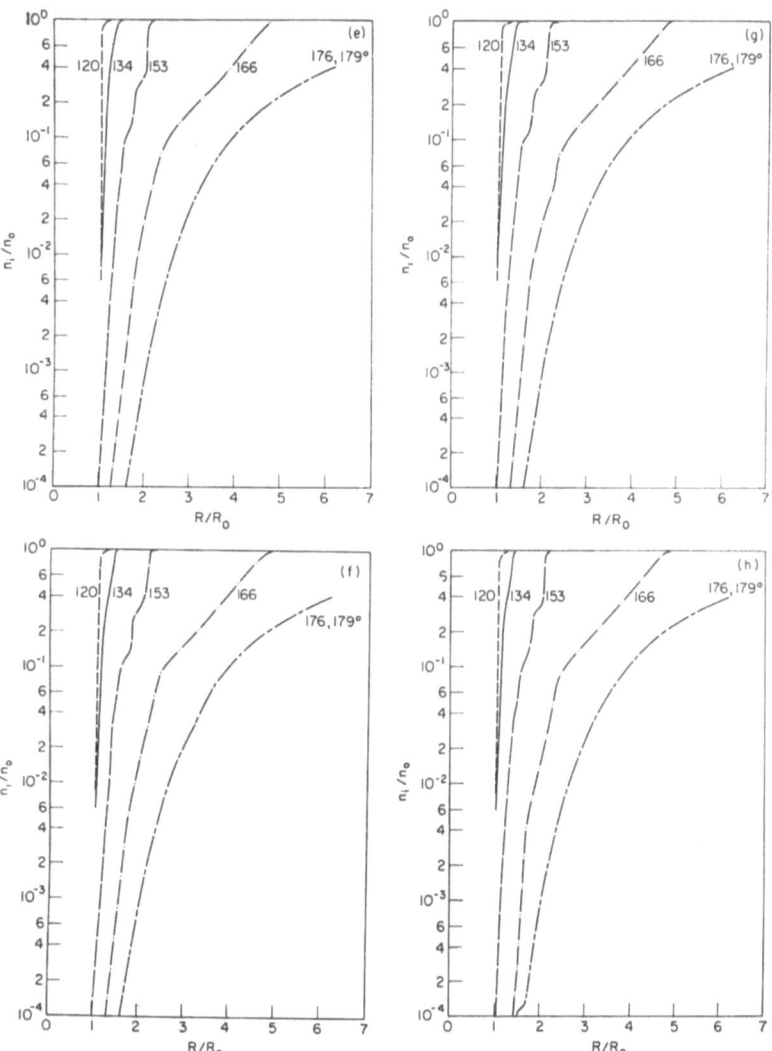

Fig. 14.   Normalized ion density vs normalized radial distance from the center of a sphere, $S = 5.6$
(Earth Parameters) for $R_0/\lambda_D = 188$ – (a) KGG and (e) LJ; 94 – (b) KGG and (f) LJ; 63 – (c) KGG
and (g) LJ; 19 – (d) KGG and (h) LJ. Note that Figures 14 (a)–(d) are identical. (See Text.)

The effect of changing speed ratio, S while other parameters are fixed, was investigated for the Martian ionosphere case. Figures 1–3 and 4–6 compare the $R_0/\lambda_D =$
$= 350$ case for $S = 19$ and $S = 13$. From comparison of these figures or comparing data
from Table II(a) and (b) it is evident that increasing S narrows and lengthens the wake.

In Figure 16 are results for the wake of a thin flat disc moving broadside to the vehicle direction of motion. By comparison with Figures 13–14(d) and (h) it is evident that
the overall length of the wake is much less for the disc than for a sphere of the same
radius and speed. It is interesting to note the extremely rapid cut off in the ion dis-

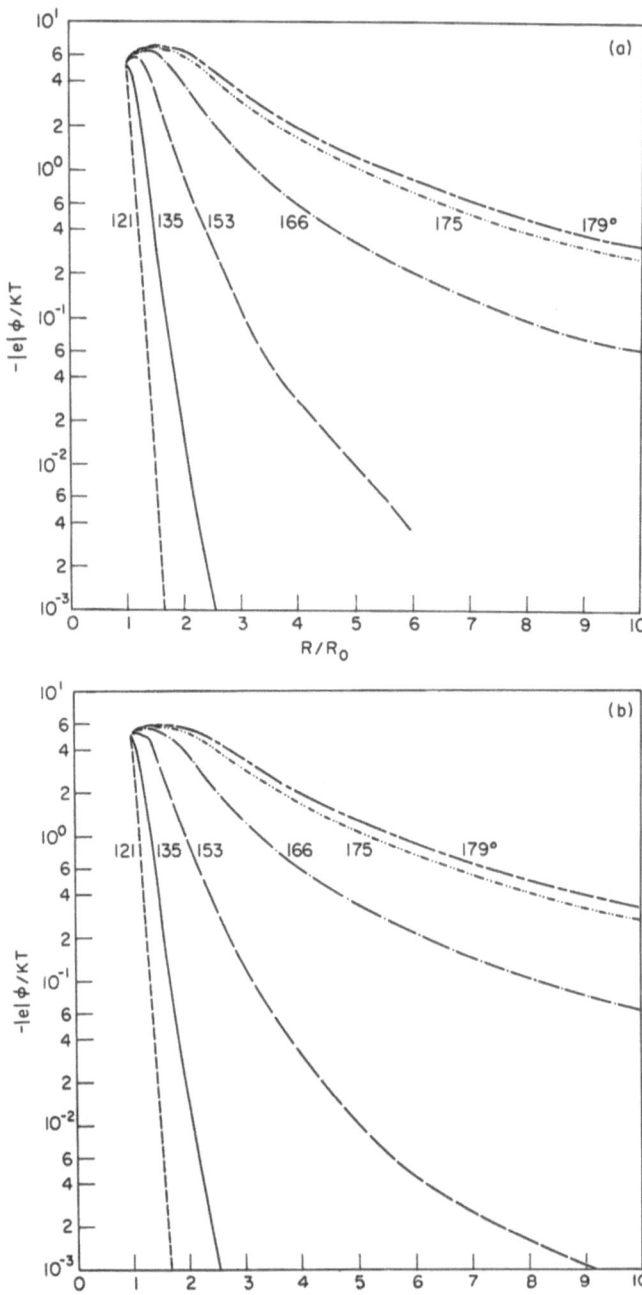

Figs. 15a–b.

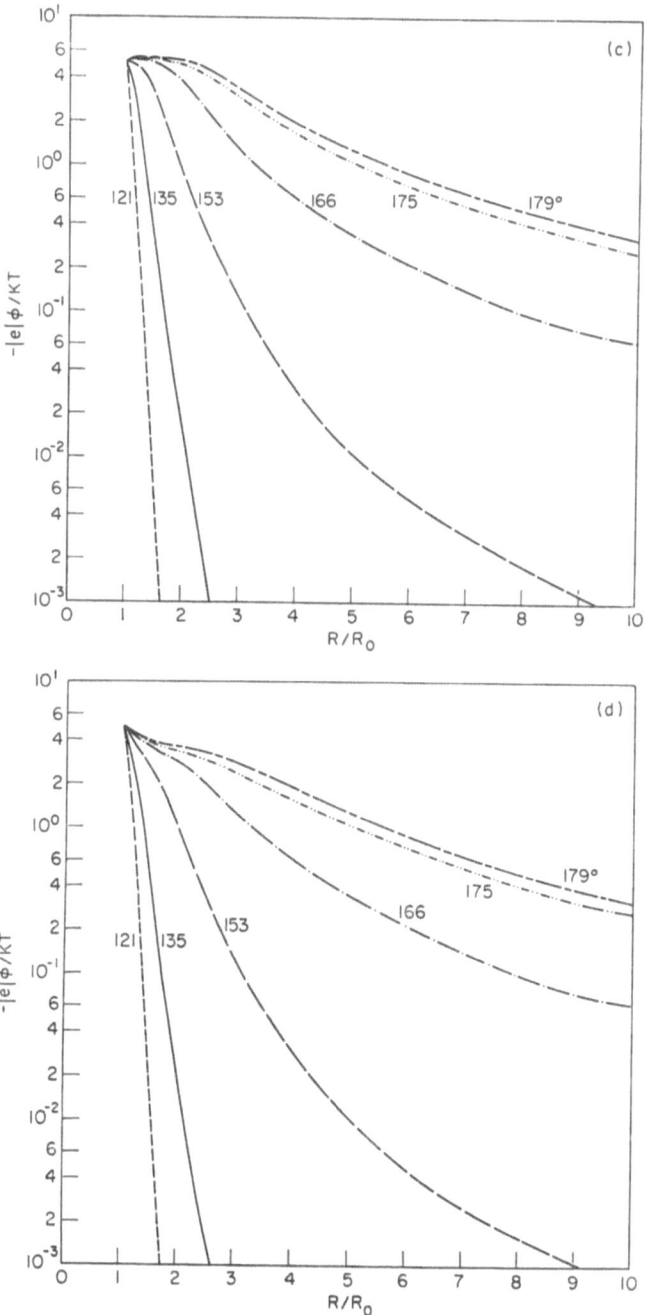

Figs. 15c–d.

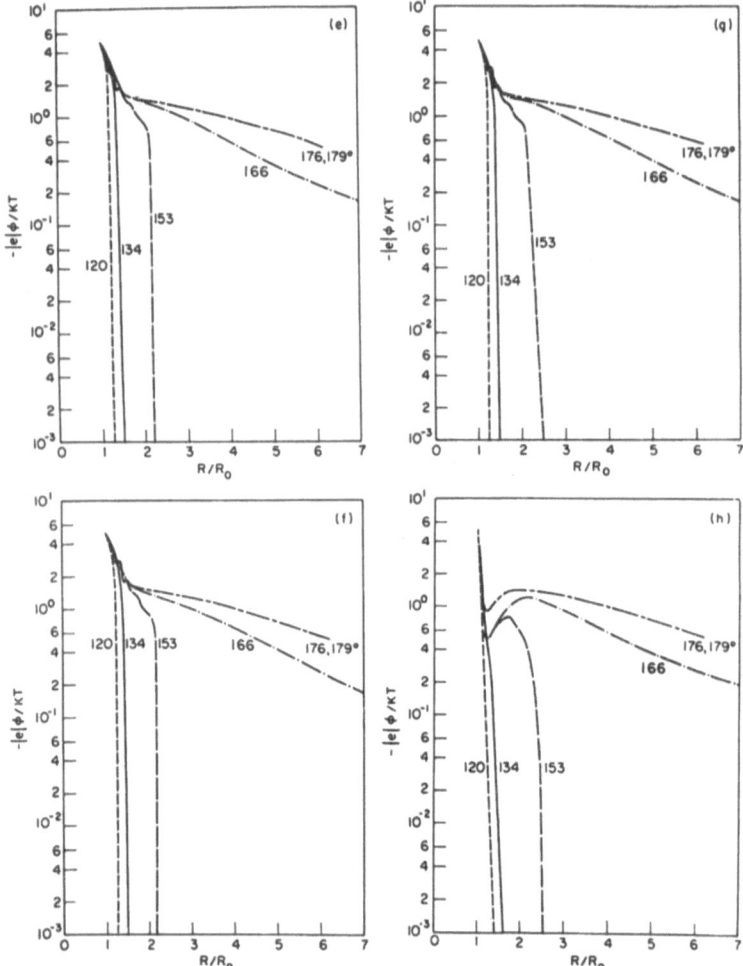

Fig. 15.   Normalized potential vs normalized radial distance from the center of a sphere, $S = 5.6$
(Earth Parameters) for R_0/λ_D 188 – (a) KGG and (e) LJ; 94 – (b) KGG and (f) LJ; 63 – (c)
KGG and (g) LJ; 19 – (d) KGG and (h) LJ.

tribution due to shadowing by the sharp edge of the disc although the effects of
electrostatic pull on the ions is evident since the sharp drop in ion density occurs
within the purely geometric shadow region. The drop off is far more abrupt than
for a sphere.

3. Summary

We have carried out a preliminary computational study of the perturbation of planetary ionospheres by entry bodies so as to draw attention to this far-from-negligible
phenomenon which must be considered in designing *in situ* probes of the ionospheric
properties, as well as in data interpretation and in estimating radio transmission to or

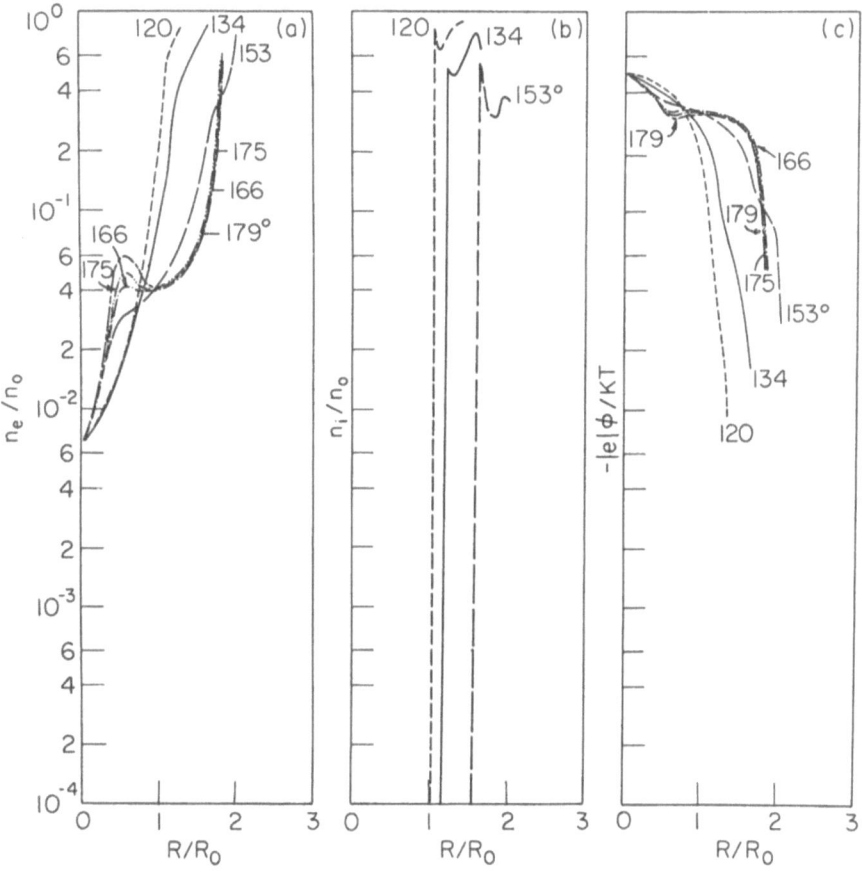

Fig. 16.   Normalized wake parameters vs normalized radial distance from the center of a circular disc, (Earth parameters) computed by Call's procedure: (a) electron density; (b) ion density; (c) potential.

from the probe. Our study has showed up graphically the lack of detailed agreement between results by different analytical-numerical approaches which have appeared in the literature. Nevertheless a few general conclusions could be made. For example, the wake behind a given vehicle is much more extensive in the Martian environment than in Earth's. In either environment, for body sizes exceeding about 20 Debye lengths (measured in the unperturbed medium), the wake length and width increases directly with body size (other parameters fixed) at distances exceeding one half the body radius multiplied by the ratio of vehicle speed to the most probable ion thermal speed. For lesser radii; i.e., within the near wake, the effect of increasing body size is much more complex involving the development of potentials well, but the detailed structure of these wells is not the same in the two theories.

Differences in the near wake structure behind flat disc and spherical vehicles of

TABLE I

Selected Data Comparisons for $S = 5.6$ (Earth Parameters).
'SAME' signifies that the number set is identical with the set above

R/R_0	R/λ_D	θ	n_e/n_0		n_i/n_0		$-|e|\phi/KT$	
			KGG	LJ	KGG	LJ	KGG	LJ
S	19	179	0.38	0.32	0.37	0.33	1	0.64
		166	0.78	1.0	0.78	1.0	0.25	0
		135	1.0	1.0	1.0	1.0	0	0
	94	179						
		166	SAME	SAME	SAME	SAME	SAME	SAME
		135						
	188	179						
		166	SAME	SAME	SAME	SAME	SAME	SAME
		135						
$S/2$	19	179	0.053	0.28	0.02	0.015	2.8	1.3
		166	0.24	0.34	0.23	0.13	1.4	1
		135	0.83	1.0	1.0	1.0	0	0
	94	179	0.023	0.28			3.8	
		166	0.23	0.33	SAME	SAME	1.5	SAME
		135	0.01	1.0			0	
	188	179	0.02				4.0	
		166	0.20	SAME	SAME	SAME	1.5	SAME
		135	1.0				0	
$S/4$	19	179	0.017	0.34	0	0	4.0	1.1
		166	0.025	0.56	0	0	3.8	0.65
		135	0.40	0.90	0.5	0.9	0.90	0.11
	94	179	0.0014	0.12		0	6.0	2.1
		166	0.0003	0.16	SAME	0.0002	5.6	1.8
		135	0.54	0.35		0.9	0.70	1.0
	188	179	0.0005	0.10		0	6.8	2.1
		166	0.0009	0.12	SAME	0.0003	6.5	1.8
		135	0.45	0.65		0.9	0.70	0.43

the same radius in the Earth's ionosphere were also investigated. The wake behind the flat body is considerably shorter than behind a sphere, and differences in the near zone structure are of course evident.

To complete a study of this type account should be taken of the effects due to wider variations in speed ratio (S) than we have considered.

Also, the effects of ion to electron temperature ratios $\neq 1$ and varying surface potentials of the entry probe due to various charging mechanisms should be considered.

Acknowledgements

The authors wish to thank G. R. Carignan, Director of the Space Physics Research Laboratory for his interest and comments. This work was supported by NASA Grant NGR-23-005-320.

TABLE II

Selected Data Comparisons. (Martian Parameters). 'SAME' signifies that the number set is identical with the set above

(a) $S = 13$

R/R_0	R/λ_D	θ	n_e/n_0 KGG	n_e/n_0 LJ	n_i/n_0 KGG	n_i/n_0 LJ	$-|e|\phi/KT$ KGG	$-|e|\phi/KT$ LJ
S/2	350	179	0.026	0.53	0.024	0.33	3.8	0.60
		166	0.97	1.0	0.96	1.0	0.050	0
		135	1.0	1.0	1.0	1.0	0	0
	100	179	0.024	0.56			3.8	0.58
		166	0.97	1.0	SAME	SAME	0.047	0
		135	1.0	1.0			0	0
				(~ SAME)			(~ SAME)	(~ SAME)
S/4	350	179	0.002	0.28	0	0	8.5	1.2
		166	0.092	0.34	0.095	0.11	2.4	1.0
		135	1.0	1.0	1.0	1.0	0	0
	100	179	0.0014	0.29			6.6	1.2
		166	0.092	0.28	SAME	SAME	2.6	0.84
		135	1.0	1.0			0	0

(b) $S = 19$

R/R_0	R/λ_D	θ	n_e/n_0 KGG	n_e/n_0 LJ	n_i/n_0 KGG	n_i/n_0 LJ	$-|e|\phi/KT$ KGG	$-|e|\phi/KT$ LJ
S/2	350	179	0.038		0.038		3.3	
		166	1.0		1.0		0	
		135	1.0		1.0			
S/4	350	179	0.002	0.36	0	0.0007	8.5	1.0
		166	0.74	1.0	0.75	1.0	0.255	0
		135	1	1.0	1	1.0	0	0

References

Call, S. M.: 1969, Ph. D. Thesis, Columbia Univ.

Fjeldbro, G., Kliore, A., and Seidel, B.: 1970, *Radio Sci.* **5**, 381–386.

Gurevich, A. V., Pitaevskii, L. P., and Smirnova, V. V.: 1969, *Space Sci. Rev.* **9**, 805–871.

Jew, H.: 1968, Ph. D. Thesis, Univ. of Michigan (Available from University Microfilms, Ann Arbor, Mich., U.S.A.).

Kiel, R. E., Gey, F. C., and Gustafson, W. A.: 1968, *AIAA J.* **6**, 690–694.

Liu, V. C.: 1969, *Space Sci. Rev.* **9**, 423–490.

Samir, U.: 1972, *Israel J. of Tech.* **10** (3), 179–188.

Samir, U. and Weil, H.: 1973, 'Electron Depletion in the Wake of Ionospheric Spacecraft – A Comparison Between Results from Langmuir Probes and Antennas', to be published.

Soffen, G. A. and Young, A. T.: 1972, *Icarus* **16**, 1–16.

Troy, B., Jr., Medved, D. B., and Samir, U.: 1970, *J. Astron. Sci.* **18** (3), 173–183.

DISCUSSION

Wrenn: Have you made any calculations of disturbance on the front side of the body which would be relevant to the operation of the planar ion trap on Viking?

Weil: No.

Polychronopulos: In view of the disparity between the quantitative results of the two described theories, what is the value of these results?

Weil: Mainly qualitative to describe general trends.

Knott: You assume a plasma temperature of 500 K for the planetary atmosphere and a ram velocity of 5 km s⁻¹ for planetary entry vehicles. Would it be possible to scale up these velocities by a factor of 100 and apply your results to the interaction of solar wind with a spacecraft?

Weil: In this case one would have to include photoeffect.

Cauffman: Since the landing will probably occur in sunlight, it would seem that photoelectrons might be an important part of the problem. Are they included in any of the models?

Weil: No, they are not.

NUMERICAL CALCULATIONS OF THE PERTURBATION
OF AN ELECTRIC FIELD AROUND A SPACECRAFT

MATTIAS SOOP*

*European Space Operations Centre (ESOC), of the European Space Research Organisation (ESRO),
Darmstadt, Germany*

Abstract. This report deals with some aspects of the electric field around a body having an insulating surface, with or without an inner conductor. Surface charge accumulation due to flow of ambient plasma electrons and emission of photo electrons is considered. The purpose of this study is to estimate the magnitude of perturbations due to the spacecraft body on the electric field measurements made at the geostationary orbit.

1. Introduction

This report deals with the perturbation of the electric field around a spacecraft arising from the surface charges that accumulate on it. The results are relevant for the ESRO satellite GEOS, which will among other things measure the electric field in space by means of spherical probes mounted at the end of 10–20 m long booms. The electric field to be measured is expected to be of the order of 10^{-4}–10^{-2} V m^{-1} (Fahleson, 1969). This means that the perturbation voltage between the probes should be less than 0.1 V and if possible less than 10^{-3} V.

The cause of the charging of the satellite surface is the inflow of ambient electrons and ions and the emission of photoelectrons on the sunlit part of the surface. The surface properties, in particular the conductivity, are important for the symmetry of the electric field perturbations.

In order to simplify the calculations the satellite is assumed to be spherical with radius R and to have two probes on opposite sides of the body at a distance L from its centre (Figure 1).

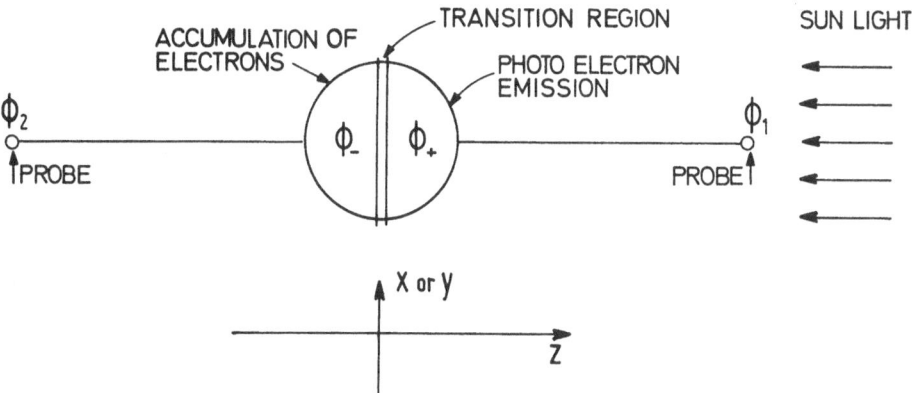

Fig. 1.   A spherical satellite with the probes aligned to give maximum perturbation. The sunlit side
is assumed to have the potential ϕ_+ and the dark side the potential ϕ_-.

* This work was performed while the author was at the European Space Research Institute (ESRIN),
Frascati (Rome), Italy.

R. J. L. Grard (ed.), Photon and Particle Interactions with Surfaces in Space, 127–136. All Rights Reserved
Copyright © 1973 by D. Reidel Publishing Company, Dordrecht-Holland

2. Physical Parameters

A discussion of phenomena causing electric field perturbations is given by Fahleson (1969) together with known or estimated data. By comparing the capacitance of a sphere with radius 0.75 m with the photo electric current or the ambient particle flux respectively one obtains the following characteristic times for charging the satellite to a steady state:

TABLE I

Physical parameters

	I(A m^{-2})	E(eV)	T(s)
Photo electron emission	10^{-5}–10^{-4}	1–10	10^{-6}
Ambient plasma electrons	10^{-6}–10^{-5}	10^3–10^4	10^{-3}
Ambient plasma ions	10^{-9}		10^{-2}
Satellite spin			1

On the illuminated part of the satellite, the dominating effect will be the photo electron emission, which will tend to charge the surface to a positive potential close to the maximum emission energy.

On the dark side, the ambient electrons will charge the surface negatively unless the surface conductivity is high enough to short-circuit it to the sunlit side, or if secondary emission is sufficiently high to keep the dark side from charging. Negative potentials of up to 10^4 volt have actually been observed (DeForest, 1972).

3. Insulating Body

A previous report (Soop, 1972) gave the results of calculations taking into account the photo-electrons but not the ambient plasma, for the two cases of conducting and insulating surfaces. In these calculations, the trajectories of a few thousand negative charges were calculated in a two-dimensional rotationally symmetric space around the satellite.

The satellite surface becomes positively charged by the electron emission, and after a transient time of the order of microseconds a steady state is attained when the surface charge is high enough to attract the subsequently emitted electrons. Using the relatively low photo emission rate given in Table I it is found that the charge on the satellite is 3–5 times greater than the charge of the electron cloud, so the satellite-electron forces dominate the electron-electron forces.

The same program is now used, including the ambient electrons, but with a lower energy and higher density than in the physical case in order to keep the computing time within reasonable limits.

The distribution function for the plasma electrons is taken to be rectangular, with a cut-off energy of 36 eV; the distribution function for the photo-electrons is shown in Figure 2.

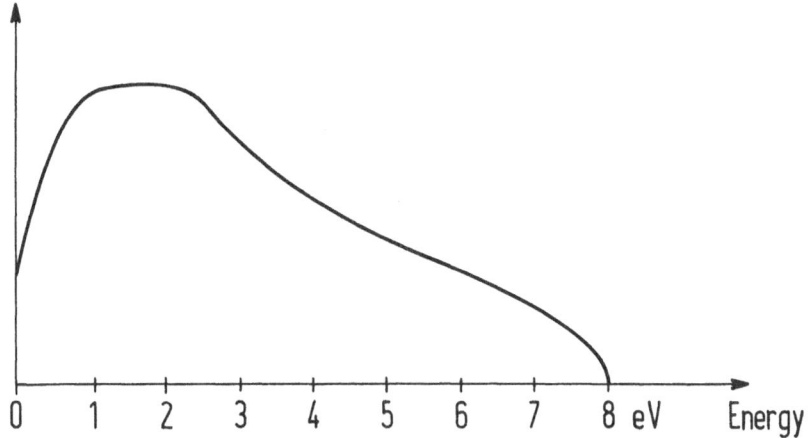

Fig. 2. Energy distribution function for photo electrons used in simulation.

TABLE II

Parameters used in calculation

	$I(\text{A m}^{-2})$	$E_{\max}(\text{eV})$	$T(\text{s})$
Photo electrons	4×10^{-5}	8	2×10^{-6}
Plasma electrons	4×10^{-6}	36	9×10^{-5}

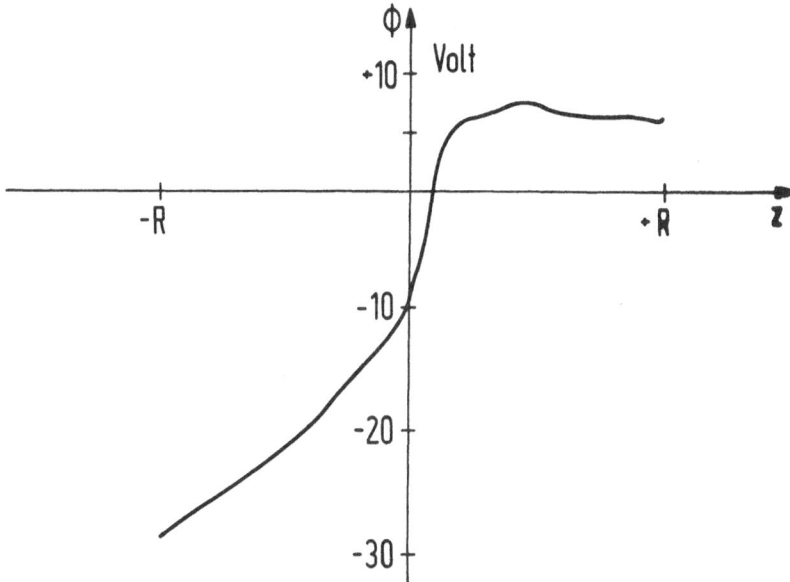

Fig. 3.   Electric potential on the satellite surface during simulation for the geometry of Figure 1.

The satellite is assumed to be made of a homogeneous insulating material in order to show the accumulation of negative charge on the dark side.

On the illuminated side the photo emission soon attains a stationary state with a constant potential which is hardly influenced by the ambient electrons. Figure 3 shows the surface potential plotted against the projection of the surface point on the Sun-satellite axis. After 2276 time steps corresponding to a time of 0.17 ms the dark side had not yet attained a stationary state in spite of the relatively high influx of electrons. It seems likely that negative charges continue to accumulate on the dark side of the satellite until the whole surface is close to the voltage corresponding to the cut-off energy of the ambient electrons.

Unfortunately there was no opportunity to continue these calculations, but if more work is done along these lines it would be advisable to use a Monte-Carlo technique with different weights on the ambient electrons and the photo-electrons. The effect of the space charges can be neglected with respect to surface charges, and then it is not necessary to simulate the evaluation of the system in time. One might also obtain useful results with a simple model where the electron trajectories are not computed in detail. It is, however, possible to estimate that the perturbation voltage $\phi_1 - \phi_2$ between the probes (Figure 1) is too high to be tolerable under the conditions stated.

In an idealized model one can consider a steady state where the dark side is charged negatively to a constant potential ϕ_- and the Sun-lit side to ϕ_+. Between the two sides there is a narrow transition region where the potential jumps from ϕ_+ to ϕ_-. In Figure 4 the numerically computed surface charge density corresponding to surface potentials of $\phi_+ = 8$ and $\phi_- = -36$ is plotted. The z-coordinate is the projection of the surface points on the Sun-satellite axis, and the width of the transition region is taken to be 1/16 of the satellite radius.

Figure 5 shows the equipotential lines around the satellite for this situation and Figure 6 the corresponding cloud of photo electrons and ambient electrons.

The height of the peaks in σ (in Figure 4) is strongly dependent on the width of the transition region, which will probably be narrower in the real case.

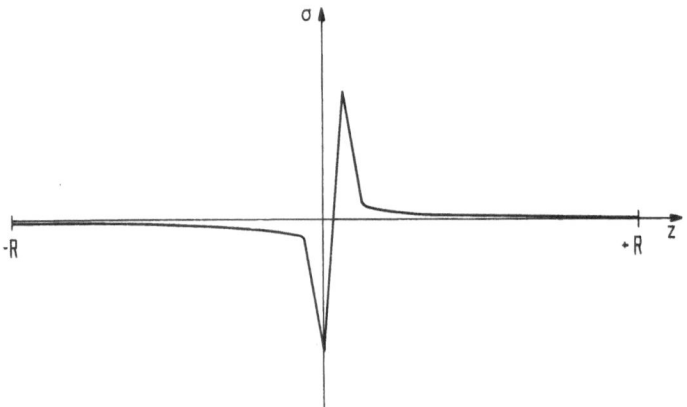

Fig. 4.   Charge density on the surface of a homogeneous insulating satellite with the potential distribution of Figure 1.

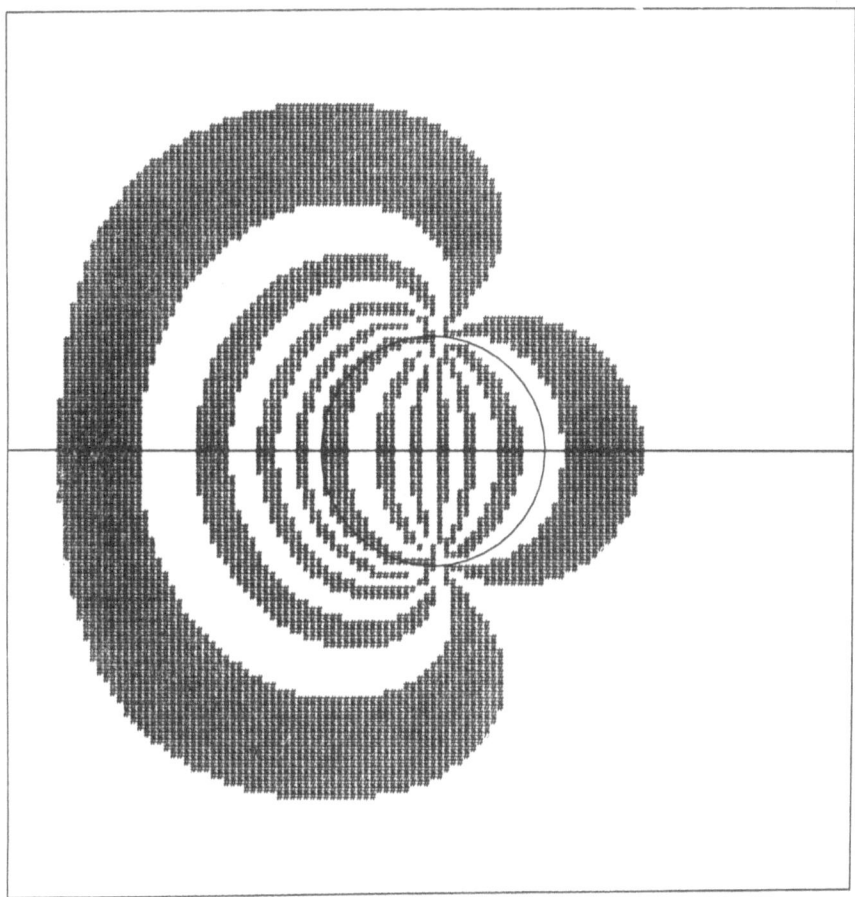

Fig. 5.   Electric field around the satellite caused by the charge of Figure 4.

The spin of the satellite is slow compared to the time required to set up the surface charge and can be neglected.

The highest perturbation potential between the probes occurs when the probes are aligned along the Sun-satellite axis, and for this case one can show analytically or numerically that when the satellite radius, R, is small compared to the distance, L, of each probe from the satellite the perturbation voltage is (see Figure 1)

$$\phi_1 - \phi_2 = (\phi_+ - \phi_-)\frac{R^2}{L^2}. \qquad (1)$$

4. Conducting Body Covered by an Insulating Layer

In the previous section we saw that a satellite made of insulating materials can be

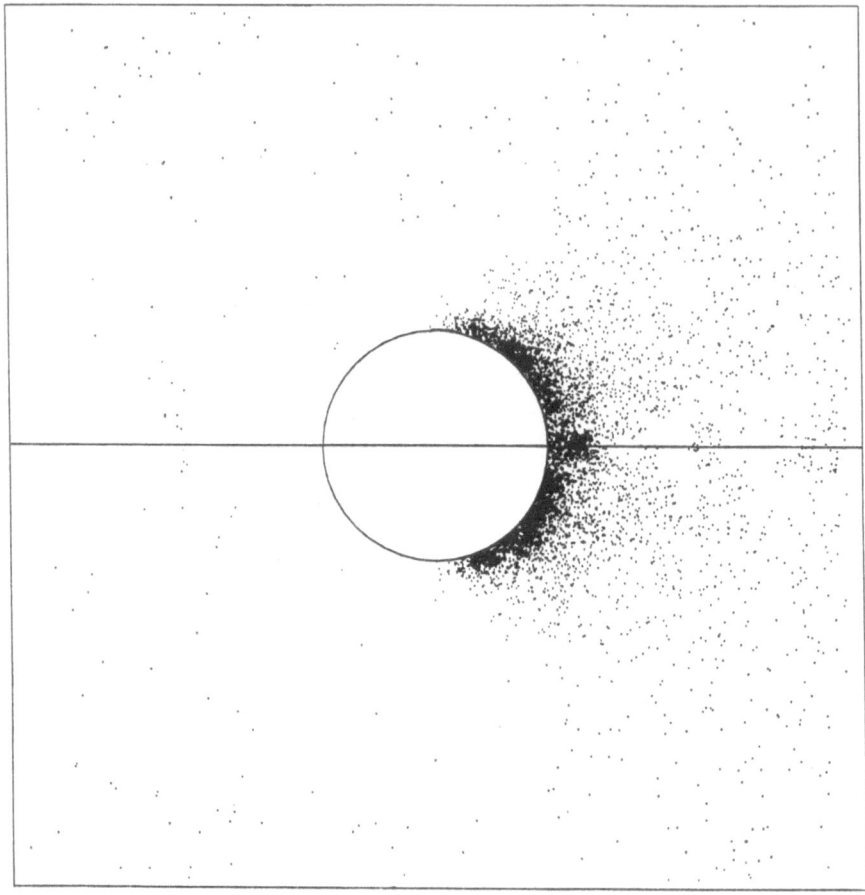

Fig. 6.   Cloud of ambient and photo electrons around a satellite corresponding to the electric field
shown in Figure 5.

charged to give great potential variations on the surface. In a practical situation, the
satellite is mainly made of metal, and the insulating material is the glass cover of the
solar cells which are mounted over a large part of the satellite surface. It would thus
be useful to estimate how a metallic body covered by a few mm thick insulating layer
will be charged when exposed to the same electron inward and outward fluxes.

Consider a metallic sphere with radius $R(1-\delta)$ covered with an insulator of thickness $R\delta$. A unit charge placed on the outer surface will give rise to a redistribution of
the charges on the metallic surface which is equivalent to a mirror charge $-(1-\delta)$
located at a distance of $R(1-\delta)^2$ from the centre (Figure 7). Since the metallic sphere
has a floating potential and total charge equal to zero, one also gets a charge $+(1-\delta)$
at the centre. Using Green's function we can now express the potential at a point $\bar{r}$
on the surface of the insulating cover as a function of the charge density $\sigma(\bar{r}')$ of the

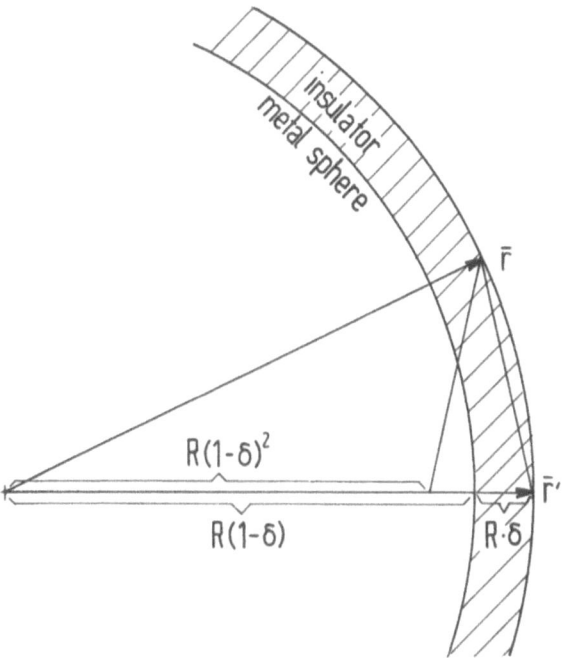

Fig. 7. Section of a metallic satellite covered with a thin insulator of thickness $R \cdot \delta$.

source point $\bar{r}'$

$$4\pi\varepsilon_0\phi\left(\bar{r}\right) = \iint\limits_S \left[\frac{1}{|\bar{r}-\bar{r}'|} - \frac{(1-\delta)}{|\bar{r}-(1-\delta)^2\bar{r}'|} + \frac{(1-\delta)}{R}\right] \sigma\left(\bar{r}'\right) dS$$

$$= (1-\delta) \iint\limits_S \left[\frac{1}{|\bar{r}-\bar{r}'|} - \frac{1}{|\bar{r}-(1-\delta)^2\bar{r}'|}\right] \sigma \, dS +$$

$$+ \delta \iint\limits_S \frac{\sigma}{|\bar{r}-\bar{r}'|} \, dS + \frac{1-\delta}{R} \iint\limits_S \sigma \, dS . \qquad (2)$$

Now, if δ is very small, the first integral on the right hand side of (2) will get contributions mainly when $|\bar{r}-\bar{r}'| \lesssim 2R\delta$. If σ varies only slightly over this region, this integral can be approximately evaluated as

$$\sigma\left(\bar{r}\right)\left(1-\delta\right) \int\limits_0^\infty \left[\frac{1}{t} - \frac{1}{\sqrt{t^2+4\delta^2R^2}}\right] 2\pi t \, dt = 4\pi R\sigma\left(\bar{r}\right)\left(1-\delta\right)\delta . \qquad (3)$$

Here we have used the integration variable $t=|\bar{r}-\bar{r}'|$ and assumed that the surface of the sphere around the point $\bar{r}$ can be regarded as an infinite plane compared to the length $2R\delta$.

If we now put $1 - \delta \approx 1$ we get from (2) and (3)

$$\phi(\bar{r}) = \frac{\delta}{\varepsilon_0} \left[R\sigma(\bar{r}) + \frac{1}{4\pi} \iint\limits_S \frac{\sigma(\bar{r}')}{|\bar{r} - \bar{r}'|} \, dS \right] + \frac{1}{4\pi\varepsilon_0 R} \iint\limits_S \sigma \, dS \tag{4}$$

On the right hand side of (4), the only term which does not contain a factor δ is the last one. This term however, does not depend on $\bar{r}$ and gives only a uniform potential to the whole surface. Only the first two terms can give appreciable potential differences over the surface. The quantity $\delta \cdot \sigma$ must remain finite for small δ, which means that a thin insulating layer on the metal sphere must have a high surface charge density in order to create a given surface potential distribution.

If we denote

$$W(\bar{r}) = \delta \cdot \sigma(\bar{r}), \tag{5}$$

the potential between two points on the surface, $\bar{r}$ and $\bar{r}_0$, can be expressed as

$$\phi(\bar{r}) - \phi(\bar{r}_0) = \frac{R}{\varepsilon_0} \left[W(\bar{r}) - W(\bar{r}_0) \right] +$$

$$+ \frac{1}{4\pi\varepsilon_0} \left[\iint\limits_S \frac{W(\bar{r}')}{|\bar{r} - \bar{r}'|} \, dS - \iint\limits_S \frac{W(\bar{r}')}{|\bar{r}_0 - \bar{r}'|} \, dS \right]. \tag{6}$$

Now we can let $\delta \to 0$ and we then get the auxillary condition

$$\iint\limits_S W \, dS = 0 \tag{7}$$

in order that the last term in (4) remains finite.

By defining

$$Q = \iint\limits_S \sigma \, dS \tag{8}$$

we can write the potential of (4) as

$$\phi(\bar{r}) = \frac{R}{\varepsilon_0} W(\bar{r}) + \frac{1}{4\pi\varepsilon_0} \iint\limits_S \frac{W(\bar{r}')}{|\bar{r} - \bar{r}'|} \, dS + \frac{Q}{4\pi\varepsilon_0 R}. \tag{9}$$

From Equation (9) and condition (7) one can now in principle determine the re-normalised surface charge density $W(\bar{r})$ and Q for a known surface potential distribution $\phi(\bar{r})$.

It should be noted that the limit $\delta \to 0$ is only a mathematical abstraction which corresponds to an infinite surface charge density. For practical purposes it is a useful approximation when $\delta \ll 1$ because it gives a convenient relationship between ϕ and W.

In Figure 8 is shown the computed values of W corresponding to the potential distribution shown in Figure 1. In this case the surface charge does not show the same

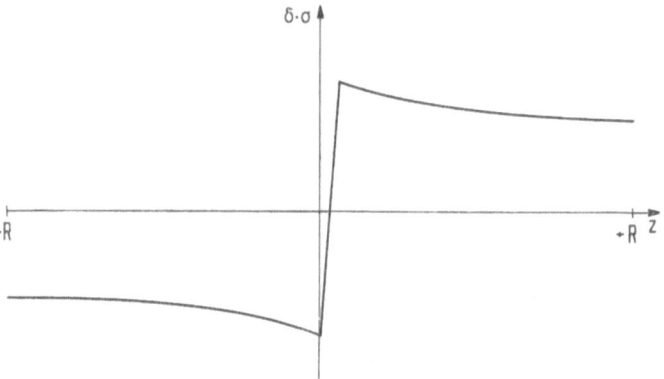

Fig. 8.   Charge density on the insulating surface of a spherical metal satellite with the potential distribution of Figure 1.

peaks near the transition region as in Figure 4 because of the short range potential described by the first term of the right hand side of (9).

In order to calculate the voltage between two symmetrically arranged probes mounted at a distance L from the satellite one can write approximately for $L \gg R$

$$\phi_1 - \phi_2 = \frac{2p}{4\pi\varepsilon_0 L^2},\tag{10}$$

where the equivalent dipole moment is

$$\bar{p} = \int \bar{r}\, dq,\tag{11}$$

where $\bar{r}$ is a vector from the centre of the spacecraft to the charge dq.

In (11) we get contributions from the charge dq at the positions $\bar{r}$ and the mirror charge $-(1-\delta)\, dq$ at the position $(1-\delta)^2 \bar{r}$

$$\bar{p} = \int \left(1 - (1-\delta)^3\right) \bar{r}\, dq \to 3 \iint_S \bar{r} W(\bar{r})\, dS, \quad \text{when} \quad \delta \to 0.\tag{12}$$

One gets from (10) and (12) a relation like (1) when the probes are aligned along the Sun-satellite axis

$$\phi_1 - \phi_2 = (\phi_+ - \phi_-)\frac{R^2}{L^2}\,\gamma,$$

where γ is a dimension-less quantity that remains finite when $\delta \to 0$. In our numerical case of Figure 8 we get $\gamma \approx 1$, so no essential change has taken place compared to a homogeneous insulating satellite.

5. Conclusions

A satellite with an insulating surface may charge to negative potentials as high as

10^4 V on the dark side, whereas the Sun-lit side will stay at a positive potential of a few volts. For a satellite radius $R = 0.75$ m and boom length 20 m, the perturbation measured by the probes may thus be of the order of

$$\phi_1 - \phi_2 = \frac{R^2}{L^2} \times 10^4 = 14 \text{ V}$$

which is far too high if DC electric field measurements are to be carried out. Any small irregularity of the surface geometry will cause a still higher perturbation. It makes no improvement if we have only a thin insulating layer on top of a conductor. It would be interesting to make computations with a surface that is partly insulating and partly conducting, but it is not expected to improve the perturbation by an order of magnitude.

The best way to avoid the high negative potential on the dark side seems to be to make the whole surface conducting and equipotential in order to ensure that an accumulation of negative charge anywhere on the surface will be compensated by the photoelectron emission on the illuminated side. If the satellite is sufficiently symmetric in shape, the only voltage perturbation is caused by the non-symmetry of the photo electron cloud (Soop, 1972), which is several orders of magnitude lower than the perturbation corresponding to the case of an insulating surface.

Acknowledgements

The author is indebted to the staff of Space Science Dept., ESTEC, Noordwijk, for valuable discussions.

References

DeForest, S. E.: 1972, *J. Geophys. Res.* **77**, 651.
Fahleson, U.: 1969, Report No. 69-36 from the Royal Institute of Technology, Stockholm, Dept. of Plasma Physics.
Soop, M.: 1972, *Planetary Space Sci.* **20**, 859.

DISCUSSION

Walker: There appears to be a singularity on the axis due to individual particles near the axis playing a disproportionate role in the calculation as they see their image on the axis.

Soop: In cylindrical co-ordinates there is a singularity on the axis for the equation of motion of a mass-point. What appears on the film, however, is a consequence of the fact that in the $r - z$ plane the point density is proportional to $n \cdot 2\pi r \, dr \, dz$ and vanishes when $r \rightarrow 0$ regardless of n. Also a particle moving along a straight line in three dimensions appears to be reflected at the axis when shown in the $r - z$ plane. Due to the low collision rate in the physical problem, the interaction of a particle with its image is negligible.

2.3. INFLUENCE OF SURFACE EMISSION ON EXPERIMENTAL MEASUREMENTS

POSSIBLE EFFECTS OF PHOTOELECTRON EMISSION
ON A LOW ENERGY ELECTRON EXPERIMENT

HELMUT R. ROSENBAUER

Max-Planck-Institut für Physik und Astrophysik, Institut für extraterrestrische Physik,
Garching, Germany

Abstract. Design considerations for the HELIOS low-energy plasma-electron instrument gave rise to a study of the possible influence of photoelectron production on the measurements.

Different sources of distortion, such as photoelectron production within the instrument, charging of the whole spacecraft, differential charging of electrically insulating parts, and the influence of the spacecraft's shape are discussed separately and methods for suppressing the different disturbances are proposed. It is shown that electron instruments can be designed such that they are insensitive to internally produced photoelectrons. The remaining distorting effects can be minimized by choosing a sphere-like symmetrical shape for the spacecraft, and by making all external surfaces electrically conductive. The latter measure is regarded to be imperative for successful measurements.

If all precautions are taken, it should be possible to measure distribution functions of solar wind electrons down to a few eV in spite of the predominant population of photoelectrons in the vicinity of the instrument.

1. Introduction

During the early years of solar wind investigation the interest of both the theoreticians and the experimenters was focused on the positive component of the plasma. In recent years interest in solar wind electrons has increased because the understanding of the interplanetary medium as a plasma is not possible without a detailed knowledge of the electron component. The electrons carry the dominant part of both the heat flux and the electric currents. Because of their higher temperature they govern the static pressure, and details of their distribution function are important for instabilities which can arise in the plasma.

The *in situ* measurement of these electrons is obviously difficult, because their average energy is of the order of only 10 to 20 eV. Therefore they are strongly affected by fields around the spacecraft (s/c). Furthermore, high fluxes of photoelectrons are produced in and around a satellite borne instrument by the solar uv radiation. Ways must be found to distinguish between the plasma- and the much higher photoelectron fluxes.

That the measurements are actually difficult to perform correctly became clear when the first attempts were made. The results from different satellites and instruments differed widely. Only the measurements of the Los Alamos Group, performed aboard the Vela satellites yielded believable results down to rather low energies (Montgomery *et al.*, 1968; Montgomery, 1972).

Because of the obvious difficulties, a study of the possible sources of distortions of electron measurements was carried out when, in 1969, an electron experiment was designed for the HELIOS mission. It was found that photoelectrons were the major source of the difficulties.

Although the example 'HELIOS' is somewhat special in some aspects, it is believed

R. J. L. Grard (ed.), Photon and Particle Interactions with Surfaces in Space, 139–151. All Rights Reserved

that most of the qualitative results are also valid for other s/c in interplanetary space.

2. Expected Results Under Idealized Conditions

In order to simplify the first considerations, the expected conditions will be idealized as follows:

(1) The s/c is spherical and conducting;

(2) only radial fields exist;

(3) space-charge effects can be neglected;

(4) the instrument (mounted under the s/c's skin) is not sensitive to photons, and no photoelectrons are produced inside the instrument.

Such a s/c in the solar wind will charge slightly positively with respect to the surrounding plasma (to about $+2$ to $+10$ V) as many authors have pointed out before (e.g. Gringauz and Zelikman (1957) and Whipple (1965)).

The experimental results are in accordance with these theoretical predictions (Montgomery *et al.*, 1972). The positive charging can be easily understood by regarding the expected currents to the surface of the s/c for 0 Volt potential. These can be calculated approximately, assuming average solar wind parameters at one AU and laboratory measured surface properties (Feuerbacher and Fitton, 1972).

For HELIOS e.g. the following values obtain:

positive currents:

Protons	$I_p \approx 4 \times 10^{-7}$ A
secondary electrons	$I_s \approx 1 \times 10^{-6}$ A
photo electrons	$I_{ph} \approx 2 \times 10^{-5}$ A

negative currents:

plasma electrons	$I_e \approx 2 \times 10^{-6}$ A.

Since the photoelectron current is by far the dominant contribution at 0 potential the s/c will obviously charge positively until most of the photoelectrons have to return to the s/c, and thus the net current becomes zero. Because of the steep decrease of the photoelectron spectrum (Feuerbacher and Fitton, 1972; Grard, 1972) this will happen at a rather small positive potential.

Since the photoelectrons returning to the s/c surface as well as the plasma electrons will be detected by an electron instrument, and since they seem to be in the same energy region, it would at first seem to be very difficult to measure low-energy plasma electrons and separate them from the high flux of photoelectrons.

However, if the simplifications assumed here actually do hold true, a complete separation of the two 'kinds' of electrons should be obtained. In energy channels below the one corresponding to the probe's potential ϕ_p, an instrument looking nearly radially outwards from the s/c should measure only returning photoelectrons, because all plasma electrons are accelerated by ϕ_p and therefore will be found in energy channels $> e\,\phi_p$. In this region, on the other hand, no photoelectrons can be detected, because all photoelectrons with an energy $> e\,\phi_p$ (in the vertical component of their velocity) will escape. Since the two spectra thus measured in one energy scan, are of completely different origin, a distinct discontinuity, both in slope and in magnitude should occur at the energy channel $e\phi_p$. (See Figure 1.)

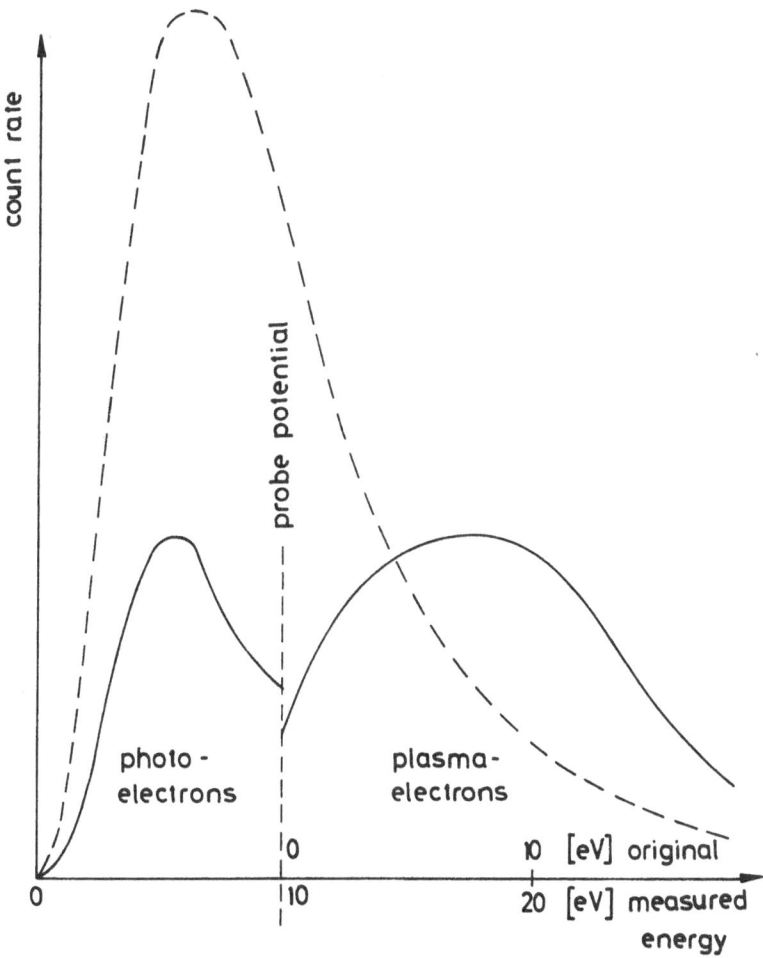

Fig. 1.   Expected electron count rates (schematical) as a function of energy for an assumed probe potential of $+10$ V. The dotted line indicates the possible effect of photoelectrons produced between the plates of the electrostatic analyser.

This means that, under the idealized conditions assumed here, the plasma electrons and the photoelectrons actually do not merge in the measured spectrum. Due to the positive charging the complete electron energy spectrum can be measured and the probe potential necessary for data interpretation can directly be derived from the same measurement.

An additional advantage of the positive charging is, that the plasma electrons with originally zero energy, arrive at the instrument with the energy $e\phi_p$ which makes them less sensitive to disturbing fields in and around the sensor. They therefore can be measured more reliably than with 0 probe potential.

Therefore it can be claimed, that under these ideal conditions the photoemission effects support plasma electron measurements, rather than disturbing them.

It must, however, be checked to what extent this simple situation applies or can be achieved.

3. Deviations from the Idealized Picture due to Photoelectrons Produced in the Instrument

Aboard most interplanetary s/c, plasma experiments are mounted such that direct sunlight will hit the entrance aperture of the instrument at least part of the time. If it is an instrument of the electrostatic deflection type such as employed by the Vela satellites or as foreseen for HELIOS, the plates can be blackened such that practically no photons can reach the particle sensor.

However, photo-electrons can be produced between the plates with the right direction and energy to reach the sensor. In the low-energy channels, they can easily outnumber the electrons coming from outside. Since the internally produced electrons are not subject to the external field of the s/c which separates photoelectrons and plasma electrons, they appear both above and below the energy channel $e\phi_p$. Because of their high flux, they can possibly smear out the break completely. This effect is indicated in Figure 1 by the dotted curve.

An additional smearing can occur due to wide view angles of the instrument. This is because photoelectrons returning to the satellite at oblique angles can have higher total energies than $e\phi_p$.

Therefore, it is necessary to design a low-energy electron instrument such that production of photoelectrons inside the experiment is well suppressed and the angle of acceptance is kept narrow.

4. Design of an Electron Sensor with Essentially no Internal Photoelectron Production

The easiest way to avoid photoelectron production inside an instrument is to mount it in a shaded part of the spacecraft. In the case of HELIOS this would, however, mean that the instrument could not measure in the plane which normally contains the solar wind flow direction and the magnetic field. It also would be impossible to mount the instrument in the s/c's equatorial plane, where the disturbing external electrical fields are expected to be minimum.

Therefore two other designs, which can be used for instruments mounted in the equatorial plane of the spacecraft were considered. The more simple one, shown in Figure 2, where the entrance of the energy analyzer is shaded by a properly formed aperture configuration in the satellite's skin, was not chosen, because the look direction would not be within the desired plane.

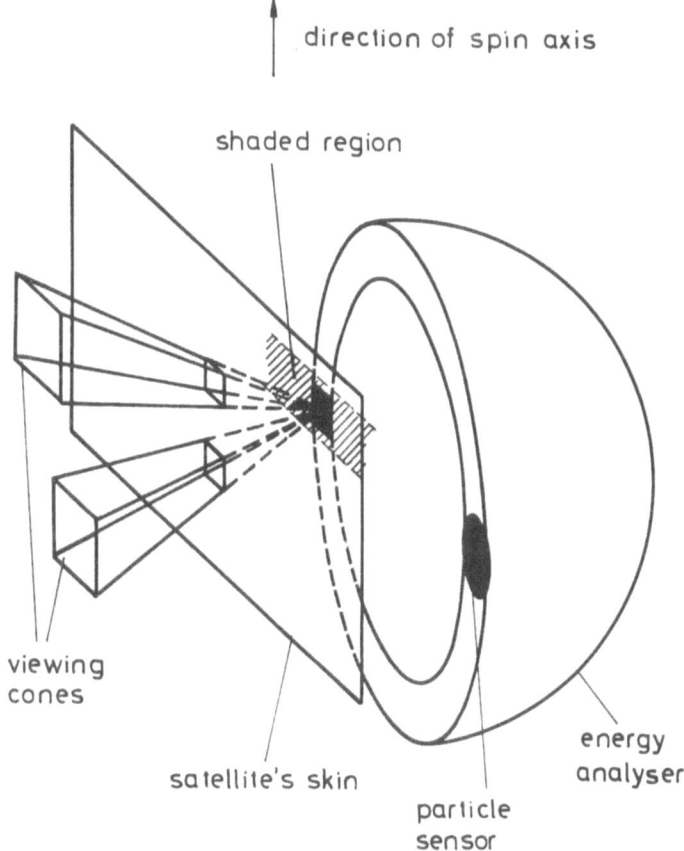

direction of spin axis

shaded region

viewing
cones

satellite's skin

particle
sensor

energy
analyser

Fig. 2.   Schematic view of an electron instrument mounted such that essentially no photoelectrons
can be produced between the plates of the energy analyser.

The instrument selected for HELIOS is schematically shown in Figures 3 and 4. It has a field of view exactly situated in the satellite's equatorial plane. In order to have the entrance of the energy analyser in shadow, the particles are first deflected 30° out of the equatorial plane by a plane electrostatic analyser. The field of view is internally limited such that all material, which could emit photoelectrons is outside the viewing cone of the energy analyser (Figure 4).

5. Disturbances due to Space Charge Effects

In addition to the radial external field, already assumed in Section 2, 'natural' outer fields will be produced by space charge due to photoelectrons on the illuminated side of the satellite and due to a proton wake effect on the side pointing away from the Sun.

For the electron instrument on HELIOS the polar components of this field can be

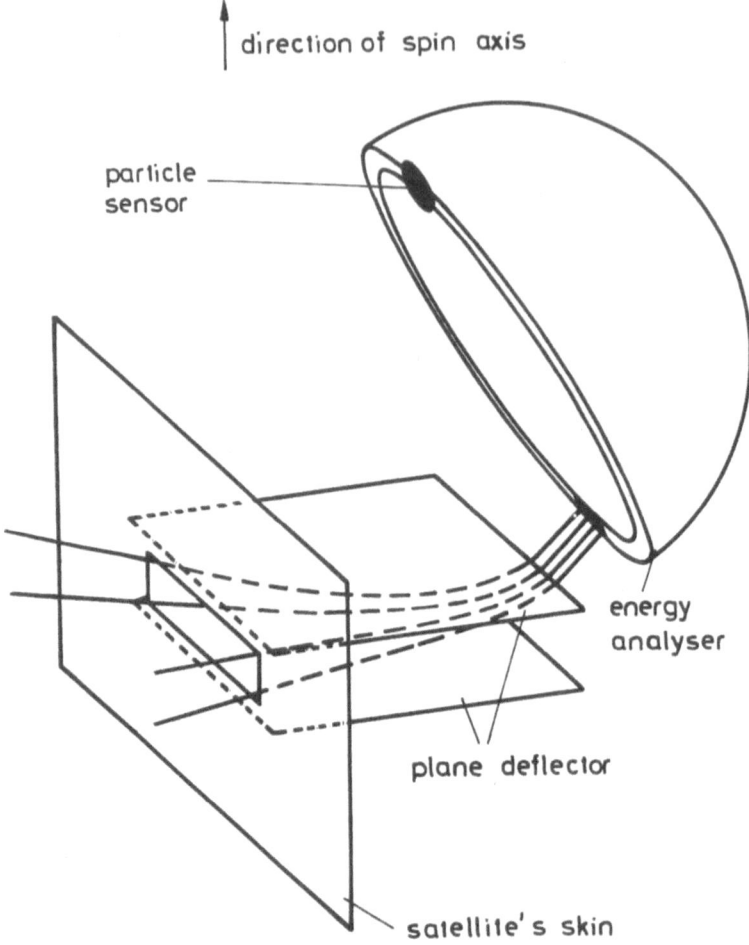

direction of spin axis

particle sensor

energy analyser

plane deflector

satellite's skin

Fig. 3.   Schematic view of the electron experiment developed for HELIOS. The particles are deflected out of the region, which can be hit by direct sunlight, by means of deflector plates.

neglected since the instrument looks out through the space charge cloud in the symmetry plane. Estimates show that also the azimuthal components are so weak that they can only slightly decrease the accuracy of the directional information in the lowest energy channels.

The radial components become effective only if a region with negative potential is generated. In this case, the photoelectrons are observed up to an energy as high as the potential difference between the spacecraft and the point with the largest negative potential in front of the instrument, and the plasma electron spectrum is cut at its low energy end by just the amount of the negative potential (compare Figure 1 and Figure 5). Consequently the break in the spectrum is no longer a measure of the s/c potential, but indicates its upper limit only.

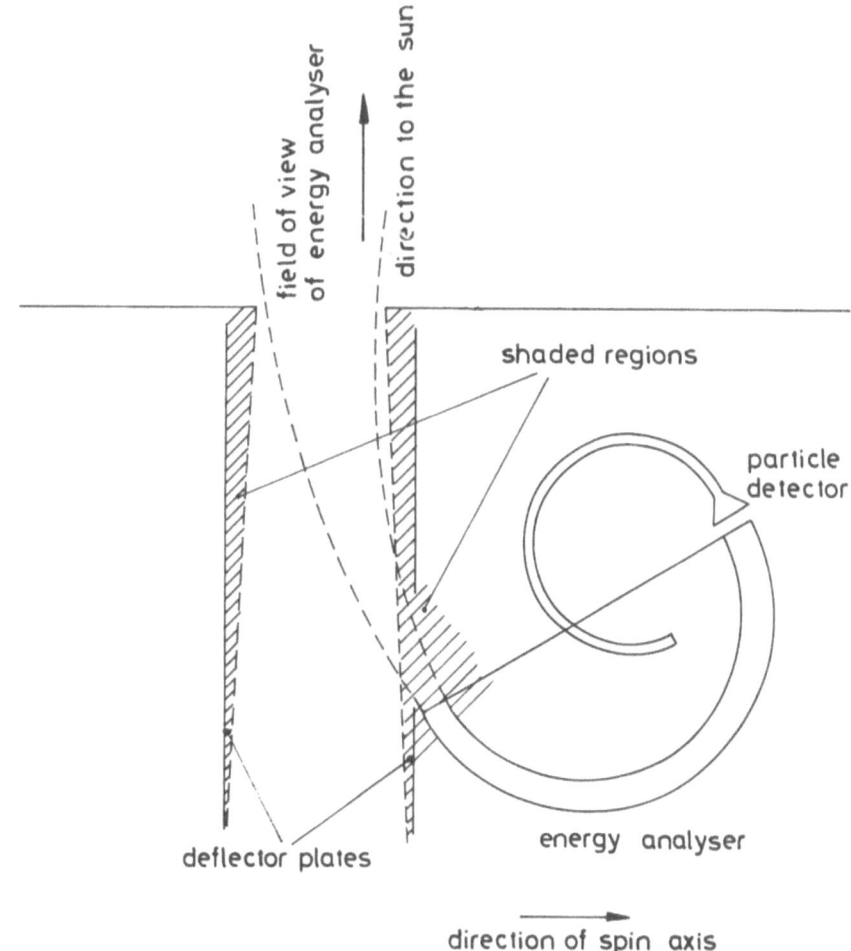

Fig. 4.   Cross section through the HELIOS electron instrument.

There are efforts under way to calculate this effect by numerical plasma simulation methods (Voigt *et al.*, 1971).

6. Influence of the Spacecraft's Shape

A sperically shaped s/c with a uniform conductive surface would be the ideal vehicle for a low-energy electron experiment. No modulation of the potential could occur due to rotation, and no differential charging would disturb the measurement. Also the 'natural' outer fields produced by space-charge would be minimum and their effects could possibly be calculated rather accurately.

The success of the electron measurements carried out aboard the Vela satellites may possibly partially be due to the favourable shape of these vehicles (picture in Montgomery *et al.* 1972).

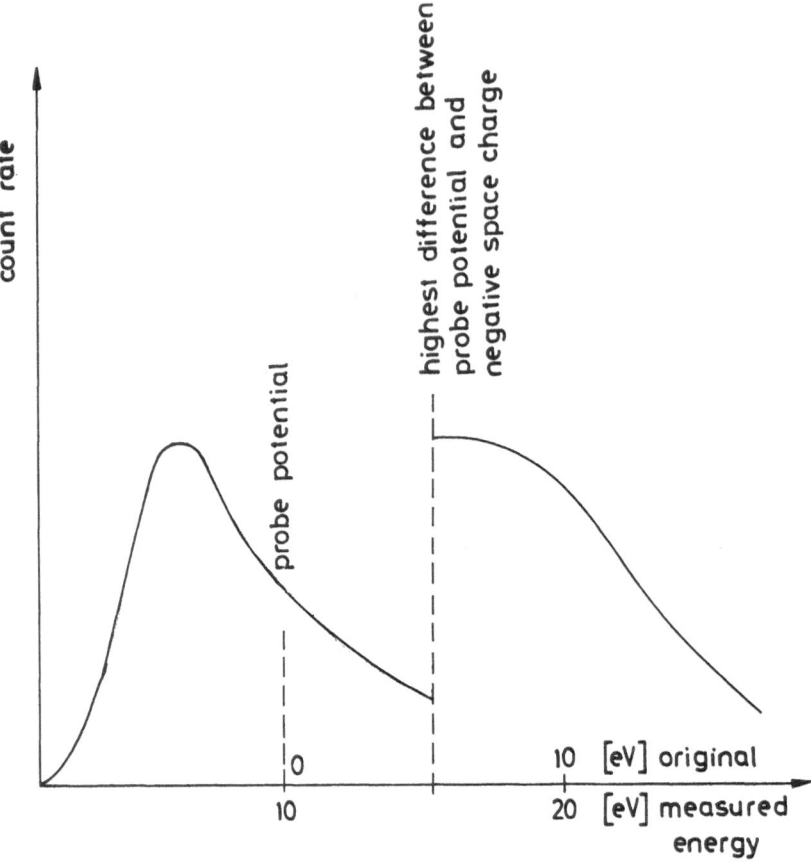

Fig. 5.   A negative space charge cloud shifts the expected break in the spectrum to higher energies.
(Compare with Figure 1.)

On the other hand, the shape of HELIOS (Figure 6) is obviously very unsatisfactory for this type of measurement. The particles, on their way to the instrument, have to pass between the two conical solar generators and can there be easily affected by surface potentials. Since a photoelectron cloud will fill the space between the cones on the sunward side, the space charge effects also will be severe.

Together with the long booms and the extended antenna, the field configuration around this *s/c* will be so complicated, that all attempts to eventually correct measurements with respect to distortions by these fields will be extremely difficult.

7. Disturbances due to Insulating Surfaces

7.1. INSULATING SURFACES IN SUNLIGHT

A large fraction of the illuminated surface of nearly all spacecraft (except some deep space vehicles, which have radioactive generators), is covered by solar cells.

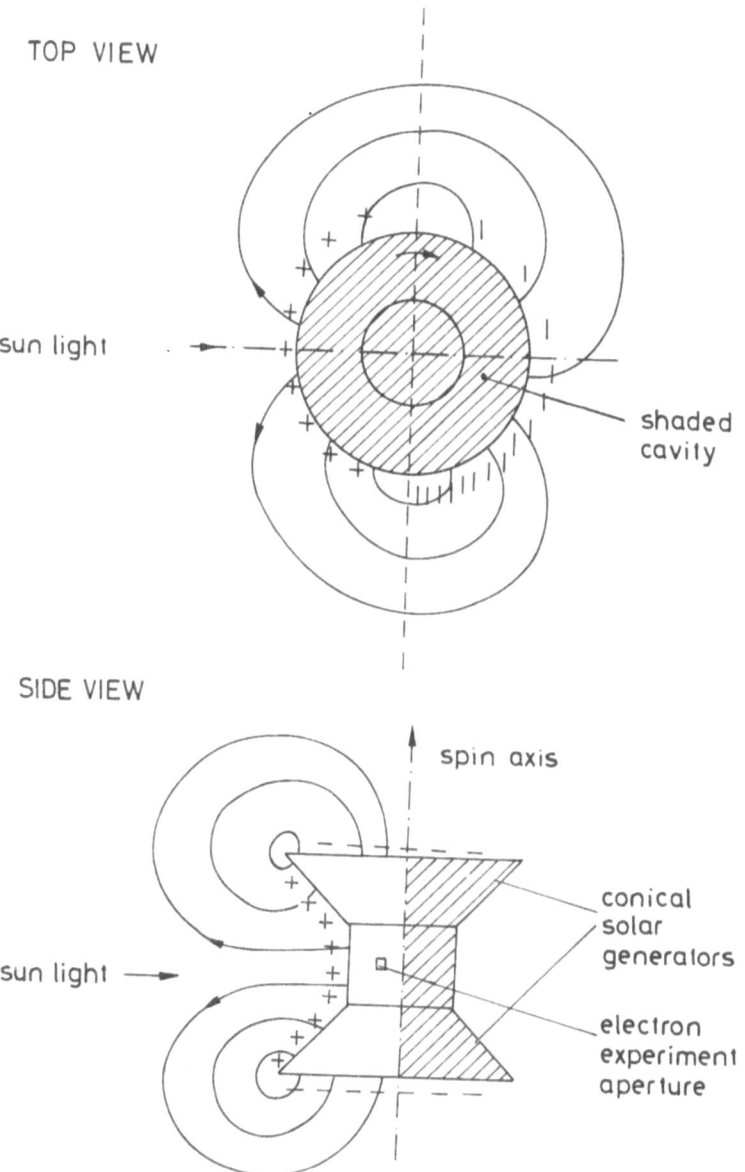

Fig. 6.   Simplified view of the HELIOS spacecraft with electric field configurations expected due to partial charging of insulating surfaces.

In the case of the HELIOS spacecraft, no conducting material at all is visible to sunlight, because the normally metallic surfaces are covered with quartz second surface mirrors for thermal reasons.

Exposed to sunlight, all elements of the surface will charge positively till the current balance condition is fulfilled. The charging will be slightly different over this surface because of different surface properties of the materials, and it will vary with time due to variations of surface properties, solar wind conditions, and solar UV radiation.

If we deal with rotating s/c's such as HELIOS, part of the surface will alternatively be illuminated and shaded.

In this case, the capacitance between the surface of the insulating coating and the spacecraft body plays a dominating role. Typical capacitances are $10-20$ pF cm^{-2}. With typical photoelectron saturation currents of $\sim 3 \cdot 10^{-9}$ A cm^{-2}, these capacitances will be charged to an equilibrium potential rather quickly (~ 0.1 s) on the sunward side. On the shaded side, however, the typical negative currents (plasma electrons plus photoelectrons returning from the illuminated to the shaded side) are much lower. Therefore, the surfaces will discharge slowly and, in the case of HELIOS (at 1 AU and under average solar wind conditions) the potential is expected normally not to fall below 0 V. When HELIOS is close to the Sun and both the plasma density and the photoelectron density become high, and probably also the electron temperature increases, the surfaces may be discharged to as much as -20 V. In this case the aperture of the instrument is shielded against low-energy electrons, and only the high-energy tail of the distribution function can be measured. In addition to this, strong azimuthal electric fields will be generated (Figure 6) which will not only falsify the directional information, but also focus or defocus the electrons depending on the azimuthal angle. Since this effect, of course, is strongly energy dependent, the measured spectra will be severely distorted (Figure 6). Since the distorting fields are not known quantitatively, interpretation of low energy particle data collected under these conditions will be essentially impossible.

7.2. INSULATING SURFACES IN SHADOW

Surfaces of electrical insulators which are permanently in shadow will charge to a relatively high negative potential, because at positive or zero potential essentially no positive current source is available to compensate for the rather high flux of plasma electrons. At high enough negative potentials the electron flux decreases, and a higher percentage of secondary electrons, and protons which are bent into the shaded regions are able to establish an equilibrium.

For HELIOS the estimated potential of the shaded areas in the cones (Figure 7) is between -10 and -100 V, depending on the plasma parameters. Obviously this effect again will focus the electrons and thus further distort the results.

7.3. THE COMPLETELY INSULATED SPACECRAFT

If, as in the case of HELIOS, the s/c body is practically completely covered with electrically insulating material, the potential of the body and of the instrument is not defined at all. Such a spacecraft is of no use for low-energy particle measurements and appropriate measures must be taken to avoid this situation.

7.4. POSSIBLE MEASURES TO AVOID DISTURBANCES DUE TO INSULATING SURFACES

Contrary to direct uv-disturbances of the measurement, which can be avoided by

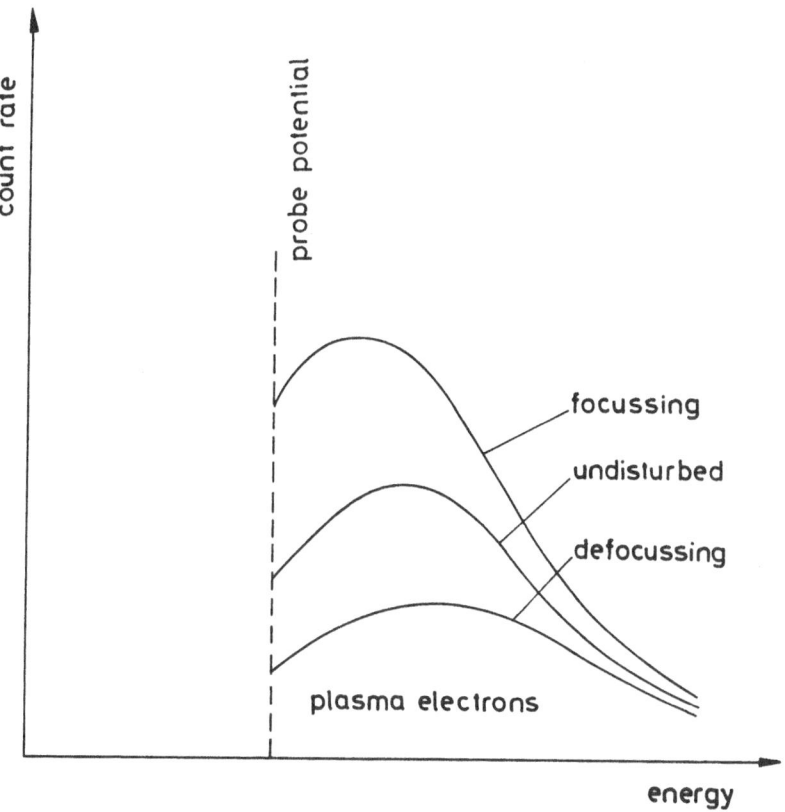

Fig. 7.   The principal expected effect of focusing and defocusing electric fields on the measured
energy spectra of plasma electrons. (Both, density and temperature results are falsified.)

proper design of the instrument by the experimenter, the disturbances by differential
charging of the s/c surface can only be reduced by appropriate design and manufacture
of the s/c. Unfortunately making a s/c 'electrostatically clean' is not yet as common a
procedure as making it 'magnetically clean'.

An obvious way to overcome the above mentioned difficulties is to make all the
surfaces of the s/c conductive.

Normally this can easily be achieved for the shaded parts. The 'thermal coating'
found there, can either be replaced by polished metallic surfaces, if thermal emission
is to be avoided, or by electrically conducting black paint, if radiation is desired.

If the shaded parts are conducting and the illuminated parts are still insulating, the
s/c main body will adopt the negative equilibrium potential described in Section 7.2.
In this case ϕ_p cannot unambiguously be determined and therefore even the interpretation of the measured high-energy tail of the spectrum is probably impossible.

In order to allow for low-energy electron measurements, some kind of electron
emitter must be provided which is electrically connected to the s/c body. This can

either be an active electron gun, which stabilizes the potential very accurately if the electrons are monoenergetic, or a sufficiently large ($>10\%$ of the shaded part) photoelectron-emitting area.

Preventing the charging of illuminated insulating areas, typically solar cell cover glasses, is somewhat more difficult. Such surfaces must be covered by a 'conductive coating' (CC) which is nearly 100% transparent in the spectral region of the sunlight important for the energy production in the solar cells. The transparency must be maintained under space conditions, i.e. UV- and particle irradiation and, in the case of HELIOS, extremely high temperatures ($\approx 200\,^\circ$C). The coating of each cover glass must be electrically connected to the s/s's main body, therefore proper contacting techniques must be established.

This problem has however been solved. A first successful attempt was made with Explorer 31 and now both the layer (indium oxide) and the contacting technique have been developed and qualified for extreme conditions in the HELIOS project.

This coating would not only help against differential charging due to photoeffects, but it would also effectively shield the ac electrical fields produced by voltage jumps in the solar generator, which occur permanently when solar cell strings pass the terminator. In this way CC would also support the 'plasma wave experiments' on satellites.

In spite of the multiple advantages for the scientific mission, it is not yet clear, whether CC will be applied on the HELIOS spacecraft.

8. Conclusion

Even though photoelectron emission on spacecraft interferes with solar wind electron measurements in several ways, there seems to be a good chance to overcome these difficulties and obtain good electron data down to essentially zero energy. This can only fully be accomplished if the instrument is designed appropriately. An additional necessary condition for success, however, is to have an 'electrostatically clean' spacecraft. Since the technical problems involved have been solved in preparation for the HELIOS mission, this technique will hopefully be applied in future missions as a matter of course.

Acknowledgements

The author is indebted to Dr M. D. Montgomery, who is presently a guest at the Max-Planck-Institut für extraterrestrische Physik, Garching, for very valuable discussions and a reading of the manuscript.

References

Feuerbacher, B. and Fitton, B.: 1972, *J. Appl. Phys.* **43**, 1563.
Grard, R. J. L.: 1972, ESTEC Tech. Working Paper No. 663.
Gringauz, K. I. and Zelikman, M. K.: 1957, *Uspekhi Fiz. Nauk* **63**, 239.
Montgomery, M. D., Bame, S. J., and Hundhausen, A. J.: 1968, *J. Geophys. Res.* **73**, 4999.

Montgomery, M. D.: 1972, in C. P. Sonett, P. J. Coleman, and J. M. Wilcox (eds.), *Solar Wind*, NASA SP-308, p. 208.

Montgomery, M. D., Asbridge, J. R., Bame, S. J., and Hones, E. W.: 1973, this volume, p. 247.

Voigt, G.-H., Schröder, H., Abromeit, C., Köneman, B., Seehusen, J., and Isensee, U.: 1971, Jahresbericht, Physikzentrum Technische Universität Braunschweig.

Whipple, E. C.: 1965, Thesis, Wheaton College, Ill.

THE EFFECTS OF PHOTOELECTRON EMISSION
ON A MULTIPLE-PROBE SPACECRAFT
NEAR THE PLASMAPAUSE*

DAVID P. CAUFFMAN

Plasma Research Laboratory, The Aerospace Corporation, El Segundo, Calif. 90245, U.S.A.

Abstract. Measurements of the potential difference between a pair of symmetrical, spherical, floating probes (14-cm diam, 5-m baseline) flown on the $S^3 - A$ (Explorer 45) spacecraft (launched November 1971, with apogee 5.2 R_E, perigee 222 km, and inclination 3.6 deg) have shown a complicated response to sheath potentials in the region near and inside the plasmapause. Typically, with increasing altitude (1) the sunward sphere will first assume a more negative potential than the antisunward sphere, (2) the polarities of the probes in the region $L = 2$ to 4 may switch, sometimes several times, until finally, (3) the detector output may saturate ($\Delta\phi > 125$ mV) with the sunward sphere more negative again.

In order to explain the above observations a model has been created in which the system is divided into four floating plasma probes: two identical spheres and the sunlit and dark sides of the spacecraft body. The interaction between the probes is computed considering space charge effects, currents due to the emission and exchange of photoelectrons, and currents from the ambient plasma. The four floating potentials are calculated by requiring current balance for each probe and solving by iteration the resulting four coupled, non-linear equations. The results of the model indicate that the $S^3 - A$ measurements described above may be explained in terms of the fact that one sphere is exposed to a greater portion of the sunlit side of the body.

This study has found that both space charge and current balance effects must be considered in order to explain the observed sheath potentials. Sheath potentials asymmetries are sensitive to ambient temperature and density, relative photoemission intensities of the spheres and spacecraft body, and the conductivity of the probe surfaces. The effects become significant for Debye lengths greater than about one-tenth of the sphere-body separation.

At the present time there is no reliable way to compensate for the effects of sheath potentials. The only way to make good floating probe measurements is to separate the probes by distances very large compared to the shielding length.

1. Introduction

Recent attempts to make floating probe measurements in the solar wind and in the Earth's magnetosphere at altitudes greater than several thousand kilometers have revealed complex relationships between the elements of multiple-floating probe systems (Cauffman and Gurnett, 1972; Heppner, 1972; Maynard and Cauffman, 1973). The floating potentials of two probes immersed in a plasma cease to be independent when the characteristic shielding length of the plasma becomes comparable to the distance between probes. Since these shielding lengths often exceed the lengths of the longest appendages flown to date (45 m on IMP-I) (Heppner, 1972), it is of considerable interest to examine the nature of the mutual interactions in a multiple-floating probe system.

One probe may affect another in several ways:

---

* Work done in part while the author was a NAS/NASA Research Associate at Goddard Space Flight Center, Greenbelt, Md. 20771, U.S.A.

(1) The presence of a charged object induces a space charge potential (which has been modified from the free-space value by the plasma medium) at the position of the conjugate probe.

(2) The plasma sheath that partially shields the charged object modifies the local plasma density, and perhaps temperature, at the conjugate probe.

(3) The physical presence of the object screens some of the ambient particles that would otherwise hit the conjugate probe.

(4) The primary probe may emit particles (notably photoelectrons or secondary electrons), some of which will be intercepted by the conjugate probe.

Thus, one probe in the sheath of another has its floating potential changed, due to alterations in its current balance and in the local plasma potential. The effects are, of course, mutual.

In the practical application of floating probes in experiments at high altitudes, the value of the floating potential is difficult to determine (e.g., see DeForest, 1972). A simple method of overcoming this problem for measurements of DC electric fields has been to measure the potential differences between identical floating probes extended on booms from opposite sides of a spacecraft. This technique has been discussed by Fahleson (1967) and reviewed by Cauffman and Gurnett (1972). It works well when probe separations are much greater than shielding lengths. If the sheaths of the spacecraft and probes were spherically symmetric, the technique would work even when the sheaths overlap significantly. Unfortunately the modes of sheath interaction detailed above do not all have symmetrical effects. The most notable problems are the following:

(1) Photoelectrons are emitted only from the sunlit facets of a probe, creating anisotropies in the radial number density and energy distributions of particles in the sheath. This effect disappears during eclipse of the satellite.

(2) If the conductance of various areas of the probe or spacecraft surface differs, then the different areas may have different floating potentials, depending upon their solar exposure and surface properties. The inhomogeneities in the floating potential of the probe surface will create anisotropies in the radial dependences of the sheath parameters.

In this paper, we shall present some experimental evidence on the nature of sheath interactions and explain these results based on a model that includes the effects of photoelectron asymmetries.

2. S^3 Measurements

In November 1971, $S^3 - A$ (Explorer 45) was launched into an eccentric equatorial (3-deg inclination) orbit with a 5.2 R_E apogee and 222-km perigee. The spacecraft experiment complement includes a pair of symmetrical floating probes designed to measure DC electric fields in regions inside the plasmapause. The sensors are 14-cm diam spheres extended on insulated booms on either side of the spacecraft such that their center-to-center separation is 5 m. A stub boom protrudes from the outer side of each sphere to maintain translational symmetry with respect to the Sun. The spheres have been coated with an aqueous carbon solution to lower their photoemission. Most of the surface of the satellite is covered by the nonconducting glass faces of solar cells. The electric field antennas are oriented perpendicular to the spacecraft spin axis, which lies approximately in the ecliptic plane about 60 deg from the solar direction.

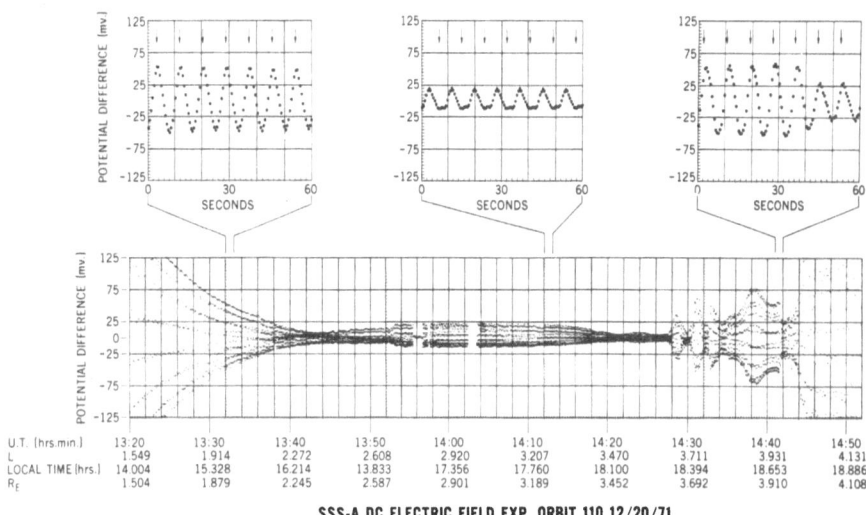

SSS-A DC ELECTRIC FIELD EXP. ORBIT 110 12/20/71

Fig. 1. Data from the $S^3 - A$ DC electric field experiment on orbit 110 outbound, showing the two types of sheath potential effects (center and right expanded inserts) including the region of saturation, and also showing the region (left expanded insert) where the measurement of electric fields is free from sheath effects.

Figure 1 shows data from the S^3 DC electric field experiment for part of an out-bound pass from $R_E = 1.5$ to 4.1. These data were taken in synchronism with the spacecraft roll. The data may be classified into three regions, shown expanded in the inserts. The first region, typified by the left insert, has a decreasing amplitude, sinusoidal waveform resulting from measurement of $V \times B$ and any ambient electric fields in the spin plane. Beyond $L \cong 2.5$, the contribution from $V \times B$ is less than a few mV. The signal which is observed at higher altitudes and which is shown expanded in the center and right-hand inserts, is due to potentials in the spacecraft sheath. The distortion of the waveform from a sinusoidal shape indicates that ambient electric fields are not the cause of these signals. There is no correlation of the sheath-induced signals with the direction of the satellite wake. At the far right ($\sim 14:42$ UT) of Figure 1, the experiment output saturates at a point identified as the plasmapause (Maynard and Cauffman, 1972). The arrows in the inserts indicate a reference point in the spacecraft roll. When the waveform maximum is simultaneous with the roll start (right-hand insert), the sunward probe has a more negative potential than the antisunward probe. When the waveform minimum is simultaneous with the roll start (center insert), the antisunward probe is the more negative one. Typically, with in-creasing altitude, the sheath potentials begin to affect the probes first such that the sunward sphere is more negative. The spheres then may switch polarities (sometimes several times). Finally, just prior to and during saturation, the sunward sphere again becomes more negative.

3. A Model of Probe Interactions

The results concerning probe potential difference polarity described in Section 2 were not expected. Signals correlating with the sun direction were anticipated because of the unavoidable existence of photoemission on only one side of the spacecraft, which deposits extra electrons on the sunward sphere. However, it was not foreseen that the sunward sphere could become the more positive sphere.

An attempt has been made to explain the observations with a model of the spacecraft system comprising four floating plasma probes. Figure 2 schematically shows the physical model, in which probes 1 and 2 are the sunward and antisunward spheres, respectively, and probes 3 and 4 are the sunlit and dark hemispheres of the spacecraft, approximated by a spherical shape. Each probe is assumed to be a conductor.

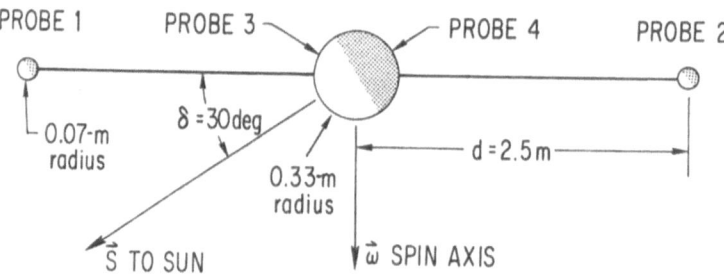

Fig. 2. Schematic diagram of the 4-probe model, consisting of a sunward sphere (probe 1), and antisunward sphere (probe 2), the sunlit spacecraft hemisphere (probe 3), and the dark spacecraft hemisphere (probe 4).

A rigorous model would involve trajectory calculations for each particle. The charge and current densities thus calculated would give, via Maxwell's equations, electric and magnetic fields. When inserted in the equations of motion of the particles, these fields would self-consistently give the positions and momenta of the particles. Even using modern computers and considering only the tenuous plasmas of distant space, there are too many particles involved to permit numerical trajectory calculations. There is no choice but to somehow average the particle behavior. We shall try to find analytic functions to represent the behavior of classes of particles. Specifically, the approach we have taken is the following:

(1) We assume 'reasonable' analytic functions for the currents to each probe as functions of probe potential and of plasma parameters.

(2) We require the sum of the currents to each probe to be zero.

(3) We assume a 'reasonable' radial dependence of the space charge potential near each probe.

(4) We require that the probe floating potential be a linear sum of the 'local plasma potential' (space charge due to other probes) and the 'current-balance potential' (such that the currents to the probe sum to zero).

The simultaneous, nonlinear equations thus obtained are solved by iteration. To test the 'reasonableness' of the functions assumed, we can use functions with some theoretical justification, and test the predicted consequences against reality.

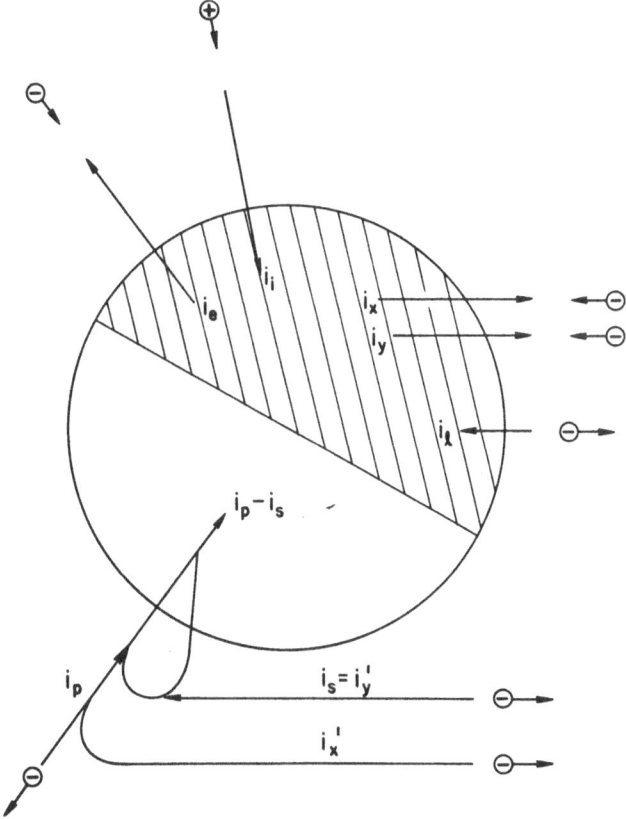

Fig. 3. Schematic sketch of the probe currents considered in the model. The symbols are identified in the text.

Figure 3 shows the currents to the probe considered in the model. These include:
(1) Ambient electron current (i_e).
(2) Ion ram current (i_i).
(3) Current due to escaping photoelectrons (i_p).
(4) Current due to photoelectrons escaping from other probes which are intercepted by this probe (i_x).
(5) Current due to photoelectrons in the sheath which are intercepted by another probe $(-i_s)$.

(6) Current due to photoelectrons in other probes' sheaths which are intercepted by this probe (i_y).

(7) Electrical leakage current (i_l).

At the present time, no attempt has been made to include currents due to secondary emission caused by energetic particle bombardment. Each current is a product of geometric factors, 'undeflected' current density, and potential dependence factors. The latter are critical elements of the model. A generalized function Ξ has been defined that is applicable to currents to a spherical probe. Figure 4 shows this function, given by

$$\Xi(\phi, \phi_0, \Xi_{max}) = \exp(\phi/\phi_0)$$

for $\phi/\phi_0 \leqslant 0$, and by

$$\Xi(\phi, \phi_0, \Xi_{max}) = \Xi_{max} - (\Xi_{max} - 1) \exp(-\phi/(\phi_0(\Xi_{max} - 1)))$$

for $\phi/\phi_0 \geqslant 0$ where ϕ is the accelerating potential and $e\phi_0$ is the most probable energy of the particles, which are assumed to have a Maxwellian velocity distribution. If $\Xi_{max} = \infty$, then, for $\phi/\phi_0 > 0$, it follows that $\Xi = 1 + \phi/\phi_0$. This branch of the function is the acceleration function derived by Mott-Smith and Langmuir (1926) for spherical probes. This form of the function is used in the model for ion and electron currents from the ambient plasma, which constitute an infinite source. Values of $\Xi_{max} \neq \infty$ are used to obtain the expressions for photoelectron currents since the supply of photo-electrons is limited.

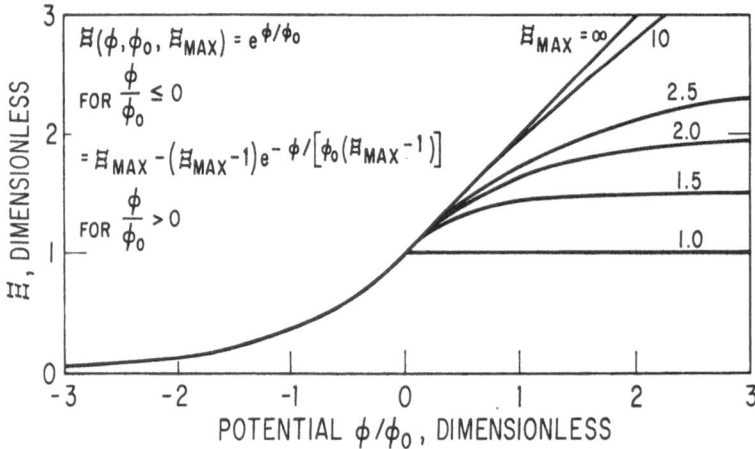

Fig. 4.   Current attraction/repulsion function, Ξ, applicable to spherical geometries (see text).

To evaluate the space charge potential at the location of one probe, due to the presence of another, we need to assume a function giving the shape of the radial potential decrease. We have assumed

$$\phi(r) = \phi(R) \frac{R}{r} \exp - \left(\frac{r - R}{\lambda}\right),$$

where R is the probe radius and λ is the characteristic shielding length. This function has theoretical justification when the floating potential is small compared to ϕ_e, ϕ_i, or ϕ_p.

Numerous other assumptions were necessary in obtaining tractable analytic functions. These include the following:

(1) Sheaths are infinite and overlapping.

(2) Ion velocity ≪ satellite velocity ≪ electron velocity.

(3) Trapped orbits are unpopulated.

(4) No ambient electric or magnetic fields exist.

(5) Sheath electric fields are radial.

(6) Photoelectrons have Maxwellian velocity distribution functions (cf. Grard, 1972).

(7) Booms are ignored.

The collective effect of the approximations and assumptions included in the model is to restrict confidence in the results to trends. The model should nevertheless be useful in order-of-magnitude estimation, and in the determination of the importance of various parameters.

The model is currently being refined. Suggestions are welcome. One area of the model that requires attention is the physics that occurs at the terminator, the boundary between the light and dark areas of a probe. The solution has presently been attempted only for the antenna axis near the sun direction. Eventually, we expect to publish a more thorough presentation of the elements of the model.

4. Results

An example of the preliminary results of the model calculations is shown in Figure 5. The potential difference between the small spheres is shown plotted as a function of the ratio λ_D/d of the Debye shielding length to the sphere-body separation distance, for several plasma energies. Electron number density increases to the left; equivalent altitude increases to the right. This figure should be examined for relative trends only; magnitudes should be ignored. The details of the figure can be altered by the choice of parameters. Any detailed conclusions should appropriately wait until the model is in a final form. However, some general conclusions can be made now and are relevant to our discussion.

In Figure 5, it is clear that the magnitude of the sphere potential difference is very dependent on both ambient density and temperature, and is 'organized' by the parameter λ_D/d. The sheath effects disappear for high densities. At low temperatures, the sunward sphere remains more negative than the antisunward sphere. At higher temperatures, the sunward probe may become more positive than the other; however, this is affected also by the choice of photoemissivities of the probe surfaces. We have also observed that, at least for low electron temperatures, the sheath effects are smaller if the spacecraft surface is a conductor (probes 3 and 4 treated as one). Irregularities occur in the curves of Figure 5 because the signs of the floating potentials

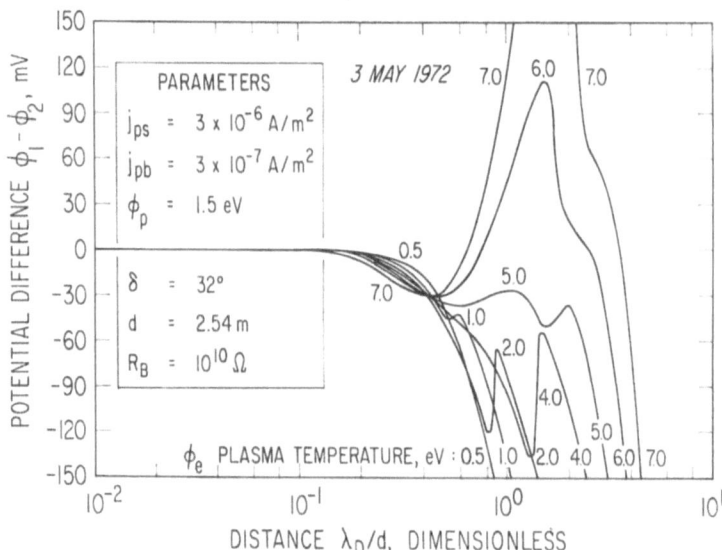

Fig. 5.   Preliminary results of the model calculations for an insulating spacecraft surface, illustrative
of trends in the dependence of probe potential difference on plasma parameters.

of the four interdependent probes change from negative to positive at different values
of λ_D/d, and at these points the shapes of the current functions change (see Figure 4).

Although various parameters in the model (e.g., the photoemission from the spheres,
photoemission from the body, coupling between the two parts of the body, energy of
the photoelectrons, and temperature of the plasma ions compared with the electrons)
are known only within reasonable limits and can be adjusted to vary the details of the
fit to the data, these general results of the model calculation *do* qualitatively explain the
observations of potential differences made with $S^3 - A$ at high altitudes. It should be
emphasized that these effects occur even though we have assumed complete symmetry
in the sensors themselves.

The portions of the curves for which $\phi_1 < \phi_2$ at small λ_D/d are, of course, due to the
incidence of more photoelectrons from the spacecraft body on probe 1. The condition
$\phi_1 > \phi_2$ occurs under some circumstances when sheath lengths become large because
sphere 2 encounters a more negative space charge on the antisunward side of the
spacecraft than sphere 1 encounters on the sunlit side. At even larger sheath lengths,
the attraction by sphere 1 of photoelectrons circulating in the sheath of probe 3
overrides the space-charge effect.

In attempting to understand the interactions that will occur when probe sheaths
overlap, we do not intend to imply that this is a tolerable circumstance for good probe
measurements. We do not know how to compensate for sheath effects. The only way
to make accurate measurements with floating probes in the magnetosphere is to have
the probes separated from other bodies by distances much larger than the shielding
length.

Acknowledgements

The DC electric field experiment on S^3 was built by Dr N. C. Maynard, to whom I am grateful both for permission to use the data, and for many helpful discussions.

This work was done in part while the author was a NAS/NASA Research Associate at Goddard Space Flight Center, Greenbelt, Md. Since July 1972, the work has been supported by the U.S. Air Force under contract F04701-72-C-0073.

References

Cauffman, D. P. and Gurnett, D. A.: 1972, *Space Sci. Rev.* **13**, 369.
DeForest, S. E.: 1972, *J. Geophys. Res.* **77**, 651.
Fahleson, U.: 1967, *Space Sci. Rev.* **7**, 238.
Grard, R. J. L.: 1972, Internal Working Paper No. 663, ESTEC, Noordwijk, Netherlands.
Heppner, J. P.: 1972, in *Proc. of the Symposium on Critical Problems of Magnetospheric Physics*, Madrid, Spain, p. 107.
Maynard, N. C. and Cauffman, D. P.: 1973, *J. Geophys. Res.* in press.
Mott-Smith, H. M. and Langmuir, I.: 1926, *Phys. Rev.* **28**, 727–763.

DISCUSSION

Feuerbacher: What do you believe is the importance of having neglected the booms?

Cauffman: The effect of the sheaths of the booms is potentially quite important. However, at the present time I have not attempted to calculate its magnitude.

Grard: Are spherical sensors better than cylindrical ones?

Cauffman: It is still difficult to say. Spherical probes should not have variations in the floating potential with aspect angle, and are easier to treat theoretically. On the other hand, cylindrical probes – that is, long wires – can usually be put further from the spacecraft for a given weight. I believe the differences may not be critically important at high altitudes because shielding lengths may exceed the dimensions of the conductive portions of either type of probe, and because, as I have pointed out today, the effects of the asymmetric photoelectron cloud around the spacecraft body may be the most important consideration on destroying symmetry. Both types of probe should give the same result if the probes are far enough from the spacecraft. I would like to see an experiment flown to test this comparison by having both types of sensors on the same vehicle.

Pedersen: The effects from photoelectron emission discussed in this paper must depend on size of the satellite and of boom length.

Cauffman: Yes, this is correct. The results shown here refer to the situation on the S^3 spacecraft where relatively short booms were used.

THE INFLUENCE OF PHOTOELECTRON AND SECONDARY ELECTRON EMISSION ON ELECTRIC FIELD MEASUREMENTS IN THE MAGNETOSPHERE AND SOLAR WIND

R. J. L. GRARD, K. KNOTT, and A. PEDERSEN

*Space Science Dept., European Space Research and Technology Centre,
Noordwijk, The Netherlands*

Abstract. The feasibility of electric field measurements by the double sphere technique both inside and outside the magnetosphere is critically reviewed. In particular, influences resulting from photo-emission, secondary electron emission and the plasma environment are analyzed. It is concluded that a double sphere aerial, 50–100 m long, can measure dc electric fields as low as 0.1 mV m^{-1}, in the solar wind and during conditions with ambient electron current density below 10^{-6} A m^{-2} in the magnetosphere, provided certain precautions are taken regarding probe symmetry. In the magnetosphere, for current densities larger than 10^{-6} A m^{-2}, the space charge of photoelectrons escaping from the satellite will give rise to potential asymmetries near the probes, and an electron gun is required for control of the satellite potential and reduction of space charge asymmetries.

1. Introduction

Satellite experiments have provided scientists with a good mapping of the DC magnetic field in the magnetosphere and interplanetary space. At the same time the charged particle population and AC electric and magnetic fields have been investigated in detail. Compared with the data obtained in these disciplines, DC electric field measurements have been collected to a much smaller extent, and in most cases by indirect methods.

This situation is mainly due to experimental difficulties encountered in a direct measurement. Nevertheless, more data on DC electric fields appear highly desirable in order to understand the origin, acceleration, and temporal and spatial density variations of the particle population. In fact, electric field data may be the key to an explanation of the dynamics of the Earth's magnetosphere (Mozer and Manka, 1971; Axford, 1969).

In the ionosphere the low energy and high density plasma causes only a small offset voltage between a spacecraft and its environment. Under such conditions rocket experiments using the double sphere technique (Fahleson, 1967) have been carried out showing good agreement with Ba-cloud observations (Fahleson *et al.*, 1971). Successful satellite measurements have been performed only at middle and higher latitudes (Maynard and Heppner, 1970; Cauffman and Gurnett, 1971). At lower latitudes the magnitude of electric fields in the frame of reference of the Earth is too small to be detected from low orbiting satellites. Spacecraft on highly eccentric orbits – such as the Imp satellites – have been equipped with long antennas for the measurement of DC electric fields, but the data are still in the stage of analysis.

R. J. L. Grard (ed.), Photon and Particle Interactions with Surfaces in Space, 163–189. All Rights Reserved

Geostationary satellites offer the advantage of measuring the DC field in the frame of reference of an Earth observer. However, results from a particle experiment on ATS 5 indicate that – unless special precautions are taken – electric field measurements will be subject to strong interference (DeForest *et al.*, 1971); the difficulties arise from the plasma environment, which is of low density and high temperature. There is experimental evidence from simultaneous electron and proton measurements on ATS 5 that the satellite floating potential can be as high as 10 kV negative with respect to space potential. Recently, AC electric field experimenters have considered the possibility that spurious AC phenomena might be produced by high voltage discharges on the satellite (Whipple, 1972). Electrostatic charging of satellites will obviously also influence low energy particle measurements.

The purpose of this paper is to review the feasibility of DC electric field measurements with a double sphere antenna at the geostationary altitude and on highly eccentric orbits with apogee in the solar wind. The electrical state of a body is determined by its plasma environment, and by the photoemission and secondary emission properties of its surface. In the first part of this paper the particle fluxes to and from satellite and probe surfaces are established; the resulting current-voltage characteristics of the body can then be derived, allowing one to define its potential in a particular space environment. It is then emphasised that certain conditions on antenna length, and satellite and probe symmetries must be met in order to ensure the reliability of the measurements.

A number of problems related to DC electric field measurements in space have already been surveyed by Fahleson (1971). The more detailed and quantitative study given in the present paper has been largely motivated by the availability of new information on the magnetospheric plasma and by a better knowledge of probe surface properties.

2. The Plasma Environment

2.1. THE IONOSPHERE

The ionosphere is a region where, under sunlight conditions, the electron random current to a surface element is generally higher than photoelectron and secondary electron current. As the equivalent plasma temperature is of the order of 0.1 V, the floating potential of a body can be expected to be negative and of the same order of magnitude. The fact that the Debye sheath is very thin also facilitates electric field measurements with a double probe. Even if experiments are carried out at high latitudes during periods of strong precipitation of energetic electrons and relatively low ambient plasma density, the equilibrium potential remains at a small negative value (Fahleson *et al.*, 1971).

2.2. THE PLASMASPHERE

In the plasmasphere the plasma density is of the order of 10^2–10^3 cm^{-3} (Carpenter, 1966) and the differential electron flux is peaking at energies below 100 eV (Shield and Frank, 1970). In this plasma environment electron fluxes and photoelectron emission fluxes are of the same order of magnitude and, depending upon the conditions in the plasmaphere, photoelectron or ambient plasma fluxes can dominate. In the transition region close to the ionosphere, the ambient plasma flux predominates for most of the time, whereas the photoelectron flux is larger near the plasmapause.

2.3. Outer Magnetosphere

This portion of space is limited by the plasmapause and the magnetopause. Contrary to the plasmasphsre, the outer magnetosphere does not normally corotate with the Earth. The plasmapause constitutes the interface between high density cold plasma on one side and low density hot plasma on the other side. Figure 1a (after Shield and Frank, 1970) gives examples of quiet time particle spectra in the outer magnetosphere, viz. inside the plasmasheet at its inner edge and at the so-called electron trough, at 5.3 and 8.8 R_E. The quiet time spectrum at the geostationary orbit may be obtained by interpolation between these two curves and is shown in dashed line.

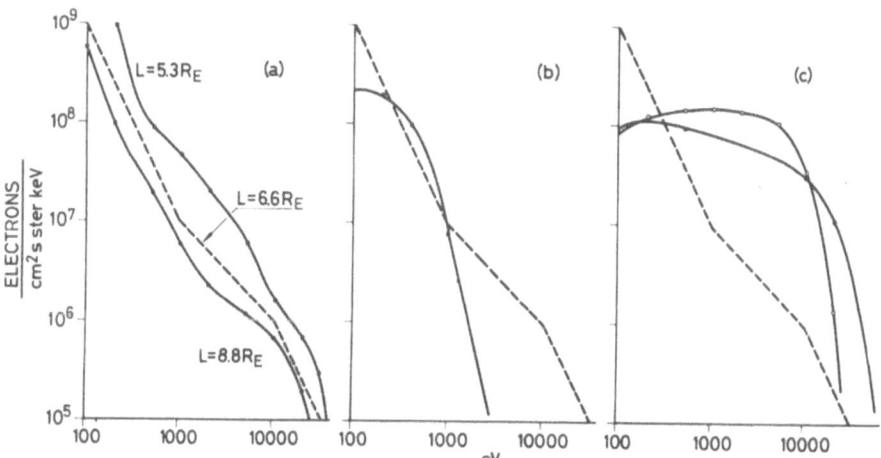

Fig. 1. (a) Electron energy spectra measure dat $L = 5.3 R_E$ and $L = 8.8 R_E$ by Shield and Frank (1970). The spectrum for $L = 6.6 R_E$ should fall between these two spectra and be given by the dashed line. (b) and (c) Possible deviations of the electron energy spectrum from typical conditions (dashed line) at the geostationary orbit during different phases of a substorm (DeForest and McIlwain, 1971).

Deviations from the quiet time spectrum have been reported by DeForest and McIlwain (1971) and are shown in Figures 1b and c. It can be seen that especially in the energy range above 1000 eV strong variations in the particle flux can be expected during a substorm. A further conclusion drawn by DeForest (1972) from ATS 5 measurements is that during substorms the contribution of particles with energies below 50 eV constitutes only 1% of the total flux, whereas at other times the flux of low energy electrons can dominate.

Particle flux measurements obtained from the geostationary spacecraft ATS 5

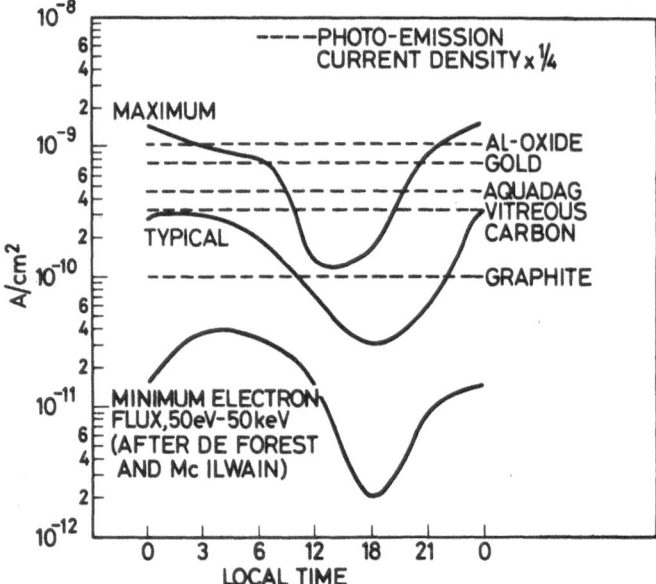

Fig. 2.   Comparison between electron current densities in the geostationary orbit and ¼ of photo-current densities for typical satellite and probe surfaces. The factor ¼ is related to the fact that for a sphere the projected area for photoemission is ¼ of the total area.

(DeForest and McIlwain, 1971) are shown in Figure 2 as a function of longitude. It can be seen from these curves that the particle flux can exceed the photocurrent from certain materials during disturbed conditions.

2.4. SOLAR WIND AND MAGNETOSHEATH

The particle population between the bow shock and the magnetopause is influenced by the interaction between the solar wind and the magnetosphere. This region is called the magnetosheath. Outside the bow shock the plasma is streaming radially away from the Sun. Parameters relevant to the solar wind and the magnetosheath under quiet conditions are summarised in Table I (Hundhausen, 1970).

TABLE I

Solar wind and magnetosheath parameters under quiet time conditions (Hundhausen, 1970)

	Solar wind	Magnetosheath
Electron and ion density (cm^{-3})	5	15
Flow speed (km s^{-1})	320	250
Flow direction	Radial	20° deflected
Proton temperature (°K)	5×10^4	10^6
Electron temperature (°K)	1.5×10^5	5×10^5

3. Surface Emission

3.1. PHOTOEMISSION

The interaction between photons and a surface results in the emission of photoelectrons when the photon energy exceeds a certain threshold given by the difference between the Fermi and the vacuum levels. The maximum energy of a photoelectron is given by the difference between photon energy and threshold energy. The energy spectrum of the photoelectrons and the photoyield depend upon the photon energy spectrum and the nature of the surface.

The knowledge of the solar photon energy distribution is essential for the study of photoemission in space. The measured flux energy spectrum $S(w)$ per $(s\ m^2\ eV)$ is shown in Figure 3 against the photon energy w in eV; these data have been reported in the literature by a number of experimenters and have been gathered by Walbridge (1971). The continuous component of the solar spectrum is approximated by a histogram, and its magnitude decreases by nearly six orders of magnitude between 4 and 12 eV. The accuracy and resolution of the measurements are good in the lower energy range, but are much less satisfactory above 15 eV.

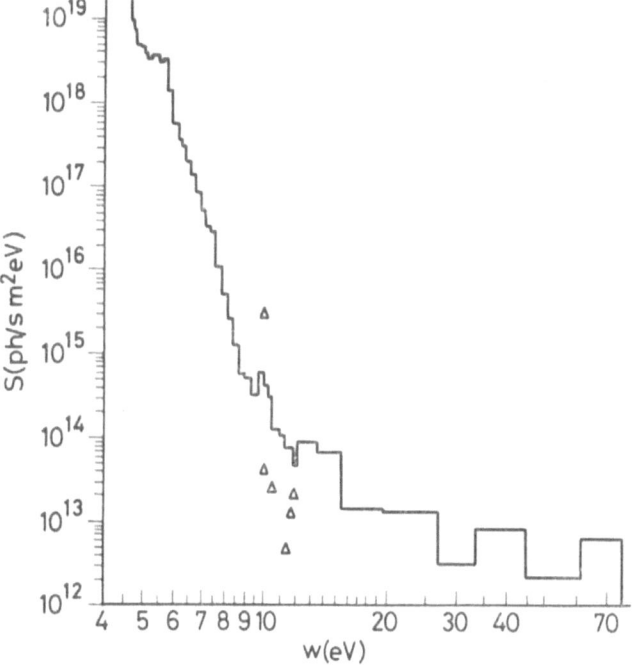

Fig. 3.   Measured solar photon energy spectrum.

3.2. SECONDARY ELECTRON EMISSION

Secondary electron emission occurs when electrons of sufficiently high energy impact on a solid. Some of the primary electrons undergo elastic collisions and are reflected without energy loss; most electrons lose energy by inelastic scattering and are re-emitted with lower energies. The secondary electron yield δ, which is defined as the ratio between the numbers of secondary and primary particles, may be higher than unity in a certain primary energy range. In this case electrons in the solid are excited and subsequently emitted. The latter mechanism may be compared with emission of electrons by photons. A certain similarity between secondary emission and photo-electric emission may also be seen from the fact that the energy spectra of emitted electrons are quite similar. More than 80% of all secondaries are emitted with energies below 20 eV.

In the literature the secondary electron emission of a certain material is usually described by the following parameters:

E_1 the lowest energy for which δ is equal to one,
E_2 the highest energy for which δ is equal to one,
E_{max} the energy where δ has a maximum,
δ_{max} the secondary yield at E_{max}.

Attempts have been made in theoretical treatments of secondary emission to derive an equation giving the yield curve as a function of the primary electron energy. Sternglass (1950) proposed the following expression, which seems to be in good agreement with experimental results:

$$\delta(E) = 7.4\, \delta_{max}\, (E/E_{max})\, \exp\left[-2\,(E/E_{max})^{1/2}\right]. \tag{1}$$

Parameters for secondary emission from a number of satellite and probe materials are given in Table II. In general, the secondary electron emission rate from a surface is roughly one order of magnitude lower than the photoemission rate, and is therefore important only for bodies not irradiated by the Sun.

TABLE II

Secondary emission properties of various materials (Gibbons, 1966; Hachenberg *et al.*, 1959) and the expected equilibrium potentials when exposed to various particle spectra

Material	Secondary emission properties		Equilibrium potential when exposed to particle spectra shown in Figure 1		
	E_{max}	δ_{max}	a	b	c
Gold	800 eV	1.45	− 800 V	− 134 V	< − 3000 V
Aluminium	300 eV	0.97	− 2150 V	− 625 V	< − 3000 V
Aluminium with oxide coating	300 eV	2.60	+ 5 V	+ 5 V	< − 3000 V
SiO₂	420 eV	2.50	+ 5 V	+ 5 V	< − 3000 V
Aquadag	350 eV	0.75	− 2140 V	− 760 V	< − 3000 V
BeCu	300 eV	2.20	+ 5 V	+ 5 V	< − 3000 V
BeCu activated	400 eV	5.00	+ 5 V	+ 5 V	+ 5 V

4. Electrical Properties of a Body in Space

4.1. CHARGING MECHANISM

The potential of a body with respect to its immediate environment is determined by the

fluxes of charged particles to and from its surfaces:
- incoming charged particles from the ambient plasma,
- photoelectrons escaping from the surface,
- secondary electrons emitted from the surface.

The polarity of the potential depends on the relative importance of the various flux contributions, and its magnitude can be determined from the current-voltage characteristic (I−V curve). The floating potential is defined as the voltage for which the sum of the currents listed above is equal to zero. Generally, when the floating potential is negative, its magnitude is determined by the energy distribution of the ambient plasma electrons; on the contrary, when the floating potential is positive, its value is a function of the energy of the photoelectrons.

One of the characteristic features of the cloud of *trapped* photoelectrons is that the total charge in the cloud is smaller than the total charge carried by the body, and the ratio of cloud to body charges varies with the size of the body and the intensity of photoemission. Referring to a computer simulation performed by Soop (1972), it can be shown that for a conducting sphere of 1.5 m diam, which is a typical dimension for a satellite, the ratio between charges in the cloud and on the sphere is of the order 0.3. For a probe with a diameter of a few cm, the charge in the cloud and any charge asymmetries associated with it can be neglected and the field has therefore, to a very good approximation, spherical symmetry (Coulomb field).

4.2. CURRENT-VOLTAGE CHARACTERISTICS

4.2.1. *Photoelectron Emission*

The photoelectron yield $Y(w)$, i.e. the number of emitted electrons per incoming photon for a given photon energy w, has been measured in the laboratory (Feuerbacher and Fitton, 1972). Figure 4a, for example, shows the curve $Y(w)$ for vitreous carbon. Multiplying the yield by the solar energy spectrum given by Figure 3, we obtain the differential flux of photoelectrons as a function of photon energy.

$$H(w) = S(w) Y(w),\qquad(2)$$

which is shown in Figure 4b.

In order to establish the current-voltage characteristic of a body for which photoemission if the dominating charging mechanism, the energy distribution of photoelectrons must be known. such measurements have been obtained for monochromatic light by Feuerbacher and Fitton (see Figure 4c). The photoelectron distribution under solar irradiation $p(\psi)$ can then be found by multiplying the energy spectrum for monochromatic light $f_w(\psi)$ by the differential photoelectron flux $H(w)$, and integrating with respect to the photon energy:

$$p(\psi) = \frac{1}{J_p} \int_0^\infty f_w(\psi) H(w) \, dw,\qquad(3)$$

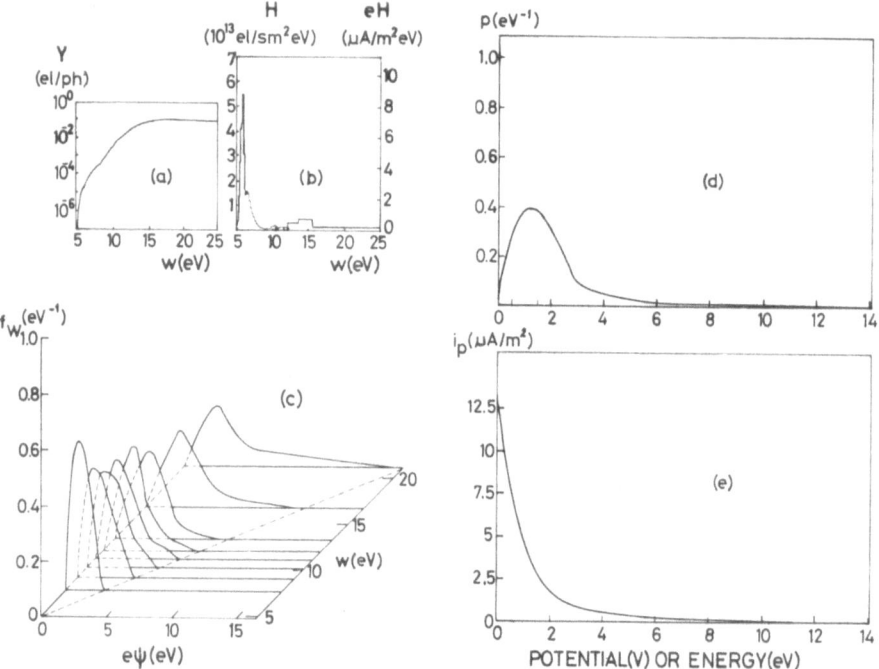

Fig. 4.   Experimental results and derived quantities on photo-electric properties of vitreous carbon, viz.: (a) Photoelectron yield. (b) Differential flux of photoelectrons vs solar photon energy. (c) Photoelectron energy distribution for different wavelengths. (d) Normalized differential flux of photoelectrons under solar irradiation. (e) Current-voltage characteristic for vitreous carbon when irradiated by the Sun.

where J_p, the saturation photoelectron flux, is introduced for normalisation purpose. The total normalised differential flux $p(\psi)$ for vitreous carbon is given in Figure 4d.

The photoelectron differential flux can be converted into a plot giving the photo-electron current density as a function of the *positive* potential of the body from which the electrons are emitted; more details on the required computations are given by Grard (1972); Figure 4e gives the characteristic curve for a planar surface made of vitreous carbon. Similar characteristics have also been computed by Grard for many other materials; a summary of these results is given in Table III.

4.2.2. *Secondary Electron Emission*

The determination of the $I-V$ curve for a surface in a plasma without solar irradiation requires the knowledge of the energy distribution of primary electrons and the yield curve for the secondary electrons (Equation (1)). The potential of the surface, V, is negative when the secondary flux is less than the primary flux, which is the case to be considered here. The primary electron current is then given by:

$$i_e(V) = e\pi \int_V^\infty (\mathrm{d}J/\mathrm{d}E)\,\mathrm{d}E, \qquad (4)$$

TABLE III

Experimental data on photoemission

Material	Aluminium oxide	Indium oxide	Gold	Stainless steel	Aquadag	LiF on Au	Vitreous carbon	Graphite	Average
Saturation current density (μA m^{-2})	42	30	29	20	18	15	13	4	21
Mean kinetic energy (eV)	1.33	1.35	1.40	1.42	1.17	2.24	1.51	1.48	1.42
Most probable energy (eV)	0.87	0.87	0.88	0.98	0.90	0.87	1.17	1.04	0.94
Max conduct./unit area planar source ($\mu\mho$ m^{-2})	45	35	30	21	21	10	13	4	23
Electron vol. dens. near the surface (10^6 m^{-3})	1670	1260	1110	770	750	460	500	150	830
Shielding distance (m)	0.17	0.20	0.22	0.26	0.24	0.42	0.34	0.60	0.25

where e is electron charge, and dJ/dE, the differential flux of primary electrons, is assumed to be isotropic. Examples for differential electron spectra are shown in Figure 1. Figure 2 shows the primary electron current for $V=0$, i.e. for a surface at space potential.

The secondary electron current is calculated by multiplying the differential energy spectra for primary electrons with the yield function $\delta(E)$ given by (1). In addition, the primary electron energy spectrum must be modified to take into account the retarding effect due to the fact that the surface potential V is negative. The secondary current density is then given by:

$$i_s(V) = e\pi \int_0^\infty \delta(E) \left[dJ(E + eV)/dE \right] dE. \tag{5}$$

For a moderately negative surface potential the primary ion flux is much smaller than the primary electron flux; for high negative potentials, however, it will increase substantially due to focussing effects. Figure 5 shows, as an example, $I-V$ characteristics for a negatively biased gold surface in a particle environment such as that shown in the dashed line on Figure 1, which is expected to correspond to average conditions at the geostationary altitude (Knott, 1972).

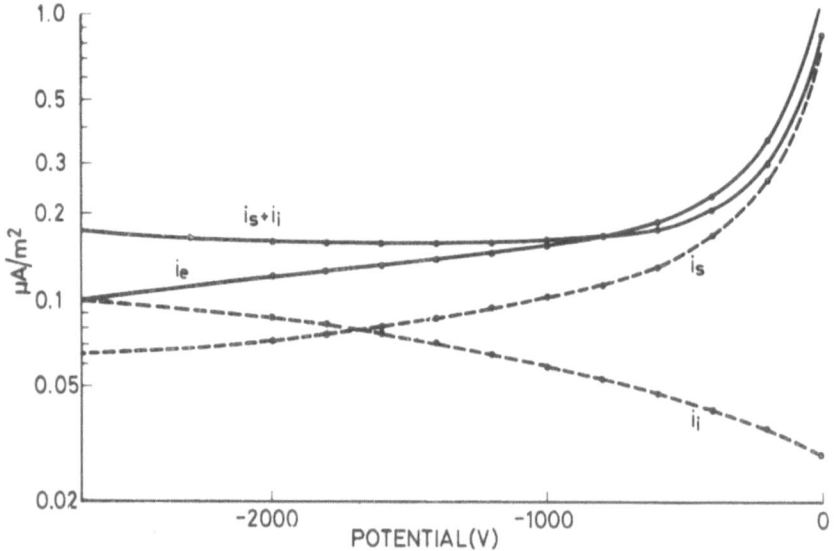

Fig. 5. Current-voltage characteristic of a gold surface when exposed to the particle environment at 6.6 R_E without solar irradiation.

4.3. FLOATING POTENTIAL

For simplicity we shall consider satellites and probes with spherical geometries. We first assume that the surface of these bodies is conducting. The case of an insulating surface is then discussed; this is of practical importance for spacecraft covered with solar cells and booms made of insulating material.

In the following treatment it is assumed that the space charge of photoelectrons escaping from the body (not to be confused with trapped photoelectrons) is negligible. This analysis is always valid for probes with radii of the order 0.1 m or less. In the case of larger objects, such as a satellite, the space charge of escaping photoelectrons plays a more complex role, which will be dealt with in Section 6.4.

4.3.1. *Conductive Sphere in Sunlight*

The effective photoemission area of a sphere is that of its cross-section πr_0^2, where r_0 is the radius of the sphere. It is assumed that photoyield does not change with photon angle of incidence, which leads to a small underestimate of photoemission. The maximum photocurrent density I_p is emitted when the body is at the plasma potential or negative with respect to the plasma, i.e. when all photoelectrons escape and none are forced to return to the body. The area for collection of ambient particles with an isotropic distribution is $4\pi r_0^2$, i.e. four times that for photoemission. This means that if the saturation photocurrent density I_p is four times the absolute value of saturation current density $|I_e + I_i|$ from the ambient plasma, the body is at plasma potential, and is negative for smaller values of I_p.

The ratio S between the area for photoemission and the total area of the surface will differ from 4 in the case of other geometries. A cylinder with a length equal to its diameter will be characterised by $S = 4.7$ when the direction of the Sun is perpendicular to its axis and $S = 3.8$ when the direction of the Sun makes an angle of $45°$ with its axis.

Figure 2 shows the current densities of electrons with energies between 50 eV and 50 keV at the geostationary orbit as a function of local time, after DeForest and McIlwain (1971). In the same figure $\frac{1}{4}$ of the magnitude of the saturation photoelectron current density is indicated by horizontal lines for various materials (Feuerbacher and Fitton, 1972; Grard, 1972). It is clear that most metals will be floating at positive potentials, whereas the potential of carbon materials may reach negative values during times of maximum electron fluxes. Vitreous carbon, in particular, is interesting as a probe material because its work function is very uniform (Trendelenburg *et al.*, 1970); as a consequence photoemission is also uniform over the entire surface of the probe; the importance of this feature for measurements on a spinning satellite will be discussed in Section 5.3.

The floating potential Vf is determined by equating to zero the sum of photo- and secondary electron, and plasma ion and electron currents $(i_p + i_s + i_i + i_e)$. In the magnetosphere i_s and i_i are normally at least one order of magnitude smaller than i_p and i_e, and can therefore be neglected. Figure 6a shows that there is only one potential V_f where the photocurrent i_p is balanced by the ambient electron current i_e. It should be noted that for the low voltages considered here i_e does not vary because it is constituted by electrons with a much higher energy than that of the photoelectrons.

In the solar wind the same body will float at a different potential, as demonstrated in Figure 6b. The photoemission current can be assumed to remain constant, since

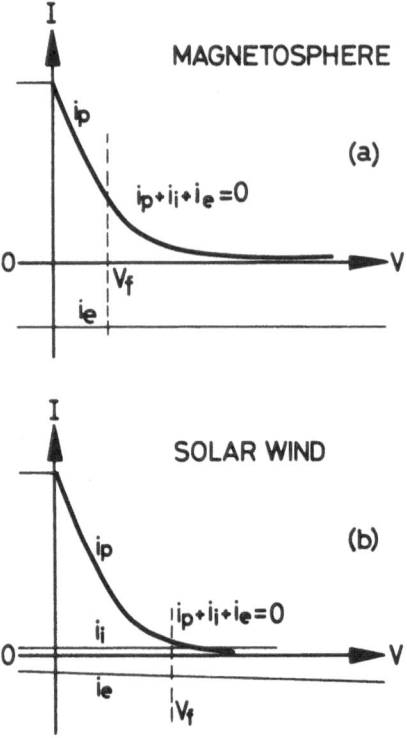

Fig. 6.   Determination of floating potential by equating photoelectron and plasma electron fluxes for: (a) Magnetosphere. (b) Solar wind.

we consider spacecraft orbiting the Earth. The ambient electron current is much smaller than any of those shown in Figure 2, and because their thermal energy is only of the order of 10 eV, the increase in ambient electron current with increasing positive bias is significant in the voltage range of interest. The positive ion current is not negligible because protons are streaming with the solar wind velocity relatively to the body. This ion current density is approximately $0.25 \ \mu A \ m^{-2}$, which is about half the electron saturation current density. Because the drift energy of the protons is of the order of 1 keV, the corresponding flux does not change over the voltage range shown in Figure 6b. For a given body, the floating potential V_f will be more positive in the solar wind than in the magnetosphere.

4.3.2. *Surfaces in Shadow*

In the absence of solar irradiation the floating potential reaches a negative value which is such that the ambient electron current i_e is exactly compensated by the secondary electron current i_s and the ambient ion current i_i; Figure 5 describes this situation for a gold surface under quiet conditions at the geostationary orbit. It is seen in this figure that the net current collected by the body is zero for a potential of approximately − 800 V. It should be noted that without the influence of secondary emission the floating potential would be at least 3 times this value.

The exact floating potential of a body in eclipse depends on both the secondary emission properties of its surface and the primary particle spectra. Table II contains computed values of the equilibrium potential for different types of surface and primary particle spectra. It can be concluded that surfaces with exceptionally high secondary emission yields, like e.g. activated BeCu, will stay at a small positive potential under practically all conditions encountered at the geostationary orbit. In the solar wind particle energies are too low to cause any significant secondary electron emission.

For satellites flying in sunlight the danger of having high voltage built up exists if insulating surfaces are exposed to space. The charge accumulated on these surfaces, when oriented away from the Sun, cannot be conducted to the front side and released as photocurrent. This means that the shadow side for a body in the magnetosphere could go to negative values of the order of 10 V in the solar wind and 1000 V in the magnetosphere. This situation, which will render electric field and low energy particle measurements useless, can be expected for a spacecraft with solar cells with glass cover made of silicon material. If the frames of the solar cells form a metallic network around the satellite, this will obviously reduce the asymmetry of the field around such a spacecraft. However, a better solution is to make use of a conductive coating on the spacecraft. A layer with a surface resistance of the order 10^5 Ω is sufficient in most situations to bring the voltage difference between Sun and shadow to less than 1 V. The technology to cover solar cells with a thin conductive coating has been recently developed (Köstlin and Atzei, 1973).

4.4. PROBE BIASING

From Figure 2 it can be seen that some of the probe materials which seem best suited for electric field experiments have such a low photoelectron yield that they may have a negative floating potential in the magnetosphere of the order of 1 kV. It would probably be impossible to carry out electric field measurements under these conditions. In principle secondary electron emission can help in this situation, if the yield is larger than one for an appreciable portion of the ambient electron population. Then the floating potential is maintained positive at a few volts because secondary electrons have an energy distribution similar to that of photoelectrons. The secondary electron yield for carbon materials is, unfortunately, less than unity for primary electron energies typical of the magnetosphere and of the solar wind.

The potential of the probe can, however, be biased to a positive value by means of a current source which drives an electron current i_b from the probe to the spacecraft (Figure 7a). The current balance of the spacecraft is achieved by increasing the ion current collected by the satellite body, and the photoelectron and secondary electron currents emitted from its surface. If the area of the conductive surface of the spacecraft is not sufficiently large, it is possible to insure the stability of the satellite potential by making use of an electron gun. The potential of the probe can be controlled by

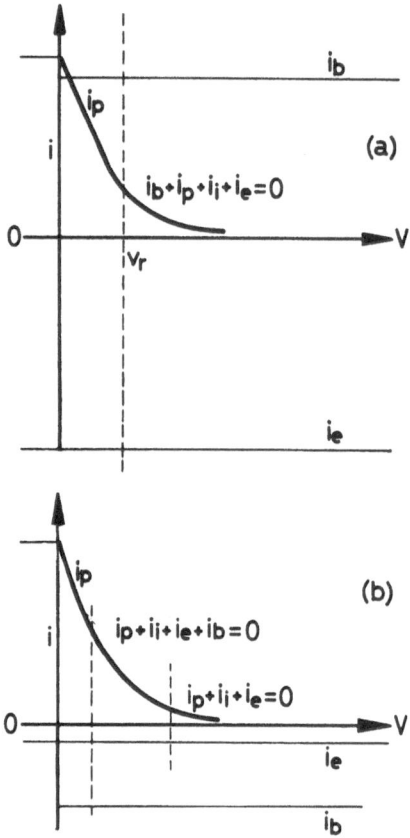

Fig. 7.   (a) Current i_b employed to bias a probe at a positive potential. (b) Current bias used to adjust the working point on the current voltage curve.

varying the intensity of the current i_b. The probe potential can be biased, for example, above the space and satellite potentials. When the ambient electron current i_e is very small, it is also possible to reduce the magnitude of the probe positive potential by reversing the polarity of the biasing current i_b, as shown in Figure 7b.

5. Symmetry Requirements for a Double Probe System

5.1. CONFIGURATION OF THE ANTENNA

The following discussion is relevant to a double probe system mounted on a spinning satellite as shown in Figure 8 and operated in sunlight. The electric field component perpendicular to the spin axis is measured as a spin modulated differential signal between the two sensors. In this chapter we will consider symmetry requirements for the sensors and the supporting booms. Influence resulting from asymmetry of the spacecraft will be dealt with in the next chapter.

The basis for such a discussion is the $I-V$ curve shown in Figure 6. The working

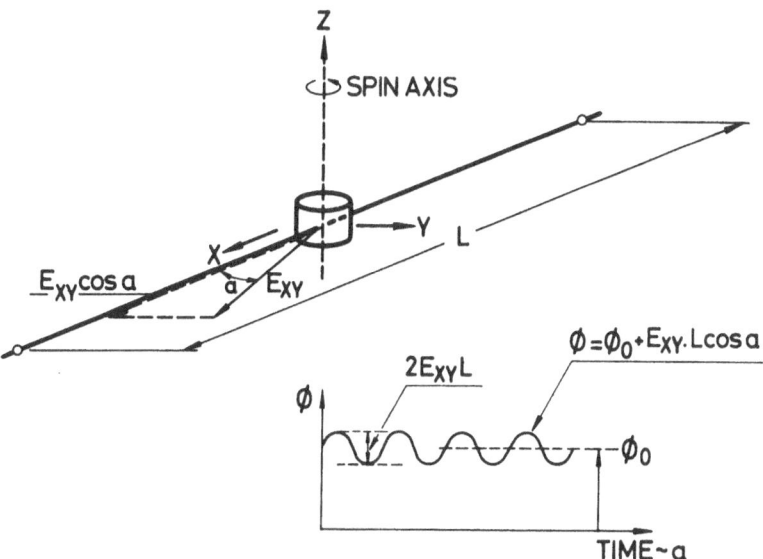

Fig. 8. Principle of measurement of DC electric field component in a plane perpendicular to satellite spin axis by modulation technique, which removes uncertainties about DC-offset between probes.

point on this $I-V$ curve is a function of the ambient electron and ion flux and of the photoelectron flux as shown in the previous paragraph. In order to minimize the effects due to discrepancies between the currents collected by the two probes, their potential should change as little as possible for a given current variation, i.e. dV/dI should be small. We assume that the ambient electron and ion energy is high and that the corresponding fluxes do not vary over the part of the $I-V$ curve which is of interest.

In this discussion it is considered that a sensitivity of 0.1 mV m^{-1} must be achieved, and it is assumed that the distance between the two probes is of the order of 50–100 m, which means that the smallest differences in potential to be measured are of the order of 5–10 mV.

5.2. SENSOR GEOMETRY

The ambient plasma current collection is proportional to the total area of the body and is not affected by its rotation, provided the energy distribution function of the particles is isotropic. However, the photoelectron current is proportional to the cross-section area of the body perpendicular to the Sun direction. Thus, the equilibrium potential of a probe varies with the satellite rotation since in general the cross-section area of the probe is a function of its attitude with respect to the Sun.

A change ΔI in one of the components involved in the current balance of the probe causes a variation of its floating potential given by

$$\Delta V = \Delta I \, dV/dI, \qquad (6)$$

or

$$\Delta V = V_0 \Delta I / i_p, \tag{7}$$

where

$$V_0 = i_p \, dV/dI \tag{8}$$

is practically independent of the probe potential and is approximately equal to 1 V for all materials, provided it can be assumed that the photoelectron energy distribution is Maxwellian (Grard, 1972).

Considering that a variation ΔA of the probe cross section A gives rise to a photoelectron current fluctuation $\Delta I = \Delta i_p$, Equation (6) becomes

$$\Delta V = V_0 \, \Delta i_p / i_p = V_0 \, \Delta A / A. \tag{9}$$

Cylindrical geometry is technically simple, but does not represent the optimum solution. A misalignment $\Delta \theta$ between two cylinders will cause a difference in floating potential.

$$\Delta V = V_0 \, \Delta \theta / \tan \theta, \tag{10}$$

where θ is the angle made by the axis of one of the cylinders and the Sun direction; ΔV is infinite for $\theta = 90°$, but is still equal to 17 mV for $\theta = 45°$ and $\Delta \theta = 1°$.

Spherical sensors do not give rise, in principle, to any interference signal at the spin frequency and are therefore to be preferred. A sphere with a radius $r_0 = 5$ cm has a typical resistance of the order of $10^9 \, \Omega$ in sunlight, and there is no major problem in designing an amplifier with a higher impedance in order to measure the differential signal with no attenuation. In order to keep the uncertainty in probe potential below 1 mV, it can be seen from (9) that the requirement to be fulfilled is $2\Delta r_0 / r_0 < 10^{-3}$; this means that for $r_0 = 5$ cm the probe must be manufactured with a nominal accuracy Δr_0 better than 25 μm, which is feasible.

5.3. UNIFORMITY OF PROBE SURFACE PROPERTIES

If the probes on a spinning double sphere antenna have non-uniform surface properties, the photo-emission from each sphere will be a function of the spin angle. Consequently the potential of the probes may vary during the spin period to the extent that the spin modulated differential signal between two probes caused by the ambient electric field is disturbed or lost. It is therefore important to assess this unwanted effect caused by non-uniform probe surfaces. In order to illuminate this problem, it is useful to outline the different behaviour of a double probe in the ionosphere and in the magnetosphere.

Figure 9a illustrates how an aerial, when operated in the ionosphere, is influenced by a difference in work function between the two probes. Here the photocurrent can be neglected with respect to the large fluxes of ambient electrons and ions. To single out the effect resulting from differences in work function, we assume that the electric field in the frame of reference of the double probe is zero. Because of the larger flux of ambient electrons, the surfaces of both probes are floating at negative potentials V_{S_1} and V_{S_2} with respect to the plasma. If we connect the probes to a voltmeter with an input impedance larger than that of the probe, a differential signal $\Delta V_F = V_{F_1} - V_{F_2}$

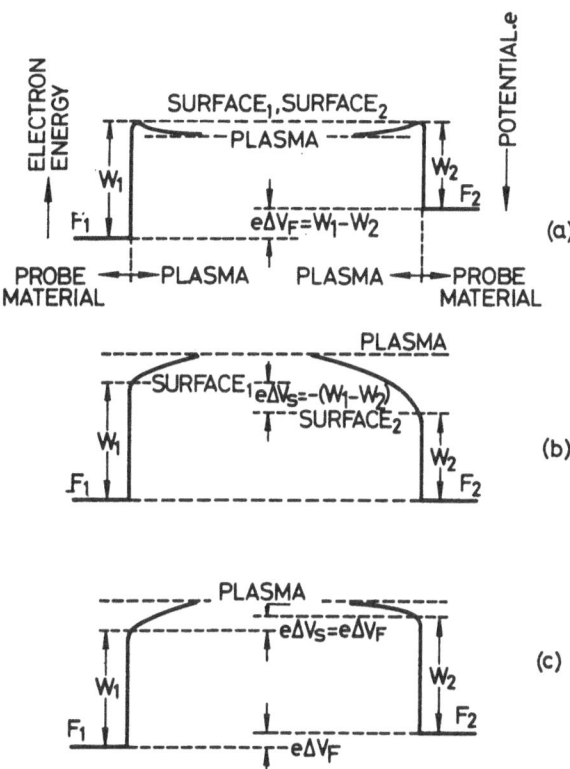

Fig. 9.   (a) The potential relationship between Fermi level, surface and plasma for a double probe in the ionosphere. The quantities F_1 and F_2 are the Fermi levels of probe No. 1 and No. 2 resp. A voltmeter between the probes will read $\Delta V_F = (W_1 - W_2)/e$. (b) The same diagram for a double probe in the magnetosphere or solar wind, in which case the surface is floating at a positive potential with respect to the plasma because photoemission dominates. Ideal probe material (escape function for photoelectrons unity for both probes) will cause no differential signal even if $W_1 \neq W_2$. (c) The same as (b), but both work functions are identical, and escape function for probe 2 is different from unity. This will give rise to differential signal $\Delta V_F = \Delta V_S$.

will be observed, because the voltmeter in effect measures the difference between the two Fermi levels V_{F_1} and V_{F_2} of the two probes. If the fluxes of ambient ions and electrons are isotropic, ΔV_F is a DC signal which does not vary with spin angle and does not disturb the spin modulated signal resulting from an ambient electric field.

If we consider the same double probe in the magnetosphere or in the solar wind, the situation is completely different because the flux of photo-electrons is larger than the flux of ambient electrons, which means that the probe surface has a positive floating potential. Figure 9b shows the relationship between plasma, surface and Fermi levels for a double probe. We assume again that the electric field in the frame of reference of the double probe is zero. Furthermore we consider that we have an isotropic flux of ambient electrons of relatively high energy. The ambient electrons are in balance with the fraction of the emitted photo-electrons, which have sufficient energy to escape from the surface; this current balance determines the floating potential and is illustrated in Figure 6. Figure 9b is meant to illustrate that in this case electrons escaping from the probe and going to infinity must overcome the work function plus the potential difference between the probe and the plasma. Considering that the photoelectron energy distribution is the same for the two probes and assuming that no energy is lost during bulk or surface transition, it follows that the potential difference between Fermi level and undisturbed plasma is the same for both probes. As a result no signal is measured with the voltmeter in the absence of ambient electric field, even when the two probes have different surface potentials ($V_{S_1} \neq V_{S_2}$).

In reality, the barrier that a photo-electron must pass before escaping from the surface of a probe is larger than the work function because of surface scattering, which depends on surface crystal structure and impurities. In Figure 9c we have illustrated this effect by taking two probes of equal work function W_1. Probe No. 1 is an ideal probe with no photoelectron scattering, whereas No. 2 has some scattering, i.e. the escape function for photoelectrons is less than unity. This leads to a differential signal on a voltmeter between the probes $\Delta V_F = \Delta V_S$.

We are only concerned about relative variations of photoemission during a spin period. Absolute differences of photoemission between the two probes will only give rise to known DC offsets, which can be subtracted from the measurements. This means that in terms of surface uniformities only variations in the integral of the escape function over the surface of the probe, illuminated by the Sun, are important.

After investigation of several probe materials it has been concluded that vitreous carbon has very favourable surface characteristics (Trendelenburg *et al.*, 1970). Vitreous carbon is a hard black glass-like material obtained by thermal degradation of polymers at 1800°C. Its density is 1.47 g cm^{-3} and its electric resistivity is 5×10^{-3} Ω cm. The point-to-point relative variation of the work function is only ± 3 mV compared to ± 50 mV for metals such as aluminium. The unusual properties of vitreous carbon are due to its glass-like surface without crystal structure and its low acceptance of surface contaminants; the latter strongly influences the escape function. Vitreous carbon can be exposed to air and still keep the same surface properties. Measurements made with incident photon energy approximately 1 eV above the photoelectron cut-off, have shown a maximum variability of the photoyield of 8% over a typical scale of 1 mm or less (Trendelenburg *et al.*, 1970). Integrated over the surface this may give rise to fluctuations of probe potential of a few mV during a spin period. This variation, which must be caused by non-uniformity of the escape function, seems to set the limit for the sensitivity of a double probe. However, more exact measurements of the variability of the photoemission over surfaces must be made before a quantitative judgement can be made. This work is currently being pursued.

5.4. BOOMS AND GUARD CYLINDERS

The booms on which the probes are mounted will cause a difference between the shadowed areas on the two spheres. This effect can be minimized by mounting the probes as shown in Figure 10. We assume that the cylindrical booms protrude over a length equal to twice the sphere diameter and that the Sun shines perpendicularly to the satellite spin axis; a differential signal will be observed when the angle between the satellite-Sun line and the boom axis is less than 11°, because one probe is shadowed by the full boom length and the other only by a boom tip.

Decreasing the radius of the boom is an advantage, but it seems difficult to reduce the boom shadow area to less than 4% of the probe cross-section area. The resulting spikes observed in the differential signal between the two probes will occur twice per rotation; they will have peak values of approximately 40 mV, and will interfere with the measurements for 22° of each half rotation. This effect can be reduced if the protruding stubs are lengthened or if the spin axis is tilted by a few degrees. In fact, a tilt of 11° eliminates this effect completely since the boom shadows are then the same on both spheres.

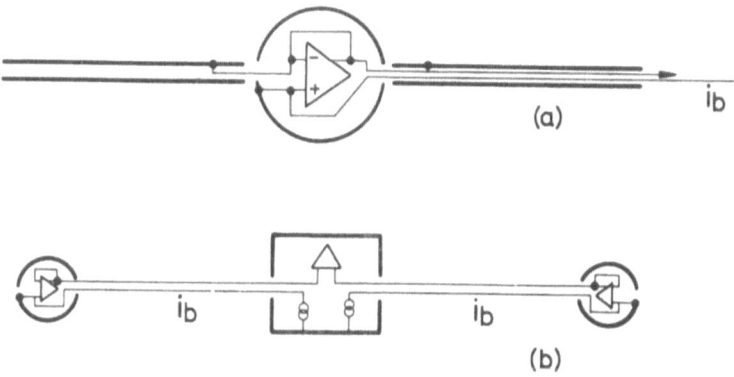

Fig. 10. (a) Probe configuration planned for a future experiment on the ESRO GEOS satellite. Probe and cylindrical guard rings are made of vitreous carbon. Preamplifiers are mounted inside spherical probes. (b) System for injecting bias current into probes.

The shadow of the satellite body on one of the probes will also give rise to strong differential signal occurring twice per rotation. If the length of each boom is 40 m and the radius of the spacecraft is 1 m, a tilt of the spin axis of the order of 1.5° will suppress this effect. If interferences due to shadows cannot be avoided, it is preferable to chop the resulting spikes by means of an on-board electronic system before presenting the signal to the telemetry.

In order to insure that current loops between probe and boom do not give rise to differential signals, it is recommended to mount cylindrical elements made of the same material as that of the probe on both sides of the spherical sensor. If these guard cylinders are held at the same potential as that of the sphere, by means of a bootstrapping circuit, they will prevent currents between the probe and the boom. A system making use of a preamplifier located inside the sphere and providing the bootstrapping for the guard cylinders is presented in Figure 10.

6. Asymmetries Caused by the Presence of Spacecraft

6.1. SURFACE POTENTIAL OF THE SATELLITE

It has already been shown in Chapter 4 that the use of insulating surfaces must be avoided, as these may charge up to highly negative potential in the absence of photo-emission. Similarly, biased probes or electrodes on the main satellite body must be arranged symmetrically with respect to the electric field sensors. The reason for this requirement is that the Debye length in the magnetosphere and interplanetary space is so large that the fields produced by biased surfaces fall slowly with distance from the satellite; in other words, plasma screening is not as effective as in the dense plasma of the ionosphere. This symmetry requirement, however, is not difficult to meet; for example, a sphere of 10 cm diam placed 2 m away from the satellite body will cause a differential signal of less than 0.1 mV m^{-1} if the booms are longer than 20 m.

6.2. TRAPPED PHOTOELECTRON CLOUD

Soop (1972) has carried out a computer simulation programme for a conductive sphere illuminated from one side in vacuum, in which case no photoelectrons leave the cloud after equilibrium has been reached; in the presence of plasma, however, the inward flux of ambient particles must be balanced by an equal flux of photoelectrons escaping from the cloud. Therefore, the vacuum case corresponds to an upper limit for the trapped photoelectron density. This study has demonstrated that photoelectrons tend to orbit to the shadow side of the sphere, and only a small asymmetry is caused by the excess of charge existing on the illuminated side.

The differential signal due to the asymmetry of the photoelectron cloud around conductive sphere of radius $r_s = 0.75$ m in vacuum can be described with the following example taken from the paper of Soop (1972):

$$\Delta V = 0.6 \, (r_s/r)^2 \text{ V}, \tag{11}$$

where r is radial distance from centre of sphere; the floating potential was assumed to be 4.3 V and the saturation photoelectron current density was taken equal to 4×10^{-9} A cm^{-2}.

The strength of the interference corresponding to the model given by (11) is illustrated in Figure 11 by curve (a); this can be compared with the amplitude of signals produced by ambient electric fields of 0.1 and 1 mV m^{-1}. The distance of the probes from the centre of the spacecraft is shown along the horizontal axis.

6.3. DIFFERENTIAL CURRENT PRODUCED BY ESCAPING PHOTOELECTRONS

The plasma current impinging on the satellite surface releases an equal amount of photoelectron current from the stationary photocloud. While the plasma current to the satellite is isotropic, the released photocurrent is anisotropic with a maximum in the direction of the Sun. We assume that the satellite has a spherical shape; the direction from the centre of this sphere to the Sun is the axis of symmetry of this system. Furthermore, the considerations concerning the escaping electrons are confined to distances r large with respect to the radius of the satellite r_s, so that the satellite can be seen as a point source. Under these circumstances we assume the following angular flux model for the outgoing electron beam:

$$j(\beta) = \frac{k+1}{2^k} \frac{i_{ps}}{4\pi} (1 + \cos \beta)^k,$$ (12)

where the angle β is measured with respect to the symmetry axis of the system oriented toward the Sun; i_{ps} is the current escaping from the photoelectron cloud which surrounds the spacecraft. The parameter k characterizes the focalisation of the electron flux in the direction of the Sun; it can be checked that for $k=0$, which corresponds to the isotropic case, the angular flux is independent of β and equal to $i_{ps}/4\pi$.

The fraction of the satellite photoelectron current collected by a small probe of radius r_0 located at a distance r from the satellite centre is given by:

$$\Delta I = j(\beta) \Delta \Omega,$$ (13)

where

$$\Delta \Omega = \pi r_0^2/r^2,$$ (14)

is the solid angle under which the probe is seen from the spacecraft.

We assume that the flux of photoelectrons escaping from the cloud surrounding a spherical body at floating potential is equal to the incoming flux of ambient electrons. and we consider furthermore that the latter is proportional to the area of this body Combining Equations (12)–(14), it is then seen that the photoelectron current flowing from the spacecraft to the probe, ΔI, can be expressed as a function of the photoelectron current emitted by the probe i_p by the following relation:

$$\Delta I = \frac{k+1}{2^{k+2}} \frac{r_s^2}{r^2} i_p (1 + \cos \beta)^k.$$ (15)

Considering now an aerial made of two probes located at a distance r from the satellite centre, but in opposite direction, it is found by inserting (15) into (7) that for $k \neq 0$ the measured differential signal has a maximum,

$$\Delta V = \frac{k+1}{4} \frac{r_s^2}{r^2} V_0,$$ (16)

when the antenna is parallel to the Sun direction.

It can be noted that, whereas the angular flux model given by (15) is strongly influenced by the choice of the focalisation parameter, the maximum of the interference signal is not much affected by the value of k. For example, replacing $k=3$ by $k=7$ only changes the magnitude of ΔV by a factor of 2. The interference caused by the collection of satellite photoelectrons by the probes, represented by curve (b) in Figure 11, is plotted as a function of distance for $k=3$, $r_s=1$ m and $V_0=1$ V.

The combined effect of the interferences given by (11) and (16) is illustrated by curve (c) in Figure 11. It is seen that the level of the signal corresponding to an electric

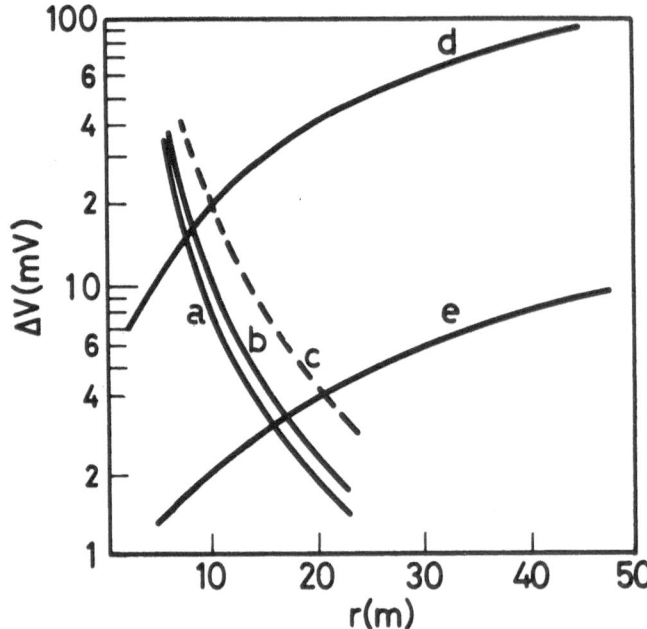

Fig. 11. Interference signals obtained from double probe measurement as function of boom length due to: (a) Photoelectrons trapped in the vicinity of the spacecraft. (b) Current due to collection of photoelectrons escaping from the spacecraft. (c) Sum of (a) and (b). Indicated in the same diagram are amplitude of signals corresponding to (d) ambient electric field of 1 mV m⁻¹, (e) ambient electric field of 0.1 mV m⁻¹.

field strength of 0.1 mV m^{-1} exceeds that of these interferences, provided the booms are longer than 20 m.

6.4. SPACE CHARGE EFFECT PRODUCED BY ESCAPING PHOTOELECTRONS

In the preceding chapter the influence of escaping photoelectrons was calculated in terms of an additional current to the probe, which is in the sunward direction from the satellite. In equilibrium the flux of photoelectrons leaving the spacecraft is approximately equal to that of the incoming ambient electrons; it follows that the density of the escaping photoelectrons near the spacecraft is necessarily larger than that of the surrounding plasma. Therefore the escaping photoelectrons represent a space charge which may influence the potential of the probes. A full treatment of this problem is beyond the scope of this paper; however, we shall make use of simplified models in order to evaluate the magnitude of this effect.

We can, for example, consider in a first approach that the escaping photoelectrons travel radially away from the satellite and that they can be represented by a number of radial line currents from the sunlit portion of the spacecraft; the extent of this space charge is assumed to be given by the plasma Debye length, λ_D. For high ambient electron current densities, of the order of 10^{-5} A m^{-2}, the density of escaping photoelectrons can reach a value of 100 cm^{-3} at the surface of the spacecraft. It is easily

found that such a space charge would create a voltage drop of 1 V across a slab 1 m
thick. As photoelectrons escape with energies of the order 1 eV, their orbits are bent
in such a way that these particles tend to diffuse with spherical symmetry as shown in
Figure 12b and c.

The potential profile corresponding to the case of spherical symmetry is schematically shown in Figure 13. Depending upon the values of the parameters which describe
the model, the spacecraft may be floating at a positive or negative potential. The potential profile may be monotonous or, on the contrary, present a minimum; the
existence of this minimum has been demonstrated by Guernsey and Fu (1970), and
Schröder (1973) who have performed self-consistent treatments of this problem.
In practice, the potential minimum adjusts itself so that the correct number of photoelectrons can escape to balance the flux of ambient particles.

The potential profile around a body is represented by curve (a) of Figure 13 when
the space charge of the photoelectrons escaping from its own surface is negligible.

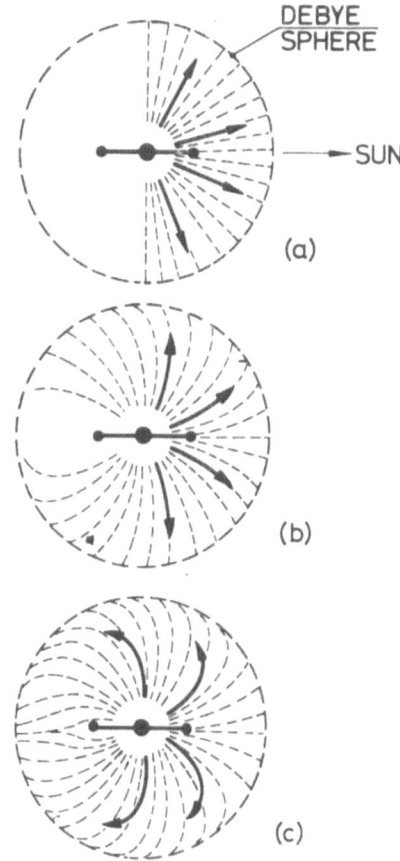

Fig. 12.   Models for the spatial distribution of photoelectrons escaping from the spacecraft: (a)
Radial line currents. (b) Deflection of trajectories due to space charge effects.
(c) Spherically symmetric model.

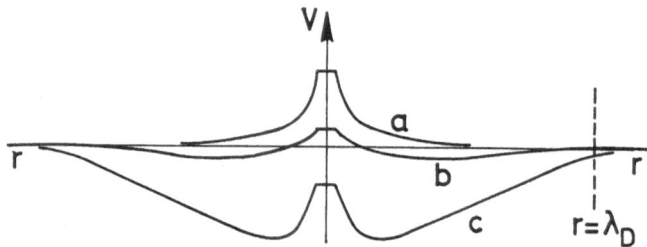

Fig. 13.   Effect of the space charge of the escaping photoelectrons on the potential distribution in the vicinity of a spherical body.

It can be shown that this is the case for a body of radius r_0 when the following condition is fulfilled:

$$r_0 \ll \lambda_D (V_e/V_a)^{3/4}, \tag{17}$$

where eV_e and eV_a are the mean kinetic energy of the photo- and ambient electrons respectively. Numerical applications yield $r_0 \ll 0.5$ m in the outer magnetosphere and $r_0 \ll 1.5$ m in the solar wind. The space charge effect of the escaping photoelectrons is therefore negligible for small probes, but could be of importance for larger bodies such as satellites.

Another feature of the escaping photoelectrons is illustrated in Figure 14; these particles tend to escape along the magnetic field lines. This effect should be particularly marked in the outer magnetosphere where the gyroradius of 1 eV electrons is of the order of 10–30 m.

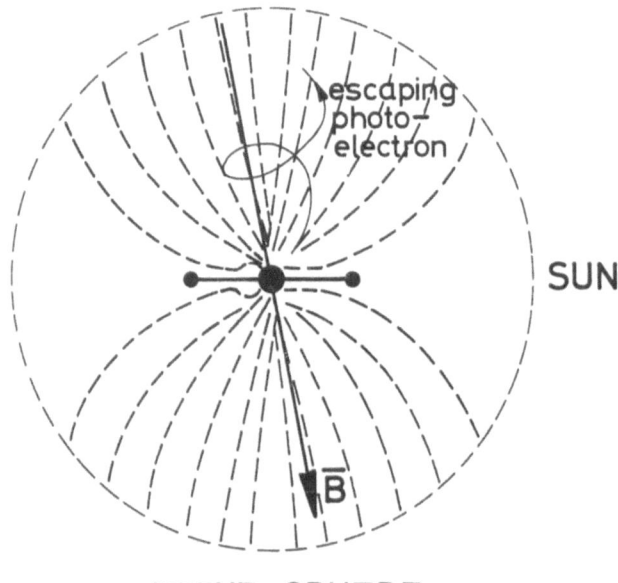

Fig. 14.   Influence of an ambient magnetic field on the spatial distribution of the escaping photoelectrons.

It is therefore extremely difficult to assess analytically the degree of symmetry of the escaping photoelectron flux, but this estimation might be done with the help of a computer. This problem is currently under investigation, but for the time being one is reduced to making assumptions on the angular distribution of this flux.

Using, for example, the model given by (12) with $k=1$ and considering a spacecraft with a radius of 1 m, it can be found that for typical solar wind conditions (ambient plasma density of 1 cm^{-3}, mean kinetic energy of 10 eV, and Debye length of 25 m) DC electric fields can be measured with an accuracy of 0.3 mV m^{-1} using probes mounted on booms only 10 m long.

In the outer magnetosphere, however, the ambient plasma energy is much higher. The flow of escaping photoelectrons is correspondingly increased, and the associated space charge extends at much larger distances, since the Debye length reaches a value of 250 m for typical ambient electron density and energy of 1 cm^{-3} and 1 keV, respectively. Assuming that the flux of escaping photoelectrons through the sunward hemisphere exceeds by only 10% the flux through the anti sunward hemisphere, it is found that a differential signal of the order of 100 mV is measured between two probes located 50 m away from a spacecraft of radius 1 m. It is therefore questionable whether fields of 1 mV m^{-1} can be measured in the magnetosphere, in particular during substorms with a boom length of 50 to 100 m. Increasing the boom length in order to substract the probes from the influence of the spacecraft in the magnetosphere does not seem very realistic; each boom should then have a length of several λ_D, i.e. 500 to 1000 m.

Naturally, one should always adopt booms as long as technically feasible (at least 50 m); however, in addition, it is necessary to use a gun which can emit electrons at high energy (1 keV) along the magnetic field lines. The energy and flux of these electrons must be accurately controlled by the experimenter, in order that the current contained in this beam exactly balances the incoming current of ambient particles. Strictly speaking, the electrons emitted by the gun will still give rise to a differential signal between the probes if the spin axis of the spacecraft is not oriented along the magnetic field; this interference, however, will be much less than for the case of freely escaping photoelectrons, the difference being that electrons emitted by the gun represent a smaller space charge because of their higher energy.

7. Conclusion

The optimization of DC electric field measurements made with a spinning satellite, using the double sphere technique, depends critically on five basic requirements, viz.:

(a) The sensing probes must be spherical.
(b) These probes must have uniform surface properties.
(c) The satellite surface must be conducting.
(d) The length of the booms must be sufficient.
(e) The spacecraft must be fitted with an electron gun.

Boom alignment considerations lead to requirement (a). Misalignments of the order of 1°, which are unavoidable in any conceivable long boom technology, are sufficient to produce error signals between two cylindrical probes which exceed the values required for electric field measurements inside the magnetosphere.

Concerning requirement (b), it is important that the probes have uniform photo-emission characteristics so that the floating potential of a probe does not change as a function of the spin angle and in this way give rise to an erroneous spin modulation signal.

Requirement (c) must be fulfilled in order to avoid any asymmetric potential distribution on the satellite surface, in particular that caused by electrostatic charging due to high energy plasma particles. These particles determine the surface potential of the dark side of the satellite if its surface is insulating. If the surface is conductive, these electrons are conducted to the illuminated side where they are released as photoelectrons.

As mentioned under requirement (d), the booms must be as long as possible. Whereas a probe separation of 20 m may be perfectly adequate for measurements in the solar wind, a boom length of 50 to 100 m is a bare minimum in the outer magnetosphere.

It has been shown, however, that it is possible to reduce the space charge effect of the photoelectrons escaping from the spacecraft by using a gun emitting a controlled flux of high energy electrons. This final requirement is especially important for measurements in the outer magnetosphere where the Debye length always exceeds the dimension of booms which are at present technically feasible.

References

Axford, W. I.: 1969, *Rev. Geophys.* **7**, 421.
Carpenter, D. L.: 1966, *J. Geophys. Res.* **71**, 693.
Cauffman, D. and Gurnett, D.: 1971, *Space Sci. Rev.* **13**, 369.
DeForest, S. E. and McIlwain, C. E.: 1971, *J. Geophys. Res.* **76**, 3587.
DeForest, S. E.: 1972, *J. Geophys. Res.* **77**, 651.
Fahleson, U.: 1967, *Space Sci. Rev.* **7**, 238.
Fahleson, U., Fälthammar, C.-G., Pedersen, A., Knott, K., Brommundt, G., Schumann, G., Haerendel, G., and Rieger, E.: 1971, *Radio Sci.* **6**, No. 2, 233.
Fahleson, U.: 1971, *Proc. of the Colloquium on the ESRO Geostationary Magnetospheric Satellite*, ESRO SP-60, 249.
Feuerbacher, B. and Fitton, B.: 1972, *J. Appl. Phys.* **43**, 1563.
Gibbons, D. J.: 1966, in A. H. Beck (ed.), *Handbook of Vacuum Physics*, Vol. 2, Pergamon Press.
Grard, R. J. L.: 1972, ESTEC Internal Working Paper No. 663 (to be puslished in *J. Geophys. Res.*).
Guernsey, R. L. and Fu, J. H. M.: 1970, *J. Geophys. Res.* **75**, 3193.
Hachenberg, O. and Braner, W.: 1969, *Adv. Electronics Electron Phys.* **11**, 413.
Hundhausen, A. J.: 1970, *Rev. Geophys.* **8**, 729.
Knott, K.: 1972, *Planetary Space Sci.* **20**, 1137.
Köstlin, H. and Atzei, A.: 1973, this volume, p. 333.
Maynard, N. and Heppner, J.: 1970, in B. McCormac (ed.), *Particles and Fields in the Magnetosphere*. Reinhold Book Co., p. 247.
Mozer, F. S. and Manka, R. H.: 1971, *J. Geophys. Res.* **76**, 1697.
Schröder, H.: 1973, this volume, p. 51.
Shield, M. A. and Frank, L. A.: 1970, *J. Geophys. Res.* **75**, 5401.
Soop, M.: 1972, *Planetary Space Sci.* **20**, 859.

Sternglass, E. J.: 1950, *Phys. Rev.* **80**, 925.
Trendelenburg, E. A., Fitton, B., Page, D. E., and Pedersen, A.: 1970, *ELDO/ESRO Scient. and Techn. Rev.* **2**, 1.
Walbridge, E.: 1971, High Altitude Observatory, NCAR, Boulder, Colorado, Internal Report.
Whipple, E. C., Jr.: 1972, 'Validity of Low Energy Particle Collection Techniques in the Magnetosphere', presented at the 17th Gen. Assembly of URSI, Warsaw.

DISCUSSION

Wrenn: There is much discussion about the reason for discrepancies between Langmuir probes and incoherent scatter radar measurements. Carlson and Sayers propose an explanation in terms of a drift of work function of probe surfaces; can you comment on the effect of such drifts on your technique?

Knott: The effect is due to contamination, and it appears to be negligible for vitreous carbon surfaces.

Wynn-Roberts: What choice of boom surface material will help to minimise the interaction between probe or boom?

Knott: The considerations are similar to those for the satellite, i.e. an insulating surface is undesirable because this would cause charge build-up. On the other hand, a conducting boom surface is also undesirable since this would cause a short circuit. Therefore, the best solution is a 'sensitive' boom surface, e.g. a semiconducting surface or series of conducting rings.

2.4. PARTICLE ENERGY DISTRIBUTIONS AROUND SPACECRAFT

CHARGED PARTICLE DISTRIBUTION IN THE NEAREST VICINITY OF IONOSPHERIC SATELLITES– COMPARISON OF THE MAIN RESULTS FROM THE ARIEL I, EXPLORER 31 AND GEMINI-AGENA 10 SPACECRAFT

URI SAMIR

*Space Physics Research Laboratory, University of Michigan, Ann Arbor, Mich., U.S.A.
and Dept. of Environmental Sciences, Tel-Aviv University, Ramat-Aviv, Israel*

Abstract. Electron and ion measurements performed by probes mounted on the Ariel I, the Explorer 31 and the Gemini-Agena 10 spacecraft which reflect on some aspects of the interaction between a spacecraft and its ionospheric plasma are compared. A partial picture of charged particle distributions around surfaces in space is thereby obtained. Results from the above satellites are compared among themselves and the degree of agreement/disagreement is demonstrated. Comparisons of results with some wake models are included, and the degree of agreement is discussed.

It is found that at the closest vicinity to the spacecraft surface and for space plasmas with high hydrogen concentration (H^+) Explorer 31 results agree with the Gurevich *et al.* wake model. For situations were O^+ is the major ionic constituent the agreement excludes the entire region of maximum rarefaction. Comparison of results from the spherical ion probe on the Explorer 31 satellite indicate that plasma neutrality in the wake region is reached at fairly close distances ($r \simeq 2R_0$) for conditions that prevail in the ionosphere at altitudes of 400–700 km. Comparing Ariel I and Explorer 31 results for the electron angular distribution close to the surface of the satellite with results from the Agena 10 maneuver shows good agreement. This is not in line with *a priori* expectations. The degree of agreement between the Explorer 31 results for the electron angular distribution in the closest vicinity to the satellite surface shows reasonable agreement with the Liu-Jew wake model. This is in a way surprising due to shortcomings of the theory and limitations of the comparison procedure itself. Results of both ion electron fluxes from the Explorer 31 satellite yield quantitatively the net current around the satellite. Wakes created by bodies of different size and surface potentials are quantitatively examined.

1. Introduction

Perturbations to the ambient plasma caused by rapidly moving spacecraft in the ionosphere have been studied theoretically since the beginning of the space era (Jastrow and Pearse, 1957). However, only with the launch of the Explorer 8 and Ariel I satellite were there attempts to analyze data in order to gain some experimental information relevant to problems involved in the interaction between a spacecraft and its natural environment. It is well known that the interaction is mutual and the phenomena involved are coupled due to effects on both the spacecraft and the plasma in the spacecraft vicinity. The potential of the spacecraft is subject to charging mechanisms such as photoemission, magnetic field line guiding, energetic particle bombardment, particle emissions etc. The mathematical treatment is quite complicated because the effects involved are coupled. Resorting to assumptions which simplify the mathematics often bring the treatment outside the realms of physical reality and practical interest.

The experimental *in situ* information available at present is fragmentary in

R. J. L. Grard (ed.), Photon and Particle Interactions with Surfaces in Space, 193–219. All Rights Reserved

nature and no complete and comprehensive description can be given to particle concentrations, (N_e and N_+) potential (ϕ) and charged particle temperature (T). However, some of the Explorer 31 and Gemini-Agena 10 data were analyzed having the above problems in mind.

The purpose of this paper is to review the main available results and demonstrate the degree of their agreement with some theoretical models and to indicate directions for future experimental and phenomenological studies. No detailed discussion will be given regarding experimental methods and techniques. The brief experimental section presented here is intended to show mainly the spacecraft configurations, probe locations and experimental objectives. More experimental details are given in *Proc. IEEE* **57**(6) (1969). Since the results treated herein are mainly from the Ariel I and the Explorer 31 satellites (i.e., between 400 and 3000 km) the discussion is restricted to the flow regime that satisfies ((1)–(3)):

$$V_T(+) \leqslant V_S \ll V_T(e) \tag{1}$$

$$\lambda_D \leqslant R_0 \ll L_{e,\,+} \tag{2}$$

$$R_L(e) \ll R_0 < R_L(+), \tag{3}$$

where: $V_T(+)$ and $V_T(e)$ are the thermal velocities of ions and electrons respectively, V_S is the spacecraft velocity, $L_{e,\,+}$ is the electron-ion mean free path, R_0 is the spacecraft radius which for practical purposes implies that an effective radius (R_0) is assigned to the satellite cross-section and any radial distance from the center of the body is defined with respect to this dimension, λ_D is the Debye length. $R_L(+)$ and $R_L(e)$ are the ion and electron Larmor radii respectively. Table I gives some relevant plasma parameters for the altitude range 400–3000 km.

TABLE I

Some relevant plasma parameters in the altitude range 400–3000 km

Altitude (in km)	$\left\lvert\dfrac{N_{e,+}}{N_n}\right\rvert$	$\left\lvert\dfrac{V_s}{V_T(+)}\right\rvert$	$\left\lvert\dfrac{V_T(e)}{V_s}\right\rvert$	$\left\lvert\dfrac{L_{e,+}}{R_0}\right\rvert$	$\left\lvert\dfrac{\lambda_D}{R_0}\right\rvert$	$\left\lvert\dfrac{R_L(+)}{R_0}\right\rvert$	$\left\lvert\dfrac{R_L(e)}{R_0}\right\rvert$
400	5×10^{-3}	~ 6	~ 37	10^3	4×10^{-3}	~ 5	$\sim 4 \times 10^{-2}$
800	5×10^{-2}	~ 4	~ 37	10^4	$\sim 10^{-2}$	~ 7	$\sim 5 \times 10^{-2}$
1200	10^{-1}	≈ 3	~ 37	10^5	4×10^{-2}	~ 8	$\sim 6 \times 10^{-2}$
3000	≈ 1	≈ 1	50	$\sim 3 \times 10^5$	6×10^{-2}	~ 8	$\sim 1 \times 10^{-1}$

The values are calculated considering: $V_s = 8 \times 10^5$ cm s^{-1}.
N_n = density of neutral particles.

2. Experimental (The Probes and Scope of Measurements)

Figure 1 is a diagrammatic representation of the three spacecraft showing the probe location on each spacecraft. Table II gives details with regard to these spacecraft.

As seen from Figure 1 and Table II both the Ariel I and the Explorer 31 included as part of their scientific instrumentation planar electron probes mounted flush on the

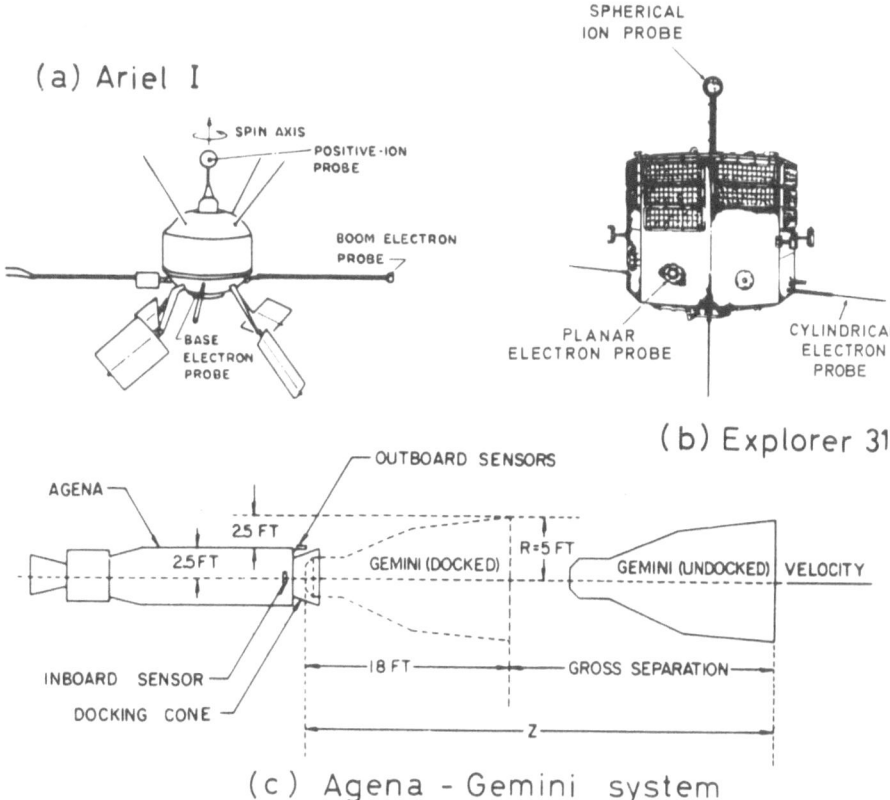

Fig. 1. Diagrammatic representation of the three spacecraft showing the spacecraft geometry and probe location.

surfaces of the satellites. Each such probe consists of a circular disc (acting as a current collector) surrounded by a guard ring to reduce sheath edge effects. The Ariel I had also a similar probe mounted on a boom of length $L=5\,R_0$. Ariel-I and Explorer 31 had also spherical ion probes (see Figure 1 (a), (b)) mounted on short stems on the spin axes.

The spherical probe consists of a spherical collector surrounded by a concentric spherical grid whose potential is adjusted to prevent electrons reaching the probe. Results from a planar multigrid retarding potential analyzer (Donley in *Proc. IEEE* **57** (6), 1969) have also been used. Although not seen clearly in Figure 1 (b) its location on the surface is similar to that of the planar electron probe mentioned above. The data from an ion mass spectrometer (Hoffman in *Proc. IEEE* **57** (6), 1969) has not yet been analyzed for the purposes of this study. Some preliminary results from this probe were published in Hoffman (1967), and will not be further discussed in the present paper. The Explorer 31 carried also two cylindrical probes (Brace and Findlay in *Proc. IEEE* **57** (6), 1969). These protrude radially 46 cm from the satellite surface and thus yield electron data at a distance approximately $2\,R_0$ from the center of the satel-

TABLE II

Some details on the Ariel I, Explorer 31 and Gemini-Agena 10 spacecraft

Item	Ariel I	Explorer 31	Gemini-Agena 10 system
Date of launch	April 1962	November 1965	July 1966
Apogee (km)	1200	3000	1400
Perigee (km)	400	500	300[a]
Inclination (°)	54	80	29
Satellite diameter (cm)	≈ 60	≈ 70	See Figure 1(c)
Experimental technique/method	Retarding potential Druyvesteyn-Modulation First and second derivatives.	Retarding potential (i) Langmuir (ii) Druyvesteyn RPA (multigrid)	Multigrid collectors, RPA Langmuir
Probe location	On satellite surface and two booms of different length.	On satellite surface and on a boom (one probe).	On the surface of the Agena
Measured quantities	N_e; N_+; T_e; T_+; ϕ_s; M_+	N_e; N_+; T_e; T_+; ϕ_s; M_+	$I_e(N_e)$, $I_+(N_+)$, T_e, ϕ_s

Suffixes (e) and ($+$) refer to electrons and ions respectively:

N = number density (concentration);
T = temperature;
M_+ = ionic mass;
ϕ_s = satellite potential.
[a] The data presented in this paper was taken at an altitude of ≈ 400 km.

lite. From measurements of the above probes on both satellites we were able to obtain information regarding $I_e^*(N_+^*)=f(r=R_0, 2 R_0, 5 R_0; \theta)$, $\phi=f(r=R_0, \theta)$, $T_e=f(r=R_0, \theta)$ and to a lesser extent $I_+^*(N_+^*)=f(r=R_0, 2 R_0; \theta)$.

The Gemini-Agena 10 spacecraft was a two-body system, consisting of the Gemini manned capsule and the unmanned Agena target vehicle. The probes on the Agena-Gemini system were planar multigrid collectors located on the Agena docking cone adaptor (see Figure 1 (c)). The two outboard probes, one electron and one ion were mounted adjacent to the Agena skin with their aperture normals at right angles to the Agena longitudinal axis. The inboard ion probe was mounted looking out the docking cone along the Agena axis. The Gemini-Agena system is shown in Figure 1 (c) in a docking/undocking configuration. More experimental details are given by Medved (1969) and Troy (1969).

2.1. EXPERIMENTAL RESULTS

2.1.1. *From the Ariel I Satellite* (Samir and Willmore, 1965, 1966; Henderson and Samir, 1967)

The results presented in this paper are of the electron and ion angular distributions only. The results for electron temperature and plasma potential variations with angle of attack are not included.

(1) Figure 2 shows the variation of the electron current ratio $[I_e(\text{base})/I_e(\text{boom})]$

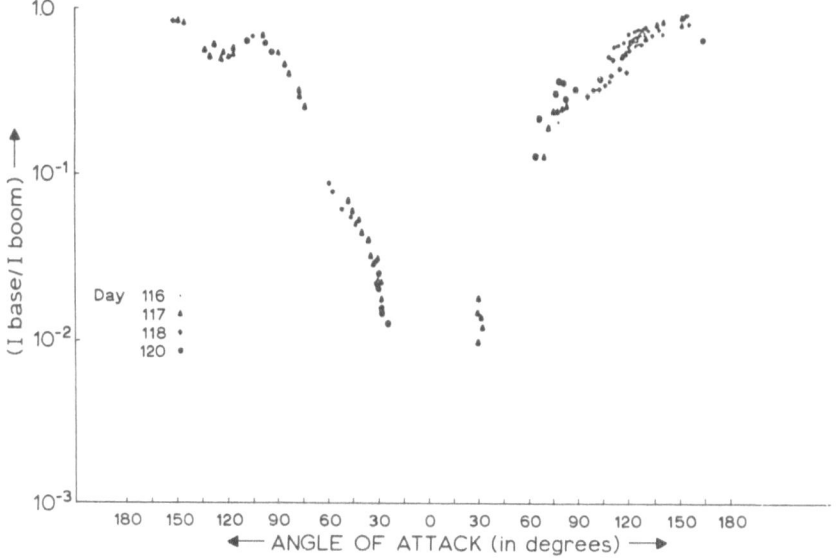

Fig. 2.   Variation of the electron current ratio $[I_e(\text{base})/I_e(\text{boom})]$ with angle of attack (θ). Ariel I data from days 116, 117, 118, 120 in 1962.

* The asterisk indicates normalized currents and densities.

with angle of attack (θ) at a distance $r = R_0$ (i.e. closest vicinity to the satellite surface). It should be noticed that the normal to the boom probe surface was parallel to the spin axis and anti-parallel to the normal to the base probe (see also Figure 1 (a)). The data points on this plot made use of measurements from four days (days 116, 117, 118, 120 in 1962), in the altitude range 400–700 km, where O^+ is the major ion constituent. It is seen that the wake zone is significantly depleted of electrons, the current at $[\theta(\text{wake})]$ being 10^{-2} times the current collected by the boom probe.

Figure 2 shows also that there is no enhancement ahead of the body ($\theta \rightarrow 180°$) and

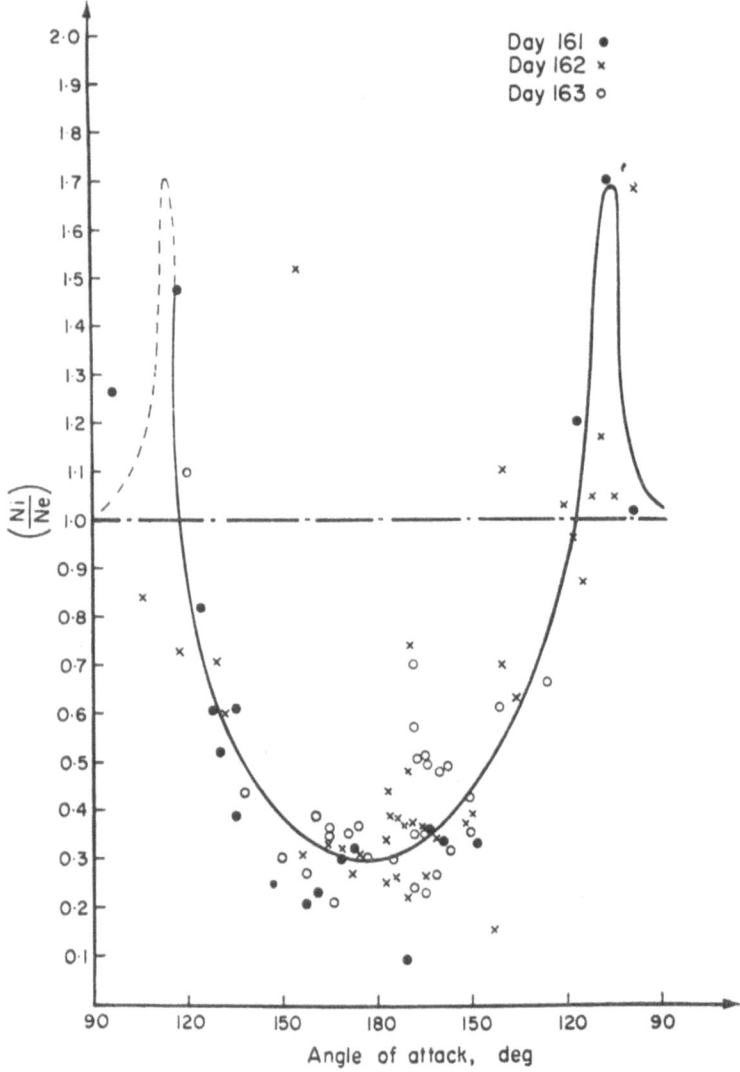

Fig. 3.   Variation of $[N_+/N_e]$ with angle of attack (θ). Ariel I data from days 161, 162, 163 in 1962. After Samir and Willmore (1965).

that the current collected by the boom-probe is apparently not significantly reduced when in its own wake. It can therefore be stated that

$$\left[\frac{I_e \text{ (base)}}{I_e \text{ (boom)}}\right] \quad \text{at } [\theta \text{ (wake)}] \text{ is} \quad \left[\frac{I_e \text{ (wake)}}{I_e \text{ (ambient)}}\right].$$

Similar results were obtained for other days, and the above statement was discussed in more detail elsewhere (Samir and Willmore, 1965; Willmore, 1970).

(2) Figure 3 shows the variation of $[N_+/N_e]$ with angle of attack for several passes on three days (161, 162, 163 in 1962). Similar results were obtained for other days. It is seen that when the ion probe looks into the wake of the satellite $(\theta > 150°)$ going towards $180°$ an ion depletion is observed. The ion depletion does not appear great as the electron depletion. This can be understood since the spherical probe on its stem may be beyond the region of minimum density immediately behind the satellite.

(3) A detailed investigation of electron probe characteristics near the edge of the wake pointed towards the possibility that strong oscillations occur in that region (Samir and Willmore, 1965). The frequency of these oscillations varies from 2.7 to 3.7 kc s^{-1} indicating the excitation of ion-plasma waves. This has been discussed by Gurevich *et al.* (1970).

(4) Using data from the boom probe we have mapped the region of disturbance caused by the satellite motion at a distance $r = 5 R_0$ from the center of the satellite. The method of analysis used here focused on examining the electron accelerating region ('plateau level') of the (dI_e/dV) characteristics (Henderson and Samir, 1967). This differs from the method of analysis used earlier. Figure 4 shows the normalized electron current versus angular position of the boom probe (ϕ) for various angles of attack (θ). It should be noticed that for different values of θ, the boom probe samples the wake of different parts of the satellite. For example, the curve for $\theta = 84°$ shows the wake structure due to the main body of the satellite. At $\theta = 60°$ the probe is in the wake created by the ion probe and for $\theta \leqslant 36°$ and $\theta > 156°$ the boom probe does not pass through any wake region. As seen from Figure 4 the minimum value of

$$\left[\frac{N_e \text{ (wake)}}{N_e \text{ (ambient)}}\right]_{r = 5 R_0} \approx 0.5 \pm 0.1.$$

This study also shows the width of the wake of both the main body and the ion probe to be larger than the geometric wake (Henderson and Samir, 1967). Figure 5 is similar to Figure 4 but presents the data with higher resolution in a limited range of θ. It is within this region that our main interest lies. It is seen that the wake produced by the main body has an enhancement in its center. Similar behaviour is seen for the wake produced by the ion probe at $\theta = 60°$. This study indicated that the wake due to ion probe is very similar in width and depth to that of the main body although the ratio of the linear dimensions of the main body of the satellite and the ion probe is about 6. It should be noted that the ion probe was biased 6 V negative with respect to the main body, which itself was between 0 to 1 V negative with respect to the ambient plasma

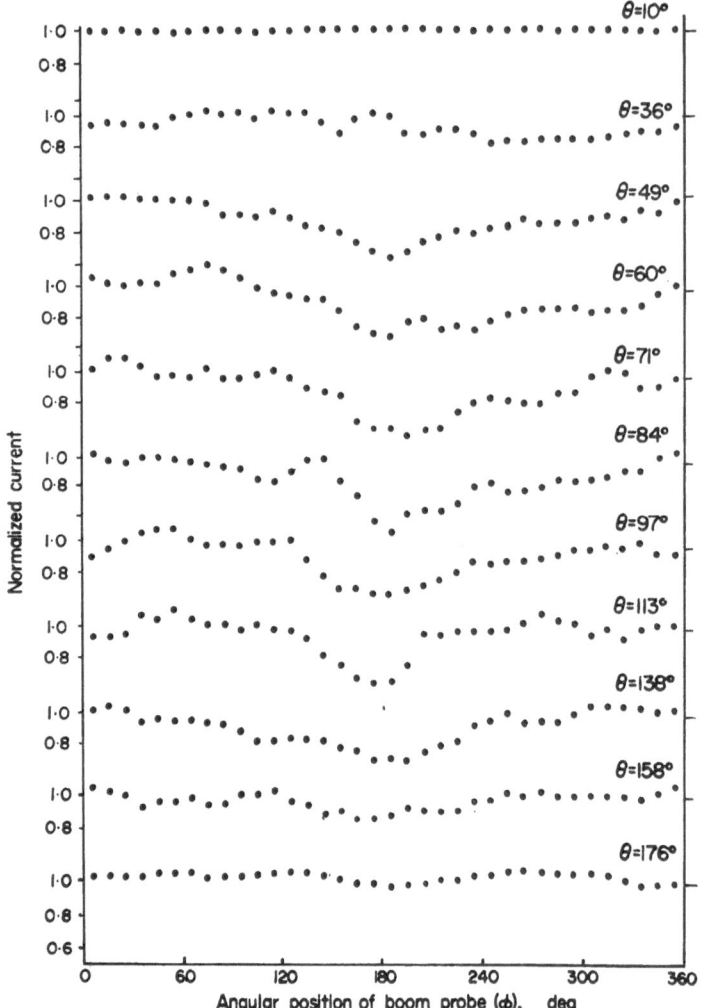

Fig. 4.   Normalized electron current versus angular position of the boom probe (Φ), for various angles of attack (θ) averaged over $\Delta\Phi = 30$ intervals to give a better resolution. After Henderson and Samir (1967)

(Samir and Willmore, 1966). Also, the ratio $[r/R_0]$, where r is the distance from the boom probe to the center of the relevant body is 5 for the main body and 33 for the ion probe, and $[R_0/\lambda_D]$ is 10 for the main body and 1.7 for the ion probe.

2.1.2. *From the Explorer 31 Satellite* (Samir and Wrenn, 1969; Samir *et al.*, 1972, 1973; Samir, 1970; Miller, 1972)

Since the spin axis of the Explorer 31 was maintained perpendicular to the orbital plane we do not have $[N_+/N_0]$ in the wake (see Figure 1(b)).

(1) Figure 6 presents an example showing the variation of the electron current (I_e)

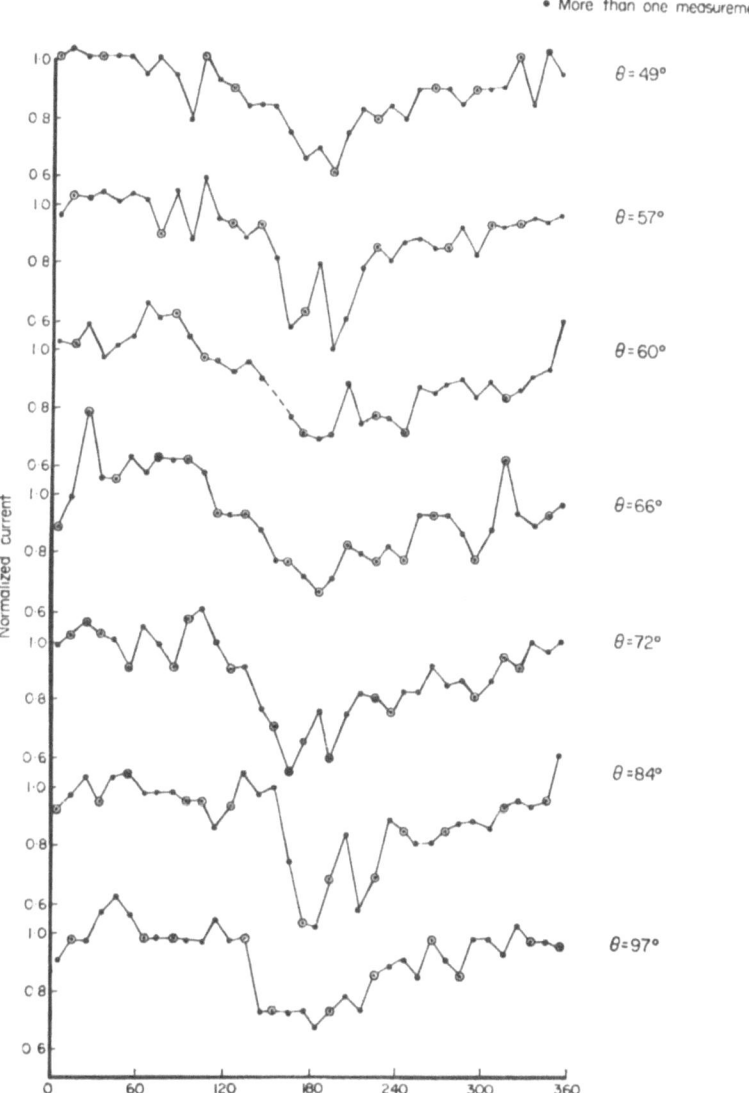

Fig. 5.   Normalized electron current versus angular position of the boom probe (Φ), for several angles of attack (θ) averaged over $\Delta\Phi = 10°$ intervals to give a better resolution. After Henderson and Samir (1967).

with angle of attack. It is seen that

$$\delta_e = \left[\frac{I_e(\text{wake})}{I_e(\text{front})}\right] \approx \left[\frac{N_e(\text{wake})}{N_e(\text{front})}\right] \quad \text{varies from} \quad 0.01$$

at perigee altitudes (see Figure 6(a)) to $\delta_e \approx 1.0$ at perigee altitudes (see Figure 6(d)).

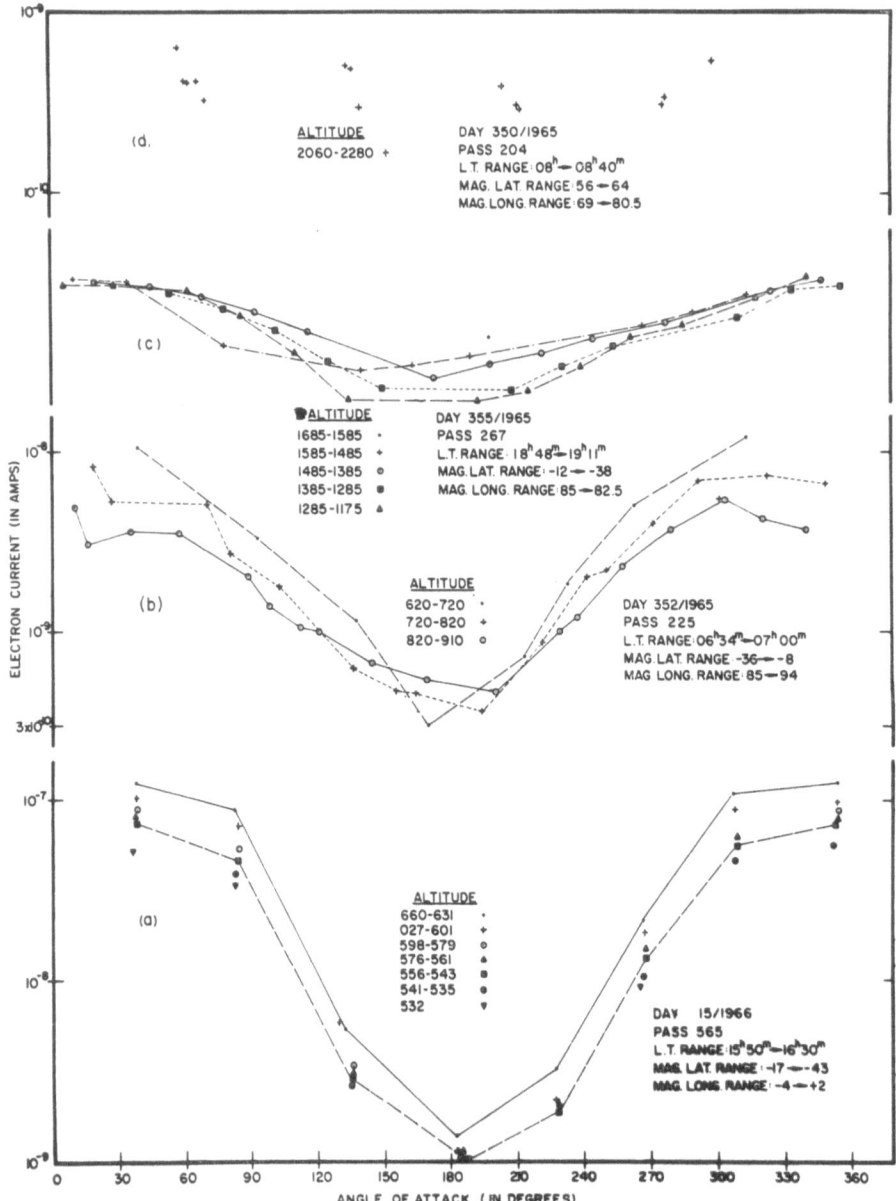

Fig. 6.   Variation of electron probe current (signal current in Amperes (Wrenn, 1969)) at space potential with the angle of attack. Explorer 31 data from days 350, 352, 355 in 1965 and day 15 in 1966.
                                    After Samir and Wrenn (1969).

Further discussion on δ_e are given elsewhere (Samir and Wrenn, 1969). It is seen that
the depth of the electron wake has its smallest value at perigee altitudes and approaches
unity on reaching apogee. The results of electron current $= f$ (angle of attack) obtained from the Explorer 31 confirm the earlier results (Samir and Willmore, 1965,

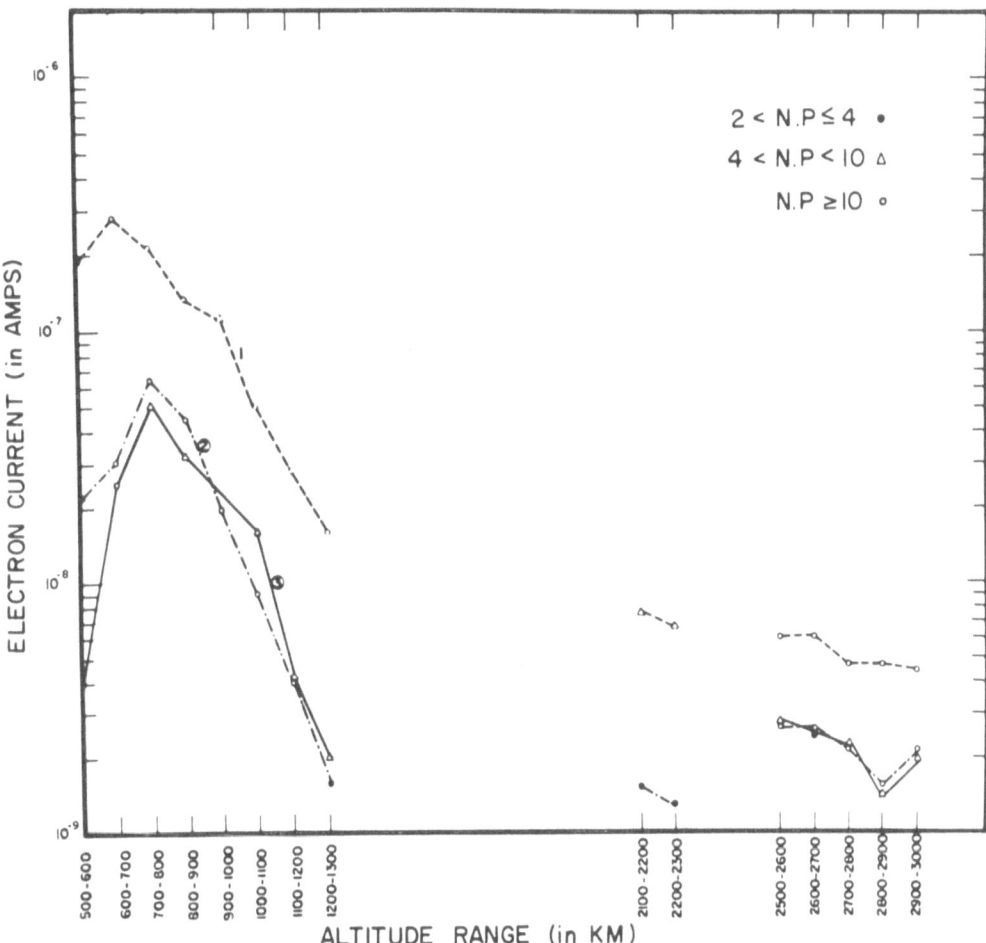

Fig. 7. Variation of $[(I_e)_{AV}]$ with altitude for $H = 100$ km altitude intervals. Explorer 31 data for days 40–101 in 1967. N.P. indicates the number of measurements used to obtain a point on each curve. Curve 1 represents $[I_e(\text{front})]$ for $\theta = 0° \pm 30°$; Curve 2 represents $[I_e(\text{wake})]$ for $\theta = 180° \pm \pm 15°$; Curve 3 represents $[I_e(\text{wake})]$ for $\theta = 180° \pm 30°$.

1966) obtained from the Ariel I satellite in the altitude range 400–700 km. Figure 7 presents a more complete average picture of I_e variations with altitude (for $H = 100$ km altitude intervals). Curve 1 in this figure represents $[I_e(\text{front})]$ for $\theta = 0° \pm 30°$ curve 2 represents $[I_e(\text{wake})]$ for $165° \leqslant \theta \leqslant 195°$ and curve 3 represents $[I_e(\text{wake})]$ for $150° \leqslant \theta \leqslant 210°$.

(2) The variation of δ_e with $[M_+]_{AV}$ was also studied. This was done by examining I_e and $[M_+]_{AV}$ with altitude using the spherical ion probe measurements (Samir and Wrenn, 1969). A large number of measurements was used in order to calculate δ_e. The idea was that if $\delta_e = f$ (altitude) then it has to be a function of the basic plasma para-

meters

$$[R_0/\lambda_D], \quad \left[\frac{T_e}{T_+}\right], \quad \left[\frac{V_s}{\sqrt{2KT/M_+}}\right], \quad \left[\frac{e\phi_s}{KT_e}\right].$$

We have restricted ourselves to the study of

$$\delta_e = f\left[\frac{V_s}{\sqrt{2KT/M_+}}\right]$$

for which the major contribution should be due to $\delta_e = f(M_+)$. We do not claim that the other parameters are necessarily of less importance, but in studying

$$\delta_e = f\left[\frac{V_s}{\sqrt{2KT/M_+}}\right]$$

the dependence on $[M_+]$ is the most significant variation.

Figure 8 presents an example of $\delta_e = f(M_+)$ for several passes. It is worthwhile noting that the relation $\delta_e = f(M_+)$ may furnish (in principle) a way to estimate the average ionic mass using the depletion of electron current in the wake as a diagnostic tool. It could be possible to use this approach to estimate $[M_+]_{AV}$ in planetary missions. Although we have not presented here the earlier Explorer 8 data (Bourdeau and Donley, 1964) which pointed towards an order of magnitude discrepancy with the Ariel I results, we believe the discrepancy to be removed after re-examining the data as shown in Samir (1970). The basis for the examination was the variation of $\delta_e = f(M_+)$.

By way of summary: (1) The more recent Explorer 31 results for $I_e(N_e) = f(\theta)$ at $r = R_0$ confirm the earlier Ariel-I results showing the ratio $\delta_e \approx 10^{-2}$ when O^+ is the major ionic constituent. (2): The Explorer 31 results yield the quantitative dependence of the electron depletion in the wake on $[M_+]_{AV}$.

(3) Analyzing measurements from the cylindrical probes on the Explorer 31 (Brace and Findlay, 1969) yields the electron currents and densities across a cylindrical collector which spans the range of 23–46 cm from the satellites surface. Compared with the electron results reported earlier from the Ariel I satellite and the planar electron probe on the Explorer 31 (Wrenn, 1969; Willmore, 1970) the cylindrical probe results do not yield $N_e = f(\theta)$ at a point (r, θ) in the satellite's wake. Figure 9 (after Miller, 1972) show $N_e = f(\theta)$ in the altitude range 520–720 km, similar to Figure 3 shown in Samir et al. (1972). Figure 10 is a reproduction of Figure 3 in Samir et al. (1972).

Although Figure 9 is plotted rather differently, it is seen that the ratio of current (density) in the wake $(\theta = 180°)$ to that in the front $(\theta = 0°)$ is similar to that shown in Figure 10 for the appropriate altitude range.

Figure 9 shows the degree of symmetry around the $\theta = 180°$ (i.e., maximum wake) position. Both Figures 9 and 10 as well as the earlier Ariel I results display assymetries that are (at least partly) due to magnetic field effects. This was discussed earlier by Samir et al. (1972) and recently by Miller (1972). Miller suggests that the angular

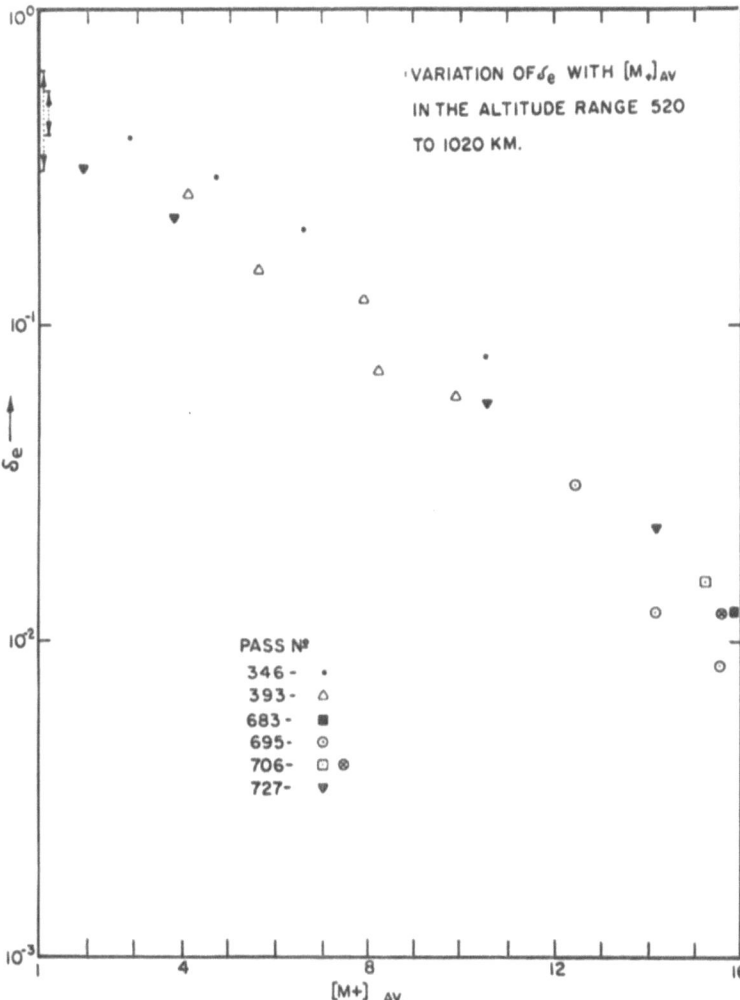

Fig. 8. Variation of δ_e with $[M_+]_{AV}$ in the altitude range 520–1020 km. Explorer 31 data for six passes. After Samir *et al.* (1972) and Samir (1970).

variation of electron density (or current) around the Explorer 31 is a linear combination of angle of attack and magnetic field (**n, B**) effects. While in principle it is definitely correct that both effects act simultaneously it is not clear to the author of this paper that a linear combination is the complete answer to the total angular variations. This last statement of ours is supported by theoretical wake studies (see e.g., the Gurevich *et al.* (1969) review paper) which display complicated density contours even when the treatments ignored magnetic field effects.

(4) If it is at all justified to combine the Ariel I ion results (see Figure 3) with the electron results from the cylindrical probes on Explorer 31 then it follows that at a distance of about $r \approx 2 R_0$ (from the center of the satellites) $N_+ \simeq N_e$ and the axial

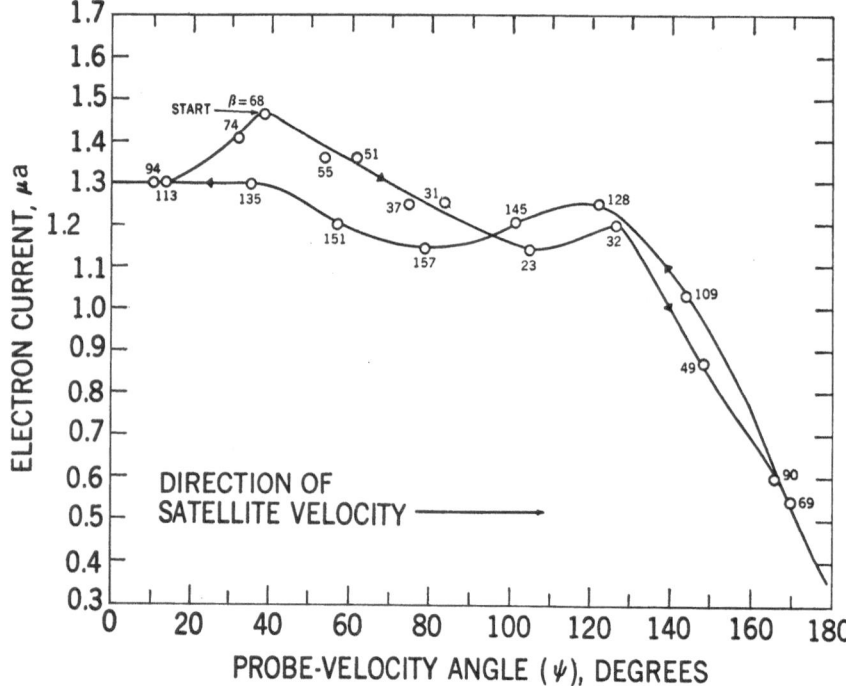

Fig. 9.   Variation of electron current (in μA) with angle of attack (θ). Explorer 31 cylindrical probe data (Brace and Findlay, 1969; Miller, 1972). Note: here ψ designates the angle of attack rather than θ as in the rest of the present paper. The numbers near each point indicate the angles with respect to the magnetic field.

extent of the region of maximum particle depletion in the wake is shorter than several authors believe (Singer, 1965). It should be born in mind that while the Ariel I satellite was nearly spherical, the Explorer 31 satellite is an octagonal parallelpiped with linear dimension given in Table II. Also, the experimental technique and method of analysis regarding electron observations on the two satellites (and the ion observations from the spherical probe) are different.

(5) Miller (1972) points out that $I_e = f(\theta)$ shows an electron current enhancement near 120°. Inspection of $[N_+/N_e] = f(\theta)$ (see Figure 3) also indicate that such an enhancement may be real. It is interesting in this context to compare with Alpert et al. (1965) where possible reasons for such a structure is discussed.

(6) Recently, ion measurements made by a retarding potential analyzer RPA (see Table II Proc. IEEE 57(6), 1969) that was mounted in the equatorial plane of the Explorer 31 satellite, with its aperture normal situated perpendicular to the satellite spin axis were used in order to study the angular distribution of ion flux (Samir et al., 1972b). Figure 11 shows the normalized particle fluxes (both ion and electron) variation with the angle of attack (θ). As could be expected $N_+(\theta) \ll N_e(\theta)$, the inequality becoming greater as one goes towards larger angles θ (i.e., towards the maximum wake position). It is unfortunate that at present there are no other ion observations to

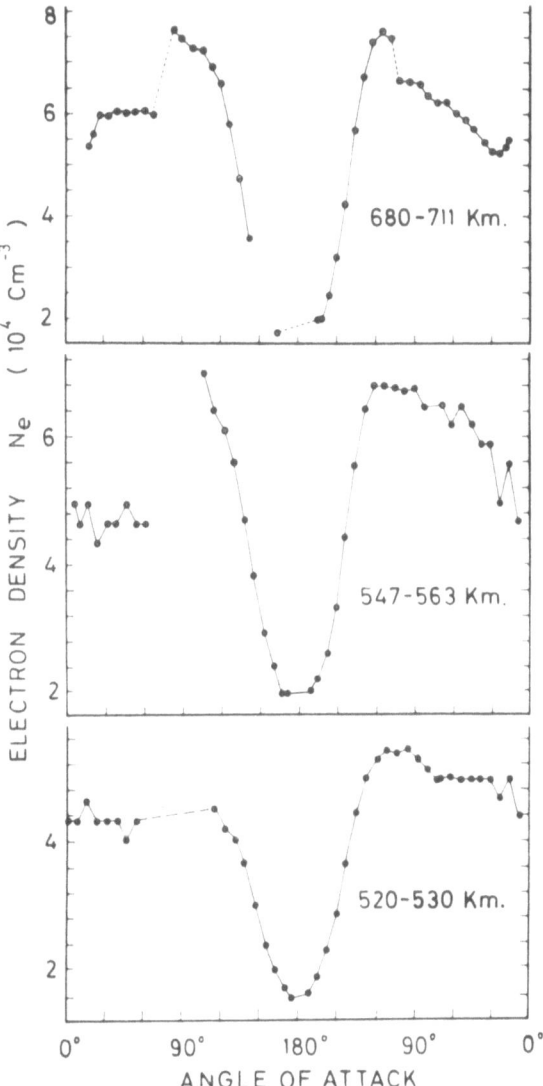

Fig. 10. Variation of electron density (N_e) with angle of attack (θ). Explorer 31 cylindrical data (Brace and Findlay, 1969) at three altitude intervals. Also Samir *et al.* (1972).

compare the above results with. An appropriate comparison could be with Hoffman's data (Hoffman, 1969). This effort will be pursued.

2.1.3. *From the Gemini-Agena System* (Troy *et al.*, 1970; Medved, 1969; Troy, 1969)

The main objective of the Gemini-Agena 10 (two body) system was to attempt an axial mapping of the wake produced by the Gemini capsule. However, the Agena itself performed some maneuvers and Figure 12 shows the results for I_e and $I_+ = f$

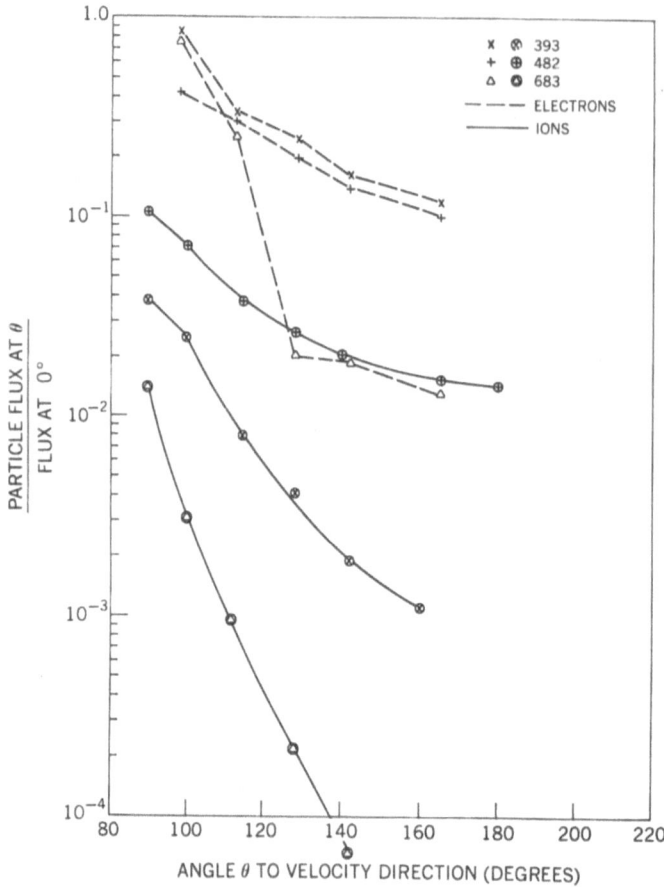

Fig. 11.   Variation of normalized ion and electron currents with angle of attack (θ) for $90° \leqslant \theta \leqslant$ $\leqslant 180°$. Explorer 31 data for three passes. Pass 393 in January 1966 in the altitude range 700–930 km, pass 482 in August 1966 in the altitude range 900–600 km and pass 683 in January 1966 in the altitude range 570–520 km. After Samir *et al.* (1972b).

(time) for a 180° yaw maneuver. The yaw as shown brought the outboard probes from position *A* where they looked forward, to position *B*, where they looked into the shadow of the docking cone (see also Figure 1 (c)). During this rotation, the probes passed through a region of maximum particle depletion produced by the main body of the Agena. It is seen that the electron current was reduced by about two orders of magnitude between position *A* and *B*, whereas the outboard ion current decreased by at least four orders of magnitude (the current value of 5×10^{-11}A is the lower limit on the electrometer).

It is interesting to note that the amount of the electron current depletion observed in the wakes of the Ariel I, Explorer 31 and the Agena is similar (i.e., $\delta_e \approx 10^{-2}$), even though the Agena vehicle differs considerably from both the Ariel I and the Explorer 31 geometry (see Figure 1 and Table II).

MAXIMUM ELECTRON AND ION CURRENT DEPLETION IN THE WAKE OF THE
AGENA – TARGET – VEHICLE
(AFTER TROY, MEDVED AND SAMIR)

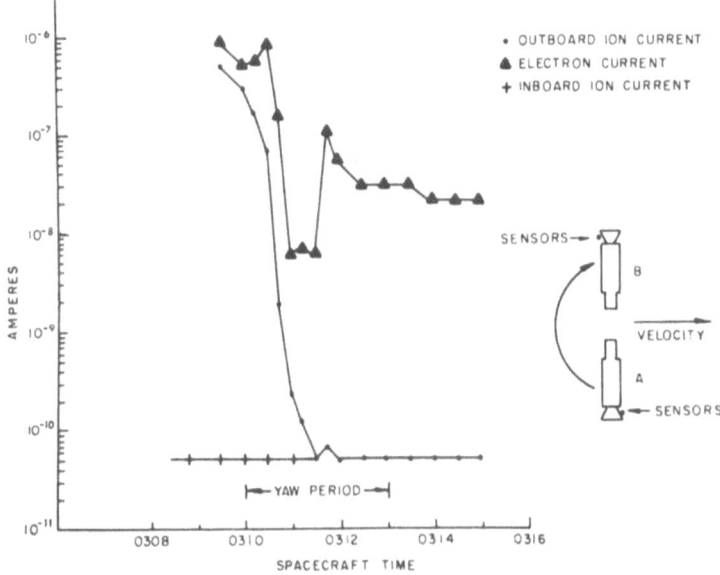

Fig. 12.   Variation of electron and ion currents with time for a 180° yaw maneuver performed by the Agena vehicle. After Troy *et al.* (1970).

This result may be interpreted as showing that the structure of the wake may be more significantly controlled by factors such as $[M_+]$ potential and $[R_0/\lambda_D]$ ratios rather than by the geometry of the body even in the very close vicinity of the spacecraft (see also Samir and Weil, 1973)).

In-situ observations from other spacecraft. Several investigators have used antennas on on rockets in order to map the wake structure behind the rockets. Of these we cite Oya (1970) and Stone *et al.* (1969). An attempt to map out the wake of the OGO II satellite was made by Weil and Yorks (1970). A comparison of the above results was performed recently by Samir and Weil (1973).

2.2. EXPERIMENT – THEORY COMPARISONS

This part of the paper presents an up-to-date comparison between our main experimental results and some near wake models. Part of the comparisons presented here are scattered throughout our earlier papers (Samir and Willmore, 1965; Henderson and Samir, 1967; Samir *et al.*, 1972).

2.2.1. *Using Ariel I Measurements*

Figure 13 presents a comparison between typical Ariel I results (when O^+ was the

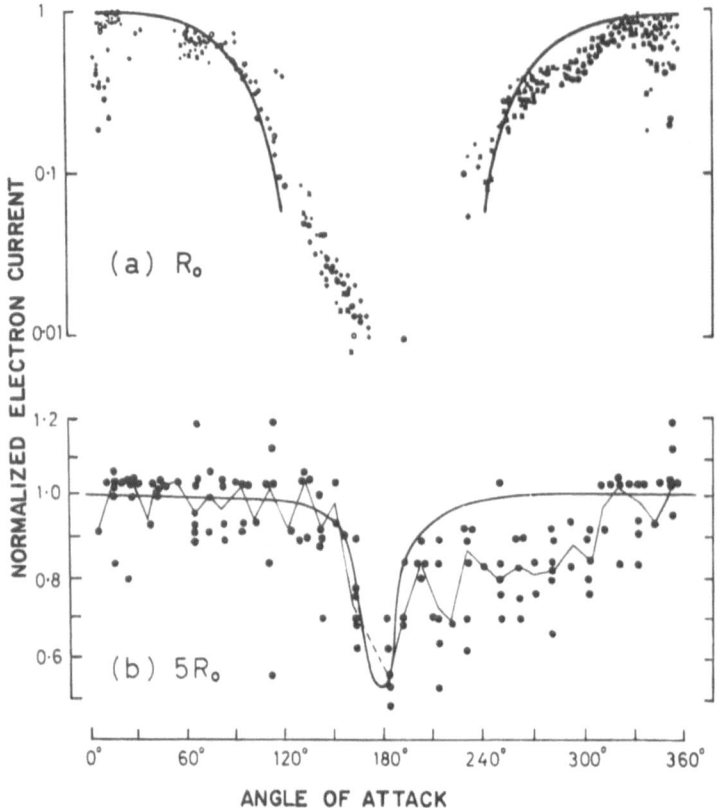

Fig. 13.   Variation of normalized electron current at different distances from the center of the Ariel I satellite. (a) Shows the variation of $[I_e(\text{base})/I_e(\text{boom})]$ at $r = R_0$ (Henderson and Samir, 1967). (b) Shows the variation of $[I_e(\text{normalized})]$ with angular position of the boom probe at $r = 5R_0$ (Henderson and Samir, 1967). The thick lines represent the Gurevich theoretical wake model (Gurevich et al., 1969; Al'pert et al., 1965).

major ionic constituent of the ionosphere) and Gurevich et al. wake models (Gurevich et al., 1969; Al'pert et al., 1965). Our results, as shown in Figure 2 of this paper, are in line with the results in Henderson and Samir (1967). As a matter of fact the experimental results shown here in Figure 13 are taken from Henderson and Samir (1967), and Figure 13 is the same as Figure 9 of Samir et al. (1972).

(1) From Figure 13 it is seen that there is a very satisfactory agreement between our normalized electron current results at $r = 5\,R_0$ and the Gurevich et al. (1969) and Al'pert et al. (1965) theoretical wake model at this distance. The continuous thick line that represents the results of the theoretical calculations is based on considering ions as neutrals i.e., neglecting the influence of the potential field on the ion motion for

$$\mu = \left[\frac{V_s}{V_T(+)}\right] = 3.75\,.$$

This agreement implies that at $r = 5 R_0$ (from the center of the satellite) the influence of the potential field is insignificant to the precision of the data.

In Henderson and Samir (1967) we carried out the above comparison. Our statement in Henderson and Samir (1967) to the effect that the experimental results are in disagreement with the Gurevich (Al'pert et al., 1965) model at $r = 5 R_0$ is incorrect and due to a mistake considering the wrong Mach number. Our comparison (Henderson and Samir, 1967) with Sawchuk's model (Sawchuk, 1963) is however correct and our results are higher by a factor of 1.5 than expected by Sawchuk (1963). The comparison with Sawchuk (1963) is of interest since this wake model was used by Weil and Yorks (1970) as a tool in their experimental mapping of the wake of the OGO II Satellite.

(2) In comparing the Ariel I results at $r = R_0$ with Gurevich et al. (1969) and Al'pert et al. (1965) we find a satisfactory agreement in the angle of attack ranges $0° \leqslant \theta \leqslant 120°$ and $240° \leqslant \theta \leqslant 360°$. A detailed account on the theoretical expressions used to compute the thick curves in Figure 13 are given in Gurevich et al. (1969).

(3) Gurevich et al. (1969) discusses our prediction (Samir and Willmore, 1965) regarding the existence of ion plasma oscillations in the wake. He claims that the instability discussed in Gurevich et al. (1969) leads to the generation of plasma oscillations with frequencies of the order of 5 kc s^{-1} for $N_e = 10^5 \text{ cm}^{-3}$ and $M_+ = 16$ which is in agreement with Samir and Willmore (1966). The problem of plasma oscillations was also undertaken in several papers by Liu and discussed in his review paper (Liu, 1969).

2.2.2. Using Explorer 31 Measurements

(1) Gurevich et al. (1969) compared his wake model with our Explorer 31 results (Samir and Wrenn, 1969) for the situations where O^+ was the major ionic constituent of the ionosphere. An agreement similar to that shown in Figure 13(a) was obtained. Namely, agreement within the angular range of $0° \leqslant \theta \leqslant 120°$ and $240° \leqslant \theta \leqslant 360°$. However, the comparison of the Explorer 31 results for conditions where the hydrogen concentrations $N(H^+)$ is significant i.e., at higher altitudes (see Figure 6 for electron current variations with altitude and Figure 8 for $\delta_e = f(M_+)$), covered the entire range of $0° \leqslant \theta \leqslant 360°$. Figure 14 shows the experiment-theory comparison for four altitude intervals in the altitude range 620–1285 km. For the altitude interval 620–720 km the relative hydrogen concentration $n(H^+)(=N(H^+)/[N(H^+)+N(O^+)])$ where $N(H^+)$ and $N(O^+)$ are the actual particle concentrations, is 0.23*, for the altitude range 720–820 km $n(H^+) = 0.45$, for the altitude range 820–910 km $n(H^+) = = 0.62$ and for the altitude range 1175–1285 km $n(N^+) = 0.94$. It is seen that the best agreement between experiment and theory is obtained for the situation of largest hydrogen concentration in the ionosphere. This could have been qualitatively anticipated since $\mu = [V_s/V_T(+)]$ is significantly smaller compared with μ at lower altitudes where O^+ is the dominating ionic constituent. Gurevich obtained his results (presented by

---

* It was assumed (Gurevich et al., 1969) that the ionospheric plasma consists of H^+ and O^+ ions only.

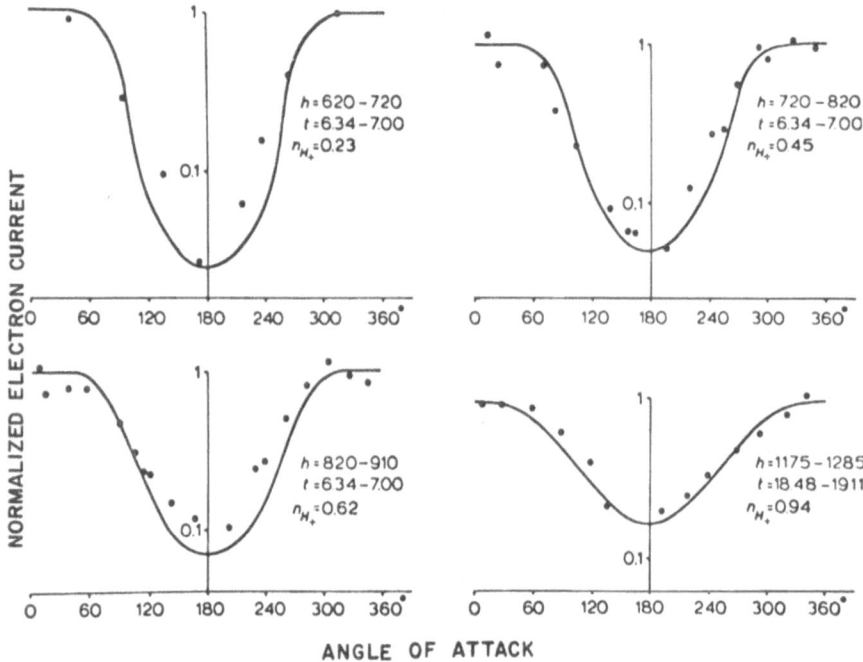

Fig. 14.   Variation of normalized electron current with angle of attack (θ) for different values of relative hydrogen density. Explorer 31 data. After Gurevich *et al.* (1969) and Samir and Wrenn (1969).

the thick lines in Figure 14) using the expression:

$$\frac{N(\theta)}{N(O)} = N(O^+)\left[\frac{1 + \mathrm{erf}(\mu(O^+)\cos\psi_0\cos\theta)}{1 + \mathrm{erf}(\mu(O^+)\cos\psi_0)}\right] +$$
$$+ N(H^+)\left[\frac{1 + \mathrm{erf}(\mu(H^+)\cos\psi_0\cos\theta)}{1 + \mathrm{erf}(\mu(H^+)\cos\psi_0)}\right], \qquad (1)$$

where:

$$\mathrm{erf}(x) = \frac{2}{\sqrt{\pi}}\int_0^x \exp(-t^2)\,dt$$

μ = Mach number of the specific ionic constituent

ψ_0 = angle associaed with the probe location on the satellite skin and its prox-
          imity to the region of the maximum rarefaction (Gurevich *et al.*, 1969),
which is actually a neutral particle approximation calculation. Gurevich *et al.* (1969)
state that for $n(H^+) \geqslant (0.1–0.2)$ the electric field hardly influences the motion of the
oxygen (O^+) ions. And since in their calculation the different ionic constituents are
non interacting it implies that for $n(H^+) \geqslant n$ (n = a certain lower limit for the percentage of hydrogen) it is the hydrogen ions that control the ion population of the wake.

Gurevich *et al.* (1969) used experimental results (Samir and Wrenn, 1969) to produce a curve comparing the measured normalized current ($[I_e(\text{wake})/I_e(\text{front})]$) vs $n(\text{H}^+)$. The degree of agreement between experiment and theory is satisfactory (see Figure 28 in Gurevich *et al.* (1969) and Figure 8 of the present paper).

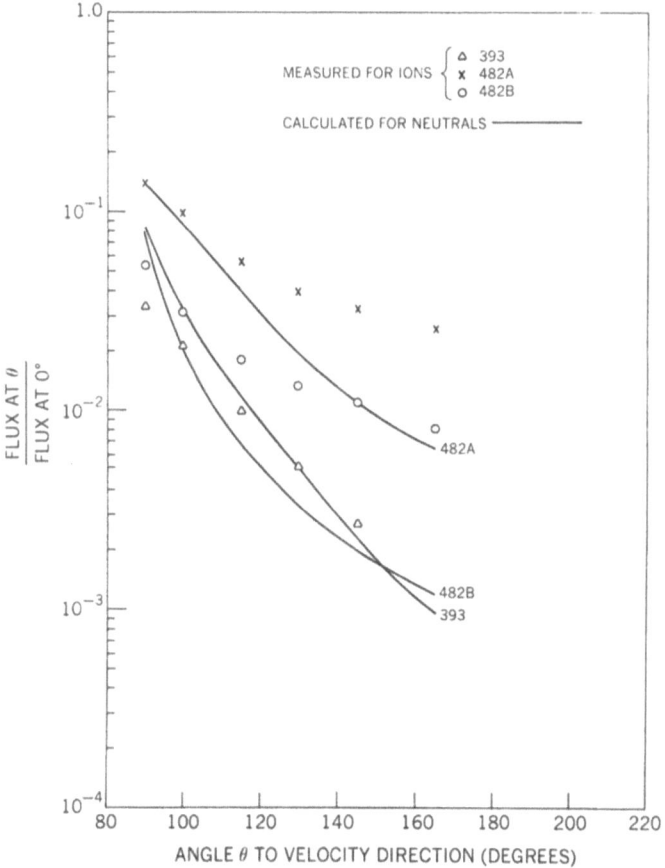

Fig. 15. Normalized ion fluxes ($[I_+(\theta)/I_+(0)]$) vs angle of attack for two passes (pass 393 and 482). Explorer 31 data. The thick line shows the theoretical computation for ions $\equiv$ neutral (Samir *et al.*, 1973).

(2) Samir *et al.* (1973) analyzed some ion probe measurements (Donley and Maier in *Proc. IEEE* **57** (6), 1969) to obtain the angular distribution of ion fluxes around the Explorer 31 (see Figure 11). The results were compared with a simple neutral approximation wake model. Figure 15 shows the degree of agreement of experiment-theory between the normalized ion fluxes vs angle of attack (θ), for two passes. The measurements of these two passes were taken in the altitude range 630–910 km. The measurements for these two passes showed H^+ and O^+ to be the two ionic

constituents and

$$2000 \leqslant T_e \leqslant 3400 \,(\text{K})$$
$$7.1 \leqslant V_s \leqslant 8.0 \,(\text{km s}^{-1})$$
$$0.71 \leqslant \phi_s \leqslant 0.95 \,(\text{Volts-negative})$$
$$2.1 \times 10^4 \leqslant N_+ \leqslant 3.4 \times 10^4 \,(\text{cm}^{-3})$$
$$9.3 \leqslant [M_+]_{AV} \leqslant 15.1.$$

For a two ion species plasma $n(\text{H}^+) = [16 - [M_+]_{AV}]/15$ hence for these two passes $0.06 \leqslant n(\text{H}^+) \leqslant 0.3$ most of the $n(\text{H}^+)$ values being centered around 0.25. The better agreement between experiment and theory (calculation for neutrals in Figure 15) being obtained for $0.2 \leqslant n(\text{H}^+) \leqslant 0.3$. This is in general accord with Gurevich *et al.* (1969) as discussed in the previous section. It should however be noted that the expression used by Samir *et al.* (1973) although similar to (1) (Gurevich *et al.*,1969) is not identical. The main difference being in the thermal motion component that appears in Samir *et al.* (1973) but does not appear in (1) (Gurevich *et al.*, 1969). For further details see Samir *et al.* (1973).

(3) Electron measurements from the flush mounted planar Langmuir probe (Wrenn, 1969) were used for comparison with the Liu (1969) and Liu and Jew (1968) theoretical wake model. Figure 16 shows the results of the experiment-theory comparison (Samir and Jew, 1972) for eight passes covering a wide range of plasma parameters. Generally, the $[V_S/V_T(+)]$ parameter varies in the range 0.85 to 5.6, the $[R_0/\lambda_D]$ parameter varies in the range 1.9 to 43, and the parameter $[e|\phi|/KT_e]$ varies in the range 3.8 to 5.7. The detailed information for each pass is given in Figure 16.

The electron data comparison given earlier in this part of the paper used the Gurevich *et al.* (1969) quasineutral theory. The comparison yielded satisfactory results for the angular range $0 \leqslant \theta \leqslant 120$ and $240 \leqslant \theta \leqslant 360$ when $N(\text{O}^+)$ is available in significant quantities and covering the entire angular range for larger $n(\text{H}^+)$ values. The present comparison covers the range $120 \leqslant \theta \leqslant 180$; thus it includes the maximum wake zone near to the satellite omitted by Gurevich. Also, the Liu (1969) and Liu and Jew (1968) wake model is not restricted by quasi-neutral limitations, and presents in a way a self-consistant approach. We are now comparing experiment with a theory that does not ignore the influence of the potential field on the ions but on the other hand assumes the electron density to obey Boltzmann's relationship (i.e. $N = N_0 \exp(-eV/KT)$). This last assumption is obviously open to doubt, though the theory considers the free stream plasma to be in a bithermal equilibrium with T_e not being necessarily equal to T_+. Using the Boltzmann expression for the electrons is obviously a weak point of the theory especially due to the existence of a negative potential barrier in the wake zone. For more detail on the above comparison indicating its limitations see Samir and Jew (1972). The degree of agreement of this experiment-theory comparison is demonstrated in Figure 16. The agreement is better than could perhaps be expected *a priori* due to the limitations on the way of comparison and the shortcomings of the theory itself. Table III shows in greater detail the measured and computed values of the normalized density (current) in the maximum wake zone (i.e. at $\theta = 165°$).

(4) The result from the Agena 10 maneuver Troy *et al.* (1970) regarding the value of $[I_e \text{ (normalized)}] = f \text{(maximum wake position)}$ is not in accord with $I_e = f(\theta)$ for different values of $[R_0/\lambda_D]$ (e.g. Kiel *et al.*, 1968). Due to the limited amount of ana-

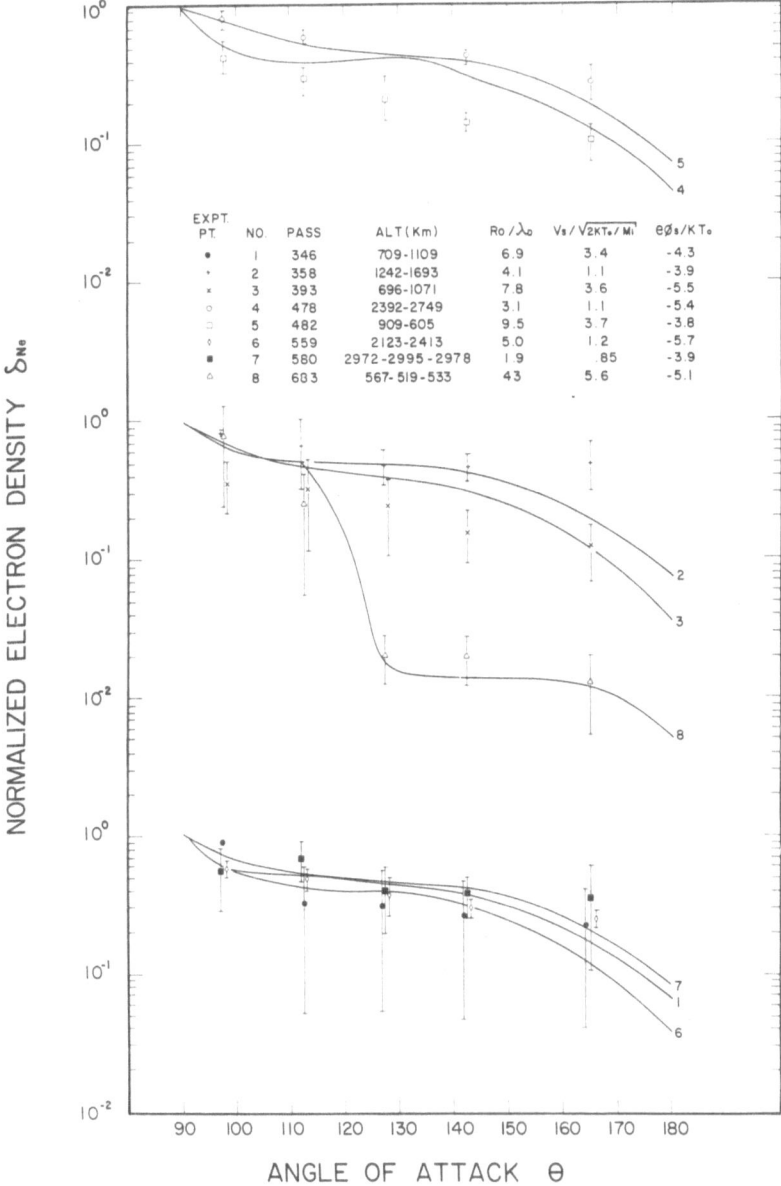

EXPT. PT	NO.	PASS	ALT(Km)	R_0/λ_D	$V_s/\sqrt{2KT_e/M}$	$e\phi_s/KT_e$
●	1	346	709-1109	6.9	3.4	-4.3
·	2	358	1242-1693	4.1	1.1	-3.9
×	3	393	696-1071	7.8	3.6	-5.5
○	4	478	2392-2749	3.1	1.1	-5.4
□	5	482	909-605	9.5	3.7	-3.8
◇	6	559	2123-2413	5.0	1.2	-5.7
■	7	580	2972-2995-2978	1.9	.85	-3.9
△	8	603	567-519-533	43	5.6	-5.1

NORMALIZED ELECTRON DENSITY δ_{Ne}

ANGLE OF ATTACK θ

Fig. 16. Variation of normalized electron current with angle of attack for eight passes. Explorer 31 data. The thick line represents the computation based on the theoretical model of Liu (1969) and Liu and Jew (1968). After Samir and Jew (1972).

TABLE III

Charge depletion in the wake zone ($\theta > 120°$)
of the Explorer 31 satellite

Pass No.	$\left[\dfrac{N_e(\text{wake})}{N_e(\text{front})}\right]_{\text{measured}}$	$\left[\dfrac{N_e(\text{wake})}{N_{eo}}\right]_{\text{computed}}$
346	$(2.2 \pm 1.8) \times 10^{-1}$	1.6×10^{-1}
358	$(5.1 \pm 2.0) \times 10^{-1}$	1.9×10^{-1}
393	$(1.2 \pm 0.5) \times 10^{-1}$	1.2×10^{-1}
478	$(2.8 \pm 0.8) \times 10^{-1}$	1.3×10^{-1}
482	$(1.0 \pm 0.3) \times 10^{-1}$	1.9×10^{-1}
559	$(2.5 \pm 0.4) \times 10^{-1}$	1.1×10^{-1}
580	$(3.6 \pm 2.5) \times 10^{-1}$	1.9×10^{-1}
683	$(1.3 \pm 0.7) \times 10^{-2}$	1.2×10^{-2}

lyzed measurements available to us (only one maneuver analyzed) and due to possible uncertainties as for the θ_i definition etc. we do not feel that it is appropriate (at this stage) to claim a disagreement between experiment and theory.

2.2.3. *About Some Theoretical Wake Models and Theory-Theory and Experiment-Theory Comparisons*

Although the amount of experimental information available from *in situ* observations is meager and fragmentary in nature this is not the case for the present state of theoretical studies. Many wake models exist each attempting to solve the same system of equations (i.e. the Boltzmann-Poisson system of equations) by using different mathematical methods and approaches subject to various physical assumptions.

Table IV lists assumptions used in some theories. The four assumptions shown are of major importance. Often authors use applicable basic assumptions but assign them to unrealistic body sizes (i.e. $[R_0/\lambda_D] \lesssim 1$). The extension of the theories to include the case of interest namely $R_0/\lambda_D \gg 1$ is not always simple or even possible without major modifications. Often very high negative potentials are attributed to satellites in the ionosphere. Although such an assumption simplifies the mathematical treatment, no such situation actually occurs in flight. Often situations applicable to probes are used for the satellite itself. This is in most cases of ionospheric satellites rarely true. Several authors claim that a certain plasma parameter is of the most significant importance and on that basis build their theory (e.g. Kiel *et al.*, 1968. See also Gurevich *et al.*, 1969.) This approach may be misleading since at the present time we do not yet know the real relative significance of each of the plasma parameters in defining the wake structure and characteristics. Such an understanding can be achieved via theory-theory and experiment-theory comparisons. It should also be realized that in studying the zones of disturbance created by the motion of the satellite, assumption applicable to the frontal zone are invalid for the wake zone. An example for that is the importance of including the ion thermal motion in the calculation. Or considering the influence of the potential field on the ion motion. Even in the wake zone itself one has to distin-

TABLE IV

Wake theories – some assumptions used

Theory	Potential field influence on the ion motion considered	Thermal ion motion considered	$N = N_0 \exp\left(\dfrac{-eV}{KT}\right)$ assumed	Magnetic field considered
Dolph and Weil (1961)	NO	YES	NO	NO
Davis and Harries (1961)	YES	NO	YES	NO
Hohl and Wood (1963)	YES	NO	NO	YES
Al'pert et al. (1965) and Gurevich et al.[b]	NO (YES[a])	YES (NO[a])	YES[a]	NO/YES[a]
Shea (1967)	NO	YES	YES	YES
Maslenikov and Sigov (1967)	YES	NO	YES	NO
Taylor (1967)	YES	YES	YES	NO
Kiel et al. (1968)	NO	YES	NO	NO
Call (1969)	YES	NO	YES	YES
Fournier (1971)	YES	YES	NO	NO
Sawchuk (1963)	NO	YES	YES	NO
Liu (1969) and Liu and Jew (1968)	YES	YES	YES	NO

[a] Through several publications Gurevich et al. used and/or did not use the assumptions mentioned.
[b] Gurevich et al. (1962, 1965, 1966, 1968, 1969, 1970) and Gurevich and Smirnova (1970).

guish between far and near (and very near) zones. And it is not always enough just to attempt to match (or patch) different regions within the wake zone. Mathematical procedures based on simplified assumptions may be valid or adequate for drag calculations but not for precise wake calculations.

3. Conclusion

The comparison of experiment-theory presented here does not constitute a comprehensive study. Only two wake models were considered. These two models differ (see Table IV) substantially and our comparison does not include the entire range of θ, especially the range $120 \leqslant \theta \leqslant 180°$ for the comparison with Gurevich et al (1969) and Al'pert et al. (1965). The comparison of experiment-theory should include more relevant theories under a variety of plasma parameters and spacecraft geometry, size and surface material. Different degrees of agreement between theory and experiment would help to gain a better physical understanding of the structure and characteristics of wakes in spaceplasmas, by assessing appropriately the real significance of assumptions (physical and mathematical) used in the theories. This understanding will be of extreme importance in the planning of the second generation of wake and sheath experiments to be performed on shuttles. For the second generation of in situ wake and sheath measurements it would be useful to use multibody systems as well as employing the idea (see Henderson and Samir (1967) for such an example) of using target bodies as wake creators. This will make possible the comprehensive study of the axial extent of the wake as well as the study of the entire zone of disturbance created by the rapid motion of a body in space. It will obviously be most useful to have maneuverable bodies as well as passive satellites of different size, geometry and surface material, that could be launched from a shuttle/space-station platform. At present we could not have performed a systematic comparison which reflects on the structure of the wake at distances $r > R_0$. The results at $r \approx 2 R_0$, $5 R_0$ are only two points on the axis and even then the measurements used in our study were performed by probes operating on different methods and techniques. We know that measurements done by different probes lead often to different results. This subject in itself requires more study.

Acknowledgement

The author wishes to thank Mr G. R. Carignan, Director of the Space Physics Research Laboratory, University of Michigan for his interest in the study and for useful discussions. Acknowledgement is also due to Professor Z. Alterman, Head of the Department of Environmental Sciences, Tel-Aviv University (Israel) for her interest in the subject. Part of the work was supported by NASA Grant NGR23-005-320.

References

Al'pert, Ya. L., Gurevich, A. V., and Pitaevskii, L. P.: 1965, *Space Physics with Artificial Satellites*, Consultant Bureau, New York.
Bourdeau, R. E. and Donley, J.: 1964, *Proc. Roy. Soc.* **A281**, 487.
Brace, L. H. and Findlay, J. A.: 1969, *Proc. IEEE* **57**, 1054.
Call, S. M.: 1969, Columbia University, Plasma Laboratory report number 46, New York.
Davies, A. H. and Harris, I.: 1961, in L. Talbot (ed.), *Rarefied Gas Dynamics*, Suppl. I, p. 691.
Dolph, C. L. and Weil, H.: 1961, *Planetary Space Sci.* **6**, 123.
Fournier, G.: 1971, Office National D'Études et de Recherches Aérospatiales, Publication number 137, Chatillon (France).
Gurevich, A. V.: 1962, *Planetary Space Sci.* **9**, 321.
Gurevich, A. V. and Pitaevskii, L. P.: 1965, *Phys. Rev. Letters* **15**, 346.
Gurevich, A. V. and Smirnova, V. V.: 1970, *Geomagnetism and Aeronomy* **10**, 313.
Gurevich, A. V., Pariiskaya, L. V., and Pitaevskii, L. P.: 1966, *Soviet Phys.* JETP **22**, 499.
Gurevich, A. V., Pariiskaya, L. V., and Pitaevskii, L. P.: 1968, *Soviet Phys.* JETP **27**, 476.
Gurevich, A. V., Pitaevskii, L. P., and Smirnova, V. V.: 1969, *Space Sci. Rev.* **9**(6), 805.
Gurevich, A. V., Pitaevskii, L. P., and Smirnova, V. V.: 1970, *Soviet Phys. – Uspekhi* **12**, 595.
Henderson, C. L. and Samir, U.: 1967, *Planetary Space Sci.* **15**, 1499.
Hoffman, J. H.: 167, *Science* **155**, 322.
Hoffman, J. H.: 1969, *Proc. IEEE* **57**, 1063.
Hohl, F. and Wood, G. P.: 1963, in J. A. Laurmann (ed.), *Rarefied Gas Dynamics*, Vol. 2, p. 45.
Jastrow, R. and Pearse, C. A.: 1957, *J. Geophys. Res.* **62**, 413.
Kiel, R. E., Gey, F. C., and Gustafson, W. A.: 1968, *AIAA J.* **6**, 690.
Liu, V. C.: 1969, *Space Sci. Rev.* **9**, 423.
Liu, V. C. and Jew, H.: 1968, *AIAA Paper* No. 68–169.
Maslenikov, M. V. and Sigov, J.: 1969, in C. L. Brundin (ed.), *Rarefied Gas Dynamics*, Vol. 2, p. 1657.
Medved, D. B.: 1969, *Rarefied Gas Dynamics*, 6th Int. Symp., p. 1525
Miller, N. J.: 1972, *J. Geophys. Res.* **77**(16), 2851.
Oya, H.: 1970, *Planetary Space Sci.* **18**, 793.
Proc. IEEE (Special Issue) **57**(6): 1969, This special issue contains descriptions of the various experiments on the Explorer 31 and the Alouette II satellites.

Samir, U.: 1970, *J. Geophys. Res.* **75**, 855.

Samir, U. and Jew, H.: 1972, *J. Geophys. Res.* **77**, 6819.

Samir, U. and Weil, H.: 1973, *Planetary Space Sci.*, in press.

Samir, U. and Willmore, A. P.: 1965, *Planetary Space Sci.* **13**, 285.

Samir, U. and Willmore, A. P.: 1966, *Planetary Space Sci.* **14**, 1131.

Samir, U. and Wrenn, G. L.: 1969, *Planetary Space Sci.* **17**, 693.

Samir, U., Maier, E. J., and Troy, B. E., Jr.: 1973, *J. Atmospheric Terrest. Phys.* **35**, 513.

Samir, U., Wrenn, G. L., and Henderson, C. L.: 1972, in Dino Dini (ed.), *Rarefied Gas Dynamics*, to appear.

Sawchuk, W.: 1963, *Rarefied Gas Dynamics*, 3rd Int. Symp. II, p. 661.

Shea, J. J.: 1967, in C. L. Brundin (ed.), *Rarefied Gas Dynamics*, Vol. 2, p. 1671.

Singer, F. S. (ed.): 1965, *Interaction of Space Vehicles with an Ionized Atmosphere*, Pergamon Press.

Stone, R. G., Fanberg, J., and Alexander, J. K.: 1969, *Planetary Space Sci.* **17**, 1437.

Taylor, J. C.: 1967, *Planetary Space Sci.* **15**, 155 and 463.

Troy, B. E., Jr.: 1969, NASA Goddard Space Flight Center Report X-615-68-164.

Troy, B. E., Jr., Medved, D. B., and Samir, U.: 1970, *J. Astron. Sci.* **18**(3), 173.

Weil, H. and Yorks, R. G.: 1970, *Planetary Space Sci.* **18**, 901.

Willmore, A. P.: 1970, *Space Sci. Rev.* **11**(5), 607.

Wrenn, G. L.: 1969, *Proc. IEEE* **57**, 1072.

DISCUSSION

Weil: How can you use the experimental results you showed to determine which of the many theoretical works gives good results when the body shapes of the satellites differ so greatly from the simple shapes used in the theories. It would seem you could only differentiate between theories whose results differ grossly.

Samir: We can only use the data which are available and in fact we have found very good agreement with the theory of Gurevich. We have not checked as yet against many other theoretical results. However some comparisons with the Liu-Jew theoretical results should appear soon in *J. Geophys. Res.* and some ion comparisons in *J. Atmospheric Terrest. Phys.* You are of course right in stating that the comparison is not at all simple.

Walker: I wish to concur with Dr Samir that much more work is needed to check out the various theories. Many theories are not necessarily intended to apply only to conditions natural to the satellite environment, but seek to be more general. Experiments are needed to test these theories for such conditions as well.

Samir: I agree with Dr Walker that more experimental work is needed. It seems to me, however, that experiment/theory comparisons using available data are the way to go at present. Any collaboration with people who worked out theoretical models is very welcome. I believe that an experiment/theory comparison is essential also for the planning of second generation experiments.

Fredricks: Was evidence found by probes on the long boom for an ion-acoustic bow shock standing off from the body? Since the ion Mach number you stated could be as great as six, it seems to me that there should exist such a standing shock which would cause refraction of the flow around the body.

Samir: The actual Mach number in the case you refer to was 3.75. The data perhaps were not averaged on a time scale that would have revealed such a shock. Also, the shock stand-off distances may be larger or smaller than the $5R_0$ boom length, so that the probe never would have encountered the shock layer in this situation. However, even in unaveraged data, we saw no obvious evidence for the presence of a bow shock. I would not consider at this stage our observations at $5R_0$ to be conclusive of whether a bow shock type disturbance does or does not in fact exist.

Knott: Was Explorer 31 covered in solar cells? Could you comment on the performance of this satellite.

Wrenn: Only about 15% of the area of Explorer 31 was covered in solar cells. Great care was taken to ensure that the satellite did have an adequate clean conducting area because one of the mission objectives was to compare and evaluate a number of probe techniques. I would say that Explorer 31 was one of the best designed ionospheric spacecraft ever built, from this point of view.

PHOTOELECTRONS EMITTED FROM ISIS SPACECRAFT

G. L. WRENN

*Mullard Space Science Laboratory, University College London,
Holmbury St. Mary, Surrey, England*

and

W. J. HEIKKILA

University of Texas at Dallas, Box 30365, Dallas, Tex. 75230, U.S.A.

Abstract. The ISIS-1 and ISIS-2 satelites carried Soft Particle Spectrometers furnished by the University of Texas at Dallas. Observations of low energy electron fluxes are presented in which photoelectrons emitted from the spacecraft surface can be identified. The intensities and energy distributions of these fluxes are described and an attempt is made to relate them to the emission characteristics.

1. Introduction

A spacecraft in orbit can collect whatever charged particles are present in its vicinity, it can also lose electron charge by means of photoemission or secondary emission. The result is that the spacecraft surface charges up to such a potential (satellite potential) that the nett current to the surface is zero. In the plasmasphere the currents due to the thermal plasma dominate and because the electron random velocities are much larger than those of the ions, a satellite potential of order -1 V results. In the magnetosphere these currents are generally small compared to the photoemission contribution and then the satellite tends to be positive by up to a few volts. In these circumstances considerable variation can occur due to irregularity in the cold plasma population and the higher energy fluxes, e.g. auroral electrons. In an extreme case where there are no incident electrons the satellite potential must tend to $+\infty$ V, the introduction of a few incident electrons would then drastically reduce this potential but there have been observations of potentials of more than $+20$ V (Norman and Freeman, 1973). It could be that secondary emission plays a significant role in some situations (Knott, 1972) but for a sunlit satellite it is believed that such a current would normally be small compared to that due to photoemission. The satellite potential would then depend upon a balance between the flux of escaping photoelectrons and the flux of ambient electrons incident on the satellite surface, assuming ion currents can also be neglected. Grard (1973) has considered this problem. The differential flux of photoelectrons emitted under solar irradiation is given by

$$H(W) = S(W) Y(W),$$

where $S(W)$ represents the solar spectrum and $Y(W)$ is the photoelectric yield, all being functions of photon energy, W.

The total flux of photoelectrons emitted then contributes a saturation value,

R. J. L. Grard (ed.), Photon and Particle Interactions with Surfaces in Space, 221–230. *All Rights Reserved*

I_s as

$$I_s = \int_0^\infty H(W) \, dW$$

and a saturation current eI_s.

In order to understand how variations in satellite floating potential might be related to incident flux changes it is necessary to know the form of the energy distribution function, $F(E)$, of these photoelectrons emitted with energy E. A number of laboratory studies have been conducted (Feuerbacher and Fitton, 1972) but it is difficult to reliably simulate the solar radiation. Grard (1973) has employed some empirical models based on experimental measurements and determined an expected satellite potential for a specific situation at geostationary altitude using the assumption that

$$F(E) = I_s \frac{E}{(e\phi_0)^2} \exp\left(-\frac{E}{e\phi_0}\right),$$

where $e\phi_0$ is the most probable energy of the distribution and is of the order of 1 eV.

If the satellite potential is ϕ_F volts then electrons with an energy greater than $e\phi_F$ can escape and thus the escaping flux is

$$I_e = \int_{e\phi_F}^\infty F(E) \, dE$$

and equilibrium results when $I_e = I_i$, the incident electron flux. The latter may well be composed of both 'cold' and 'hot' components.

No in-flight measurements of photoelectrons emitted from spacecraft have previously been available but Langmuir probes have yielded estimates of saturation photoemission currents between 4×10^{-9} A cm^{-2} and 9×10^{-9} A cm^{-2} (Hinteregger et al., 1959; Bourdeau et al., 1966; Wrenn, 1966; Norman and Freeman, 1973).

2. Measurements

The ISIS-1 and ISIS-2 satellites carried Soft Particle Spectrometers furnished by the University of Texas at Dallas. The instrumentation is described by Heikkila et al. (1970) and Table I lists a number of pertinent parameters. Figure 1 presents an example of the data received from ISIS-1 in the form of a grey scale spectrogram and plots of total number energy fluxes. The electrons seen here are principally of ionospheric origin, these must have been produced by photoionisation in the conjugate ionosphere because the local ionosphere was in darkness, they are peaked along the field line direction.

Sun pulses are present when sunlight enters the spectrometer directly but smaller subsidiary peaks are also observed. Figure 2 gives an expansion of part of the record showing the variation in total flux and the pitch angle as measured by on-board magnetometers. The subsidiary peaks occur at a pitch angle of 90° (0° ≡ field aligned moving away from the equatorial plane

180° ≡ field aligned moving towards the equatorial plane). Consideration of the geometry of the situation in Figure 3 clearly indicates that these peaks are due to

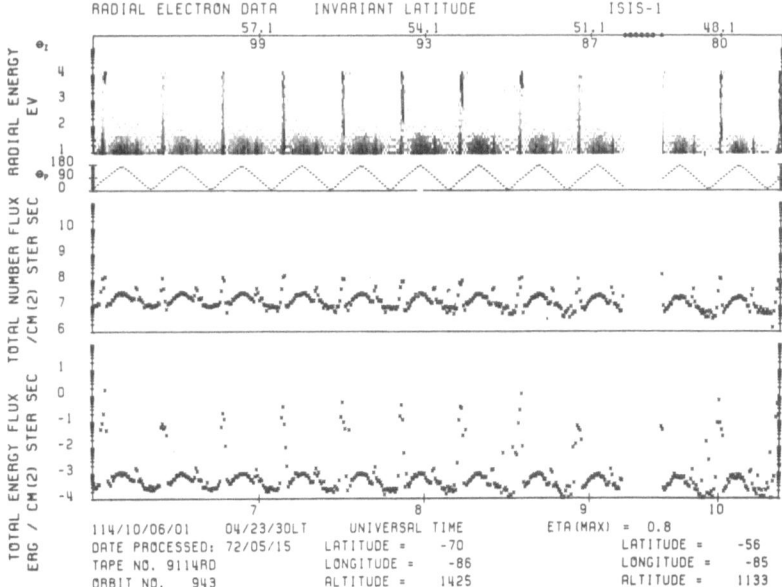

Fig. 1.   ISIS-1 Spectrogram showing photoelectron fluxes.

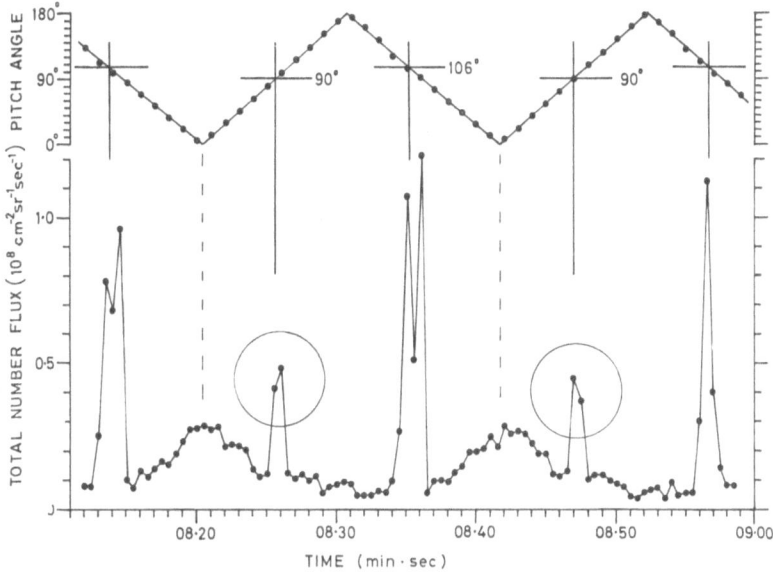

Fig. 2.   Spin modulation of total number flux with pitch angle.

TABLE I

Instrument parameters for photoelectron measurements

	ISIS-1	ISIS-2
Altitude (km)	570 → 3500	1380
Operation dates	37'69 → 299'69	106'71 →
Energy range (eV)	9.5 → 11600	5 → 13150
Energy resolution	0.8	0.25
Energy overlap	0.45	0.63
Angular ⎱ FWHM	7° × 25°	4° × 13°
Response ⎰ 1%	16° × 36°	9° × 27°
Geometric factor (cm² sr)	0.0012	0.0005
Detection efficiency	0.3 → 0.7	0.7 → 0.8
Count period (s)	0.0111	0.0111
Satellite spin period (s)	20	20

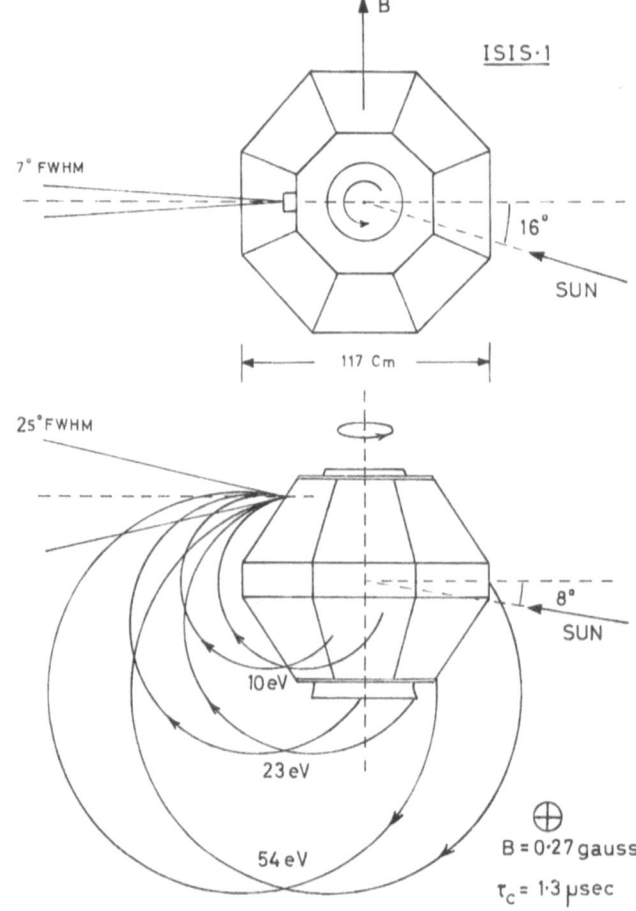

Fig. 3.   ISIS-1 orientation with respect to Sun and magnetic field on 24.4.69 at 10.08.30 UT.

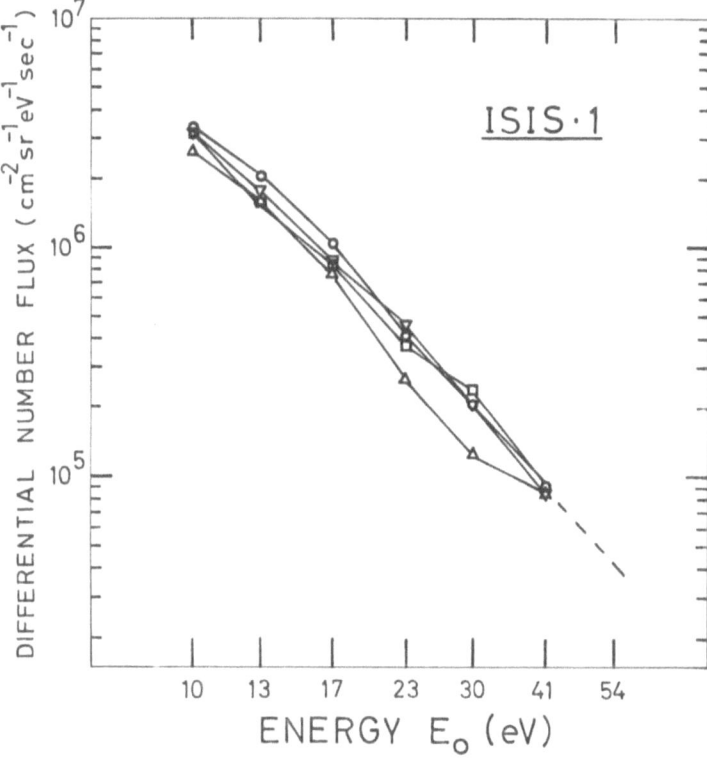

Fig. 4.   Energy spectra of photoelectrons emitted from ISIS-1.

electrons emitted from the satellite and returning due to their trajectories in the mag-netic field. The position of the solar and magnetic vectors are shown for the spin orientation corresponding to the peaks, the acceptance angles in the two planes are also illustrated. Clearly only electrons with a pitch angle of $\sim 90°$ can enter the anal-yser and trajectories for such electrons of 3 different energies are shown with incident angles of $\pm 12.5°$ which is the FWHM angular response in the meridional plane. It should be noted that the gyration period, τ_c is so small that the satellite can be con-sidered as stationary. It is immediately obvious that it is very difficult to relate the measured flux to an emission flux because of the constraints imposed by the allowed trajectories on the area and angle of emission. Further complication is introduced by the shape of the satellite and its surface variability. Having said that, Figure 4 gives the number flux spectra for the four sweeps corresponding to the circled points in Figure 2. Averaging these and subtracting the small ionospheric contribution leaves a total flux of photoelectrons with energy >7 eV as 2.4×10^7 cm^{-2} sr^{-1} s^{-1}. It must be admitted that there is some uncertainty in the energy scale deduced from a three decade exponential sweep, the suspicion is that the bottom end really corresponds to a somewhat higher energy (Wrenn and Heikkila, 1972); this actually appears to be borne out by the trajectory considerations of Figure 3.

Figure 5 presents similar ISIS 2 data again showing a peak at the 90° pitch angle position. Here there are no direct sun pulses, the collected ionospheric photoelectrons are seen to be nearly isotropic although the direct flux clearly exceeds the conjugate one. The trough at 53.52 probably results from spacecraft shadowing of incoming electrons with pitch angle $\sim 90°$. Figure 6 illustrates the satellite orientation in this case and the permitted trajectories for the smaller acceptance angles. Notice that the solar illumination is now such that the higher energies might be cut-off; Figure 7 shows the two appropriate spectra as open symbols. Subtraction of the ionospheric photoelectron contribution then yields the closed symbols and the total fluxes with energy > 5.8 eV are 4.4×10^8 cm^{-2} sr^{-1} s^{-1} ($\bullet$) and 2.4×10^8 cm^{-2} sr^{-1} s^{-1} ($\blacktriangle$).

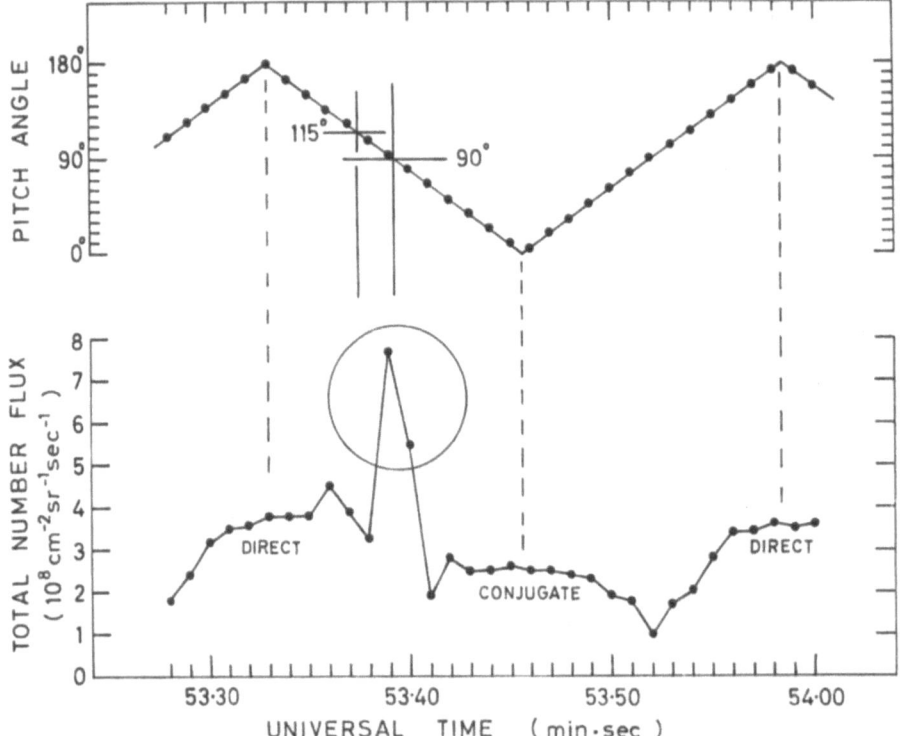

Fig. 5.   Spin modulation of total number flux with pitch angle. (ISIS-2 3.9.71 at 22.53.40 UT).

The ISIS-2 instrument incorporated a 5 V repeller screen in order to tie down the bottom end of the exponential sweep and this promotes confidence in this energy scale.

3. Discussion

Only two examples of data have been presented but they were selected because the photoelectrons of satellite origin could be clearly identified. Although the character-

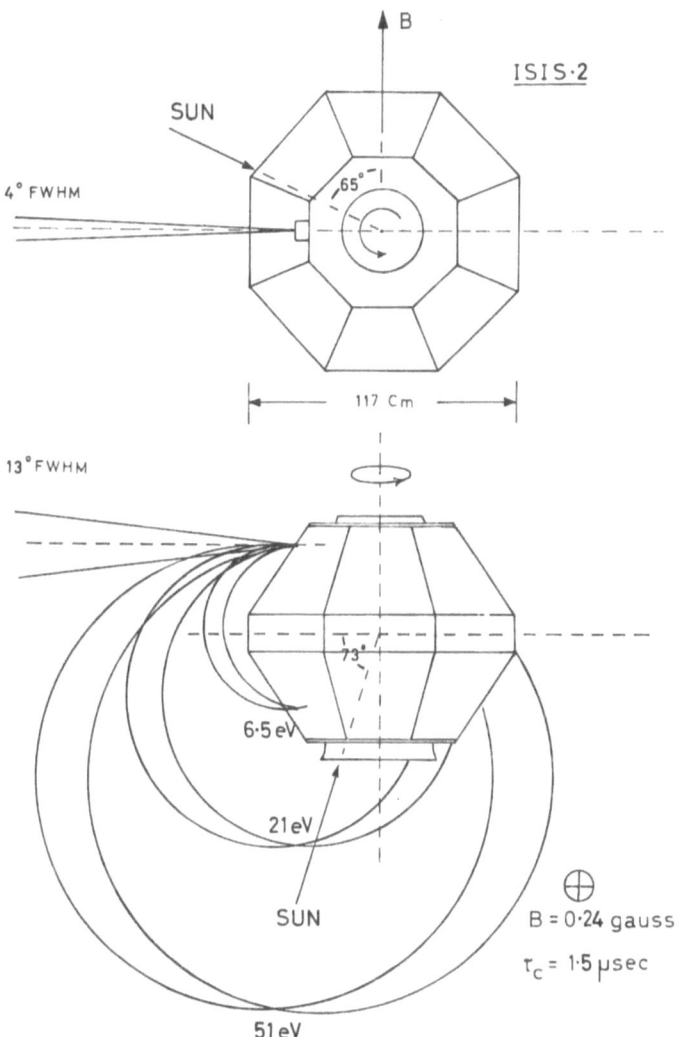

Fig. 6. ISIS orientation with respect to Sun and magnetic field at 22.53.40 UT.

istic 90° pitch angle peaks are common throughout the available data, the orientation and location of the satellites is seldom suitable for good resolution of these signals. Direct sun pulses or incident fluxes often obscure them, and since the peaks are quite sharp it is necessary for the radial detector to be in a mode with good telemetry sampling. It is useful if the satellite spin axis is perpendicular to the magnetic field, this rarely happened for ISIS-1 but ISIS-2 does operate in a cartwheel mode which is more favourable. It is really not meaningful to try to relate these measurements to the theoretical or laboratory emission fluxes in a rigorous way because of the difficulties indicated above.

The total flux estimates can be compared to the saturation flux.

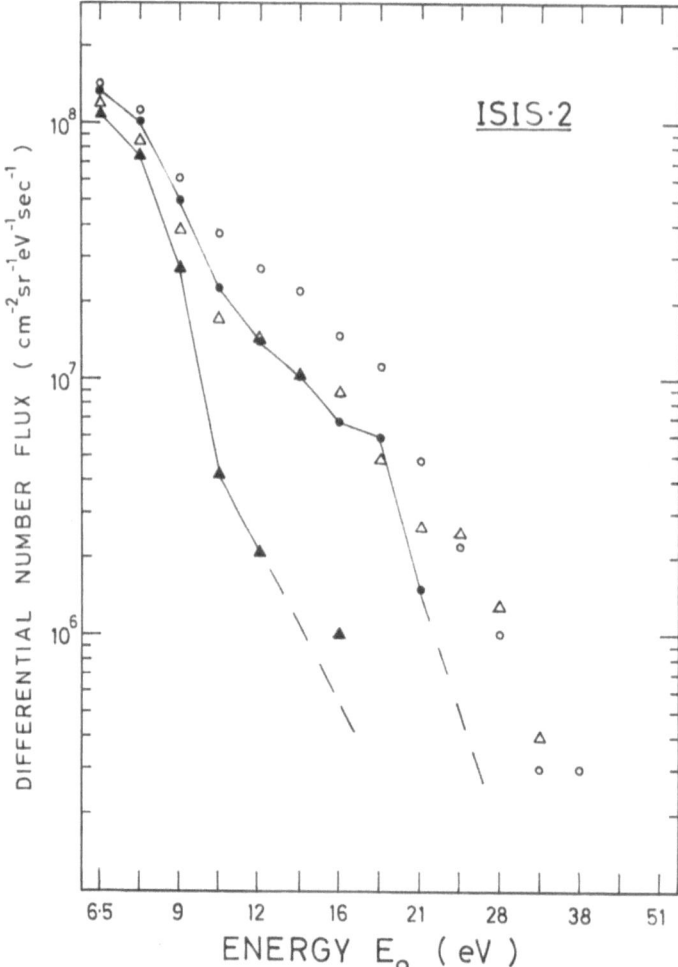

Fig. 7.   Energy spectra of photoelectrons emitted from ISIS-2.

Taking $eI_s = 5 \times 10^{-9}$ A cm^{-2} to be a typical current density gives

$$I_s = 3 \times 10^{10} \text{ electrons cm}^{-2} \text{ s}^{-1}$$
$$= 5 \times 10^{9} \text{ electrons cm}^{-2} \text{ sr}^{-1} \text{ s}^{-1}$$

assuming an isotropic emission.

The spectra shapes are clearly distorted by the variation of the effective area of emission with energy and it is not appropriate to attempt an estimation of equivalent 'temperature' for these distributions. It is interesting to compare them with the theoretical model described above. In Figure 8 the ISIS measurements are plotted with model curves calculated for $I_s = 5 \times 10^{-9}$ cm^{-2} sr^{-1} s^{-1}, taking $\phi_0 = 1$ V and $\phi_0 = 2$ V. In the circumstances the compatibility between data and theory is not unreasonable,

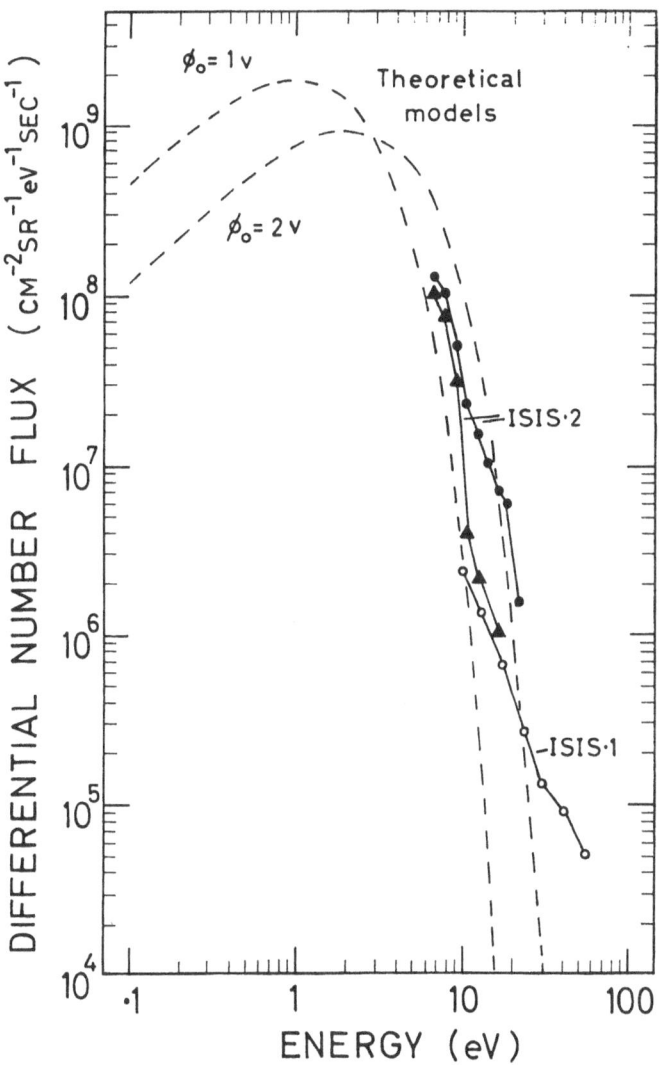

Fig. 8. Measured energy spectra compared with theoretical models with $\phi_0 = 1$ V and $\phi_0 = 2$ V.

the better agreement appears to be obtained with the higher 'temperature' but the more likely conclusion is that the distribution is not close to Maxwellian at high energies. This is in fact supported by the laboratory measurements used by Grard but it is possible that the effects of low wavelength radiation (e.g. He 304 Å) have been underestimated.

Little ISIS-2 data has so far been reduced but it is hoped that it will be possible to make a more quantitative study of the emitted photoelectrons when measurements under a variety of solar illumination conditions become available.

References

Bourdeau, R. E., Aikin, A. C., and Donley, J. L.: 1966, *J. Geophys. Res.* **71**, 727.
Feuerbacher, B. and Fitton, B.: 1972, *J. Appl. Phys.* **43**, 1563.
Grard, R. J. L.: 1973, *J. Geophys. Res.*, in press.
Heikkila, W. J., Smith, J. B., Tarstrup, J., and Winningham, J. D.: 1970, *Rev. Sci. Instrum.* **41**, 1393.
Hinteregger, H. E., Damon, K. R., and Hall, L. A.: 1959, *J. Geophys. Res.* **64**, 961.
Knott, K.: 1972, *Planetary Space Sci.* **20**(8), 1137.
Norman, K. and Freeman, R.: 1973, this volume, p. 231.
Wrenn, G. L.: 1966, Ph. D. Thesis, London.
Wrenn, G. L. and Heikkila, W. J.: 1972, *Trans. Am. Geophys. Union* **53**(4), 474.

DISCUSSION

Montgomery: What again is the source of the counting rate spike on the dark side away from the sun pulse? Does the energy spectrum and angular distribution of this spike fit this explanation? The spike seems to be quite narrow.

Wrenn: The spike is due to photoelectrons emitted on the sunlit side of the satellite, these can return to the dark side due to their orbits in the magnetic field. The angular distribution is consistent with this explanation, the acceptance angle of the instrument being quite narrow, and the energy distribution is reasonable in spite of the uncertainties involved.

ENERGY DISTRIBUTION OF PHOTOELECTRONS
EMITTED FROM A SURFACE ON THE OGO-5
SATELLITE AND MEASUREMENTS OF
SATELLITE POTENTIAL

K. NORMAN and R. M. FREEMAN

Mullard Space Science Laboratory, University College London,
Holmbury St. Mary, Surrey, England

Abstract. A Langmuir probe experiment, provided by the Mullard Space Science Laboratory and flown on the OGO-5 spacecraft, is used to measure the energy distribution of photo-electrons emitted from the gold surface of the spherical probe in sunlight. Comparison with laboratory measurements and theoretical models yields a most probable energy of 1.1 eV for the distribution. The effects of photo and secondary emission on the potential of the spacecraft are discussed and some examples are shown.

1. Introduction

A langmuir probe experiment to measure the electron density and temperature of the thermal plasma within the magnetosphere was included in the payload of the OGO-5 satellite. The experiment which has been described in more detail by Freeman *et al.* (1970), utilised a small gold-plated spherical sensor mounted at the end of a boom approximately 2.5 m from the main body and solar panels of the spacecraft. A saw-tooth ramp voltage was used to sweep the probe slowly through the local plasma potential and a small amplitude sinusoidal voltage was simultaneously applied. The amplitude of the AC component of the probe current was therefore proportional to the voltage derivative of the total current (di/dV) and was telemetered continuously.

Mott-Smith and Langmuir (1926) showed that the electron current (i) collected by a small spherical probe in a thermalised plasma, is given by Equation (1) when the probe is biassed negative of plasma potential, and by Equation (2) when it is biassed positive. i_0 is the current at plasma potential and is defined by Equation (3).

$$i = i_0 \exp\left(-\frac{eV}{kT}\right) \tag{1}$$

$$i = i_0\left(1 + \frac{eV}{kT}\right) \tag{2}$$

$$i_0 = NAe\left(\frac{kT}{2\pi m}\right)^{1/2}, \tag{3}$$

m = mass of the electron;
e = charge on the electron;
T = electron temperature;

k = Boltzmann's constant;

A = surface area of the probe;

V = potential of the probe relative to plasma potential;

N = electron density.

The positive ion contribution to the probe current may be neglected in the electron accelerating region (Equation (2)) and in the part of the retarding region (Equation (1)) close to plasma potential. Subject to this restraint, therefore, the probe response is due to the electron component of the plasma. The rms amplitude of the alternating component of the probe current (I_a) is given by Equation (4) in the electron retarding region, and by Equation (5) in the acceleration region, where it is independent of the probe potential.

$$I_a = \frac{i_0}{\sqrt{2}} \frac{eV_a}{kT} \exp\left(\frac{-eV}{kT}\right) \tag{4}$$

$$I_{a0} = \frac{i_0}{\sqrt{2}} \frac{eV_a}{kT}, \tag{5}$$

V_a = amplitude of the sinusoidal voltage.

Thus a graph of 'logI_a' vs 'V' has a slope in the retarding region which is inversely proportional to the electron temperature and a constant value in the acceleration region which is a known function of electron temperature and density. The discontinuity in the probe response at plasma potential is more clearly defined on the current derivative curve than on the total current curve (see Figures 1(a) and 1(b)) and the satellite potential is indicated by the potential of the probe, with respect to the satellite, as it passes through the local plasma potential.

The simple langmuir theory may not be applied to the experimental determination of electron temperature and density when the debye length (λ_D) in the plasma becomes comparable with the separation distance of the probe and satellite, since the probe is not in contact with the ambient plasma but is within the sheath around the spacecraft. For the OGO-5 experiment this point is reached when the density falls below approximately 10 cm^{-3} near the plasmapause boundary. However, it will be shown that a measurement of the local potential within the sheath is possible but that satellite potential calculated on this basis will be less than the actual potential difference between the satellite and ambient plasma. In summary, the experiment measures the potential of the satellite relative to the position of the probe 2.5 m away.

2. Photoelectric Emission

The probe was positioned at one end of the spacecraft and as the angle of the solar panels changed around the orbit it was sometimes shielded and sometimes exposed to sunlight. The observed response of the probe to solar illumination, compared to its characteristic plasma response as it is swept through plasma potential, is shown diagramtically in Figure 1(c) to 1(e). The first of these figures represents the variation in

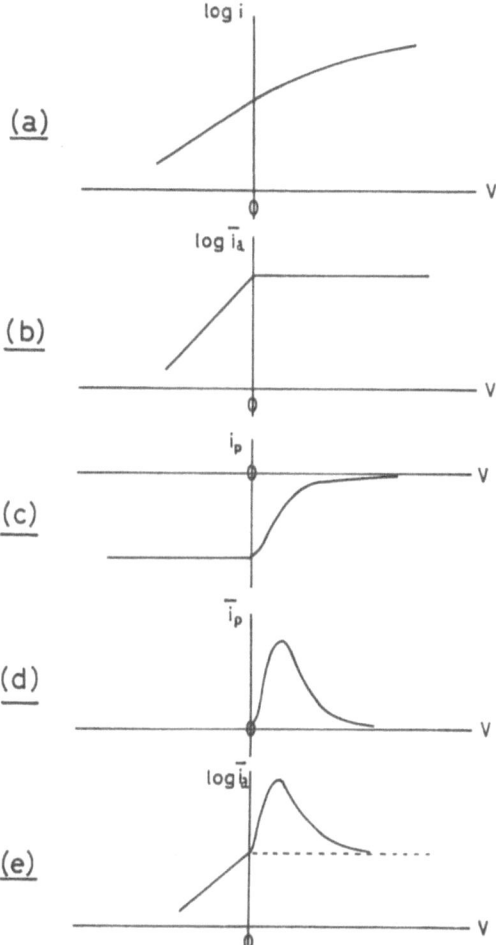

Fig. 1.   Diagramatic representation of response curves for a small, spherical langmuir probe near plasma potential: (a) logarithm of the probe current and (b) logarithm of the voltage derivative of the probe current for a thermal plasma; (c) probe current and (d) voltage derivative of the probe current for photoelectrons emitted from the probe; (e) logarithm of the current derivative with thermal plasma and photoemission contribution. All parameters are plotted against the probe voltage.

the total photocurrent from the probe surface, which is constant when the probe is negative of local plasma potential and decreases in the positive region as the increasing potential barrier prevents more of the emitted electrons from escaping the vicinity of the probe. The use of the AC technique produces the differential curve shown in Figure 1 (d) and it can be shown that it is proportional to the energy distribution of the photoelectrons leaving the surface, if the ambient plasma density is small enough to be neglected and spherical symmetry of the equipotentials is maintained near the probe.

Figure 1 (e) illustrates a probe characteristic in sunlight with some ambient plasma present. The relative magnitudes of the photo and plasma currents varied consid-

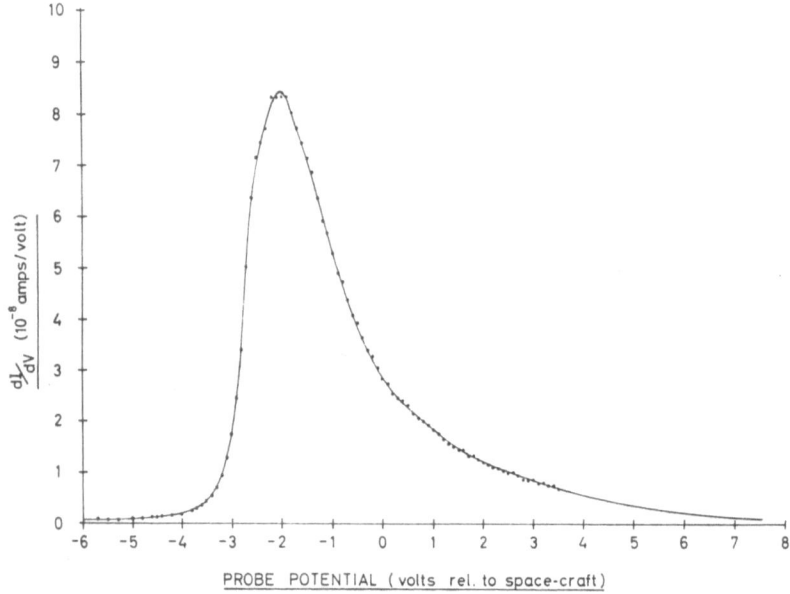

Fig. 2a.   Photoemission curve from the OGO-5 langmuir probe. The voltage derivative of the
current vs the probe voltage with respect to the satellite.

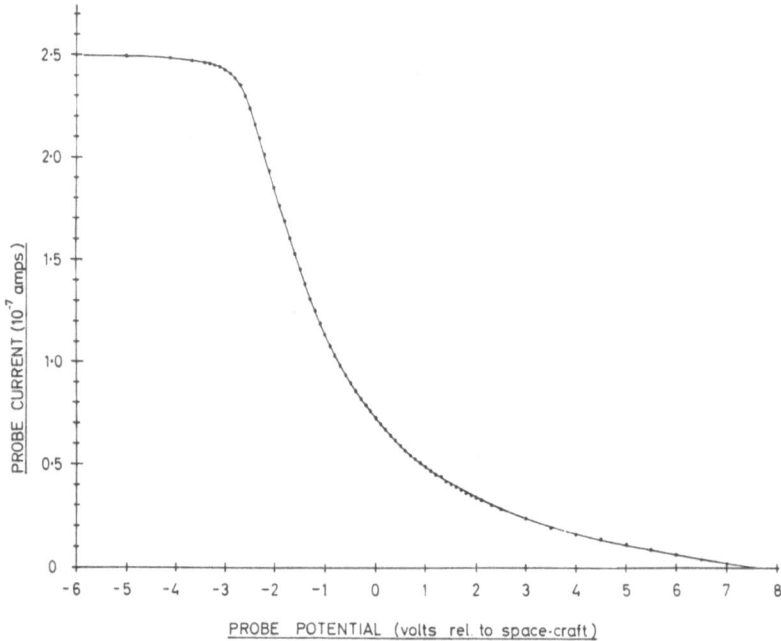

Fig. 2b.   Photoemission current from the OGO-5 probe vs probe voltage with respect to the
satellite.

erably and there were many periods when the satellite was in the outer magnetosphere; the probe in sunlight and the ambient plasma current negligible. These periods were readily recognised by the peaked photoemission response of constant maximum amplitude and the $i-V$ curve returning very close to zero at the positive end of the sweep; unlike Figure 1 (e). The shapes of the photoemission curves were practically identical and it was possible to subtract a typical curve from a composite such as Figure 1 (e) to obtain a reasonable estimate of the electron density when the peak photoemission response was several times greater than that of the plasma. The validity of the subtraction was checked by observing data as the probe went into or out of the solar panel shadow and checking that the photoemission subtraction did not produce a discontinuity in the calculated plasma density profile.

The differential photoemission curve is shown in Figure 2a and the corresponding emission current as a function of probe voltage, Figure 2b, was obtained by integrating the first curve. The total photo current from the 6 mm diam gold sphere was 2.5×10^{-7} A, and if this is averaged over the exposed normal area, a figure of $8.8 \times \times 10^{-9}$ A cm^{-2} is derived for the photoemission current density of a gold surface in sunlight.

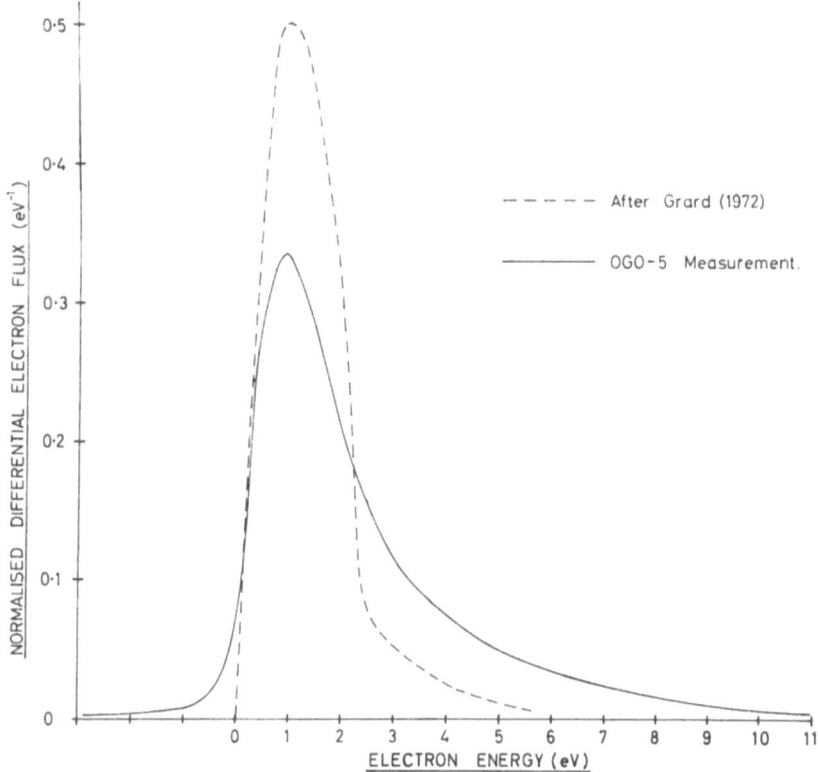

Fig. 3.   Energy distribution of photoelectrons from the OGO-5 probe (continuous line) and that due to Grard (broken line), vs arbitrary energy scale.

The energy distribution of the photoelectrons was obtained by normalising the differential curve to have unit integral over the measured energy range, and it is shown by the solid curve in Figure 3. The broken curve in the same figure shows the photoelectron energy distribution from a gold surface derived by Grard (1972) from a combination of solar spectrum measurements and a laboratory study of the photoelectric yield for gold at many different wavelengths. The latter curve was fitted to an energy distribution (P) given by Equation (6) and the most probable energy (E_0) was found to be 0.88 eV.

$$P(E) = \frac{E}{E_0^2} \exp\left(-\frac{E}{E_0}\right).$$ (6)

The two curves in Figure 3 were arbitrarily aligned so that their maxima coincided on the energy scale but clearly the langmuir probe distribution had a greater proportion of high energy electrons than that due to Grard. The theoretical distribution of Equation (6) was recomputed for different values of E_0 and it was found that the OGO-5 results corresponded quite closely to this type of distribution with a most

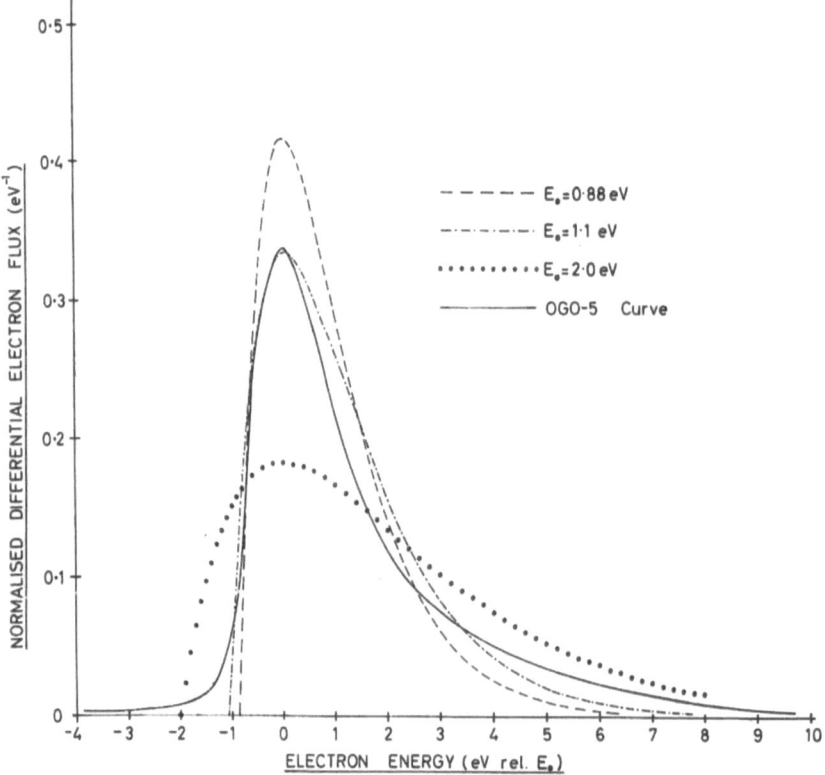

Fig. 4.   Comparison of experimental energy distribution curve for photoelectrons with three theoretical models.

probable energy, $E_0 = 1.1$ eV. Figure 4 demonstrates the comparison of the experimental curve with the theoretical expression, taking three values of E_0.

3. Satellite Potential Measurement

The langmuir probe electron current arose from three sources and as the relative magnitudes changed around the orbit, the dominant source was used to estimate satellite potential.

(1) *Thermal plasma.* Within the Earth's plasmasphere the local plasma potential was identified with the discontinuity in the shape of the probe 'current-voltage' characteristic, as is described by the langmuir theory. This defined the potential difference between satellite and plasma since the probe voltage was continuously monitored.

(2) *Photoemission from the probe.* This was nearly always the main current source, outside the plasmasphere when the probe was in sunlight. At the negative end of the potential sweep, electrons emitted from the probe surface terminate either on the satellite surface or in the surrounding plasma, but as it sweeps slowly positive, a deepening, electron potential well is created in the vicinity of the probe as it passes through the equilibrium potential at that point in the sheath surrounding the spacecraft. The photocurrent decreases at this potential which corresponds with the left hand edge of the curve in Figure 2(a). The satellite potential derived for that figure would be approximately $+3$ V and represents a lower limit.

(3) *Secondary emission from the satellite.* Low energy electrons were often detected far into the magnetosphere and in the solar wind, and were always observed when the satellite crossed the magnetosheath region. They had an approximately maxwellian energy distribution with a characteristic temperature of 10^4 K and their appearance was shown by Freeman (1973) to coincide with the increase in flux of higher energy electrons in the range 300 eV to 800 eV. They were attributed to secondary emission from the satellite surface as a result of the incidence of the higher energy electrons. The probe response was similar to that for a thermal plasma and it enabled the satellite potential to be estimated in a similar manner, but again it represented a lower limit to the actual potential. These secondary electron curves were used to determine satellite potential when the probe was shadowed by the solar panels or when the satellite was in the Earth's shadow.

The OGO-5 satellite potential followed a general behaviour pattern around its orbit which had a perigee of 1.1 Earth radii (R_E) and an apogee of 24 R_E. Close to perigee, the potential reached a minimum of -7 V to -10 V and it increased steadily as the satellite moved away from the Earth, reached 0 V at, or just beyond, the plasmapause boundary. At greater distances the potential was more variable but for much of the time it remained in the region $+3$ V to $+5$ V. The probe was swept over a 10 V range which could be stepped on command between the extremes of ± 20 V and a routine command sequence was operated around the orbit to follow the general pattern. Rapid changes in potential occurred but it was only possible to follow them if data

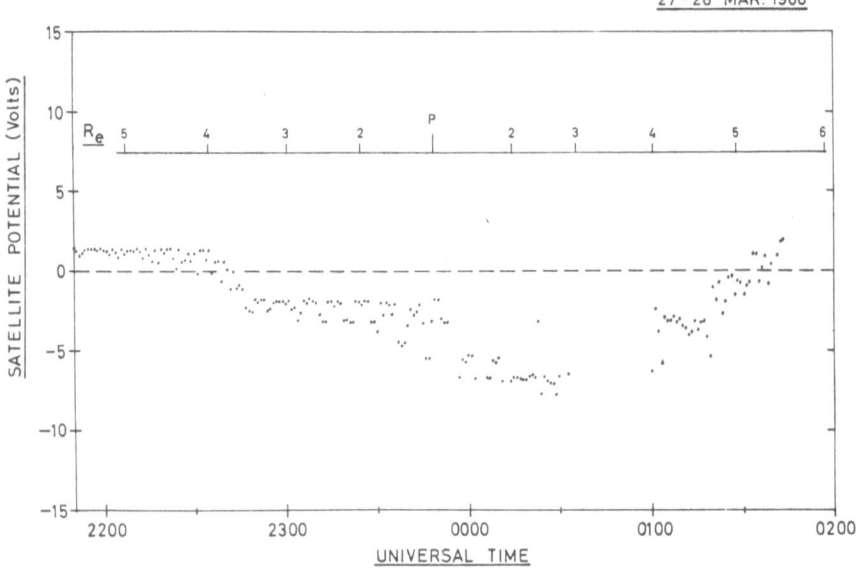

Fig. 5.   OGO-5 potential changes during a pass through perigee.

was monitored in real time, and the satellite potential remained within ±20 V. Some measurements of satellite potential are shown in Figures 5–9.

Figure 5 is taken from a satellite pass through perigee and the corresponding electron density profile showed the plasmapause boundary at 3.8 R_E inbound and 3 R_E outbound. However a plasma 'ledge' extended beyond the outbound plasmapause

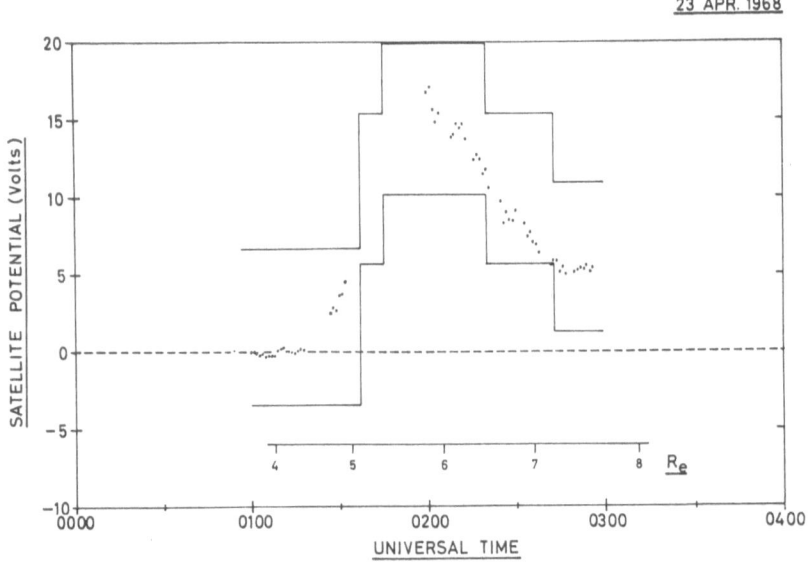

Fig. 6.   Satellite potential variation on 23rd April 1968.

boundary to 5.5 R_E. On all observed perigee passes there is good agreement between the disappearance of the thermal plasma and the point at which the satellite potential goes positive.

It was noted on some orbits that the satellite potential went much more positive than the more usual $+3$ to $+5$ V, when it was just outside the plasmasphere. On a few occasions the data was monitored and sweep range commands sent to track the changes in potential. Two examples are given in Figure 6 and 7. During the first, the satellite was outbound at 5 R_E when its potential rapidly moved more than 20 V positive. The upper and lower limits of the probe sweep range are indicated by the continuous straight lines. The potential remained outside the range of the instrument for 30 min, then steadily declined to its more usual value of $+5$ V over a one hour period. The satellite was at 0230 local time, the thermal plasma density was very low

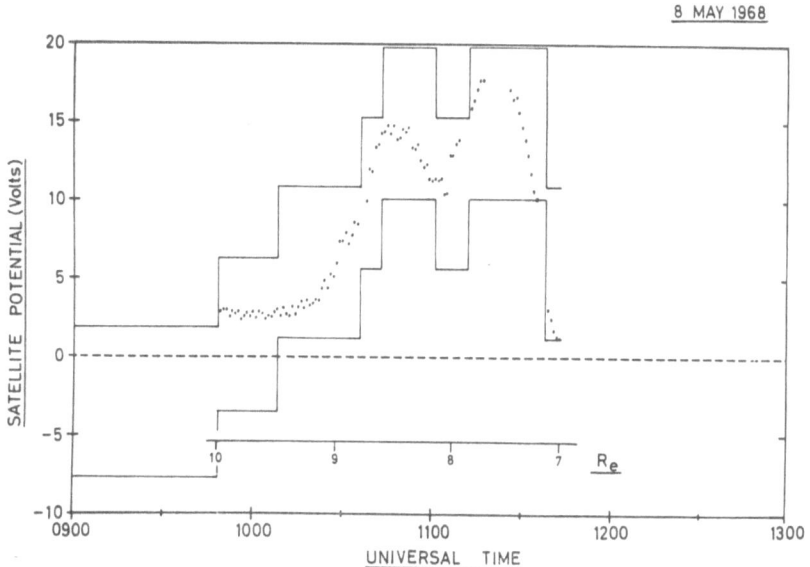

Fig. 7.   Satellite potential variation on 8th May 1968.

and no secondary emission was detected from the satellite which indicated an absence or very low flux of higher energy electrons. Under these conditions, photoemission would drive the satellite more positive than in the presence of a higher density plasma. Figure 7 illustrates a similar example on another pass when the satellite was inbound at a local time of 0730. The satellite potential on this occasion did not exceed 20 V but the plasma density and secondary emission fluxes were also very low.

There was a short period during August and September 1968 during which the satellite was eclipsed by the Earth when it was outside the plasmasphere. This gave an opportunity to study the behaviour of satellite potential in the low density magnetospheric plasma with no photoemission from the surface. Seven inbound passes crossed

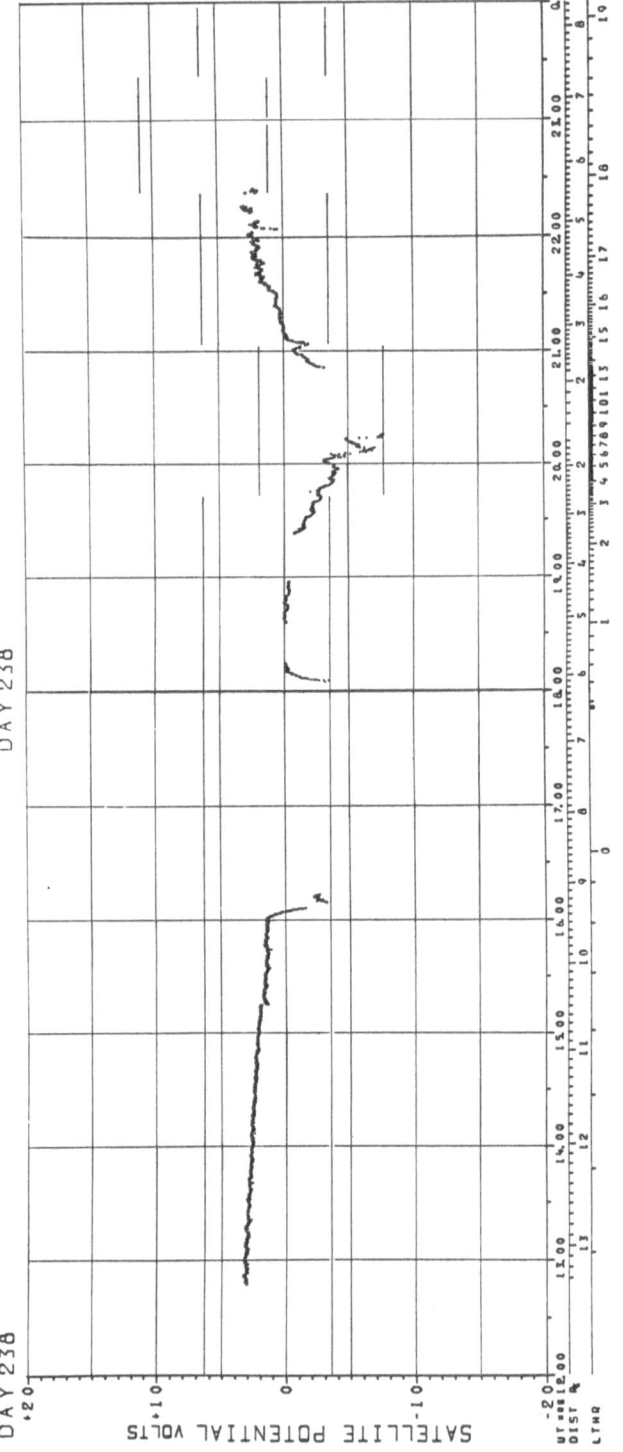

Fig. 8.   Change in OGO-5 potential during eclipse period on 25th August 1968.

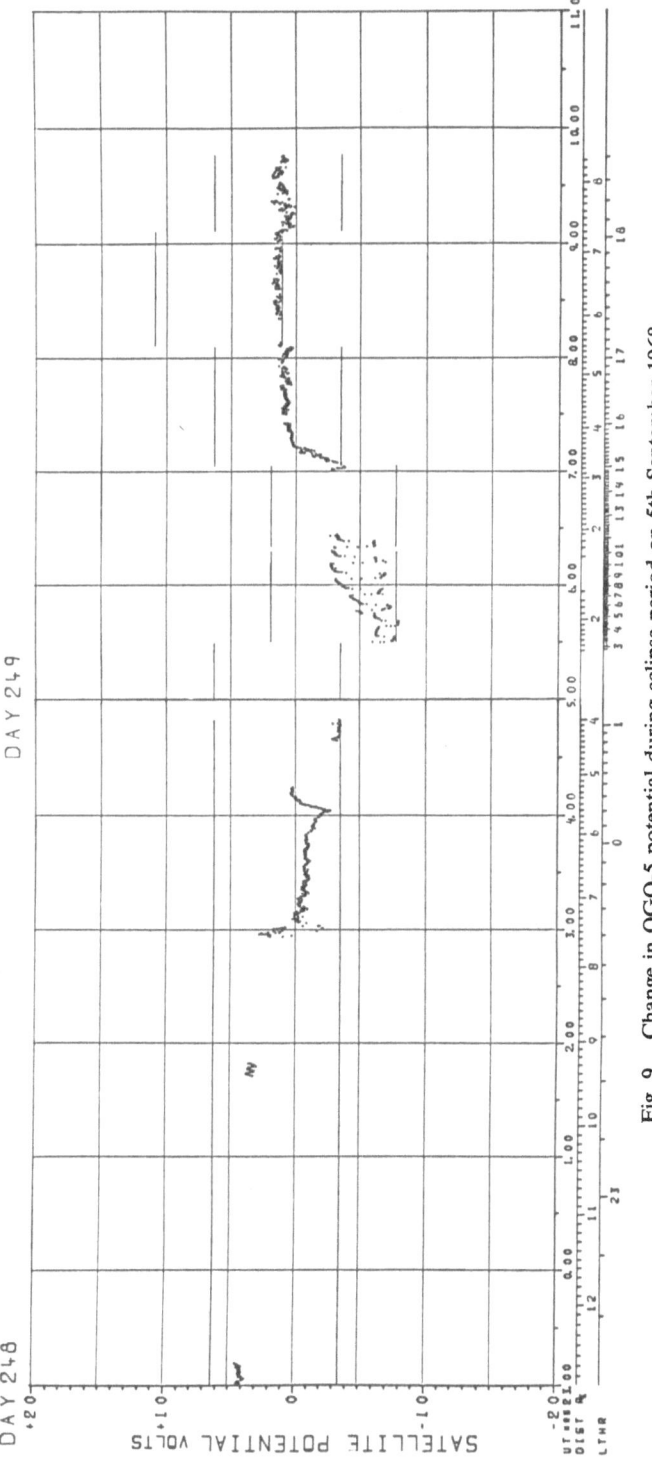

Fig. 9. Change in OGO-5 potential during eclipse period on 5th September 1968.

the Earth's shadow at geocentric distances between 6 R_E and 10 R_E but sufficient real time data was available from only two. The first was on 25 August (Figure 8) and the eclipse lasted from 1609 until 1812 UT; the second (Figure 9) was a shorter eclipse on the 5 September which lasted from 0256 until 0401 UT. Both figures show the satellite potential plotted against universal time, geocentric distance in Earth radii and local time. The sweep range was commanded on a routine basis during these orbits and at the time of the eclipse the probe sweep covered a measurement range of -3.5 to $+6.5$ V, as indicated.

Figure 8 shows that the satellite potential was a few volts positive and decreasing slowly as it approached the plasmapause. At the start of the eclipse it fell rapidly below the -3.5 V range limit and returned equally rapidly to 0 V at the end of the eclipse; shortly before the satellite entered the plasmasphere.

Inspection of the langmuir probe curves indicated that before the eclipse shown in Figure 9, the satellite potential was more positive than the top of the sweep range at $+6.5$ V. At the start of the eclipse, it dropped steeply to -1 V and remained fairly constant until near the end of the period when it began to go more negative. There was a steep 3 volt rise in potential as the satellite went into sunlight but it did not return to its pre-eclipse level because it was crossing the plasmapause boundary at this time.

The local plasma density appeared to be equally low at the start of both eclipses and although both showed a rapid decrease in satellite potential, there was a notable difference in the actual values.

4. Discussion

The derived photoemission current for gold of 8.8×10^{-9} A cm^{-2} is higher by a factor of between 1.3 and 6 than some earlier measurements made in the space environment (Bourdeau *et al.*, 1961; Hinteregger *et al.*, 1959; Wrenn, 1966). The value is probably too high because the illuminated area of the probe was treated as the projected area of the sphere normal to the Sun and implicitly assumes that the photoelectric yield efficiency is independent of the angle of incidence. Measurement of the variation in yield efficience would allow this figure obtained from a hemispherical surface to be converted properly to give the photocurrent from a normal plane surface.

The proportionality of the probe current derivative to the photoelectron energy distribution depended mainly on the spherical symmetry of the surrounding equipotentials. This would not appear to be the case when the photoelectrons are emitted from one side of the probe only and the ambient plasma density is very low. However, it was shown by Soop (1972) that the equipotentials in the vicinity of a conducting sphere in sunlight were spherically symmetrical even though the photoelectron cloud was confined mainly to the illuminated side. The more distant equipotentials would be distorted by the presence of the spacecraft when the plasma density was very low but there was no significant change in the shape of the measured photoemission curve when the plasma density was relatively high (i.e. > 20 electrons cm^{-3}). This is demonstrated by the ability to subtract the photoemission curve from a composite like Figure 1 (e) to obtain an acceptable plasma curve.

The potential of a spacecraft in the outer magnetosphere depends critically on a number of factors, the most important of which are solar electromagnetic radiation, the density and energy distribution of the local plasma and the nature and shape of the satellite surface. The OGO-5 spacecraft had a complicated surface and with the data available it was not possible to predict the behaviour of the surface potential. The results are therefore presented as examples of available measurements for this particular satellite. The examples were chosen to include the region of a geosynchronous orbit and the satellite eclipse period which were the subject of some recent publications (DeForest, 1972; Knott, 1972). The potential at 6.6 R_E was shown to vary between approximately 0 V and perhaps a few tens of volts positive when the satellite was in sunlight whilst in eclipse it could go negative to an undetermined extent. However, one eclipse measurement indicated that there was sufficient low energy plasma to maintain the potential at -1 V.

The term 'satellite potential' must be qualified for a spacecraft such as OGO-5 which did not have an equipotential surface. The measurements refer to the 'chassis ground' which was exposed at a number of places on the main body of the satellite, and on the back of the solar panels it was covered with a paint of relatively low resistivity. The rest of the satellite body was covered with aluminised mylar with the insulating side exposed.

Acknowledgements

We wish to thank Prof. R. L. F. Boyd, CBE, FRS for his interest and encouragement; Prof. A. P. Willmore who initiated the experiment; Mr J. Blades and Mr R. Wood of Pye, Ltd., Cambridge, for the electronic design; Mr P. Sheather for the mechanical design and the U.S. National Aeronautics and Space Administration for incorporating the experiment in the OGO-5 spacecraft.

References

Bourdeau, R. E., Donley, J. L., Serbu, G. P., Whipple, E. C., Jr.: 1961, *J. Astron. Sci.* **8**, 65.
DeForest, S. E.: 1972, *J. Geophys. Res.* **77**, 651.
Freeman, R. M., Norman, K., and Willmore A. P.: 1970, in V. Manno and D. E. Page (eds.), *Inter-correlated Satellite Observations Related to Solar Events*, D. Reidel, Dordrecht, Holland, p. 524.
Freeman, R. M.: 1973. Ph. D. Thesis.
Grard, R. J. L.: 1972, ESTEC Internal Working Paper, No. 663.
Hinteregger, H. E., Damon K. R., and Hall, L. A.: 1959, *J. Geophys. Res.* **64**, 961.
Knott, K.: 1972, *Planetary Space Sci.* **20**, 1137.
Mott-Smith, H. M. and Langmuir, I.: 1926, *Phys. Rev.* **28**, 727.
Soop, M.: 1972, *Planetary Space Sci.* **20**, 859.
Wrenn, G. L.: 1966, Ph. D. Thesis.

DISCUSSION

Rosenbauer: What was the surface material of the probe?

Norman: Gold, and for comparison with laboratory measurements the results of Dr Grard for gold were used.

Feuerbacher: Did you correlate the amount of electrons in the high energy tail with solar activity?

Norman: No.

Young: What were number densities when the satellite potential was very positive? What was the extent of this region in the radial direction; in local time?

Norman: Number densities were not measurable, but the spacecraft was outside the plasmasphere. Spatial extent was $\sim 1-2 R_E$ at 7–9 h local time.

Wiesemann: If you measure the photoelectrons at low plasma density, you might have no spherical symmetry of the probe sheath. Thus, it's questionable whether your characteristic really gives the energy distribution or not. This effect might possibly explain the difference between theoretical and experimental curves.

Norman: The probe sheath would not be spherically symmetrical at large distances because of the presence of the spacecraft. The effect of this asymmetry was neglected.

Weil: In some of your curves of spacecraft potential near perigee there was a sudden jump in potential near perigee and the curve was therefore very unsymmetrical. Is this because of a sudden change in the level of photoemission?

Norman: Yes. In fact, in those cases the satellite was going into eclipse just after the perigee point. I neglected to note this on the figures.

Pedersen: Are you sure that the observed energy distribution of photoelectrons near the Langmuir probe is not influenced by secondary emission from the probe or any satellite surface?

Norman: All measurements used for information about photo-electron energy distribution were taken at times when the ambient electron flux was so low that secondary emission could be neglected.

2.5. POTENTIAL OF THE SPACECRAFT SURFACE

LOW-ENERGY ELECTRON MEASUREMENTS AND
SPACECRAFT POTENTIAL: VELA 5 AND VELA 6

MICHAEL D. MONTGOMERY*

*Max-Planck-Institut für Physik und Astrophysik,
Institut für extraterrestrische Physik, Garching, Germany*

and

J. R. ASBRIDGE, S. J. BAME, and E. W. HONES

University of California, Los Alamos Scientific Laboratory, Los Alamos, N.M., U.S.A.

Abstract. Low-energy electron measurements from 10 to 1000 eV made by the Los Alamos plasma probes aboard the VELA Satellites provide a direct measure of the spacecraft potential with respect to the surrounding plasma when this potential rises above the lowest measured energy level, and in addition, give an indirect indication when the potential is smaller.

It has been found, as expected, that the spacecraft potential is a strong function of the incident electron flux from the ambient plasma and therefore depends strongly on the region in which the measurements are made (solar wind, magnetotail etc.). In addition, information concerning the dependence of spacecraft potential on photocurrent has been obtained from data gathered in the magnetotail and solar wind during eclipses of the Sun at the spacecraft. A preliminary evaluation of the data gives the following results:

(1) In the high latitude magnetotail, when the electron flux from the ambient plasma is low, the potential is positive and can rise to over 70 V.

(2) The measured relationship between the incident electron flux f_e and the satellite floating potential φ can be approximated (for f_e between 1×10^7 and 8×10^7 cm^{-2} s^{-1}) by the relation $f_e = (f_s/4)$ $\exp(-\varphi/\varphi_0)$; where f_s (the photoelectron saturation flux) $= 7.3 \times 10^8$ cm^{-2} s^{-1}, and $\varphi_0 = 13.6$ V.

(3) The spacecraft potential in the solar wind in sunlight (where $f_e \approx 3 \times 10^8$ cm^{-2} s^{-1}) is positive and is usually between $+3$ and $+5$ V.

(4) In the solar wind, but in darkness, the s/c potential is slightly negative – most probably between 0 and -2 V.

1. Introduction

A number of experimental and theoretical studies, both in space and in the laboratory, of spacecraft charging effects have been recently published (see for example Grard, 1972; Feuerbacher and Fitton, 1972; DeForest and McIlwain, 1971; and DeForest, 1972).

DeForest for example, has found direct evidence from ATS 5 measurements, that in the Earth's shadow at synchronous altitude and during the injection of substorm-associated ~ 10 keV particles, the body of the spacecraft was charged to negative potentials reaching -10 kV. As a result of these findings, the question of spacecraft charging effects has become important not only for the proper interpretation of plasma data, but also for the possible practical consequences of local discharges that evidently can occur as a result of these high potentials (Fredricks and Scarf, 1973).

The purpose of this publication is to describe the spacecraft charging effects of the

* On Leave from the Los Alamos Scientific Laboratory. Supported by the Alexander von Humboldt-Stiftung Special Program for Collaboration between Scientific Institutions in the Federal Republic of Germany and in the U.S.A.

R. J. L. Grard (ed.), Photon and Particle Interactions with Surfaces in Space, 247–261. All Rights Reserved
Copyright © 1973 by D. Reidel Publishing Company, Dordrecht-Holland

VELA 5 and 6 satellites and their relation to low energy electron measurements under the following conditions: (1) Magnetotail; outside the plasma sheet (low incident electron flux), in sunlight. (2) Magnetotail; inside the plasma sheet (varying amounts of incident electron flux), in sunlight. (3) Both 1 and 2 above in darkness (The Earth's shadow). (4) Solar wind in sunlight. (5) Solar wind in darkness (the Moon's shadow). All measurements were made between 18 and 18.5 R_E due to the Vela satellite's near circular orbits.

2. Experimental Description

Figure 1 is a photograph of the two Vela 5 satellites in their launch configuration. It can be seen that each satellite is roughly spherical in shape composed of a number of flat, triangular solar arrays. The plasma instrument is mounted on the top right-hand apex position of each satellite as it appears in the picture and can be recognized by the prominent narrow slit of the Sun sensor. A number of short whip-antennas can also be seen that are out of the direct field of view of the plasma detectors.

The directions of the Vela satellite's spin axes are actively controlled such that the spin vectors always point directly away from the Earth. It can then be easily seen from Figure 1 that in the solar wind, when the satellite is between the Sun and the Earth, the plasma instrument is on the sunlit side of the spacecraft while in the magnetotail near local midnight, the detector is in partial shade.

Details concerning the operation and design of these instruments or the similar VELA 3 instruments, as well as an outline of data reduction procedures, can be found in Bame *et al.* (1967) and Montgomery *et al.* (1970). For the purpose of describing these electron observations, we need only a few essential details:

The measurements were made with hemispherical plate electrostatic analyzers. The polar acceptance angle of each is 110° centered on the spacecraft equatorial plane, and the azimuthal resolution is 2.8° fill-width-at-half-maximum. During each rotation of the satellite (every 64 s), 16 electron and 16 proton differential energy spectra are generated. The spacing between individual spectra is 4 s (22.5°), and the measurement of each spectrum requires 0.320 s. The electron energy measurements cover the range 9.8–4150 eV (VELA 5) or 6.5–2322 eV (VELA 6) in 20 logarithmically spaced steps.

3. Observations

3.1. Magnetotail

Figure 2A presents typical energy spectra obtained in and near the plasma sheet by the VELA 6B analyzer. The curves represent the measured differential electron flux, averaged over one *s/c* revolution for 3 different 64 s periods in full sunlight. Curves 1 and 2 were obtained on two different days in the magnetotail, and the relative lack of electrons above 100 eV leads one to conclude that the spacecraft was not in the plasma sheet. The sharp downward break in the low-energy flux is interpreted as a direct indication of the satellite's positive potential relative to the surrounding medium since all higher-energy photoelectrons escape from the spacecraft. In this case the

Fig. 1. Photograph of the Vela 5 satellite pair – ready for launch.

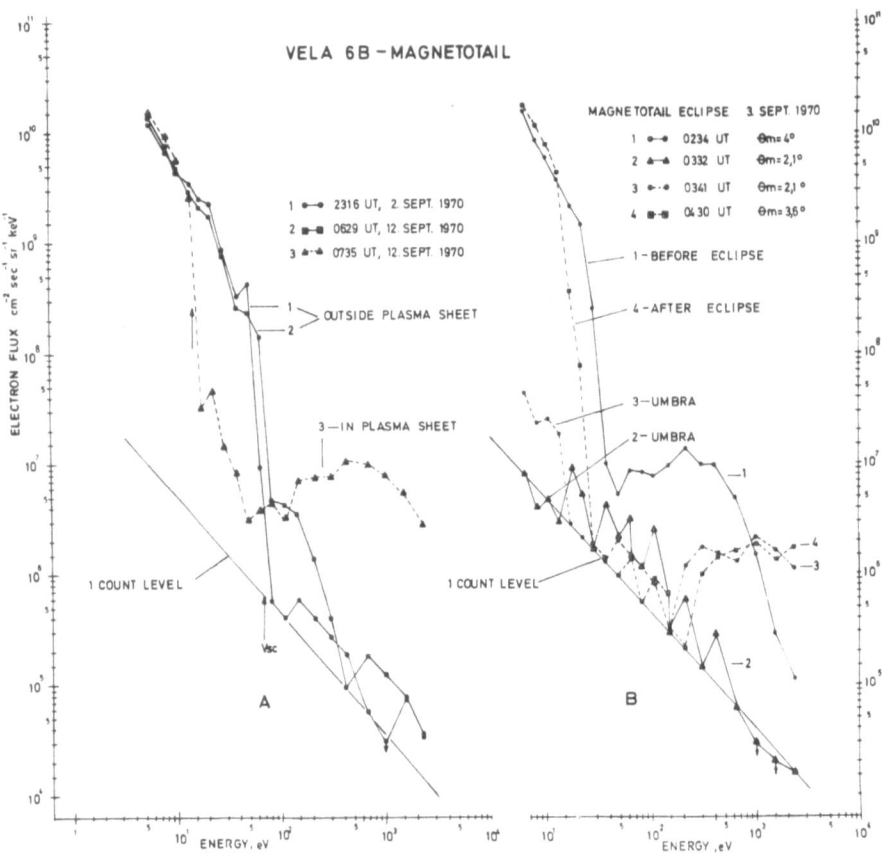

Fig. 2. Electron flux energy spectra in the magnetotail showing dependence of s/c potential on the plasma sheet electron flux, and s/c illumination. The spectra were obtained at the following positions in solar ecliptic radius, latitude and longitude: Curve A1 (18.5 R_E, 12.8°, 173°), A2–3 (18.5 R_E, 11.7°, 165.1°), B1 (18.5 R_E, 3.9°, 179,7°), B4 (18.5 R_E, −1.12°, 183,3°) the angle between the Sun-satellite line and the extendend Sun-Earth line is given by θ_m for curves B1–B4. The '1 count level' is shown in order that the number of counts represented by each point can be estimated.

satellite potential indicated by the arrow labeled V_{SC} is in the vicinity of 70 V, which is a more-or-less typical value for the so-called high latitude magnetotail. The spacecraft potential rarely exceeds 100 V, which means that at most times there is at least a small flux of incident electrons from the ambient plasma to the spacecraft that balances the photoelectron flux above ∼100 eV. This flux is usually, however, close to instrumental background and therefore difficult to measure.

Curve 3, Figure 2A, shows the changes in spectral shape that always occur when the plasma-sheet envelops the spacecraft. Upon detection of plasma sheet electrons, the break in the photoelectron spectrum moves to lower energies. As expected, the energy of the break, and therefore the satellite potential, is strongly dependent on the magnitude of the plasma sheet electron flux, that in this case was ≈5×10^7 cm^{-2} s^{-1}. The total flux integrated over all energies for curves 1 and 2 (mostly photoelectron flux) was $2 \pm 0.3 \times 10^8$ cm^{-2} s^{-1}, while the total flux obtained from curve 3 was $2.4 \pm 0.3 \times 10^8$ cm^{-2} s^{-1}. These fluxes are about equal, as they should be, if the plasma flux were simply replacing photoelectron flux.

The measured electron fluxes before, during, and after an eclipse of the sun by the earth are presented by Figure 2B. These data make it clear that the low-energy electrons observed in the plasma sheet are in fact photoelectrons. Curve 1 is a more-or-less typical spectrum obtained just before the eclipse. The satellite potential was $\sim + 32$ V, and the integrated plasma flux was $\sim 2 \times 10^7$ cm^{-2} s^{-1}. On the other hand, briefly after the start of the eclipse at 0300 UT, the flux over the entire energy range was very close to background. In the magnetotail, low energy electrons are completely absent only during eclipse periods. The fact that almost no high-energy electrons were observed comes about because the satellite during the early portion of the eclipse was outside the plasma sheet.

Curve 3, obtained later during the eclipse, shows a reappearance of plasma sheet electron flux, but this time at higher energy. However, this energization is in the wrong direction to be due to a change in s/c potential, since with the loss of photoelectron current the s/c potential should decrease. However, upon inspection of the low-energy part of Curve 3, one observes a weak flux of low energy electrons.

These low energy electrons appear and disappear in direct relation to the flux of more energetic plasma electrons, and thus appear to be secondary electrons generated by the incident 1–3 keV plasma sheet electrons. Therefore the s/c must be at a somewhat positive potential. In fact, since the flux seems to cut off sharply at about 12–15 eV, it appears that the s/c potential at this time is $\sim + 15$ V. If this interpretation is correct, it means that the flux of excaping secondary and backscattered electrons above 15 eV is enough, together with the incident proton flux, to balance the incident plasma sheet electron flux and thus hold the S/C potential positive. Another possibility, of course, is that this small flux of low energy electrons comes from secondaries generated within the detector itself, and has no meaning concerning the s/c potential. But since these 'secondary' electrons are observed to be approximately isotropic, this possibility does not seem likely. Spectrum 4 was obtained just after the s/c returned to full sunlight, and represents a typical photo- and plasma sheet electron spectrum.

Another more direct way to determine the change in spacecraft potential between sunlight and shadow from these and similar measurements, would be to associate energy shifts in the plasma sheet spectra with variations in illumination. However this method can only give a rough indication of such potential changes because the plasma sheet electron fluxes are so variable with time. On the other hand, the lack of large obvious energy shifts clearly related to changes in s/c illumination means the spacecraft potential does not shift more than a relatively small fraction of the characteristic plasma electron energy. The practical result is that spacecraft potentials appear to remain very much less than the 5–10 kV observed at synchronous altitude on ATS 5 by DeForest (1972), probably because the electron fluxes in the plasma sheet are much less and lie in an energy range where the secondary electron yield is quite high.

As has already been explained, a direct measurement of the dependance of s/c potential on incident electron flux can be obtained by simply noting the s/c potential (from the break in the low-energy electron flux spectrum) and the corresponding incident electron flux at various times when the plasma flux is clearly separated in energy from the photoelectron flux. Figure 3 is a scatter plot of results obtained from spectra similar to those of Figure 2A. The s/c potential φ is shown as a function of incident plasma electron flux f_e and it is easily seen that a rather well defined relationship is obtained between the s/c potential and incident flux.

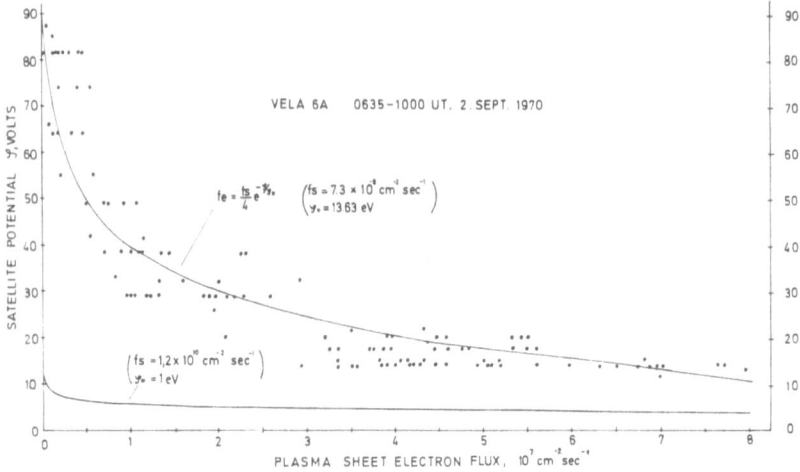

Fig. 3.   Measured dependence of the satellite potential on the electron flux from the plasma sheet.

Some, but not all, of the scatter in the points is due to the practical difficulty in estimating the s/c potential. The remaining scatter probably results from time variations in the higher energy ($\gtrsim 20$ eV) photoelectron flux due to variations in the incident treme UV-flux from the sun.

In any case, it is useful to compare the measured form of the charging characteristic over this potential range with the predictions of Grard (1972) for lower potentials. He employed measured values of the solar photon spectrum and laboratory measured photoelectron yield (accurate for energies $\lesssim 10$ eV) and assumed a simple Maxwellian energy distribution. An estimate of the s/c potential in sunlight is conceptually quite simple to make. The s/c will float at a potential such that the net current to the s/c is zero. In general, this current arises from the following sources: f_e, the electron flux from the ambient plasma; f_{pe}, flux due to escaping photoelectrons; f_{bs}, flux due to backscattered electrons; f_{se}, flux due to secondary electrons generated by incident plasma electrons that can escape the s/c potential; f_p, flux from incoming protons, and f_{sep}, flux due to secondary electrons generated by incoming protons. The corresponding integrated flux balance can be written as follows:

$$F_e = F_{pe/4} + F_{bs} + F_{se} + F_p + F_{sep}, \tag{1}$$

where $F_e \equiv f_e A$, etc., A is the surface area of the s/c, and the factor $\frac{1}{4}$ is due to the fact that the s/c is illuminated from only one side. All other fluxes in this case are assumed isotropic. Since the incident proton flux f_p, from the plasma is only about 10% of the incident electron flux f_e, the terms F_p and F_{sep} will be neglected. In addition since the s/c potential is positive, at least in sunlight, many of the lower energy secondary electrons will not escape and therefore not contribute. However, the term F_{bs} due to electron backscatter can, as discussed earlier, be a sizable fraction of the incoming plasma electrons F_e. Direct measurements at synchronous altitude (DeForest, 1972) are consistent with an average backscatter fraction of $\sim\frac{1}{2}$ while in our case it is probably even higher. However, for simplicity we will here assume a negligible backscatter fraction. In this case, we need only consider a simple balance between the flux of incident plasma electrons and the fraction of photoelectrons that can escape the s/c potential, φ:

$$f_e \simeq f_{pe}(\varphi)/4, \tag{2}$$

where $f_{pe}(\varphi)$ can be found by integrating the differential photoelectron energy spectrum $J(\psi)$ from $\psi = \varphi$ to ∞. Grard (1972) has shown that, at least for the low energy component, <5 eV, a Maxwellian model should provide good description of $J(\psi)$. tron Under this assumption f_{pe} may be expressed

$$f_{pe} = f_s \exp(-\varphi/\varphi_0), \tag{3}$$

where f_s is the saturation photoelectron flux and $e\varphi_0$ is the most probable photoelec-energy. Substituting (3) into (2) the following relation is obtained:

$$f_e = (f_s/4) \exp(-\varphi/\varphi_0). \tag{4}$$

f_s and φ_0 can be found from a fit of (4) to the directly measured relation of φ and f_e. The corresponding fitted curve and the resulting values of f_s and φ_0 are indicated by the upper smooth curve of Figure 3. The values of f_s and φ_0 valid for lower energy electrons and s/c potentials, and therefore higher ambient electron fluxes (Grard, 1972), are indicated by the lower curve. The effect of including a significant elastic backscatter fraction would simply reduce f_e, and therefore f_s, since the spectral shape would remain the same. It is emphasized that this fit is made only to parameterize the curve and not the compare with any particular physical model. The reason for the difference between Grard's and these values for f_s and φ_0 is clearly due to the presence of an enhanced high energy tail on the simple Maxwellian low-energy photoelectron distribution.

Therefore, it is seen that the photoelectron spectrum can, to first order, be represented by the sum of the two Maxwellians. The lower energy Maxwellian becomes dominant when f_e becomes comparable or greater than $f_s/4$. This means, that Grard's results should become dominent for $f_e \gtrsim 2 \times 10^8$ cm^{-2} s^{-1}. The maximum value for f_e in the data sample available for this study was not more than 8×10^7 cm^{-2} s^{-1}, so it was not possible to follow φ to lower values and thereby test the low-potential part of the relationship.

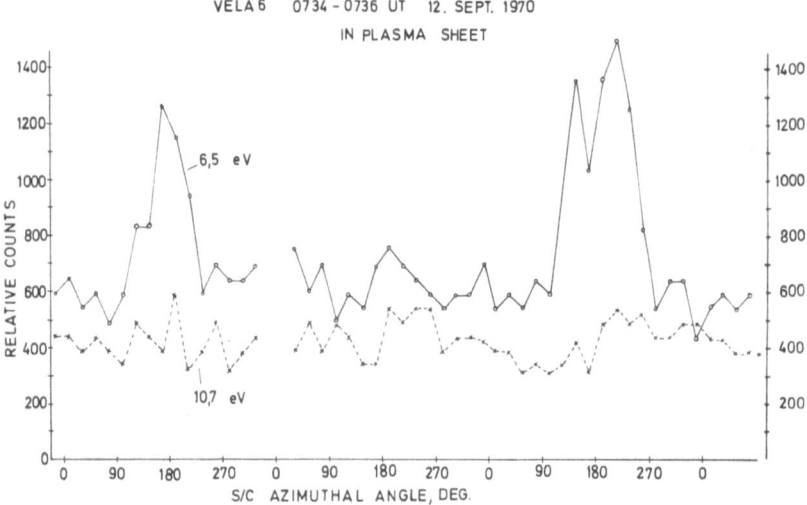

Fig. 4.   Example of a photoelectron angular distribution at selected energies in the plasma sheet.

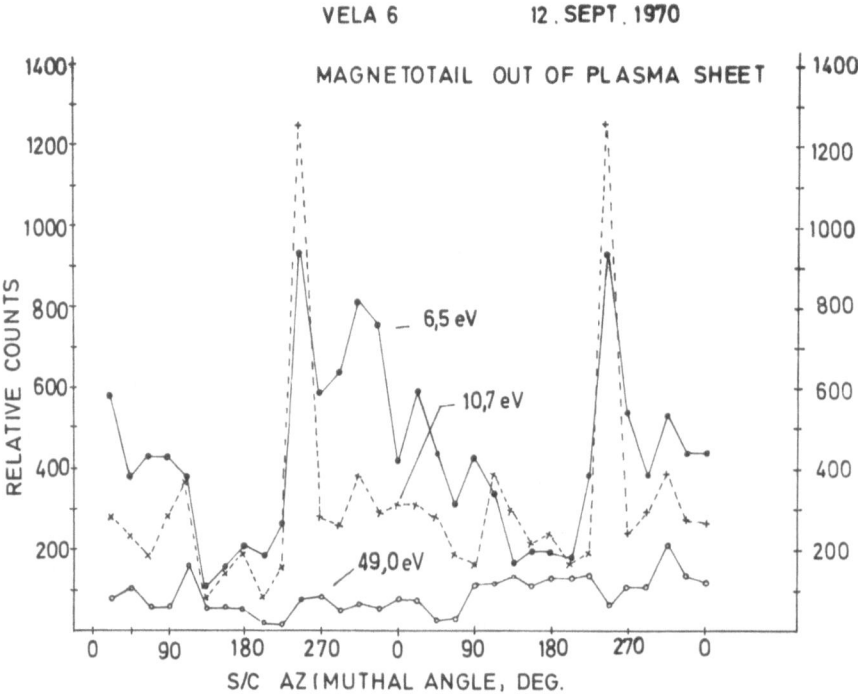

Fig. 5.   Example of an energy-dependent photoelectron angular distribution in the magnetotail outside the plasma steet.

Figures 4 and 5 respectively present typical photoelectron angular distributions measured both in and out of the plasma sheet. The data in both figures were taken when the angle between the spin axis and anti-Sun direction was $\sim 20°$. Thus the area around the detector was always at least partially in shadow. The azimuthal angle is measured with respect to that point in the s/c's rotation where the detector normal points closest to the Sun. Figure 4, obtained in the plasma sheet during a period when $f_e = 5 \times 10^7$ cm^{-2} s^{-1}, shows a strong anisotropy in the 6.5 eV channel while the 10.7 eV shows almost none. In addition the flux peaks near 180° where the detector is farthest from the Sun. When the s/c is out of the plasma sheet, Figure 5, the effect is even stronger and extends to higher energies. Again the photoelectron flux enhancements occur away from the Sun, but are shifted in azimuth with respect to Figure 4. In general the magnitude and phase of the spin modulation depend strongly on the magnitude of f_e. These modulation effects extend to higher energies and are stronger when f_e becomes less, i.e. when the s/c potential becomes more strongly positive. Even so, considerable numbers of low-energy photoelectrons are observed at all azimuthal positions even when the s/c potential reaches the most positive observed value of $\sim +100$ V.

The fact that the higher energy photoelectrons exhibit almost no angular anisotropy is not surprising since these electrons are expected to travel relatively far from the spacecraft and many of them can reach the dark side. The rather wide polar acceptance angle of the instrument makes it possible to detect electrons arriving at oblique angles ($\leqslant 35°$) to the surface. The observation of many low-energy electrons on the 'night' side of the s/c might at first seem more surprising, but this must be due to the fact the s/c potential is so screened by the photoelectrons that even 5–10 eV electrons can travel more than a s/c diameter away from the surface. Some, in addition, may be secondaries generated by incoming higher energy photoelectrons. The spin modulation of the low-energy electrons is more difficult to explain, but is probably due to differential charging of the insulated parts of the s/c surface. The potentials of these surfaces would be expected to vary with Sun angle and therefore could cause spinphased variations in counting rate at the detector. In the solar wind, such modulation effects are not seen.

3.2. IN THE SOLAR WIND

Figure 6 presents electron flux measurements obtained in the solar wind, again averaged over one s/c rotation. Curve 1, Figure 6A, is a reasonably typical solar wind electron spectrum, that was obtained shortly before the start of an eclipse by the Moon. The solar wind density was by chance only 2.5 cm^{-3} compared to an average of ~ 7.5 cm^{-3}. Because of the lower than average density in this example, the perturbation in the counting rate of the lowest energy channel due to s/c charging was somewhat larger than normal. The total electron flux for this example was $\sim 3 \times 10^8$ cm^{-2} s^{-1} and the proton bulk speed was 464 km s^{-1}. For purposes of comparison, a typical photoelectron spectrum obtained in the magnetotail and taken from Figure 2 has also been plotted (Curve 2).

VELA 5A — SOLAR WIND

VELA 6B — MAGNETOTAIL

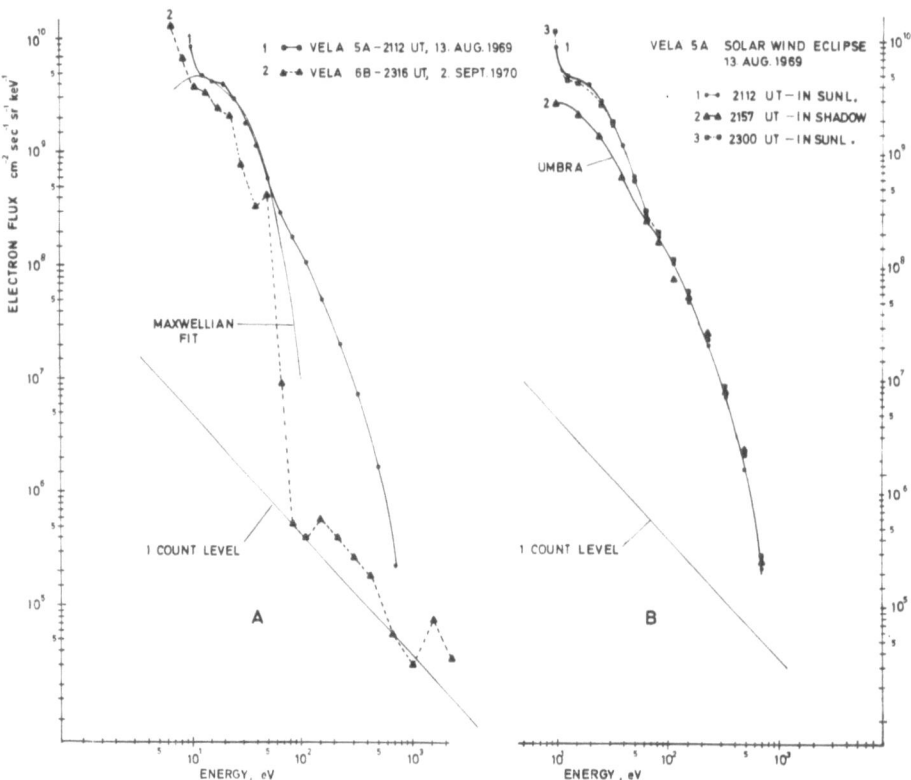

Fig. 6.   Electron flux energy spectra in the solar wind showing the dependence of the spectral shape on s/c illumination. Curve A1, from the magnetotail is included for purposes of comparison. The spectra were obtained at the following solar ecliptic positions (R, Lat., Long.): A3 (18.45, 4.9°, 26.6°); B1 (see A2, Figure 2): B3 (18.44, 8.3°, 29.0°).

The data analysis procedure used to calculate the electron plasma parameters such as density, temperature, etc. employs a 2-dimensional bi-Maxwellian fit to channel 2 through 6 to provide an estimate of the velocity distribution for energies below that corresponding to channel 2 where the measurements are not reliable (for details see Montgomery *et al.*, 1970). Since the calculated density depends rather strongly on this fit, it is sensitive to the shape of the electron flux curve for energy channels 2 through 6. Compare the fit and measured data points in Figure 6A. It can be seen that, excluding channel 1, the fit is quite good.

A general check on density measurements can be made by a comparison with the nearly simultaneously measured proton density. Although the proton density measurements near the time of the eclipse were compromised by the s/c attitude (the spin axis was pointing too close to the Sun), there is almost always good agreement between measured proton and electron densities over quite a wide range of values. Figure 7 is included as an example of this behaviour. Thus evidence is provided that, above

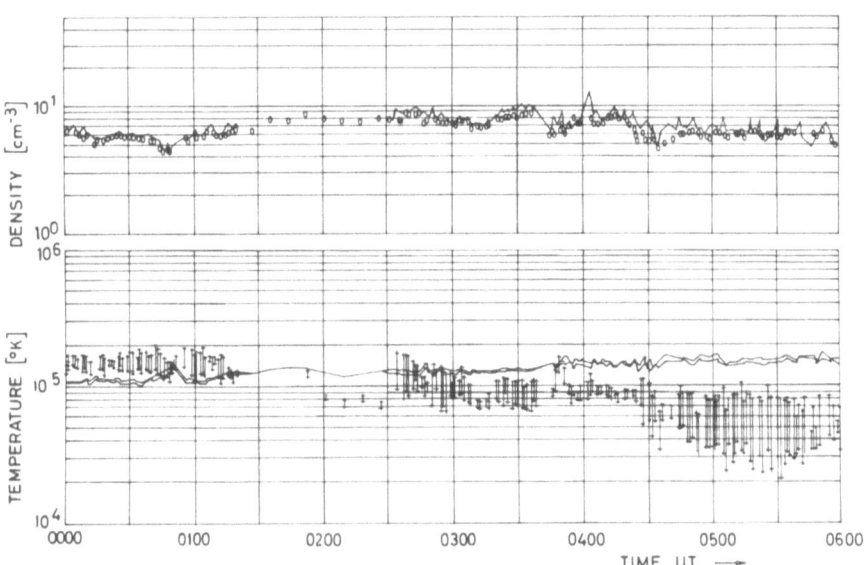

Fig. 7. Solar wind parameters measured by VELA 5 showing the general agreement between electron and proton densities. The short vertical lines are drawn joining the maximum and minimum proton temperatures, and thus their lengths represent the magnitudes of the temperature anisotropies. The two electron temperatures are indicated by the two closely spaced continuous curves. Proton density is indicated by the continuous curve while the electron density is shown by the small circles.

about 12 eV, the *shape* of incident electron flux is not significantly disturbed by *s/c* charging effects. The incident electron flux at this time was $\sim 3 \times 10^8 \text{ cm}^{-2} \text{ s}^{-1}$, and if a calculation using the results of Grard (1972) is made, it is found that the *s/c* potential should be only ≈ 2.3 V. If the disturbance in the counting rates at 10 eV is interpreted as being due to a focussing effect resulting from a positively charged *s/c*, the actual potential is probably somewhat larger than this theoretical value. On the other hand if the potential were much larger than 5 V the electron density would no longer follow variations in proton density due to simultaneous changes in *s/c* potential and distortion of the electron spectra (*s/c* charging effects are neglected for the purpose of data reduction). Thus, in the solar wind in sunlight the spacecraft potential probably is between $+3$ and $+5$ V.

Further information on the *s/c* potential can be obtained from data collected during the eclipse. Returning to Figure 6B, the changes in the electron spectra due to the absence of photoelectrons can be seen. In this case the counting rates at low energies are sharply reduced, and the apparent, derived density is about a factor of 2 less. An expected shift of *s/c* potential to a smaller value has occurred, and since the 10 eV electrons are not cut-off, or even noticeably affected, the potential must not be more negative than a few volts. In fact, it is possible by comparing flux curves (1 and 2 for example) to estimate the change in *s/c* potential as the *s/c* moves into darkness. If focussing effects are neglected (justified by chosing points at energies high compared to the estimated s/c potential) one can write the following relation between the two curves:

$$F_2(E - \Delta E) = F_1(E)\left(\frac{E - \Delta E}{E}\right)^2,$$ (5)

where F_1, (or F_2) is the measured counting rate as a function of energy E in sunlight (or shadow), $\Delta E/e$ is the negative shift of the potential (the reduction in energy of electrons having had original energy E), and $((E - \Delta E)/E)^2 = C(E, \Delta E)$ is the correction factor due to the change in phase-space sampling from an energy shift of ΔE (see Montgomery et al., 1970). The subscripts identify the corresponding curves. Using curves 1 and 2, and E in the range 20 to 40 eV, it can be seen by applying (5) that the shift between the curves is consistant with a ΔE of only 5 eV.

The solar wind observations can now be briefly summarized as follows: (1) The ≲10 eV part of the electron spectrum appears to be more distorted by s/c charging when the s/c is in sunlight than in shadow, thus the absolute value of the s/c potential is greater in sunlight. (2) The s/c potential shifts downward by about 5 V when the s/c enters darkness. (3) The s/c potential clearly does not exceed +10 V nor does it fall below −10 V. If the reasonable assumption is made that the s/c is somewhat negative in darkness, the s/c potential is most likely ∼0 to −1 V in shadow and therefore +4 to +5 V in sunlight.

The major consequence of these observations, of course, is that relatively few of the incoming solar wind electrons actually contribute to current flow away from the s/c. It is possible to roughly estimate this 'sticking' coefficient from the measurements simply by requiring a balance of electron and proton current during the eclipse. In this example the proton flux (directed flow) was 1.2×10^8 cm^{-2} s^{-1} (solar wind speed $=470$ km s^{-1}), while the isotropic electron flux varied from 3×10^8 cm^{-2} s^{-1} in sunlight to about 1.6×10^8 cm^{-2} s^{-1} during the eclipse. Taking into account the factor of ∼4 difference in total current contribution between directed and isotropic fluxes and assuming that 100% of the incoming protons are simply absorbed, it is found that $1.2/(4 \times 1.6) \approx 0.2$ of the incoming electron flux actually contributes to the current. Of course this estimate is only approximate since some secondary electrons will be generated by the incoming protons, and it is even possible that the s/c potential remains slightly positive during eclipse. It should also be mentioned that solar X-ray measurements made aboard the s/c show that the eclipse was essentially total during the period 2156–2201 UT, and the s/c was approximately 40 R_E downstream from the Moon thus far from any plasma wake region. This rather high back-scatter fraction is also consistant with the low-energy photoelectron angular distributions obtained in the magnetotail discussed in Section 3.1.

To complete the description of the eclipse and the effect of the ∼5 V swing in s/c potential on the counting rates and the derived solar wind density and temperature, the time histories of counting rates for four selected energy channels, and the corresponding density and temperature appear in Figure 8. It should be noted that during the

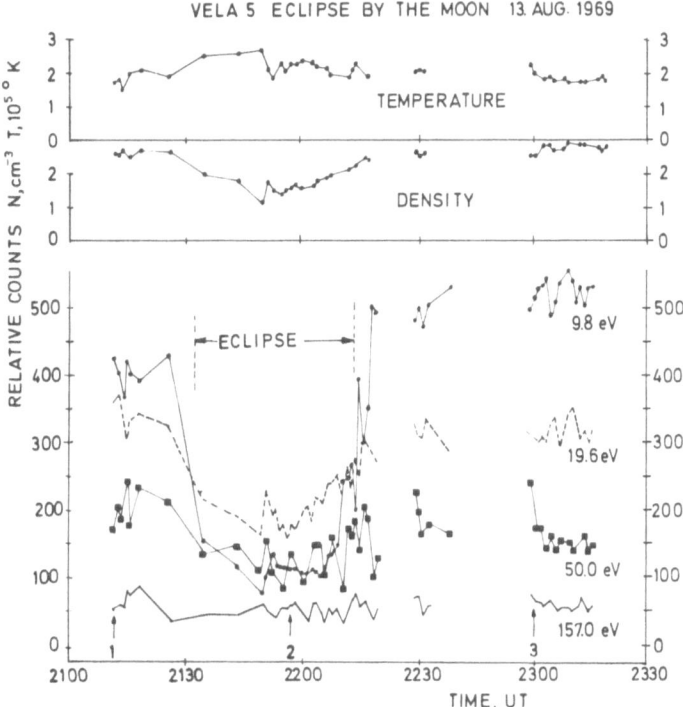

Fig. 8. VELA 5 electron temp, density, and relative counting rate for 4 selected energy channels during a solar eclipse by the Moon. Time points 1, 2 and 3 correspond to curves 1, 2 and 3 of Figure 6B. The proton bulk speed during this event varied from 470 to 460 km s⁻¹.

eclipse, the effect on the temperature is quite small while the apparent density is about a factor of 2 smaller. The widely spaced points are derived from so-called 'store' data – a much restricted one-dimensional data set. The resulting parameters are much less accurate.

4. Summary

The results concerning the electron measurements in the magnetotail are relatively clear. The relationship between the ambient plasma electron flux f_e and spacecraft potential φ can be approximated (following Grard (1972)) by the equation $f_e = (f_s/4)_e^{-\varphi/\varphi_0}$ where $f_s = (7 \text{ to } 8) \times 10^8$ cm⁻² s⁻¹ and $\varphi_0 = 13.6$ eV; for f_e in the range from $(1 \text{ to } 8) \times 10^7$ cm⁻² s⁻¹. Grard (1972) has obtained from laboratory photoelectron yield measurements $f_s \sim 1.2 \times 10^{10}$ cm⁻² s⁻¹ and $\varphi_0 \sim 1$ eV. However, this equation is expected to be valid for values of $f_e \gtrsim 1.2 \times 10^8$, i.e. low values of the s/c potential ($\lesssim 4$V).

The difference between these direct measurements and Grard's extrapolated results can then be easily explained by, and brings out the importance of, the weak high-energy tail on the photoelectron energy distribution due to solar radiation in the *EUV* and shorter wave lengths.

Data were not available at the time of this study to check the relationship over higher values of f_e and therefore cover values of φ where Grard's results should be applicable.

Angular distributions of photoelectrons measured in the magnetotail show that many low-energy photoelectrons reach the dark side of the spacecraft.

The solar wind observations appear to indicate the following:

(1) Excess counting rates in the lowest energy channel ~ 10 eV in sunlight indicate a s/c potential that is somewhat positive; probably somewhere between $+3$ and $+6$ V during periods when the incident measured ambient electron flux is ≈ 3 to 6×10^8 cm^{-2} s^{-1}, uncorrected for electron backscatter. The fact that the electron density accurately follows changes in the simultaneously measured proton density, Figure 7, means that the s/c potential is not much more than $+5$ V.

(2) Changes in energy spectra during eclipse, i.e. direct comparisons of energy spectra between sunlight and darkness, clearly indicate only a small decrease in the s/c potential of ~ 5 V. This means that during eclipse, the s/c potential was most likely only slightly negative $\simeq 0$ to -2 V. A possible explaination for such a small potential change is the high probability of backscatter of low energy plasma electrons. The resulting decrease in effective plasma electron current away from the s/c is also consistent with the higher than expected positive s/c potential in sunlight.

Since the magnitude of the s/c potential is smaller in shadow than in sunlight, the solar wind values obtained in shadow should be more accurate. In fact, the bi-Maxwellian fits to the data obtained in shadow are quite good even including the lowest energy channels. However, in spite of this, the measurements of the solar wind electron parameters are not seriously affected by the s/c potential in sunlight. It is seen that the electron density, aside from a constant, tracks the proton density over a wide range of values and the electron temperature is scarcely affected.

Since the only electron plasma parameters needed to complement the positive ion measurements in the solar wind are the temperature and higher moments of the velocity distribution (such as heat flux), and since these higher moments are not affected by a spacecraft potential of a few Volts, corrections for variations in s/c potential are not routinely carried out.

Acknowledgement

The authors thank Drs H. Rosenbauer and H. Miggenrieder of the Max-Planck-Institut für Extraterrestrische Physik and Prof. V. M. Vasyliunas of the Massachusetts Institute of Technology for helpful discussions.

This research was done as part of the VELA Nuclear Test Detection Satellite Program which is jointly sponsored by the Advanced Research Projects Agency of the Dept. of Defence and the U.S. Atomic Energy Commission.

References

Bame, S. J., Asbridge, J. R., Feldhauser, H. E., Hones, E. W., and Strong J. B.: 1967, *J. Geophys. Res.* **72**, 113. Asbridge

DeForest, S. E. and McIlwain, C. E.: 1971, *J. Geophys. Res.* **76**, 3587.
DeForest, S. E.: 1972, *J. Geophys. Res.* **77**, 651.
Feuerbacher, B. and Fitton, B.: 1972, *J. Appl. Phys.*, in press.
Fredricks, R. W. and Scarf, F. L.: 1973, this volume, p. 277.
Grard, R. J. L.: 1972, ESTEC Tech. Working Paper No. 663.
Montgomery, M. D., Asbridge, J. R., and Bame, S. J.: 1970, *J. Geophys. Res.* **75**, 1217.

ELECTROSTATIC POTENTIALS DEVELOPED BY ATS-5

SHERMAN E. DEFOREST

University of California, San Diego, Calif. 92037, U.S.A.

Abstract. Investigation of the properties of low-energy particles measured on board ATS-5 show that the synchronous spacecraft can charge to -12.000 V in eclipse, and several hundred volts in sunlight. Differential charging can produce local fields of several thousands of volts per meter in the near vicinity of the spacecraft surface. Time constants for charging can be less than a second to tens of minutes.

1. Introduction

ATS-5 was launched into synchronous orbit on August 12, 1969. Shortly thereafter, it was stationed at 105° W longitude (local midnight at 0700 UT) with orbital inclination of 2.30°. Before launch, research groups at both the University of California, San Diego and Lockheed, Palo Alto insisted that conducting collars be placed around the apertures of low-energy particle detectors. This was meant to be a partial shield against local electric fields which could affect particle counting rates. In addition, some concern was expressed about the fact that the viewing cones of two instruments looked out through a cylinder of solar cells. The general geometry is shown in Figure 1. In flight, ATS spins at about 100 rpm about the longitudinal axis. The spin axis is oriented parallel to the Earth's axis.

Within weeks it was obvious that these precautions had not been sufficient to prevent differential charging from taking place in the vicinity of the end detector. The big surprise, however, was that certain puzzling particle events could be shown to be due to the whole spacecraft charging to thousands of volts. No one else had reported seeing these kinds of potentials developed on spacecraft, although several predictions of a few volts potential had been made. Therefore, a paper (DeForest, 1972) was written which discussed this charging with emphasis on a model that had been developed to predict it. This paper will review that work, then discuss some more recent work on the differential charging, and finally present some evidence of effects which have not been satisfactorily explained yet.

2. Instrumentation

The UCSD plasma detector is shown schematically in Figure 2. Four such detectors are provided on ATS-5. They are arranged in electron-proton pairs looking parallel and perpendicular to the spin axis. The energy range from 50 ev to 50 keV is covered in 62 logarithmically placed steps. A complete scan takes about 20s.

3. Total Spacecraft Charging

3.1. ECLIPSES

The plasma environment seen at synchronous altitude is quite different from that seen

R. J. L. Grard (ed.), Photon and Particle Interactions with Surfaces in Space, 263–276. All Rights Reserved

S. E. DE FOREST

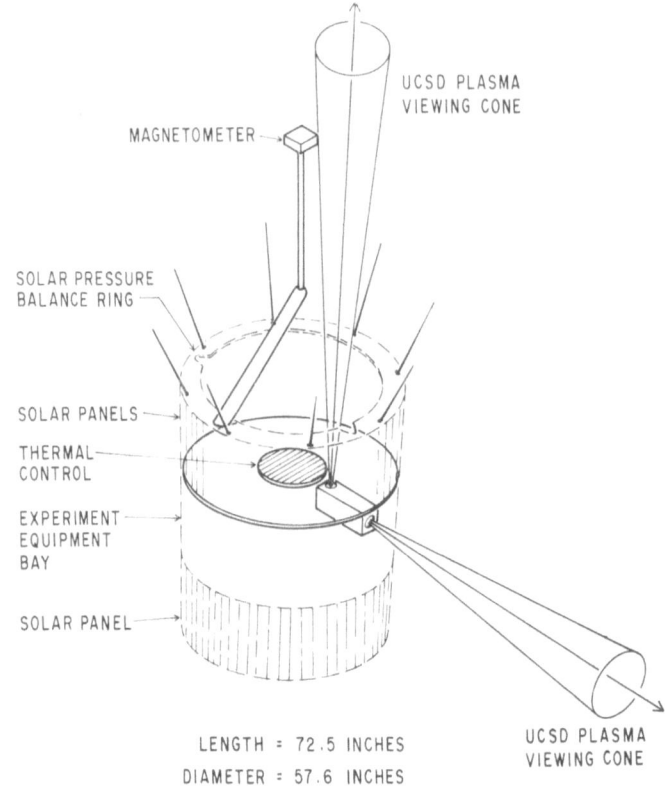

UCSD PLASMA
VIEWING CONE

MAGNETOMETER

SOLAR PRESSURE
BALANCE RING

SOLAR PANELS

THERMAL
CONTROL

EXPERIMENT
EQUIPMENT
BAY

SOLAR PANEL

LENGTH = 72.5 INCHES

DIAMETER = 57.6 INCHES

UCSD PLASMA
VIEWING CONE

Fig. 1.   Schematic representation of ATS-5

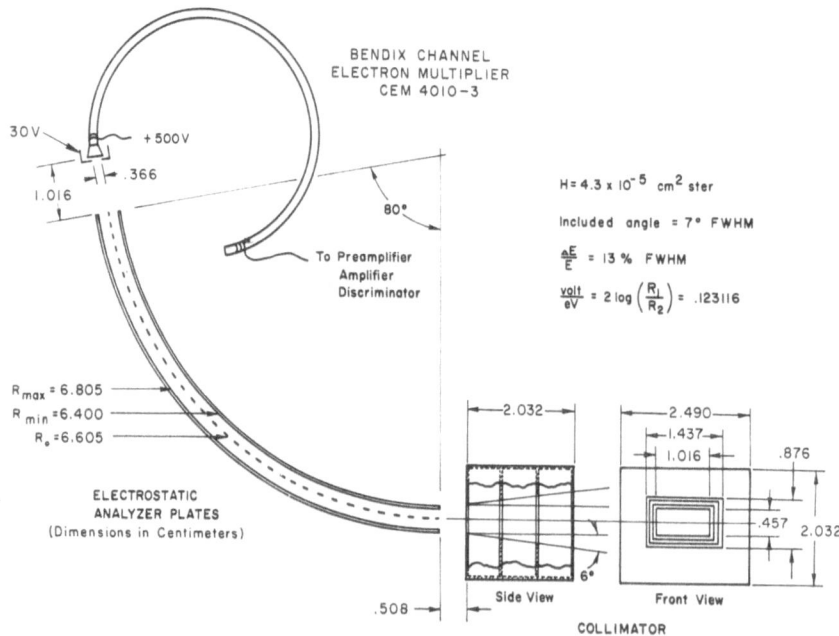

BENDIX CHANNEL
ELECTRON MULTIPLIER
CEM 4010-3

30V   + 500V

.366

1.016

80°

To Preamplifier
Amplifier
Discriminator

$H = 4.3 \times 10^{-5}$ cm^2 ster

Included angle = 7° FWHM

$\frac{\Delta E}{E}$ = 13 % FWHM

$\frac{\text{volt}}{\text{eV}} = 2 \log\left(\frac{R_1}{R_2}\right) = .123116$

R_{max} = 6.805
R_{min} = 6.400
R_o = 6.605

ELECTROSTATIC
ANALYZER PLATES
(Dimensions in Centimeters)

2.032

2.490

1.437

1.016   .876

.457

2.032

6°

.508

Side View

Front View

COLLIMATOR

Fig. 2.   Schematic representation of the UCSD plasma detector on ATS-5.

at lower altitudes (Chappel *et al.*, 1970; DeForest and McIlwain, 1971; Frank and Ackerson, 1972). Particularly in the midnight region, great variability of plasma density and temperature is found. In addition, a synchronous spacecraft goes into eclipse for periods of about $\frac{1}{2}$h ever night for a period of 3 to 4 weeks on either side of an equinox. Therefore, the satellite's response to any given plasma in both darkness and sunlight must be known.

This is demonstrated by the data shown in Figure 3. It contains 24 h of data taken on October 16, 1969. The format is a plot of energy vs time for both electrons and protons. Note that the energy scale for protons is reversed so that zero energy electrons and protons have the same origin. The gray scale is modulated by the counting rate with low rates being dark. The scale is allowed to overflow and recycle. The

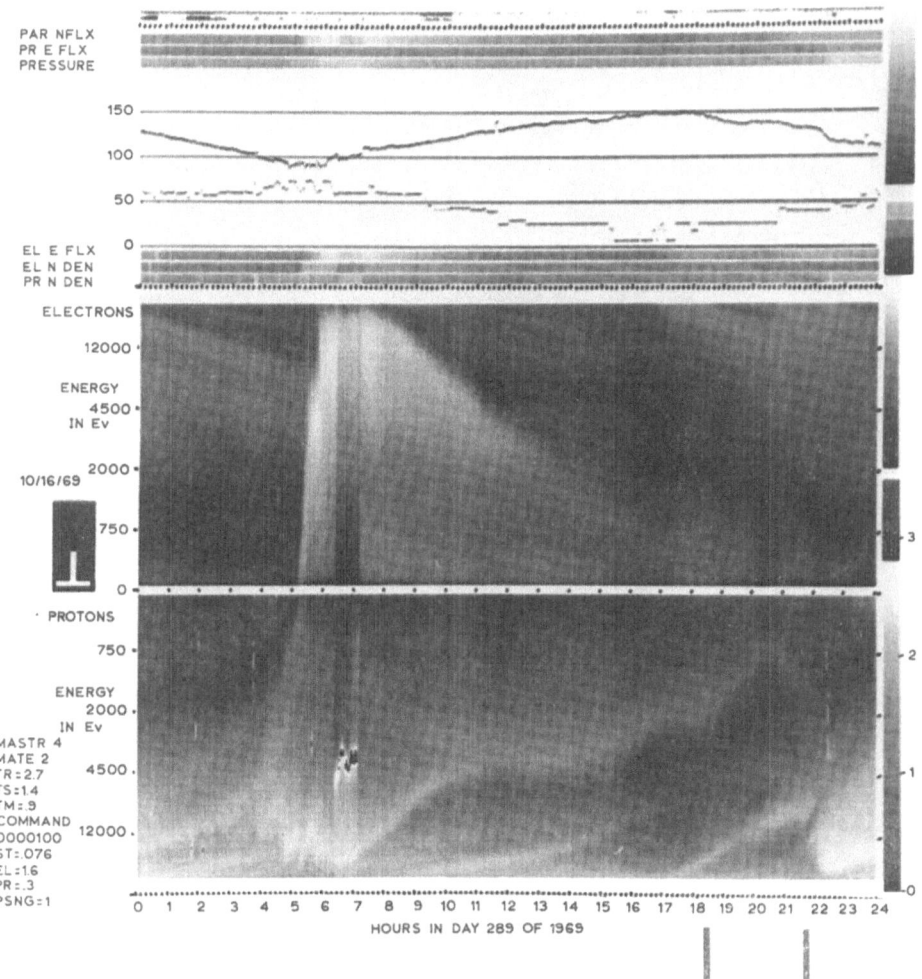

Fig. 3.   Spectrogram of data taken perpendicular to spin axis on October 16, 1969.

highest proton counting rates in the figure are black. For more details on this method
of presentation, see DeForest and McIlwain (1971).

These data (which are taken only from detectors looking perpendicular to the spin
axis) show a sudden change in character at about 0620 UT and return to what would
appear to be a normal development at 0720 UT. It is not surprising that this type of
event was first thought to be a crossing of a particle trapping boundary from closed
to open and back. One of the first papers presented on these data even suggested this
explanation (DeForest, 1969). However, the orbit calculation showed that on this
night, and every other night when such events occur, the spacecraft enters and leaves
eclipse at exactly the times of the event. For this reason, we had to assume that the
apparent changes in spectra are due to some effect local to the spacecraft.

The most obvious explanation is that ATS is charging to a negative potential of
about 4200 V in eclipse. This is shown to be the case by considering the detailed

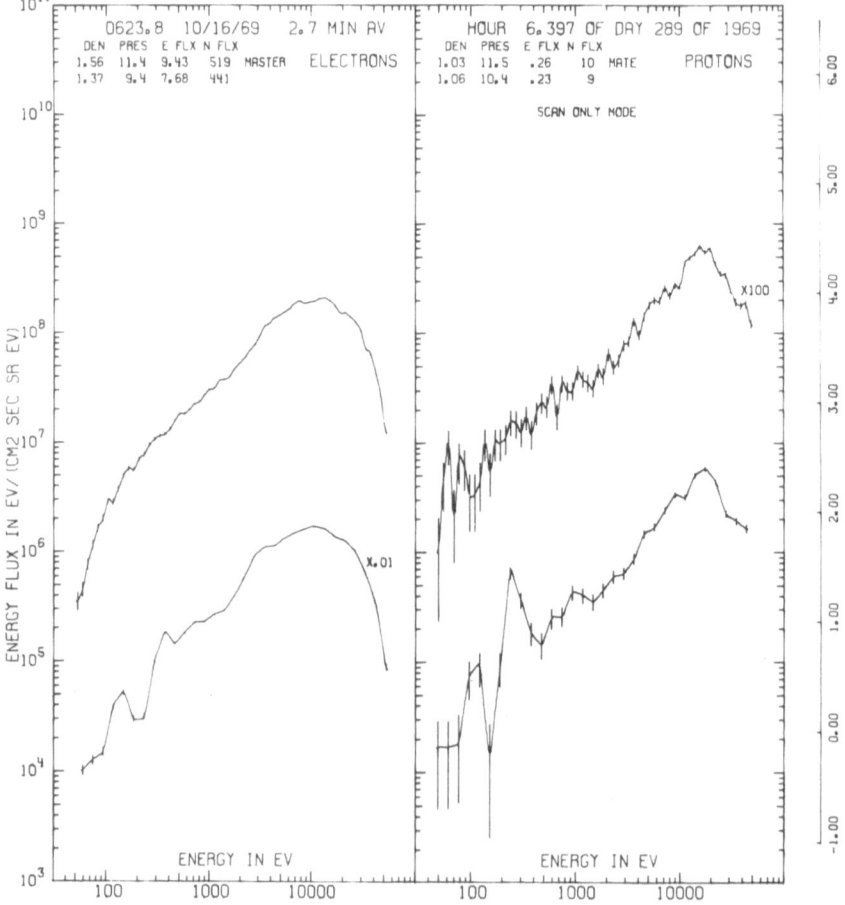

Fig. 4.   Pre-eclipse particle data on October 16, 1969 (top curves are data perpendicular to the spin
axis; bottom curves are parallel).

spectra. Figure 4 shows a cut taken through the previous figure just prior to the event. Data from both sets of detectors are shown in the figure with the perpendicular at the top. Data from the two directions have been separated by a factor of 100 for convenience. From the integrals calculated at the top of this figure, it is seen that the plasma has a density of about 1/cm³. The electrons have a temperature of about 5000 eV, and the protons have a temperature of about 10000 eV. Both species are roughly Maxwellian in shape.

Figure 5 is in the same format, but is taken from the middle of the event. Note the sharp cut-off in protons below a certain energy. In both of these figures, energy flux, which is nearly proportional to counting rate, has been plotted.

When the same data are replotted as phase space density against energy as in Figure 6, the charging hypothesis is confirmed. In this representation, the data are seen to be almost identical except for an offset of 4.2 keV, and not only are the electron and proton spectra offset in opposite directions, as they should be, but also the shift in

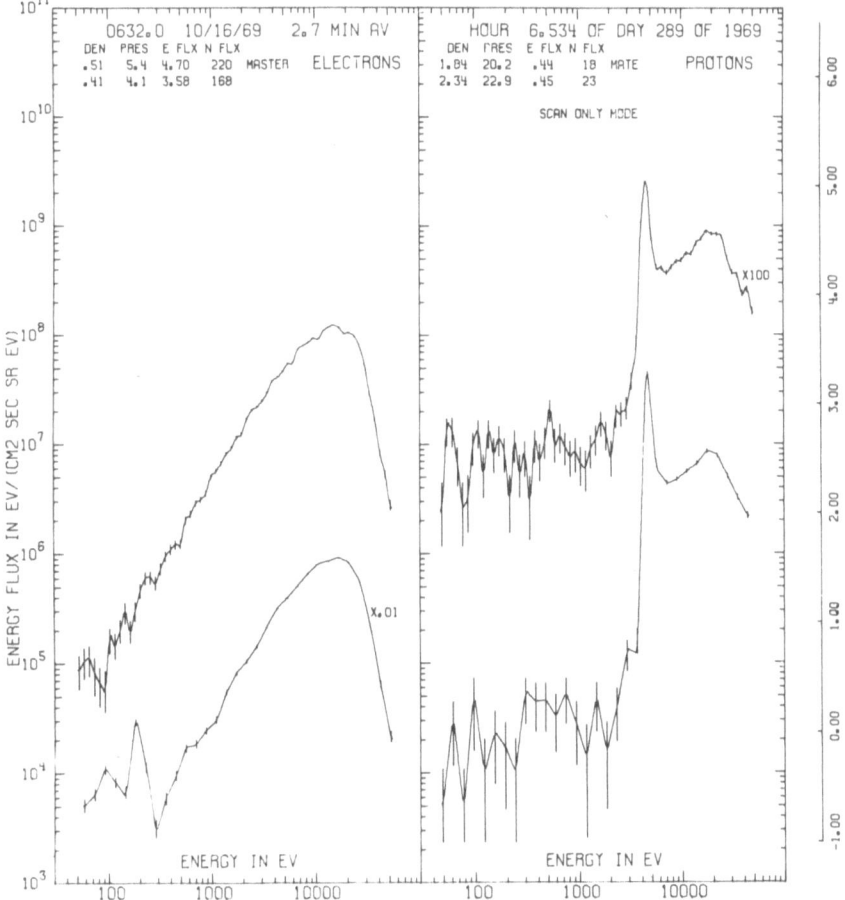

Fig. 5.   Particle data during eclipse on October 16, 1969.

Fig. 5.   Particle data during eclipse on October 16, 1969.

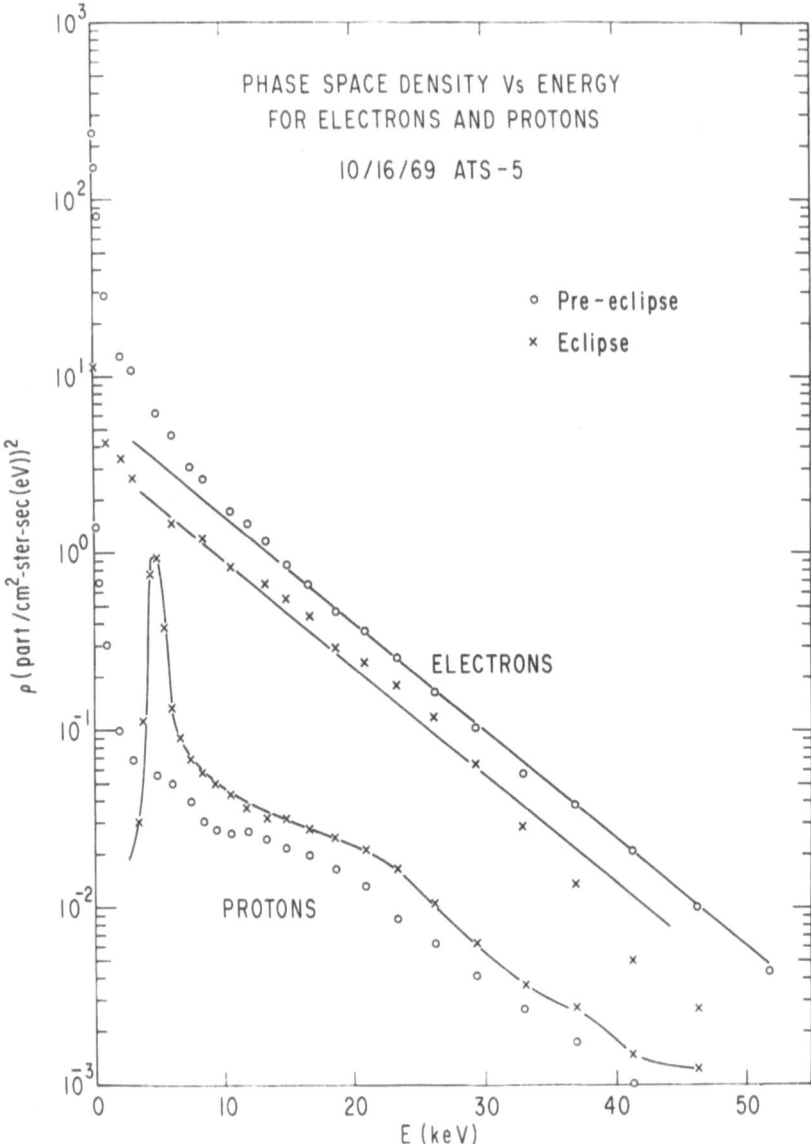

Fig. 6.   Phase space density of particles before and during eclipse.

both cases is the same amount that would have been predicted from the location of the proton cutoff. This is emphasized by the parallel solid lines drawn through the electron data at a constant separation of 4.2 keV. The largest deviation from this simple explanation is in the 30- to 50-keV electrons, but from Figure 3, we know that electron spectrum is changing fast enough in this time period to account for this discrepancy.

Other examples would show much the same story. This particular case was chosen for publication only because it has a relatively simple injection with fairly stable spectra.

ATS-5 does not charge to high potentials during every eclipse. An injection of hot plasma must precede the eclipse by not more than approximately 3 h for significant charging to take place. On the average, slightly more than half of the eclipses result in charging. On several days, injection has taken place while the spacecraft was in the shadow. The spacecraft went from almost zero potential to a very large potential within a single scan.

3.2. SUNLIGHT

ATS-5 can also charge to several hundred volts in the sunlight. All such events are confined to the midnight to dawn time sector. This might have been expected by the nature of injections at midnight, but another feature of sunlight charging is that it only happens during the winter months. This has been true from launch until the most recently analyzed data of spring 1972.

A typical charging event would be to about -150 V for a period of 5 min to an hour.

The seasonal effect must be due to the seasonal variation in the magnetosphere which causes ATS to cut through higher lines of force during the summer. Therefore, occurrence of this effect is probably a function of the location at which the spacecraft is parked. Detailed magnetospheric mappings would have to be made before a prediction of sunlight charging could be made for other longitudes.

3.3. THEORY OF CHARGING

A theory to explain these total charging events has been developed and presented in detail elsewhere (DeForest, 1972). The technique will be only summarized here.

As a first step, it was assumed that the spacecraft was made of aluminum, and then standard laboratory values were used to predict backscattered electrons (Sternglass, 1954), secondary electrons due to both incoming protons and electrons (Whipple, 1965), and photo-electrons (Hinteregger *et al.*, 1959). No corrections were made for magnetic or wake effects.

The spectra of outgoing electrons as produced by incoming particles was then calculated for several representative charging events. Keeping the outgoing spectral shapes constant, the normalization was varied to obtain an indicated zero net flux to the spacecraft. An excellent fit was found when the surface was emitting only about $\frac{1}{4}$ as many electrons of pure aluminum would.

The best value for the photo-electron flux produced by ATS-5 was obtained by considering the sunlight charging events. We calculated 8.2×10^{-10} A cm^{-2} as compared with 3×10^{-9} A cm^{-2} for pure aluminum measured on a rocket flight (Whipple, 1965).

A side effect of this calculation was to predict that the spacecraft normally is about one volt positive potential in the night region. This prediction has also been made in a recent paper by Grard (1972).

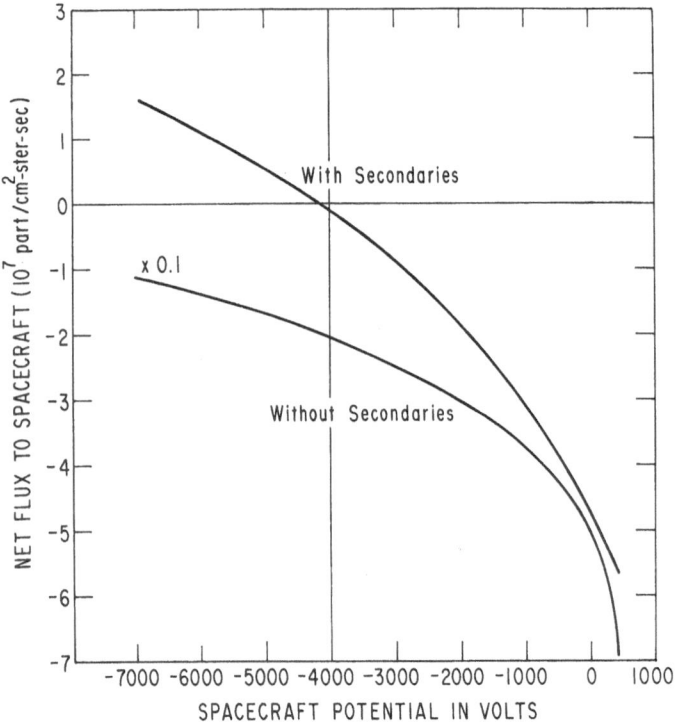

Fig. 7.   Theoretical net flux to ATS-5 vs spacecraft potential during eclipse.

Figure 7 shows the predicted net flux to ATS-5 as a function of potential using this model and the pre-eclipse data from Figure 4 with the assumption that the photo-electron flux has been turned off. The proper potential is predicted to within the resolution of the instrument.

4. Differential Charging

4.1. IN SUNLIGHT

The total charging events are spectacular oddities which have been useful in developing the theory of charging, but several types of differential charging are more important to particle experimenters in that they can happen all the time.

It will be remembered that the one set of detectors look parallel to the spin axis of the spacecraft, yet in Figure 8 a large modulation is seen in the protons looking along the spin axis. This modulation is at the spin frequency. Simultaneously, there is a cut-off in the electron spectrum. The modulation of the protons increases to higher energies as the cut-off in the electron spectrum increases. Perpendicular detectors see neither effect, but the parallel detectors see similar events on every active day from the autumnal equinox to the vernal equinox.

As was seen in Figure 1, the parallel detectors are at least partially shielded from

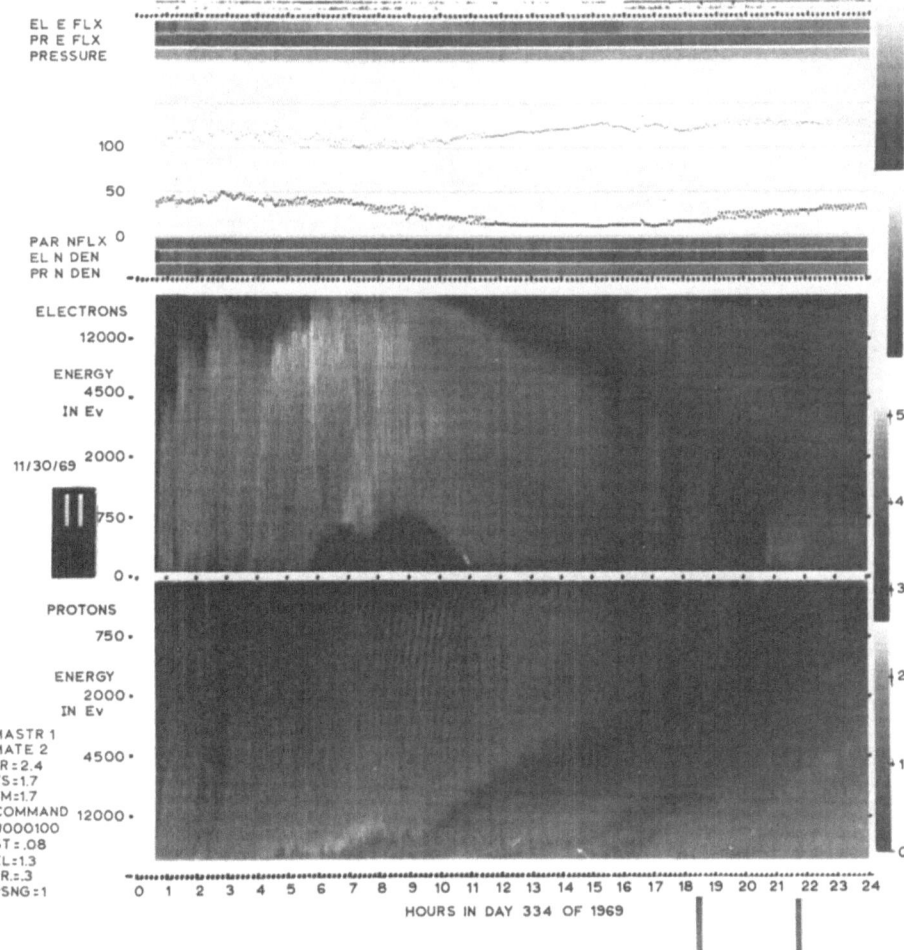

Fig. 8.   Spectrogram of data taken parellel to spin axis on November 30, 1969.

the external plasma by a ring of solar cells with several antennae and a boom on it. Initially, we thought that sunlight was being reflected down into the cavity by one of these booms. Something (most likely the thermal control unit) would then change up in the way that would be expected in the dark. Then the periodic illumination of the cavity temporarily decreases the potential. The net effect would be to produce an electric field that is dependent on the spin angle.

A better explanation can be deduced from the special event shown in Figure 9, which contains data from the parallel detector for the same eclipsing event as previously shown. A slight differential charging had taken place starting at about 0600 UT, coincident with the main injection. During the eclipse, this feature slowly disappears over about 20 min. After leaving the eclipse, the differential feature does not return immediately, but rather builds up with a similar time constant. The same general type

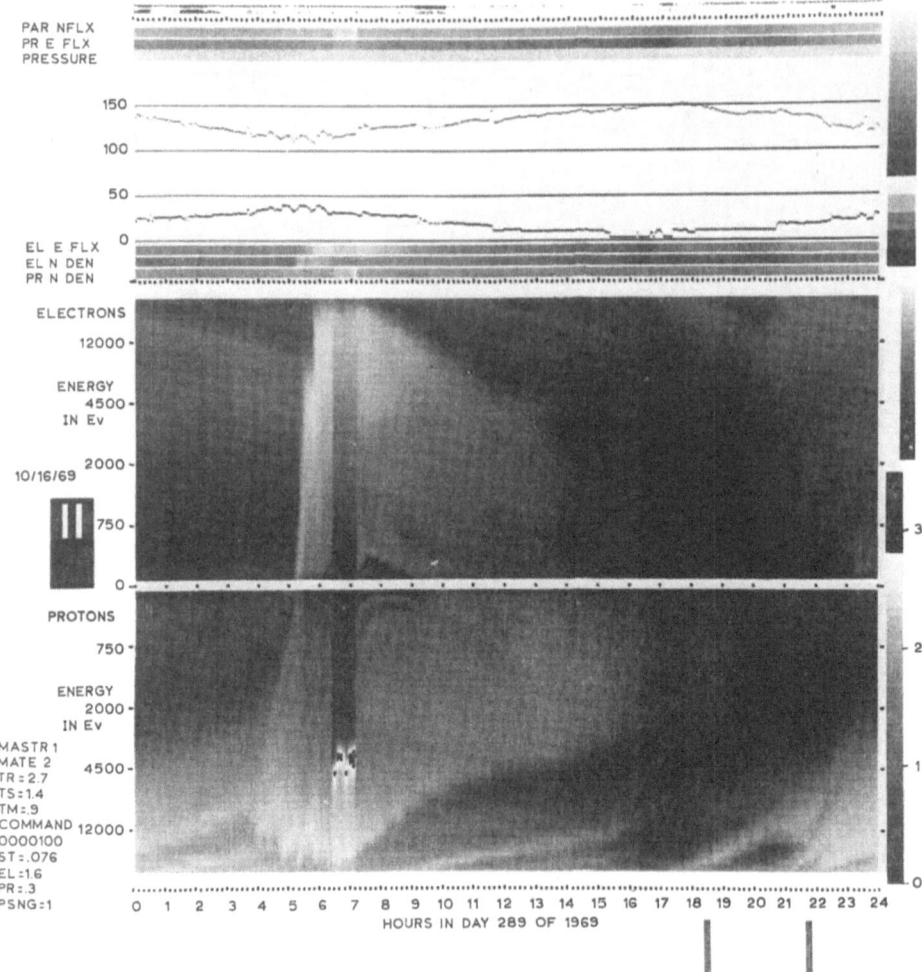

Fig. 9.   Spectrogram of data taken parellel to spin axis on October 16, 1969.

of behavior is seen on all eclipsing events when there has been differential charging before entering the shadow.

The capacitance of the spacecraft to the outside plasma is of the order of picofarads, but the capacitance formed by an oxidized layer of aluminum on top of metal can be several thousands of picofarads per square centimeter. The events shown in Figure 9 can then be explained by saying that in the dark part of the solar cell cavity something charges up in the same manner as eclipse charging. When the spacecraft enters the Earth's shadow, it charges up through an effective capacitor of about 1 picofarad. The previously charged surface then finds itself several thousands of volts above its equilibrium potential, and proceeds to discharge, but it is working through a much larger

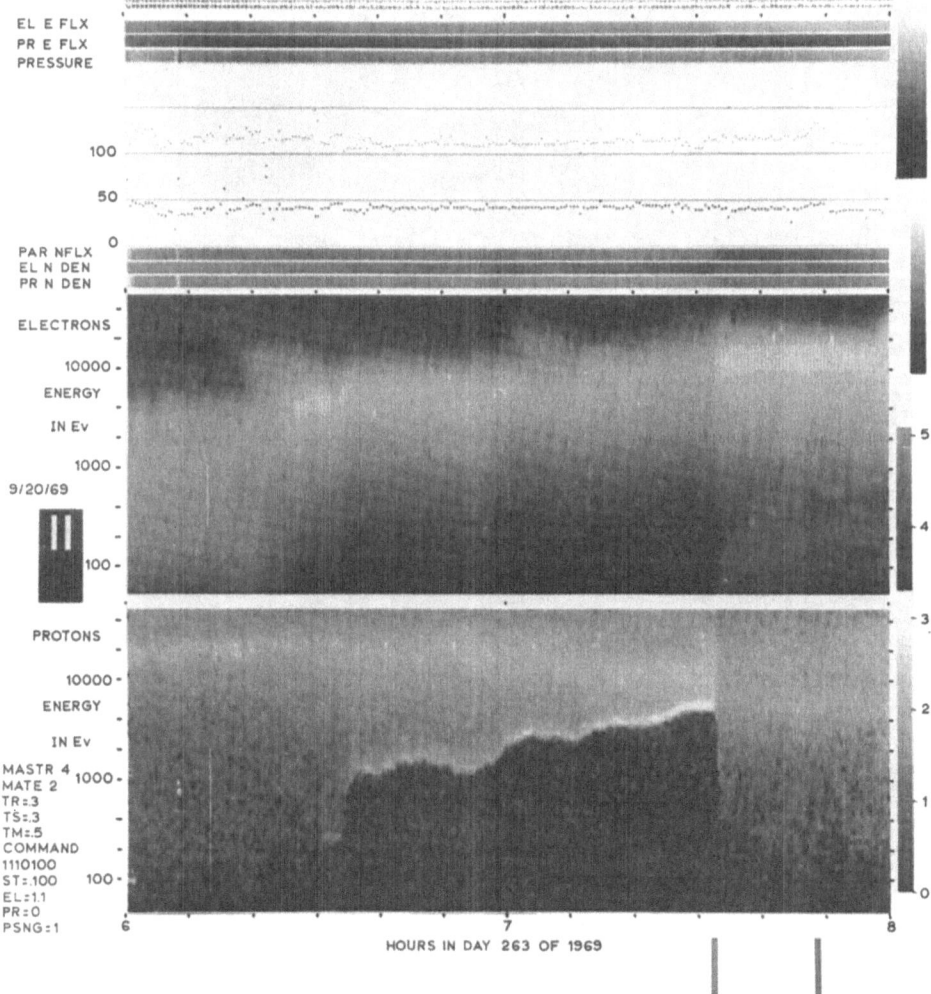

Fig. 10.   Spectrogram of data taken parellel to spin axis on September 20, 1969.

capacitance back to the main spacecraft body. Similarly, this large capacitor takes several minutes to charge up after leaving the shadow.

The proton modulation as seen in Figure 8 is then thought to be the result of a changing $E \times B$ force acting on incoming particles. During typical differential charging events, the angle between the spin axis and the local magnetic field is about 30°.

This phenomenon occurs until the annual precession of the plane of the orbit tips the spacecraft far enough to illuminate the cavity continuously.

While the perpendicular detector never sees such striking events, a non-spin modulated notch is frequently seen in the lowest energy electron channels. A possible explanation of this could be a slight charging of the solar cell covers in sunlight with a discharge

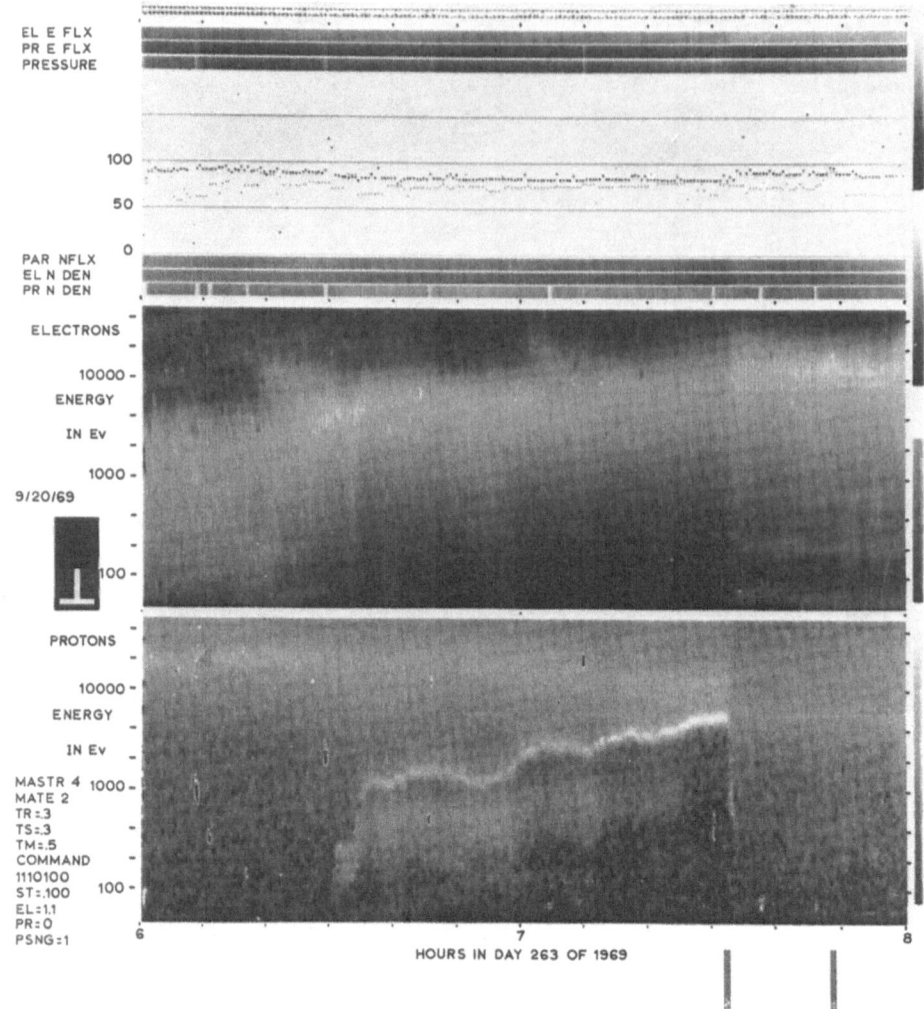

Fig. 11.   Spectrogram of data taken perpendicular to spin axis on September 20, 1969.

time long compared to a spin period (about 0.5 s). ATS-5 is a particularly simple spacecraft in exterior configuration, and that probably simplifies differential charging problems for the perpendicular detector, but differential charging might be expected on any spacecraft operating in the outer magnetosphere, depending on its configuration and orientation to the Sun. This is particularly important when measuring plasma flows by looking at modulations in a spinning detector.

The magnitudes of electric fields produced in the vicinity of the parallel detectors can be estimated to be of the order of thousands of volts per meter.

4.2. IN SHADOW

Figure 10 shows the parallel detector's response during an eclipse. No particles are

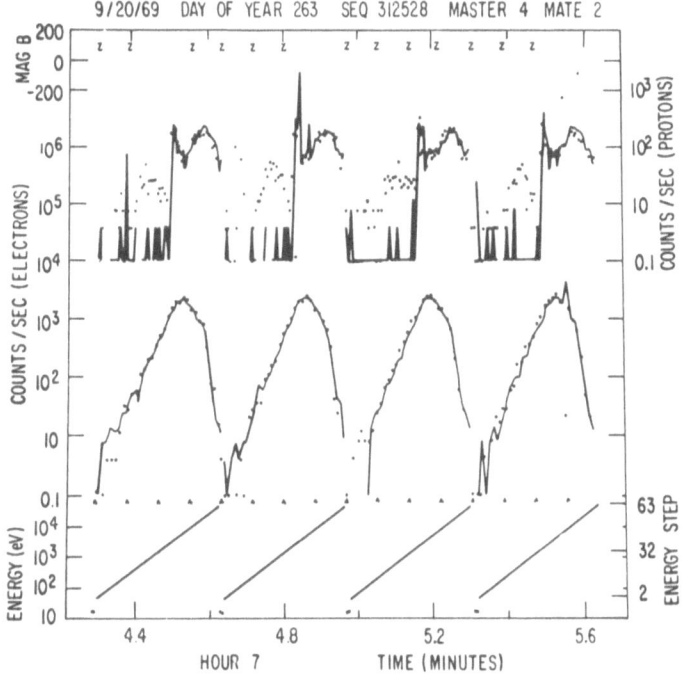

Fig. 12.   Particle data during eclipse on September 20, 1969.

seen below the cutoff energy. However, in Figure 11 we see the perpendicular data taken at the same time, and there is a definite peak in counting rate below the cutoff. Figure 12 shows four scans taken during the event. The presence of these particles is particularly difficult to explain since they do not occur on every charging (cf. Figure 5). They have too high a flux to have been created by ionizing neutrals in the vicinity of the spacecraft. Elastic scattering of ions near the spacecraft is also much too weak a source. The most likely candidate at this time seems to be particles sputtered from surfaces of the spacecraft which are not at the same potential as the main body. Data on sputtering (Strong, 1938) would indicate that the incoming fluxes are strong enough to produce these particles. However, the exact nature of these particles cannot yet be determined, and we still have no adequate explanation of why they occur sometimes and not others.

5. Summary

Many types of surface charging phenomena have been seen on ATS-5. Some have not yet been adequately explained. At this time, it is known that the spacecraft can charge to well over 10000 V in eclipse, and to several hundreds of volts in sunlight. Differential charging which produces fields of at several thousands of volts per meter have also been seen. Time constants for charging can range from a fraction of seconds to tens of minutes depending on the surface being charged.

All of these effects are serious hazards to the study of low-energy natural plasmas. Particle optics and accessibility can become hopelessly snarled in these fields.

The possibility of subsystem failure due to discharge of these potentials exists, and may already have been observed (Fredrichs and Scarf, 1972).

Acknowledgement

Many fruitful conversations with Prof. C. E. McIlwain have helped to understand these data. The work was supported by NASA Contract NAS 5-10364 and Grant NGL 05-005-007.

References

Chappel, C. R., Harris, K. K., and Sharp, G. W.: 1970, *J. Geophys. Res.* **75**, 50.
DeForest, S. E.: 1969, *EOS* **150**, 660.
DeForest, S. E.: 1972, *J. Geophys. Res.* **77**, 651.
DeForest, S. E. and McIlwain, C. E.: 1971, *J. Geophys. Res.* **76**, 3587.
Frank, L. A. and Ackerson, K. L.: 1972, *J. Geophys. Res.* **77**, 4116.
Fredricks, R. and Scarf, F.: 1973, this volume, p. 277.
Grard, R. J. L.: 1972, submitted to *J. Geophys. Res.*
Hinteregger, H. E., Damon, K. R., and Hall, L. H.: 1959, *J. Geophys. Res.* **64**, 961.
Sternglass, E. J.: 1954, *Phys. Rev.* **95**, 345.
Strong, J.: 1938, *Procedures in Experimental Physics*, Prentice-Hall, Englewoodcliffs, N.J.
Whipple, E. C.: 1965, Ph.D. Thesis, George Washington University, Washington, D.C.; also published as NASA Tech. Note X-615-65-296.

DISCUSSION

Gold: Do any effects show a relationship to the function of the transmitters? Charging can occur as a result of rectification in the plasma of any high frequency excitation of an antenna feed.

DeForest: No, no relation has been seen with the spacecraft transmitters-observations exist for periods that some of the transmitters were on or off and no change was seen in our data.

Grard: Why don't you observe any modulation at spin frequency on the measurements made with the detector perpendicular to the spin axis?

DeForest: I assume that differential charging time constants are similar for objects near both detectors. But small deviations of the parallel particles cause shadowing by the solar panel. No such objects can easily block the perpendicular particles. This can be seen from Figure 1.

Rosenbauer: Referring to the effects which you explained as particles produced on the spacecraft in the presence of high field-strength, I would like to ask if these effects decreased with the time after launch. I would like to know whether they might be due to outgassing effects.

DeForest: The effects decreased somewhat, at most by a factor of two since the launch in 1969, but I am not sure about that.

OBSERVATIONS OF SPACECRAFT CHARGING
EFFECTS IN ENERGETIC PLASMA REGIONS

R. W. FREDRICKS and F. L. SCARF

Space Sciences Dep., TRW Systems Group, One Space Park,
Redondo Beach, Calif. 90278, U.S.A.

Abstract. Recently DeForest (1972) presented direct evidence that in the local morning quadrant at synchronous altitude, electrons of energies $\sim$1–10 keV injected by motion of the plasma sheet during substorm-associated events can charge the surface of ATS-5 to large negative potentials (up to -9 kV under eclipse conditions). The highest values occur when neither cold plasmapause nor cool photoelectrons are present to neutralize the influx of hot plasma sheet electrons, and DeForest also reported that differential charging produced local E-fields of several hundred volts/meter. In this paper we present additional evidence, derived from engineering data on anomalies observed in the operation of subsystems aboard non-NASA synchronous orbiters. Flight data and laboratory simulations indicate that portions of the surface of a spacecraft not only charge to many kV (negative) during substorm-correlated events, but also suffer discharges (arcs or coronas). The large amplitude electromagnetic pulses with high frequency Fourier spectra irradiate cabling, and can cause the observed anomalous changes of state of electronics subsystems. In one spacecraft, the solar cell substrate and the aluminized mylar super-insulating blanket material have been identified as probable sources of arc or corona discharging. Samples of these materials were subjected to laboratory tests in an effort to simulate conditions of spacecraft charge during substorm/eclipse phenomena, and were found to suffer electrical breakdown under voltage gradient conditions well within the range of spacecraft potential gradients reported by DeForest. One subsystem was also tested in the laboratory and found to suffer the anomaly observed in orbit in the presence of arcing such as occurs on either solar cell substrate or superinsulation material. The correlations of a large number of spacecraft subsystem anomalies with substorm activity in the morning quadrant of the magnetosphere and the laboratory investigation of discharge phenomena associated with surface charging will be presented in detail. These charging effects and accompanying problems need not be restricted to synchronous orbits, but are to be expected in other regions containing energetic electron plasmas of sufficient density, such as the Earth's low altitude plasma sheet or the Jovian magnetosphere.

1. Anomalous Events in Geostationary Orbit

1.1. Introduction

In late 1971 it came to the authors' attention that several anomalies in the behavior of certain subsystems aboard a number of different non-NASA spacecraft in synchronous orbit had been observed. It was noted that when the occurrence times of these anomalous events were plotted against Local Time (LT), they showed a systematic preference for the local morning sector of the magnetosphere. A plot of 23 such events is shown in Figure 1.

The *exact* nature of the anomalous behaviors of the various subsystems shown in Figure 1 is of no particular importance to our discussion of their probable cause in the subsequent portions of this paper. It is sufficient to say that all events in Figure 1 involved unexpected changes of state in one or more of the electronics subsystems aboard the satellites.

The most obvious geophysical phenomena with which these observed events could be correlated are magnetic storms and magnetic substorms. During such events, it is

R. J. L. Grard (ed.), Photon and Particle Interactions with Surfaces in Space, 277–308. All Rights Reserved
Copyright © 1973 by D. Reidel Publishing Company, Dordrecht-Holland

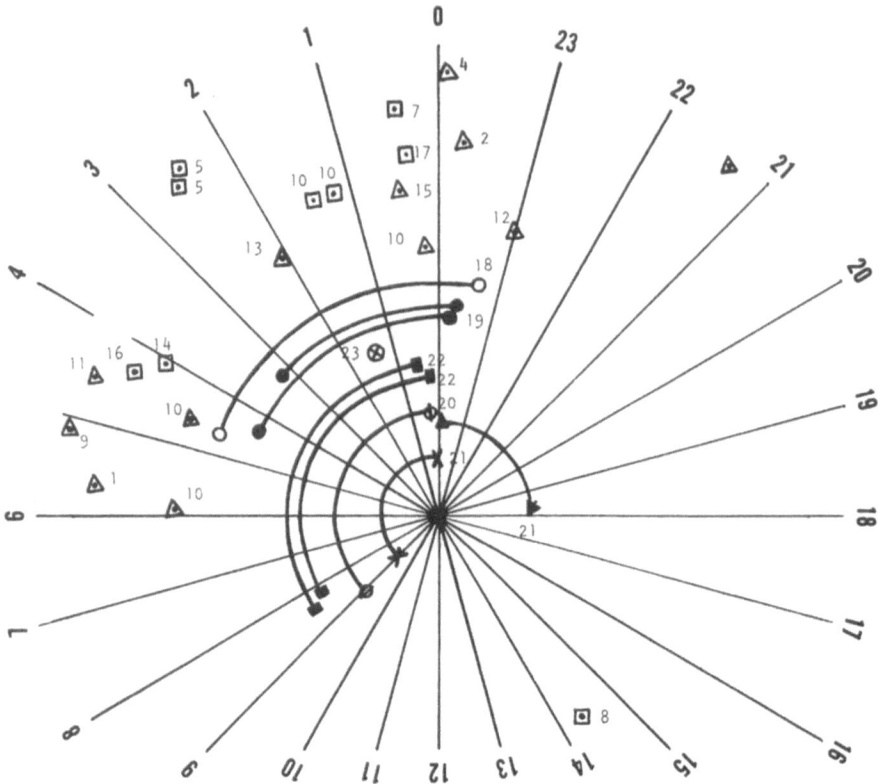

Fig. 1. Distribution of anomalies observed by several geostationary satellites. Rays are local time at spacecraft in hours, numbers identify events. Radius vectors have no significance, and differ only to allow visual resolution of events.

known that the relatively cool, dense plasmapause population at 6.6 R_E is swept away, and replaced by an injection spectrum of energized plasma sheet electrons and protons. As shown very lucidly by DeForest (1972), such events can charge spacecraft surfaces in darkness to many kilovolts negative, and even sunlit surfaces to several hundreds of volts negative, relative to plasma potential. Such charging could lead to electrical discharges (arcs) in spacecraft structures, and radiation from these arcs could cause anomalous behavior of electronics subsystems.

For this reason, we undertook a correlative study of the times of occurrence of the spacecraft events and of enhanced ground magnetic activity. First, we made correlations with the 3-hourly indices Kp and (or) ap. Then we obtained rapid-run magnetograms from various surface observatories in an effort to obtain time correlations of higher resolution than the 3-hourly indices can provide.

In Figure 1, the anomalous events numbers 1 to 17 and number 23 are discrete events, in the sense that they were observed to occur during real-time telemetry acquisitions. On the other hand, events numbers 18 to 22 occurred at some time between the end of a real-time telemetry acquisition and the beginning of another. Thus, the electronics subsystem changed state at some time during this time period. These unresolved anomalous events are depicted in Figure 1 as the solid arcs.

In the remainder of Section 1, we discuss the correlation of event times with substorm occurrence. In Section 2, we give a brief background review of observations that can be interpreted as spacecraft charging from the ATS-5 measurements. In Section 3, we develop some crude models of spacecraft differential charging, and the structural components most likely to be involved. In Section 4 we describe laboratory tests of some of these structures, and show that arc discharges may occur on them within the voltage limits of spacecraft charging charging as given by DeForest (1972) for ATS-5. The measured electromagnetic pulses from these discharges were found to be energetic enough, and to have the proper high-frequency content, to produce anomalous behavior of electronics subsystems aboard the spacecraft. A discussion of these results, and their impact on spacecraft design considerations, appears in Section 5.

1.2. CORRELATIONS WITH THE Kp INDEX

In an effort to establish a preliminary connection among the events in Figure 1 and enhanced ground magnetic activity, we compared these observed event times with the Kp index value for the 3-h period containing the anomalous event. The comparison is shown in Figure 2, where events are marked by the star-like symbol, with an arrow pointing to the 3-h Kp value for that event. Note that the large majority of these events occurred during periods having $Kp \geqslant 3_+$. One event occurred when $Kp = 2_0$, but the previous 3-h period had $Kp = 3_0$. Another event occurred when $Kp = 2_+$ but, again, followed a 24-h period for which $Kp \geqslant 3_0$. Four events occurred when $Kp = 3_-$, but values of 3_+ were observed prior to the period containing the event. Only the single event on September 2, 1971 was associated with a low Kp value of 1_-, and is probably an event not associated with substorm activity.

Thus, 22 of the 23 events in Figure 1 occurred either during 3-h periods of $Kp \geqslant 3_+$, or during periods of 3_0, 2_+ or 2_0 which followed immediately after periods of greater Kp. These latter occurrences we infer to be due to substorm-associated phenomena, or electron injection events, occurring after a previous substorm had swept out the cool plasmapause population, leaving the spacecraft in a plasma environment conducive to charging by even moderate fluxes of plasma sheet electrons with mean energies of a few keV.

1.3. CORRELATIONS WITH THE ap INDEX, AND LOCAL TIME

In Figure 3 we show the 3-hourly ap index for the month of November 1971. The darkened sections of this histogram correspond to the 6-h period during which one of the spacecraft, which observed anomalies in one of its electronics subsystems, was in the local midnight to local dawn quadrant of the magnetosphere. The periods during which the anomalous events were observed are marked by the label 'event' and an arrow.

Note that the smallest value for which an event occurred was $ap = 11$ on November

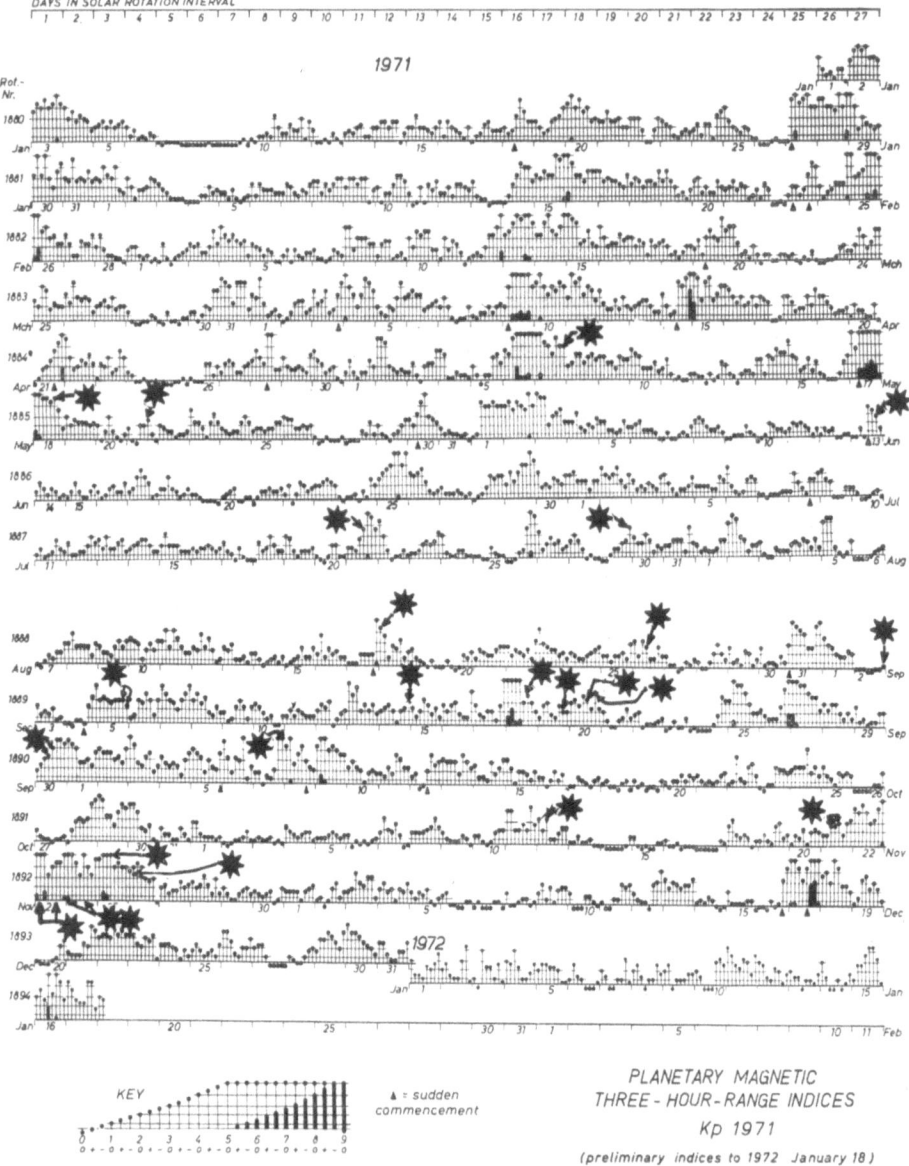

Fig. 2.   The three-hourly values of *Kp* for 1971. Occurrence of events in Figure 1 are marked by the star-like symbols and arrows.

21; *ap* was 18 on November 12, and the rest had *ap* ⩾ 32. Also note that during periods of other very large values of *ap*, but with the spacecraft *not* in the local morning quadrant, no anomalous events were observed. The only blackened period of large *ap* (= 39 on November 22) for which no event occurred coincided with a period during which the electronics subsystem had been inactivated deliberately. Similar results are obtained for time periods other than November 1971.

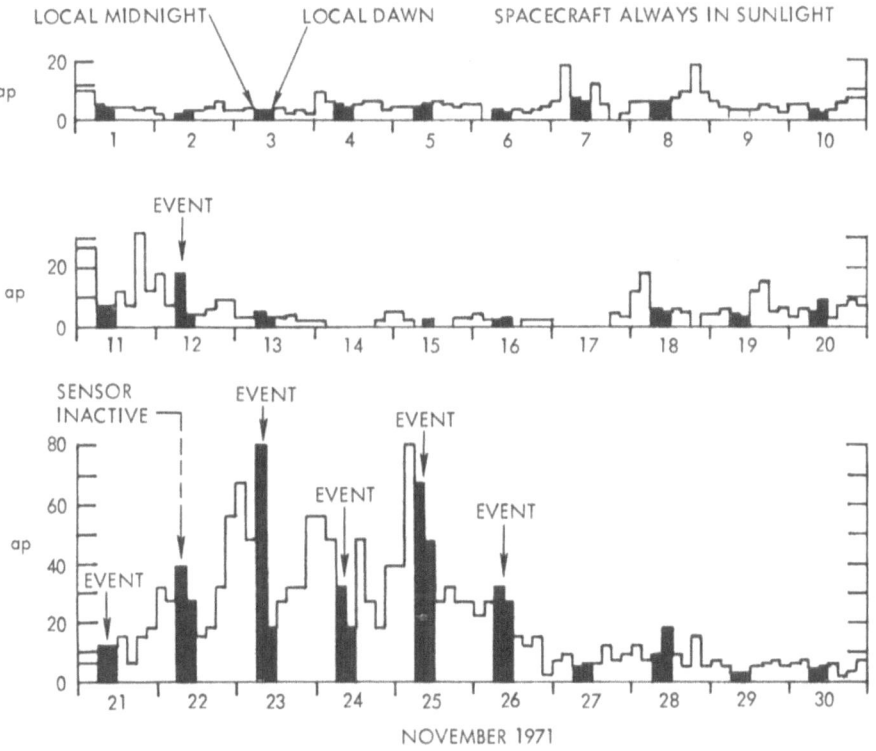

Fig. 3. Plot of three-hourly index ap for November 1971. The blackened portions are the 6-h intervals when one geostationary spacecraft was in the 0000 to 0600 LT sector.

1.4. CORRELATIONS WITH GROUND MAGNETOGRAMS

The previously discussed correlations with Kp and ap have only a 3-h resolution corresponding to the averaging time used to compute these indices. For this reason, rapid-run magnetograms were obtained from several ground observatories.

When making correlations between events observed at synchronous altitude (6.6 R_E) and at ground observatories, one must remember that magnetic substorms and their associated enhancements of ionospheric current systems (electrojets, etc.) tend to be somewhat localized, and after the onset of the main phase, complex auroral zone ionospheric phenomena move about with variable time sequences (Rostoker, 1972). The most favorable orientation of the satellite-ground observatory configuration for correlation is when the two participants lie on nearly the same L-shell and nearly the same geomagnetic longitude. For many of the events shown in Figure 1, the favorable ground stations are in northern Siberia, and at the time of this writing, the rapid-run magnetograms from these stations were not yet available to us.

As an example of a correlation with substorm activity using ground magnetograms, we have selected event number 23 of Figure 1, which occurred at 0812 UT on November 26, 1971, for a detailed discussion. In Figure 4 we show the magnetograms for the

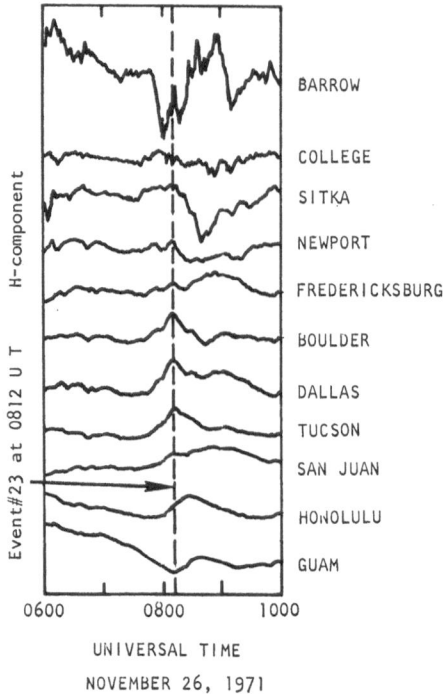

Fig. 4.   Magnetograms of the *H*-component from the stations noted at the right margin, for event
23 of Figure 1 on November 26, 1971, at 0812 UT.

period 0600 UT to 1000 UT on November 26, 1971. As one can see from Figure 3,
this period was near the end of a more extensive period of sustained magnetic activity
which began on November 22, 1971. One sees the evidence for the *H*-component bays
at first Barrow and later at Sitka. The bay at Barrow began shortly before 0800 UT,
and that at Sitka shortly after 0812 UT. The dashed vertical line at 0812 UT shows the
time at which the anomalous event was observed on the geostationary satellite in the
local early morning sector.

     To confirm that an electron injection event occurred at 6.6 R_E during this time
interval, we show in Figure 5 the energy-time spectrogram of the University of California, San Diego (UCSD) electrostatic analyzer data from ATS-5 between 0600 UT
and 1000 UT (corresponding to 2300–0300 local time for ATS-5). These data were
kindly furnished to us by Dr Carl McIlwain of UCSD. The top panel shows the components of geomagnetic field intensity in gammas, parallel to (darker trace) and perpendicular to (lighter trace) the spacecraft's spin axis. The central panel shows the
grey-coded, 20-s averages of electron flux intensity as a function of energy (ordinate)
and time (abscissa). The lowest panel is the same for protons. The analyzer axis was
perpendicular to the spin axis and in a near-equatorial plane of the satellite. (See
DeForest and McIlwain (1971) for details of their experiment and data displays.)

     From these data, one can see that near 0750 UT (or ∼0050 LT) an injection of

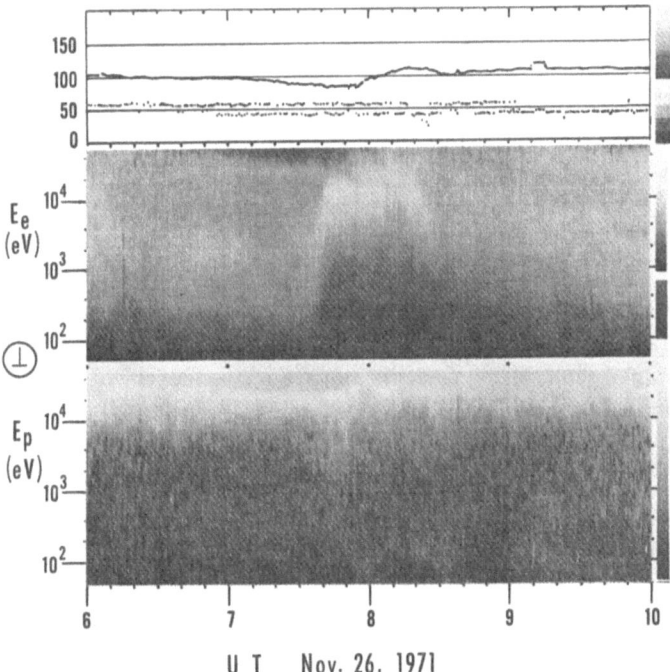

Fig. 5. ATS-5 data for November 26, 1971, showing $E_e \gtrsim 10$ keV injection event in progress at 0812 UT. Details are in test. (Courtesy of Dr C. E. McIlwain.)

electrons having $E_e \sim 10$ keV began, and continued to about 0825 UT (0125 LT). The event on the other synchronous satellite (event number 23 in Figure 1) occurred at 0812 UT. This event therefore occurred during the injection event shown in Figure 5. In fact, the very faint semicircular trace in the lowest part of the proton panel that starts at the bottom just before 0800, reaches a peak at about 200 eV near 0812, and returns to the bottom after 0830 UT, shows that this was an ATS charging event. This trace represents the thermal protons that were accelerated by the spacecraft charge, and the ATS-5 data do confirm that the charging peak occurred near 0812 UT. Thus, we have not only correlated the anomalous event with ground magnetic activity but also with an observed ~ 10 keV electron injection event measured by instruments aboard an adjacent synchronous orbiter, ATS-5.

In addition to this particular correlation, we found that 21 of the 23 events of Figure 1 are definitely associated with either substorms or magnetic storms. In 19 of these 21 cases, satellite-to-ground station geometry was sufficiently favorable that timing the event with the magnetograms presented no significant problem. The other two cases had less clear-cut time correlations, but there was general magnetic activity in the ground magnetograms during time periods containing the satellite events; however, magnetograms from the most favorably located stations were unavailable.

In summary, the 19 definite, two probable, and two questionable correlations between satellite events in Figure 1 and ground magnetograms indicating storm or sub-

storm activity establishes, in our opinion, a convincing statistic to support the hypothesis that the anomalous behaviors of electronics subsystems on these geostationary
satellites have a direct connection with substorm or storm plasma environments in the
morning sector at 6.6 R_E.

2. Observations of Spacecraft Charging

DeForest (1972, 1973) has published an excellent description of ATS-5 data which
presents evidence for electrical charging of this geostationary satellite during energetic
electron injection events, in the local morning sector, associated with magnetic storm
or substorm activity.

DeForest noted that, during such disturbed periods (in the winter months, at least)
the electron fluxes in these injection events could be sufficiently high to charge ATS-5
to between ~ -50 to -300 V in sunlight, and between -1000 V and -9000 V during
eclipses. He ascribed this large difference in the illuminated and eclipsed values of
spacecraft potential to the absence of the photoelectron current during eclipses.

The measurement of this spacecraft potential was made by the examination of
electron and proton differential energy spectra, as shown in Figure 6 (courtesy of
Dr DeForest). One notes that in this case, the electron spectrum during eclipse has

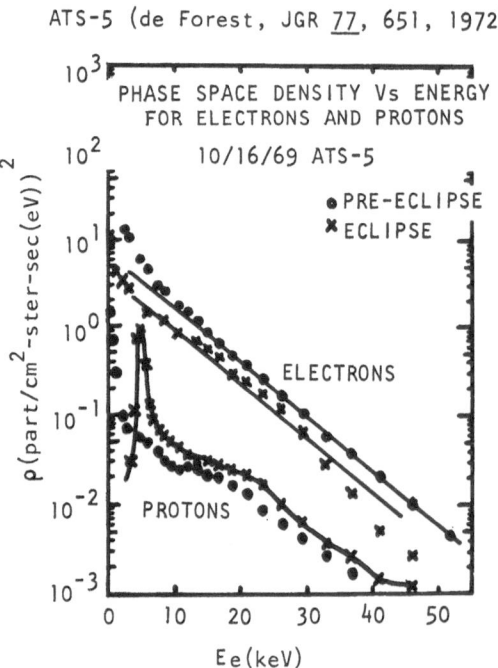

Fig. 6. Flux vs energy spectra taken on ATS-5 spacecraft during an electron injection event ($E_e \sim 5$
keV, $E_p \sim 10$ keV) showing -4.2 kV spacecraft potential shift of pre-eclipse spectra during a spacecraft eclipse (after DeForest, 1972).

been retarded by some 4200 V, while the proton spectrum has been accelerated by the same voltage, compared to the spectra taken shortly before the eclipse. Such a linear shift of equal magnitude and opposite sign must be due to the development of a -4200 V potential of the spacecraft, relative to the plasma reference. For more such results, we refer the reader either to DeForest's paper in this Symposium, or to his previously published work (DeForest, 1972).

It is our opinion that these observations of charging of the geostationary satelliet ATS-5 during substorm injection events in the local morning sector, and the correlations we have made between substorm activity on ground magnetograms and the incidence of anomalous events on other geostationary spacecraft when located in the local morning sector, lead to the conclusion that the anomalies in electronics subsystems must be either directly or secondarily caused by such spacecraft charging phenomena.

3. Models of Spacecraft Differential Charging

3.1. REVIEW OF CHARGING THEORY

A portion of a spacecraft immersed in an ambient plasma will come into electrical equilibrium with that plasma by developing surface charges of the proper sign and magnitude to reduce the *net* (surface-integrated) current between plasma and spacecraft to zero. The total current is computed from all of the partial currents contributed by the ambient electrons and ions, the backscattered electrons and ions, by secondary electrons and ions, and by photoelectrons from any illuminated areas.

The very simplest of theories will be used here, for the 'collisionless' plasma regime. In this simplest of treatments, all current carrying charged particles are considered to be Maxwellian with temperatures T_α, that is, they have distribution functions in velocity space

$$f_\alpha(\mathbf{v}) = (m_\alpha/2\pi\kappa T_\alpha)^{3/2} \exp\left(- m_\alpha \mathbf{v}\cdot\mathbf{v}/2\kappa T_\alpha\right), \tag{1}$$

where κ is Boltzmann's constant, m_α the species mass. The partial current densities then have the general form

$$j_\alpha = N_\alpha q_\alpha (m/2\pi\kappa T_\alpha)^{3/2} \int d^3\mathbf{v}\, \mathbf{v}\cdot\mathbf{n} \exp\left(- m_\alpha v^2/2\kappa T_\alpha\right), \tag{2}$$

where is the partial number density, q_α is the signed charge, and we compute the current density j_α perpendicular to a surface of unit normal $\mathbf{n}$.

Let us use a geometrical model of a cylindrical spacecraft with covered ends, define a coordinate system with $v_\parallel$ along the axis of the cylinder and $v_\perp$ normal thereto.

Then, the current density incident on the wall of the cylindrical spacecraft may be written

$$j_{\alpha\perp} = N_\alpha q_\alpha (m_\alpha/2\pi\kappa T_\alpha)^{3/2} \times$$
$$\times \int_{-\infty}^{\infty} dv_\parallel \exp\left(- m_\alpha v_\parallel^2/2\kappa T_\alpha\right) \int_{v_0}^{\infty} 2\pi v_\perp^2\, dv_\perp \exp\left(- m_\alpha v_\perp^2/2\kappa T_\alpha\right), \tag{3}$$

where $v_0 = 0$ if we expect a surface potential of zero, or an accelerating potential for particles of charge $q_\alpha = \pm e$, and $v_0 > 0$ if the surface potential is expected to retard the α-species. For the end covers of the cylinder, one has

$$j_{\alpha\parallel} = N_\alpha q_\alpha (m_\alpha/2\pi\kappa T_\alpha)^{3/2} \times$$

$$\times \int_{v_0}^{\infty} v_\parallel \, dv_\parallel \exp(-m_\alpha v_\parallel^2/2\kappa T_\alpha) \int_0^{\infty} 2\pi v_\perp \, dv_\perp \exp(-m_\alpha v_\perp^2/2\kappa T_\alpha) \quad (4)$$

for each end, with v_0 having similar meaning as before.

For a synchronous orbiter, one expects to find large, negative potentials in eclipse or in shadows, and either moderate, negative or small, positive potentials in sunlight. The large, negative potential occurs during injection of substorm electrons of some several up to perhaps 20 keV mean energies. It is *only* this substorm-associated case that we discuss here.

In eclipse or shadow, the negative surface potential causes total escape of all secondary electrons, and suppression of secondary ions. Thus the current balance equation will require currents to the unilluminated wall of the spacecraft to satisfy

$$N_e e \left(\frac{2\kappa T_e}{m_e}\right)^{1/2} \frac{2}{\sqrt{\pi}} \int_{e\Phi/\kappa T_e}^{\infty} t^{1/2} e^{-t} \, dt - N_p e \left(\frac{2\kappa T_p}{m_p}\right)^{1/2} - N_* e \left(\frac{2\kappa T_*}{m_e}\right)^{1/2} = 0, \quad (5)$$

where N_* and T_* are the number density and effective temperature of secondary electrons, respectively. Similarly, for the end currents one has for eclipsed or shadowed surfaces

$$N_e e \left(\frac{2\kappa T_e}{m_e}\right)^{1/2} e^{-e\Phi/\kappa T_e} - N_p e \left(\frac{2\kappa T_p}{m_p}\right)^{1/2} - N_* e \left(\frac{2\kappa T_*}{m_e}\right)^{1/2} = 0. \quad (6)$$

An equivalent photoelectron term $-N_{ph} e (2\kappa T_{ph}/m_e)^{1/2}$ can be added to both of these equations under illuminated conditions. The two densities N_* and N_{ph} must be computed from the appropriate yield factors for the surface materials in question, as pointed out in the works of Knott (1972) and Grard (1972). For spherical geometry, one has only a $j_{\alpha\perp}$ given by

$$j_{\alpha\perp} = 2 N_\alpha q_\alpha \left(\frac{2\kappa T_\alpha}{\pi m_\alpha}\right)^{1/2} \int_{v_0/\langle v_\alpha \rangle}^{\infty} t^2 e^{-t^2} \, dt, \quad \langle v_\alpha \rangle = \left(\frac{2\kappa T_\alpha}{m_\alpha}\right)^{1/2} \quad (7)$$

which leads to the current balance condition

$$N_e e \left(\frac{2\kappa T_e}{m_e}\right)^{1/2} \left(1 + \frac{e\phi}{\kappa T_e}\right) e^{-e\Phi/\kappa T_e} - N_p e \left(\frac{2\kappa T_p}{m_p}\right)^{1/2} - N_* e \left(\frac{2\kappa T_*}{m_e}\right)^{1/2} = 0 \quad (8)$$

to which a term $-N_{ph} e (2\kappa T_{ph}/m_e)^{1/2}$ for photoelectrons may be added, as before. If we write the saturation current density *to* the spacecraft (in the conventional sense) as

$$j_+^{(S)} = N_p e \left(\frac{2\kappa T_p}{m_p}\right)^{1/2} + N_* e \left(\frac{2\kappa T_*}{m_e}\right)^{1/2} + N_{ph} e \left(\frac{2\kappa T_{ph}}{m_e}\right)^{1/2} \tag{9}$$

dan *away* from the spacecraft as

$$j_-^{(s)} = N_e e \left(\frac{2\kappa T_e}{m_e}\right)^{1/2} \tag{10}$$

then for the planar (Equation (6)), cylindrical (Equation (5)), and spherical (Equation (8)) surfaces, the equilibrium potential Φ assumed by that surface is, within the simple theory, given by the solution to the transcendental equation of the general form

$$\Psi_g \left(e\Phi/\kappa T_e\right) = j_+^{(s)}/j_-^{(s)} = R, \tag{11}$$

where the geometry-dependent function Ψ_g is

$$\Psi_g \left(e\Phi/\kappa T_e\right) = \begin{cases} \exp\left(-e\Phi/\kappa T_e\right), & \text{for the plane} \\ \dfrac{2}{\sqrt{\pi}} \displaystyle\int_{e\Phi/\kappa T_e}^{\infty} t^{1/2} e^{-t}\, dt = \dfrac{\Gamma\left(\tfrac{3}{2}, e\Phi/\kappa T_e\right)}{\Gamma\left(\tfrac{3}{2}\right)}, & \text{for the cylinder} \\ \left(1 + e\Phi/\kappa T_e\right) e^{-e\Phi/\kappa T_e}, & \text{for the sphere.} \end{cases} \tag{12}$$

These three functions are plotted in Figure 7 as a function of the normalized energy $|e\Phi/\kappa T_e|$. Since the saturation current densities are functions of the thermal properties of the ambient plasma, of the secondary and photoelectron yield factors of the surface material, and of the direction of incidence of solar photons on the local surface, it is not true that the ratio R in Equation (11) is independent of geometry. However, if we assume that the ratio R is a constant, then the intersection of horizontal lines $R=$ constant $\leqslant 1$ with the Ψ_g yield three different values of $|e\Phi/\kappa T_e|$. Since our initial assumption was $\Phi \leqslant 0$, for the example $R=0.3$ shown in Figure 7, one obtains potentials

$$\begin{aligned} \Phi &= -1.2\,\kappa T_e/e \;\;(\text{plane}) \\ &= -1.84\,\kappa T_e/e \;(\text{cylinder}) \\ &= -2.42\,\kappa T_e/e \;(\text{sphere}). \end{aligned}$$

Because actual spacecraft surfaces are complex and inhomogeneous, the practical case of spacecraft charging is probably well beyond hope for adequate theoretical treatment. However, the above treatment does provide some guidance concerning orders of magnitude of surface potentials.

Conversely, if we apply this simple theory to the experimental results of DeForest (1972) during eclipse of ATS-5, we can estimate the ratio $R=j_+^{(S)}/j_-^{(S)}$. Let us take the example of Figure 7, where $\Phi = -4.2$ kV. According to DeForest, the density of injected protons and electrons was ~ 1 cm^{-3}, with $T_p \sim 10$ keV and $T_e = 5$ keV. If there were *no* secondary emission, then $R = (m_e T_p/m_p T_e)^{1/2} = 0.033$. In this case, for

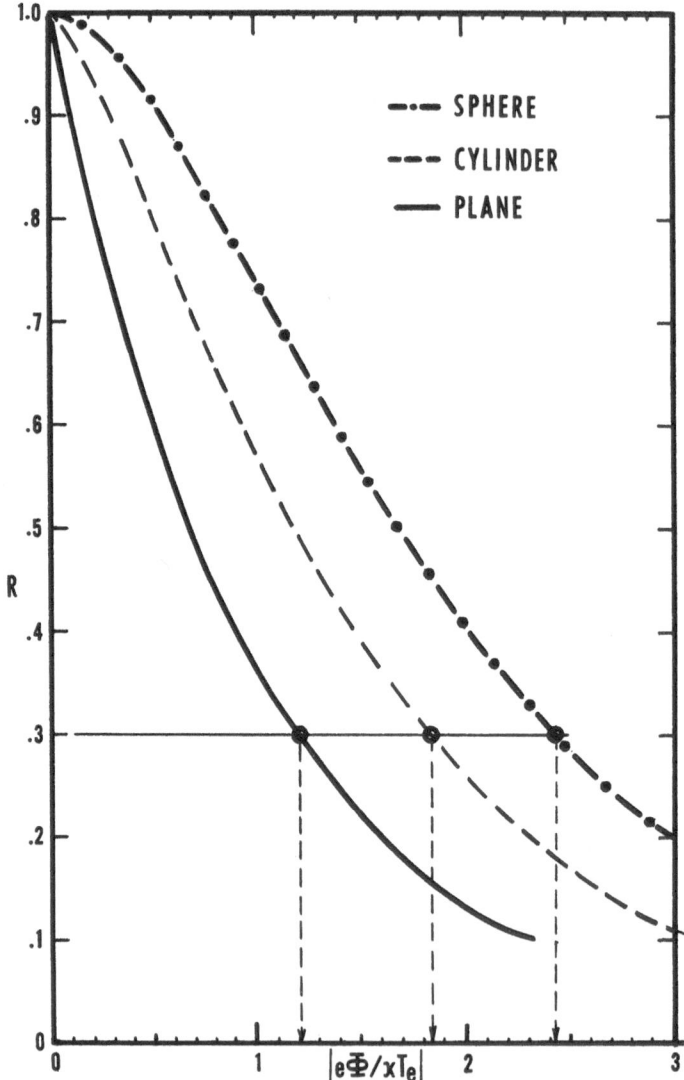

Fig. 7.   Shows charging potential $e\Phi(<0)$ as a function of the saturation current ratio $R = j_+{}^{(S)}/j_+{}^{(S)}$ for spherical, cylindrical and planar surfaces by electrons of energy $E_e \sim \kappa T_e$. Note difference in equilibrium potential for constant value of R.

the three geometries of spacecraft surface, the curves in Figure 8 yield the results

$$\Phi = -(3.4) \; (\kappa T_e/e) = -17 \text{ kV (plane)}$$
$$\Phi = -(4.4) \; (\kappa T_e/e) = -22 \text{ kV (cylinder)}$$
$$\Phi = -(5.25) \; (\kappa T_e/e) = -26 \text{ kV (sphere)}.$$

These voltages, especially that of the plane, are consistent with DeForest's comments

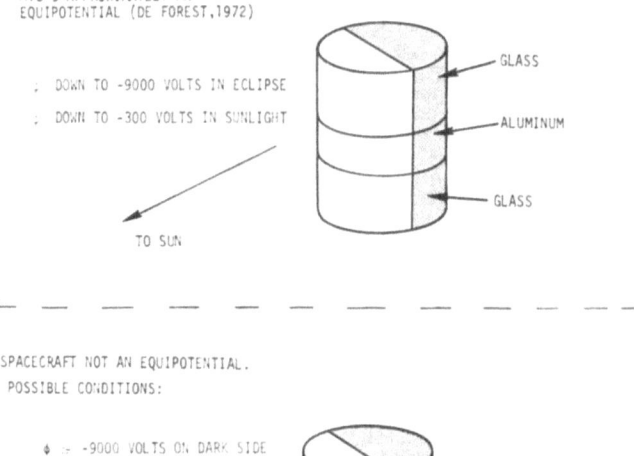

ATS-5 APPROXIMATELY AN
EQUIPOTENTIAL (DE FOREST,1972)

. DOWN TO -9000 VOLTS IN ECLIPSE

. DOWN TO -300 VOLTS IN SUNLIGHT

TO SUN

GLASS

ALUMINUM

GLASS

SPACECRAFT NOT AN EQUIPOTENTIAL.
POSSIBLE CONDITIONS:

φ = -9000 VOLTS ON DARK SIDE

φ = -300 VOLTS ON SUNLIT SIDE

TO SUN

ALL INSULATING

MATERIAL

Fig. 8.   Schematic diagrams of two different spacecraft geometries, one with a conducting equatoria
ring and another covered by (dielectric) solar arrays.

that extrapolation of his curve labeled 'without secondaries' in Figure 7 of his 1972
paper would yield a predicted potential about 3 times the measured one.

If we draw a vertical line at $|e\Phi/\kappa T_e|=0.84$, which is the value of 4.2 kV divided by
electron voltage of 5 kV, then the intersections yield

$R = 0.425$ (plane)
$R = 0.64$ (cylinder)
$R = 0.80$ (sphere).

In all geometries, then, secondary emission currents and backscattered electron currents from the surface materials of ATS-5 during eclipses must be a significant fraction
(40% to 80%) of the incident currents.

3.2. DIFFERENTIAL CHARGING

One should recall that the surfaces of different spacecraft vary widely. For example,
many spacecraft are cylindrical in shape, and have solar cells with glass covers coating
the entire cylindrical surface. Some of these are open on one or both ends, with both
dielectric and conducting surfaces bearing instrumentation exposed to both sunlight
and plasma environment. Others have one end open in this manner, and the other end
with a thermal closure surface covering it. Still others having cylindrical geometry
have varying materials (dielectrics, thermal balance surfaces, conductors, paint, open-

ings, etc.) distributed over all surfaces. Some spin at 15 rpm, some at 60 rpm, and some, including ATS-5, at as much as 100 rpm. Still others, usually with solar paddles, are attitude-stabilized, and do not spin at all. In fact, most such satellites launched so far are veritable nightmares to analyze for electrical behavior in synchronous or other orbits in the distant reaches of the magnetosphere.

As an example of the different charging characteristics of two spacecraft structures, in Figure 8 we show the idealized ATS-5 geometry quoted by DeForest (1972), and an idealized structure similar to many geostationary communications satellites. In the upper panel, the spacecraft has an aluminum (or perhaps some other conductive) band girdling its equator. Above and below this 'belly-band' the surface is covered with solar cell arrays (a dielectric surface). The ends may be open or closed by some prescribed closure material.

In the lower panel, the spacecraft is entirely covered with solar panels, that is, a dielectric material. Again, the ends may be open or closed. One immediately foresees a vast difference in spacecraft charging potential in these two cases. In the first geometry, under illumination, the conductive belly-band will be an equipotential surface, and even during substorm injection events, the effective area for photoelectron emission can be great enough to hold the potential of the entire ring to the -50 V to -300 V limits found by DeForest on ATS-5.

On the other hand, the shadowed regions of dielectric may assume a very large negative potential during these events, since photoelectron emission currents from illuminated dielectric surface areas are completely insulated from charging currents on the dark areas. If the time constant for charging is short compared to one-half the spin period, dark areas may achieve many kilovolts negative, while illuminated surfaces are at 50–300 V negative. Such *differential* charging can lead to very large potential gradients (i.e., electric fields) at points on the surface of the spacecraft. Also, if surface dielectrics are thin sheet materials mounted on conductors below, large gradients can be anticipated across the dielectric. A local gradient of many tens of kV per cm can lead to electrical breakdown of the dielectrics, with corona or arc discharges. These effects may or may not cause electronics failures and mechanical damage, depending upon the nature of spacecraft function and design.

In Figure 9 we show a schematic of a geostationary spacecraft similar to one of those involved in the observation of events in Figure 1. The top panel shows the illumination conditions during the winter solstice, the bottom panel those during a summer solstice. In winter, the open end of the cylinder is completely shadowed. Thus, only energetic electrons (nearly isotropic) during injection events will enter this end. Inside the open end are various conductive structures which electrically connect to the spacecraft common 'ground'.

In summer, the open end is partially illuminated and photoelectrons can be emitted from both the inner cylindrical surface of the solar array substrate, and from the conductors connected to the spacecraft common "ground". The spacecraft spin period is on the order of 1 s.

Thus, in these two cases, different differential charging of various spacecraft struc-

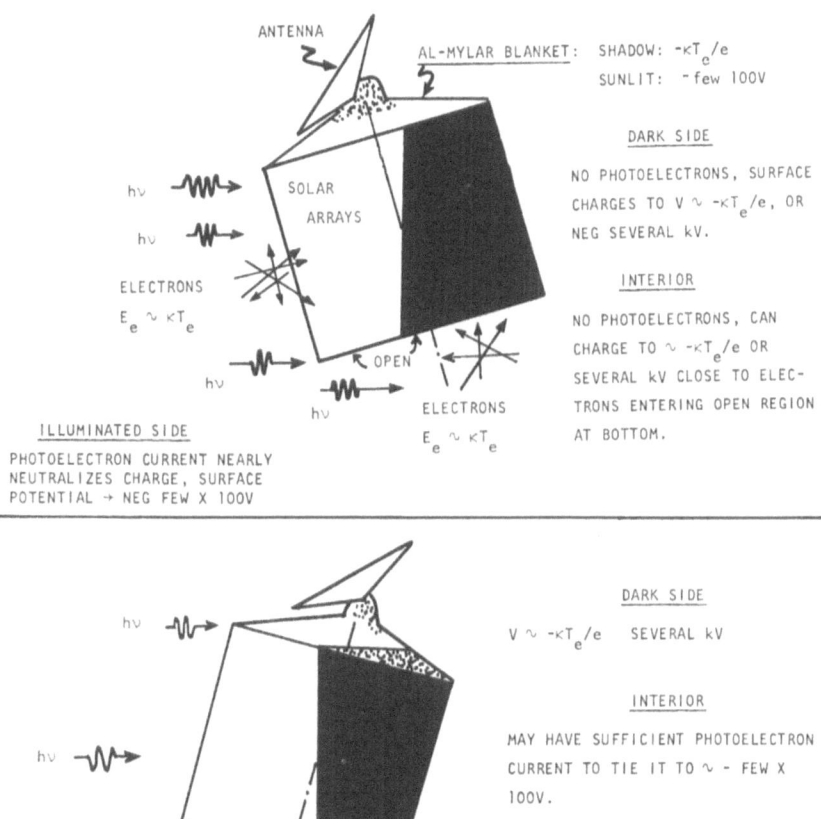

ANTENNA

AL-MYLAR BLANKET: SHADOW: $-\kappa T_e/e$
                    SUNLIT: ~few 100V

DARK SIDE

NO PHOTOELECTRONS, SURFACE
CHARGES TO V ~ $-\kappa T_e/e$, OR
NEG SEVERAL kV.

INTERIOR

NO PHOTOELECTRONS, CAN
CHARGE TO ~ $-\kappa T_e/e$ OR
SEVERAL kV CLOSE TO ELEC-
TRONS ENTERING OPEN REGION
AT BOTTOM.

SOLAR ARRAYS

hυ

ELECTRONS
$E_e \sim \kappa T_e$

OPEN

hυ

ELECTRONS
$E_e \sim \kappa T_e$

ILLUMINATED SIDE

PHOTOELECTRON CURRENT NEARLY
NEUTRALIZES CHARGE, SURFACE
POTENTIAL → NEG FEW X 100V

DARK SIDE

V ~ $-\kappa T_e/e$ SEVERAL kV

INTERIOR

MAY HAVE SUFFICIENT PHOTOELECTRON
CURRENT TO TIE IT TO ~ - FEW X
100V.

ILLUMINATED SIDE

~ - FEW X 100V

OPEN

hυ

AL-MYLAR
SHADOWS: ~ $-\kappa T_e/e$ ~ -kV
SUNLIT: ~ - FEW X 100V

Fig. 9. Schematic diagram of structure of a typical geostationary communications satellite under
two different seasonal illumination conditions.

tures can occur during the substorm-associated injection events. In the 'winter' configuration, the dielectric outer surface will charge to a relatively low negative voltage
(say ~ − 100 V) while the shadowed outer surface will charge to perhaps as much as
− 9000 V, provided the time constant is less than 0.5 s. The so-called common 'ground',
through bombardment of the open end by the keV injection electrons, can swing down
to ~ − 9000 V.

In the 'summer' configuration, the dielectric surfaces again have ~ − 100 V (sunlit)
to − 9000 V (shadowed) charging. However, provided enough exposed surface area is
available in the open end, the common 'ground' can assume the smaller, ~ − 100 V,
potential.

In either configuration, the closed end of the cylinder (which is de-spun) will have shadowed areas, at least from the antenna, if not a terminator. The closure material is multiple layers of dielectric and conductor with a dielectric outer surface. This means it is a capacitor, estimated to be some 10 to 100 pf. If discharged by an arc with a 1 μs duration, it could produce a transient power of up to 4000 W. A significant electromagnetic pulse containing high frequencies is thus possible.

The solar cell arrays around the cylindrical surface are constructed as shown in Figure 10. There are eight individual panels, each electrically isolated from the others except through solar cell string connections to the main power bus. The glass-covered solar cells are mounted on a substrate structure consisting of a fiberglass sheet of thickness 0.010 in. (0.0254 cm) bonded to a honeycomb structure 0.25 in. (0.635 cm) thick with honeycomb cell dimension ~0.25 in. and wall thickness ~0.001 in. (0.002 54 cm). A photograph of a section of this structure is shown in Figure 11. The glass-covered solar cells cover all but ~2000 cm^2 of this substrate.

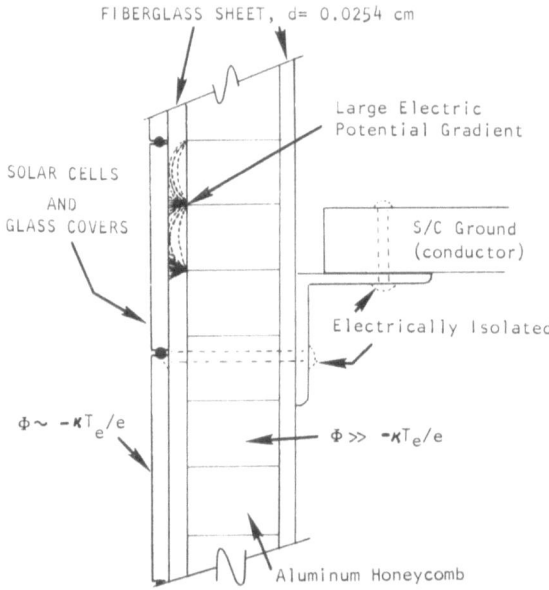

Fig. 10.   Schematic cross-section, not to scale, of the solar array structure and its mounting to space-craft main structure. See text for details.

Each panel is mounted to the spacecraft common ground structure by fiberglass brackets and insulated bolts. The capacitance of this structure is difficult to estimate, but a crude estimate is ~6 pf per panel, relative to an ambient plasma of Debye length ~50 m, or about 50 pf for the whole array of 8 panels.

Our estimate for the time constant required to charge the surface of a panel in darkness by energetic electron fluxes typically measured by DeForest (1972) on ATS-5 during substorm-associated injection events (that is, ~4 × 10^9 electrons cm^{-2} s^{-1}) to a voltage of some −5 to −10 kV is less than 0.1 s. This was based on the observation

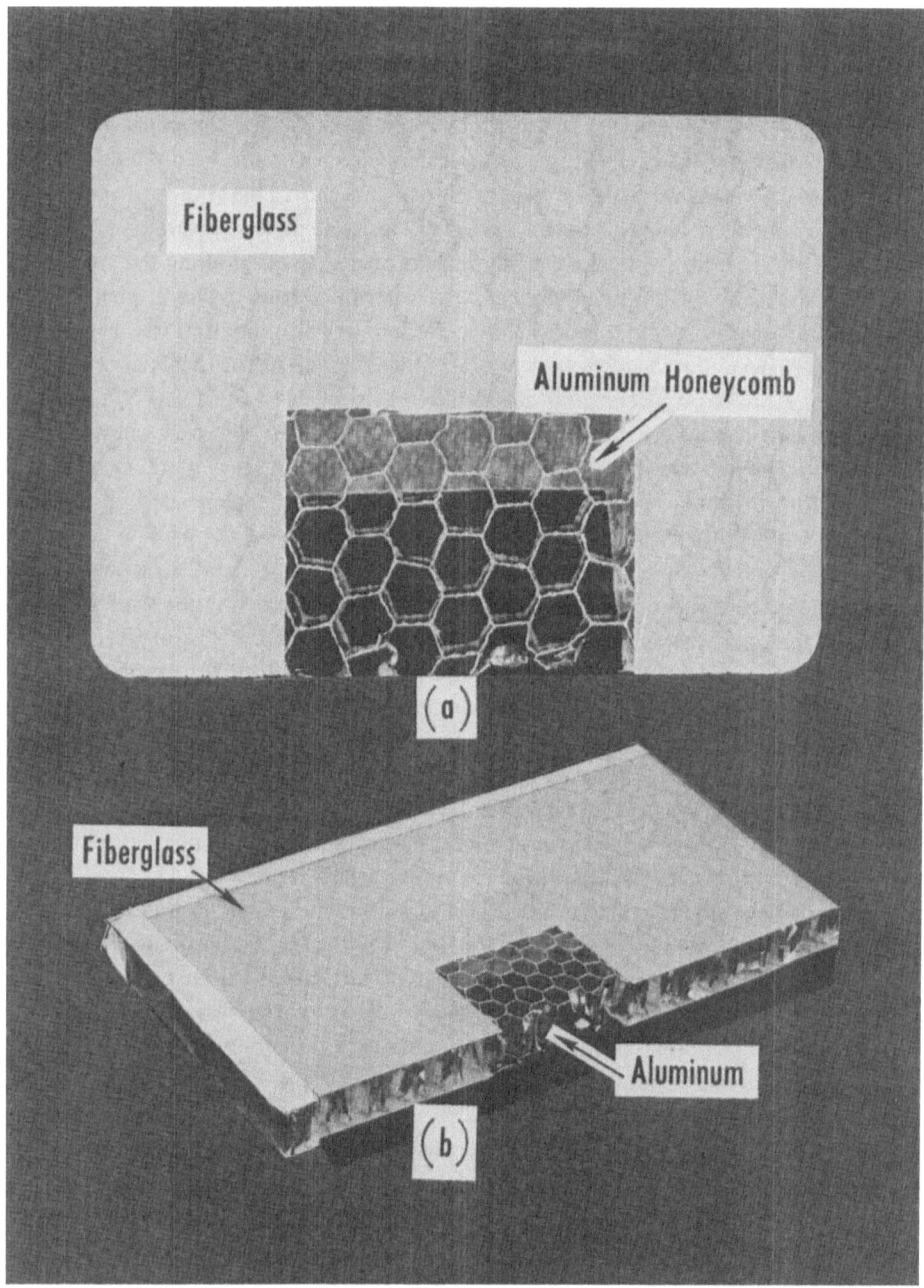

Fig. 11.   Photograph of the fiberglass-aluminum honeycomb solar array substrate material.

that, if the development of a retarding potential with time were ignored, and charge were delivered to a 50 pf capacitor by the full flux, then $Q = j_-^{(S)}(0) A \tau_{min} = C\Phi$, or $\tau_{min} = C\Phi / j_-^{(S)}(0) A$, where A is the spacecraft area and Φ the final potential. One obtains $\tau_{min} \sim 2 \times 10^{-3}$ second. If we allow a factor of 50 for the effect of surface potential retarding successively more of the electron flux with time, we obtain $\tau \sim 10^{-1}$ s.

The only experimental evidence supporting such a short charging-time constant is from an unpublished observation on ATS-5 (DeForest, private communication). As DeForest (1972) pointed out, the UCSD plasma probe pointing along the spin axis is located within a cavity, and surfaces near it can alternately charge and discharge as the probe and its nearby surfaces spin through shadow and illumination (see Figure 6 of DeForest's paper). Significant spin modulation was apparent for low-energy protons. This modulation had no detectable phase-lag with respect to solar aspect angle (De-Forest, private communication), indicating that at least in this case charging-time constant was sufficiently less than a spin period (0.6 s for ATS-5 at 100 rpm).

A question still remains concerning whether: once charge is deposited on the solar array structure, is the resistivity of the fiberglass sheet in Figure 11 low enough to allow charge to leak through it to the aluminum honeycomb conductor below? Certainly, no leakage will occur through the solar cell glass covers on time scales of a spin period, so that only the exposed area of fiberglass ($\sim 2 \times 10^3$ cm^2) is involved. We use the nominal resistivity $\varrho \sim 10^{17} \Omega$ cm for the thickness 2.54×10^{-2} cm and exposed area 2×10^3 cm^2 to obtain a resistance per panel of $R = 1.3 \times 10^{12} \Omega$. The estimated capacitance is ~ 10 pf, so that $\tau = RC \sim 13$ s. This is many times the spin period (~ 1 s) or estimated charging time ($\lesssim 0.1$ s). We conclude that the outer surface of the solar panel will hold its induced charge unless electrical breakdown through it occurs.

Another spacecraft component on which large potential gradients may develop is the thermal closure material described previously. This is made of 11 layers of very thin mylar on one side of which a thin aluminum film has been deposited. A schematic (with fewer than the actual 11 layers) is shown in Figure 12, at a terminator boundary. The dark area can be charged to many kV negative, while the illuminated area may emit enough photoelectron current to remain only a few tens to a few hundreds of volts negative. The potential distribution in this case could assume a structure similar to that in the lower panel, with gradients of ~ 50 kV cm^{-1} at a sharp terminator, directed *along* the outer surface. There is no way to determine the electrical connections among the eleven layers of dielectric and aluminum, since handling and mounting of such thin, fragile sheets can easily cause pin-hole-like connections accidentally at random points in the structure. Therefore, potential distributions *across* the blanket are not known.

We thus identified the solar array structures and the thermal blanket materials as spacecraft components on which substorm-associated charging could conceivably produce both large potentials, and more importantly, large potential gradients, and therefore selected them for laboratory high-voltage testing described in the next section. In particular, the large gradient configurations shown in Figures 10 and 12 were suspected.

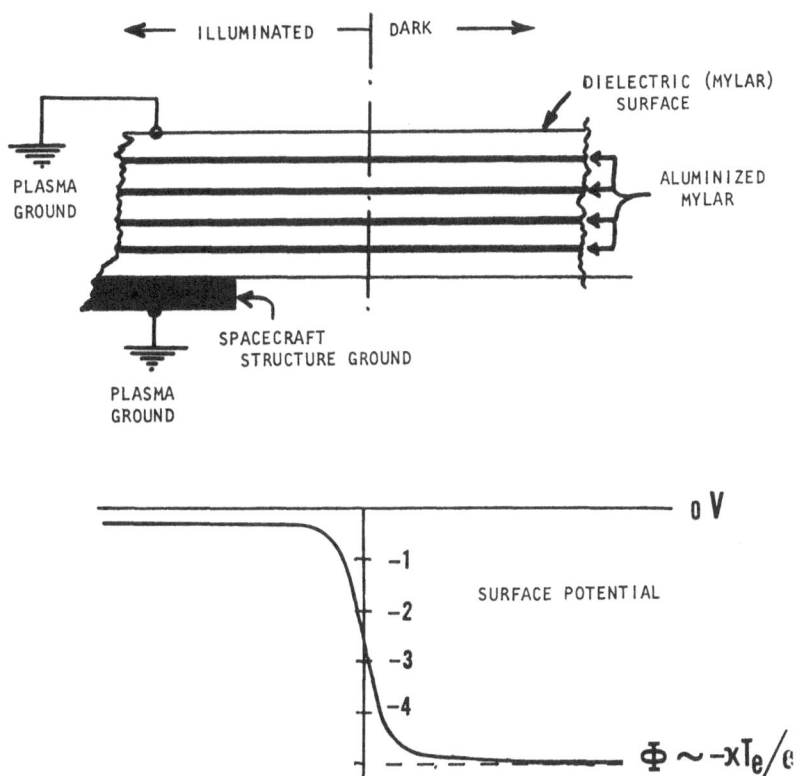

Fig. 12.   Schematic cross-section of multi-layered aluminized mylar thermal blanket material, and sketch of potential distribution possible at a terminator boundary for $\kappa T_e \sim 5$ keV.

4. Laboratory Test Results

4.1. GENERAL COMMENTS

In view of the evidence that:

(a) Geostationary satellites are known to charge differentially during substorm-associated injections of hot plasma sheet electrons;

(b) anomalous events involving changes of state of electronics subsystems on such satellites were timed with known substorm occurrence in ground magnetograms;

(c) structural components of these spacecraft were identified that could possess large potential gradients during charging events; it was decided that either the large potential gradients, or their subsequent breakdown through corona or arc discharges, was responsible for the anomalous behavior of the subsystems.

Thus, laboratory tests on a flight-spare electronics subsystem, including its inter-connecting cabling and electrical tests of the suspected structural components, were performed. We present in this section the tests and their main results.

4.2. STATIC AND LOW FREQUENCY TESTS ELECTRONICS SUBSYSTEMS

A flight-spare subsystem, including cables, was first subjected to static and low-frequency magnetic and electric fields of large magnitude to determine whether quasi-static large potential gradients or large transient low-frequency currents in spacecraft structures could cause anomalous behavior of the subsystem such as that which occurred in flight. These tests and their results are described below.

4.2.1. *Magnetic Field Tests*

It was first determined that the anomalous behavior could be induced by sufficiently large magnetic fields. A 2.25-in. diam coil of 700 turns was pulsed with a peak current of 0.5 A of 10 ms duration. The field at the center of this coil was about 50 G. When two turns of the interconnecting cabling of the electronics subsytem was wrapped around the source coil, anomalous behavior was induced.

The direct susceptibility of circuitry within the subsystems shield boxes was then tested. This susceptibility is due to change of magnetic flux within the boxes (dB/dt). It was found that over a frequency range of several hundred Hz to 1 MHz, the suscep-tibility peaked at several hundred kHz. However, the external field required was >50 G.

4.2.2. *Electric Field Tests*

Sinusoidal voltages up to 10 kV at frequencies up to 100 kHz were applied to subsys-tem boxes and cabling. These units were placed between two large parallel plates separated by 25 cm. Thus, fields up to 80 kV m^{-1} were applied. No anomalous behav-ior of the subsystem was produced by these large fields at these low ($\leqslant 100$ kHz) frequencies. We concluded that direct application of substorm charging potential gradients across boxes and cables, within the ranges estimated for such in-orbit gradients, could not cause the observed anomalies.

4.2.3. *Summary of Results*

The ranges covered by these tests, and the main results, are summarized below

> Test program ranges
> Magnetic fields: 100 Hz–1 MHz
> $\quad\quad\quad\quad\quad\quad$ 0–50 G
> Electric field: $\quad$ 100 Hz–1 MHz
> $\quad\quad\quad\quad\quad\quad$ 0–10 kV-m^{-1}
> *Main results*
> Anomalous behavior induced by:
> ~ 50 G at 100 kHz, magnetic fields
> No anomaly for $\leqslant 100$ kV m^{-1}, electric fields

4.3. HIGH FREQUENCY TESTS OF ELECTRONICS SUBSYSTEM

During the low-frequency electric field tests described above, occasional arc discharges and coronas in the test equipment occurred. These were observed to cause the anomalous behavior of the electronics subsystem under test. To verify this, a tesla coil was next employed to induce the anomalies. With the tesla coil and the electronics subsystem electrically isolated from one another, arcs were struck in air from the tesla coil to a floating conductor at varying distances from arc location to the subsystem boxes and cables.

It was found that such arcs at distances up to ~ 5 m, and corona discharges around the electrode of the tesla coil at ~ 70 cm separation, produced the anomalous behavior.

It was found that proper shielding (and high-frequency grounding of the shielding) of the interconnecting cabling eliminated arc-induced anomalies.

This, of course, suggested that high-frequency, large amplitude pulses produced by the tesla coil arc irradiated the cabling, causing spurious signals to be produced, thus initiating subsystem anomalous behavior.

A careful determination of the electric and magnetic field strengths in these pulses was then made, using B-dot and D-dot probes. It was found that most of the energy radiated by these arcs appeared in the electric component. The results were as follows: at 0.5 m arc-to-probe separation, a 950 V m^{-1} pulse with 20 ns rise-time was observed. At 1.0 m separation, the electric intensity dropped to 400 V m^{-1}. With the D-dot probe disconnected, a baseline signal induced in the measuring equipment and probe cable was less than a 50 V m^{-1} equivalent. The magnetic field component at arc-to-probe separation of 30 cm had a peak pulse amplitude of $\sim 10^{-2}$ G and rise-time ~ 20 ns.

These results indicated that arcing in structures of the spacecraft close to the cabling associated with electronics subsystem could have produced the observed anomalies in its behavior. Since the solar array substrate and thermal closure materials, identified as potential sources of breakdowns due to large potential gradients, are well within the arc-to-subsystem cabling separations used in tesla coil tests, it was indicated that these structures should be subjected to high-voltage testing, as described next.

4.4. ELECTRICAL HIGH-VOLTAGE BREAKDOWN TESTS

The purposes of high voltage tests were: first, to measure potential differences which could produce breakdown on or through the samples of materials used on spacecraft in order to ascertain whether arcs could be produced at voltage drops on the order of those possible on a geostationary spacecraft; second, to determine the degree of and type of degradation occurring as a result of arc discharges in or on the surface of the sample material; and third, to measure the frequency content and field intensity of electromagnetic pulses from such arc discharges.

The materials selected were, as stated previously, the most likely to be exposed to severe differential charging during substorm-associated injections of hot plasma in the local midnight to local dawn sector at 6.6 R_E. First, samples of the fiberglass-aluminum honeycomb substrate material were tested without any mounted solar cells; then, samples of this substrate containing solar cells and their glass covers were tested; third, samples of the multilayered aluminized mylar thermal blanket material were tested. These samples were not subjected to any special cleaning procedures.

The equipment used in these tests consisted of the following: a 20 kV, 5 mA high-voltage supply; oscilloscopes, amplifiers and meters; a vacuum system capable of $\sim 10^{-6}$ torr. All high voltage tests were performed on material samples both in air and in partial vacuums of less than 2×10^{-5} torr. Voltage was applied to various electrode placement geometries in an effort to simulate hypothetical potential gradient situations predicted for the spacecraft in-orbit charging. Potential differences were gradually increased until a breakdown occurred. If the arcing ceased, the voltage was increased to re-initiate the breakdown. If a continuous arc occurred, the voltage was decreased until the arc quenched, and then subsequently increased to determine at which the arc re-ignited.

4.4.1. *Tests of the Aluminum Honeycomb-Fiberglass Substrate*

The aluminum-honeycomb substrate tests were performed both in air at atmospheric pressure and in vacuum with pressure of 1.5×10^{-5} torr. Three configurations of electrodes were used as shown in Figure 13. In configuration (a) the voltage was applied across both fiberglass sheets. In (b) both electrodes were on one fiberglass surface and in (c) the voltage was across one sheeet only.

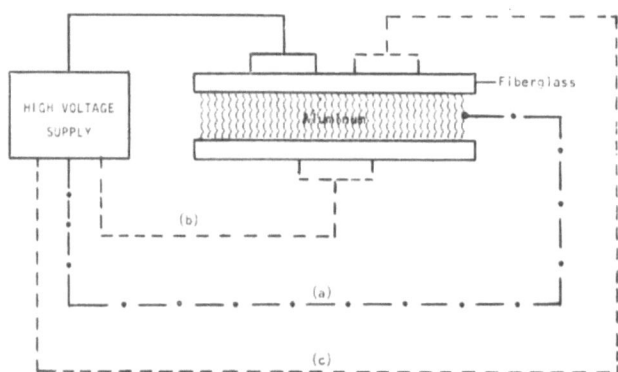

Fig. 13.   Schematic diagram of configurations used in high-voltage testing of solar array substrate samples. Broken lines show various connections of second electrode for fixed connection (solid lines) of first electrode.

When the voltage was applied in configuration (a) the fiberglass broke down at 1650 V at room temperature and atmospheric pressure and at ~ 6000 V in vacuum. In both cases, after the initial breakdown, subsequent arcs occurred at lower voltages. A continuous discharge subsequently occurred at 500 V and at 250 V in vacuum.

In configuration (b) the initial discharge occurred at ~ 3800 V in air and 7500 V in vacuum. Subsequent continuous breakdown occurred at 1000 V in air and 700 V in vacuum.

Only a vacuum test was performed on the sample in configuration (c). In this case the initial breakdown occurred at 6000 V and continuous arcing at 600 V.

In all of the above tests the occurrence of the arc produced a burn spot on the fiberglass so that the position of the ends of the arc could easily be determined. The burn spots were not necessarily near the edges of the electrodes. The arcs, however, did occur at a position where the various aluminum-honeycomb cells where joined together.

4.4.2. *Tests with Solar Cells and Cover Glass on Aluminum-Honeycomb Substrate*

The solar cell on substrate tests were performed in vacuum with the configurations shown in Figure 14. When the high voltage was applied using configurations (a) and

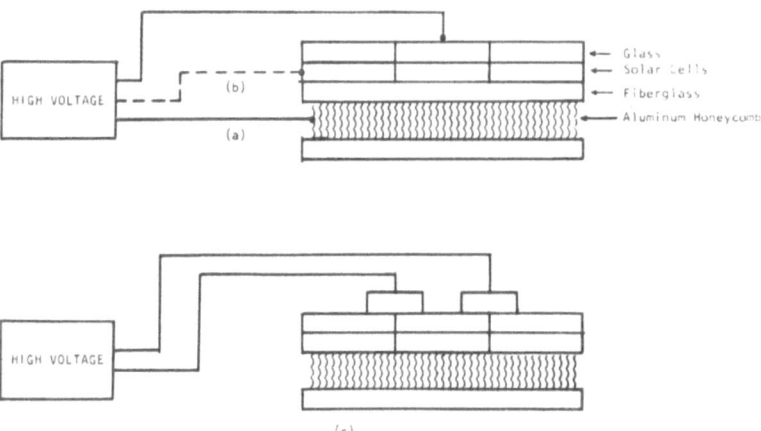

Fig. 14.   Similar to Figure 13, but showing test configurations for samples of substrate with solar cells and their glass covers mounted.

(b), no breakdown occurred below about 13000 V. When arcing did occur, the arcs ran along the surface of the glass. With configuration (c), breakdown occurred at 7000 V. The arc traveled down the separations between cover glasses and broke through the fiberglass. Subsequent arcs occurred at lower voltages, with a continuous discharge at 1000 V. Before and after each of the above tests, the current from the solar cells was checked, and no apparent degradation of the cells was observed. After the test with configuration (c), burn marks were found in the substrate fiberglass.

4.4.3. *Thermal Blanket Tests*

The tests on the thermal blankets were performed using the configuration shown in Figure 15. Initial breakdown occurred around the edges of the blanket at ~ 5000 V, depending on proximity of electrode to edge of blanket. Subsequent breakdowns required higher voltages. Aluminum was vaporized from all layers of the blanket, particularly near the edges and along the wrinkles of the thin layers.

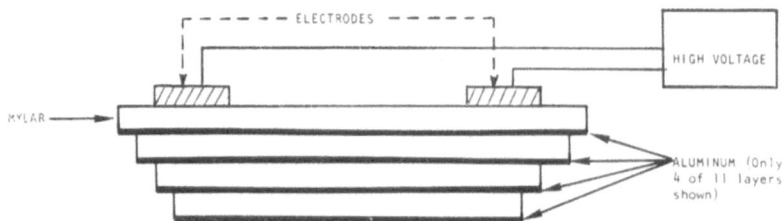

Fig. 15.   Schematic of surface potential gradient test of the aluminized mylar multi-layered blanket.
Only four of the eleven layers are schematicized.

The vaporization of the aluminum produced mechanical damage, as illustrated in Figure 16. This is just one of the eleven layers in the thermal blanket material. The left-hand section (a) is an undamaged layer prior to breakdown testing. The right-hand section (b) is the result of arcing in the blanket. Note how the vacuum sparks travel along lightning-like paths. These paths coincide with creases in the original material, where in all probability aluminum vapor was created, with concentrations in the almost microscopic valleys of these creases, allowing a metallic gaseous to occur. One can easily imagine the resultant degradation of thermal characteristics of such a superinsulating material after arcing

4.4.4. *Summary*

Table I summarizes the main results of the electrical breakdown and arc discharge tests. All of the cases summarized in this Table are possible sources of arcs resulting from potential gradients that may be achievable during substorm-associated space-craft charging phenomena in orbit.

(a)   (b)

BEFORE   AFTER

Fig. 16.   Shows undamaged (a) single layer of thermal blanket material prior to high-voltage testing and (b) after high-voltage breakdown. The damage is clearly due to vacuum arcing which vaporized aluminum, leaving transparent mylar along arc channels.

TABLE I

Arc breakdown potentials

	Breakdown potential diff. (V)	Configuration	Remarks
Thermal blanket ($<2 \times 10^{-5}$ torr)	~ 5000	Both terminals on top mylar layer	Breakdown around edges. Subsequent breakdowns require higher voltages. Test not continued past 15000 V. Aluminum vaporized on all layers of blanket, particularly along wrinkles of thin layers.
Fiberglass/Aluminum Honeycomb ($<2 \times 10^{-5}$ torr)	~ 6000 ~ 7500 ~ 6000	Across one sheet of fiberglass. Across two fiberglass sheets. Both terminals on same fiberglass surface	V decreases down to as low as 250 V after repetitive sparking. V decreases to as low as 700 V after repetitve sparking. V decreases to as low as 600 V after repetitive sparking.
Solar cells and cover glass on substrate ($<2 \times 10^{-5}$ torr)	~ 7000	Both terminals on cover glass surface. Terminals cover space between cover glass	V decreases down to 1000 V after repetitive arcing. Spark travels down cracks between cover glasses. No observable degradation of solar cells.

4.5. Electromagnetic Pulses from Arc Discharges

The high-voltage breakdown tests indicated that arc discharges could be induced by voltage gradients within the limits foreseen on the orbiting geostationary spacecraft during charging events. In order to determine whether electromagnetic pulse discharges could be responsible for anomalous behavior of the electronics subsystem, the amplitude and waveform characteristics of these arcs in actual breakdowns of samples of spacecraft component materials were measured for comparison with the characteristics of pulses from the tesla coil test arcs, which were known to induce anomalous behavior.

A sample was placed in the vacuum chamber (at less than 4×10^{-5} torr). An electrode configuration was selected to simulate the expected potential gradient condition on the spacecraft, and high-voltage applied until an arc occurred. The signal from this arc was detected by an antenna consisting of a 14 in. (~ 36 cm) long brass rod of cross-sectional dimension ~ 0.6 cm, coupled at its center to a shielded current probe.

The signal was amplified and connected by shielded cable to an oscilloscope, the trace of which was photographed. The response of this system was acceptably flat between 1 MHz to 50 MHz. The oscilloscope was placed at least 30 ft (~ 10 m) away from the vacuum system to minimize any direct coupling of pulse radiation into the oscilloscope. The system was operated with the brass antenna disconnected from the current probe to ensure absence of directly coupled arc-produced pulse signals in cabling and oscilloscope systems. Measurements were then performed at antenna-to-sample separations of 50 and 100 cm.

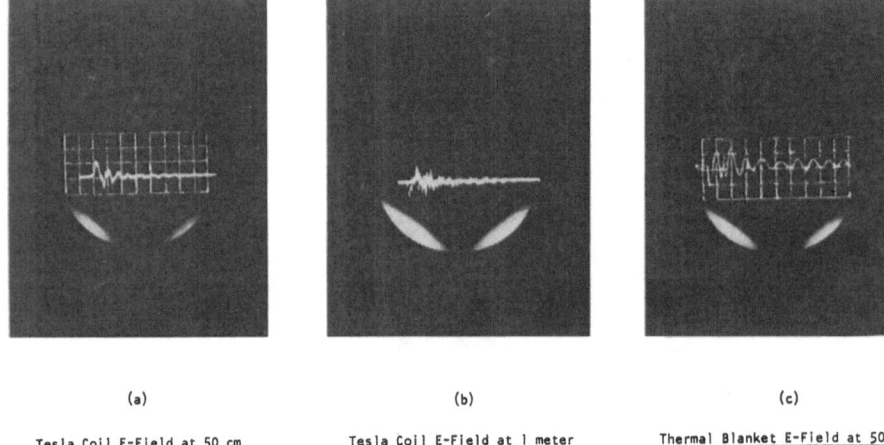

(a)   (b)   (c)

Tesla Coil E-Field at 50 cm   Tesla Coil E-Field at 1 meter   Thermal Blanket E-Field at 50 cm

Amplitude Setting = 1 V/cm   Amplitude Setting = .5 V/cm   Amplitude Setting = 0.5 V/cm
Sweep Setting = .05 µsec/cm   Sweep Setting = .05 µsec/cm   Sweep Setting = .05 µsec/cm
Calibration = 800 V/meter/cm   Calibration = 400 V/meter/cm   Calibration = 400 V/meter/cm

Fig. 17.   Shows oscilloscope electric field component of wave forms of pulses from arc discharges produced at probe-arc separations indicated, for arc sources shown. (c) has been enhanced to allow reproduction. Details are discussed in text.

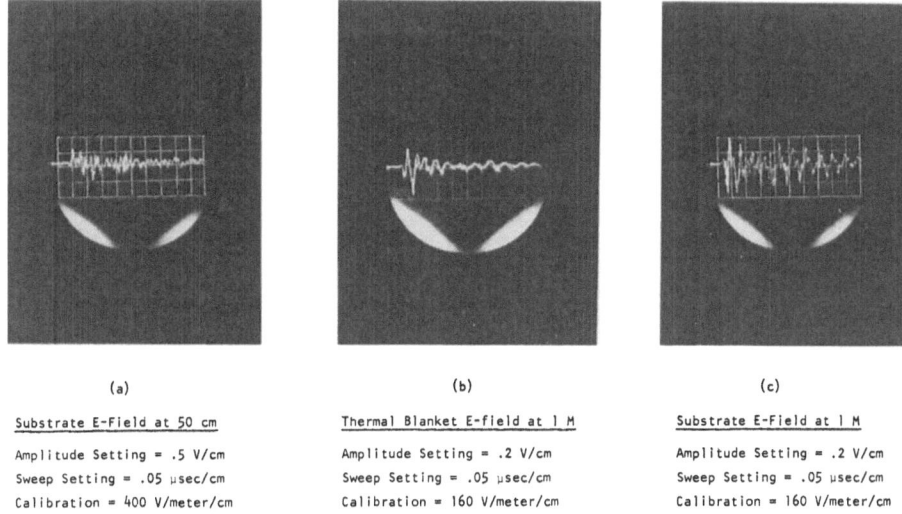

| (a) | (b) | (c) |

Substrate E-Field at 50 cm   Thermal Blanket E-field at 1 M   Substrate E-Field at 1 M

Amplitude Setting = .5 V/cm   Amplitude Setting = .2 V/cm   Amplitude Setting = .2 V/cm

Sweep Setting = .05 μsec/cm   Sweep Setting = .05 μsec/cm   Sweep Setting = .05 μsec/cm

Calibration = 400 V/meter/cm   Calibration = 160 V/meter/cm   Calibration = 160 V/meter/cm

Fig. 18.   Similar to Figure 17. Traces have been enhanced to allow reproduction.

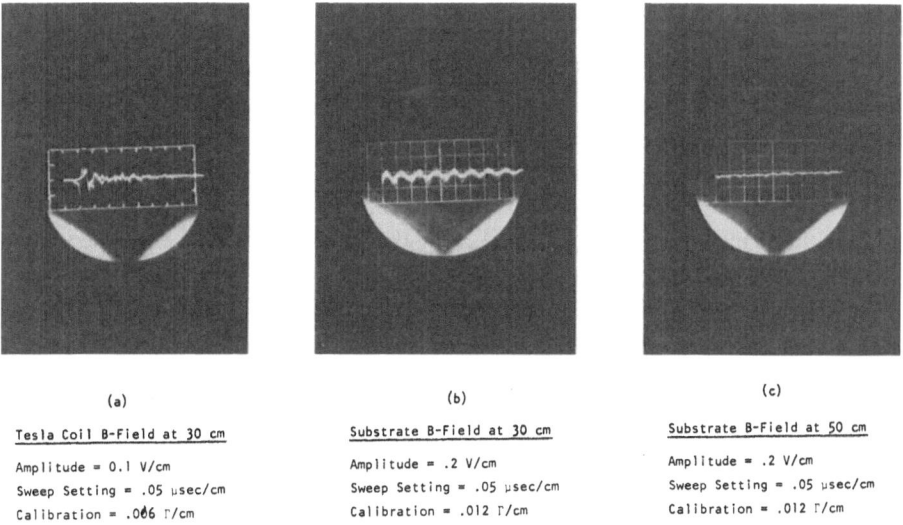

| (a) | (b) | (c) |

Tesla Coil B-Field at 30 cm   Substrate B-Field at 30 cm   Substrate B-Field at 50 cm

Amplitude = 0.1 V/cm   Amplitude = .2 V/cm   Amplitude = .2 V/cm

Sweep Setting = .05 μsec/cm   Sweep Setting = .05 μsec/cm   Sweep Setting = .05 μsec/cm

Calibration = .066 Γ/cm   Calibration = .012 Γ/cm   Calibration = .012 Γ/cm

Fig. 19.   Similar to Figures 17 and 18, for magnetic field component
of pulses produced by arc discharges.

The magnetic fields of the pulses were measured with the same system, but using a
high-frequency B-dot probe in place of the brass antenna. The same baseline checks
against direct coupling into cables and oscilloscope were made with the B-dot probe
disconnected.

Photographs of the various electric and magnetic field waveforms obtained during
these tests are shown in Figures 17, 18, 19. Because of the non-repetitive nature of

TABLE II

Maximum em radiation from arc discharges

	E			B	
	Amplitude (V m^{-1})		Rise time	Amplitude (Γ)	Rise time
	50 cm	100 cm	(ns)	30 cm	(ns)
Tesla coil (in air)	~ 950	~ 400	~ 10	$< .02$	~ 15
Thermal blanket	~ 700	~ 250	~ 15	$< .02$	~ 15
Fiberglass/aluminum					
honeycomb substrate	~ 600	~ 325	~ 10	$< .02$	~ 15

pulses from the arcs, the quality of the original photos was less than desired, and it was necessary to touch up some of the traces to produce sufficient contrast so that they could be reproduced for publication. The calibration information necessary to convert these photographed waveforms to risetimes and either V m^{-1} or gamma amplitudes is provided below the photograph.

The results of these intensity tests are given in Table II.

5. Discussion

5.1. Correlations of anomalies with substorms

The results of our study of 23 anomalies, shown in Figure 1, showed that 19 of these were definitely correlated with the *Kp* (or *ap*) indices and/or with ground magnetograms which indicate that substorms or magnetic storms were in progress during the time period including the observed event on the geostationary (or drifting near-geostationary) spacecraft. Two more anomalies have a very probable correlation with enhanced magnetic activity. Thus 21 of the 23 events can be explained as end products of the interaction of spacecraft with a substorm environment.

We feel that the correlations found represent extremely convincing evidence that these anomalies were induced by primary interaction of spacecraft with hot electrons injected into the vicinity of the synchronous orbit, with a subsequent charging of some of the spacecraft surfaces to high negative voltages with respect to others, followed by electrical breakdown involving arc and corona discharges.

5.2. Models for causes of anomalies

After examination of spacecraft structures, we identified two likely candidates for electrical breakdown and arc or corona discharge production. The two candidates were: (1) the 8 solar array substrate structures, consisting of an aluminum honeycomb sandwiched between two thin sheets (0.010″) of fiberglass, with each of the eight panels electrically isolated from the rest of the spacecraft structure, and from each other; (2) the thermal blanket material, consisting of eleven very thin layers of aluminized mylar, the whole in some sort of 'frictional' contact with spacecraft common structures via aluminum tape.

The physical model of surface differential charging of these structures was formulated, in which it was postulated that at certain points, potential gradients (electric fields) could be foreseen to achieve sufficient magnitudes to cause electrical breakdown. This model can be summarized as follows:

(a) The outer surface of the solar array structure charges to large negative voltages, say $-\Phi_0$, with respect to the aluminum honeycomb and perhaps the spacecraft common ground portions of the interior. Or;

(b) The shadowed portions of thermal blanket material charge to $-\Phi_0$, while the illuminated sections adjacent charge to a much smaller negative voltage $-\Phi_1 \gg -\Phi_0$.

(c) In either case (a) or (b), the gradient $\partial\Phi/\partial r$ can exceed the value of electric field necessary to strike corona or arc discharges. In case (a) the arc would be expected to occur through the fiberglass to the sharp edges of aluminum honeycomb below. In case (b), arcing would most likely follow paths in the creases of the layers of aluminized mylar, as vacuum arcs are wont to do;

(d) The arcs or corona discharges so produced radiate large amplitude, fast risetime electromagnetic pulses, with a great majority of the energy in the electric field component. These pulses irradiate the electronics subsystem's interconnecting cable, producing spurious signals and anomalous behavior.

These models lead to the design of specific laboratory tests of solar array substrate samples and samples of thermal blanket material.

5.3. LABORATORY TEST RESULTS

5.3.1. Samples of the solar array substrate material were tested in high vacuum ($\sim 1.5 \times 10^{-5}$ torr). The results were:

(a) Initial arc discharge occurred at ~ 6000 V between an electrode on the fiberglass surface and the aluminum honeycomb beneath. Arcing occurred near the sharp intersections of the aluminum honeycomb cells and the lower surface of the fiberglass sheet. After a short time, the voltage required to strike arcs decreased to ~ 250 V, because the spots burned through by initial discharges created carbonized paths;

(b) With the voltage applied between two electrodes on opposite fiberglass surfaces, so that the entire drop was across the whole sample, initial arcing occurred at ~ 7500 V, and shortly afterward continuous breakdown ran at ~ 700 V.

(c) High voltage breakdowns with solar cells in place did not appear to degrade solar cell output current capability.

5.3.2. Samples of the aluminized mylar thermal blanket material were tested by placing two electrodes on the same outer surface (mylar) of the samples, with the following results:

(a) Initial breakdown ocurred at ~ 3000 V. The arcing was observed to travel around the edge of the sample. Arcs apparently occurred in all 11 layers, following paths in the creases of the layers.

(b) These arcs vaporized the aluminum in the arc channels (creases) and also around the edges of the sample. This removed large areas of aluminum, leaving the areas affected transparent.

(c) After having arced for a sufficient time, the breakdown voltage increased to up to 15000 V, apparently due to the removal of the aluminum in potential vacuum arc channels, so that vaporized metal was no longer available to feed the arcs.

(d) No voltage, up to ~ 20 kV, was capable of causing a breakdown directly through the eleven layers when electrodes were placed on opposite sides of the blanket.

5.3.3. The electromagnetic pulses from these arcs were measured with B-dot and D-dot probes, with the following results:

(a) The energy in radiated pulses was found to reside largely in the electric field component;

(b) Electric field pulses of up to 400 V m^{-1} with 20 ns rise times were observed at 1 m distance from the arc. These pulses are sufficiently strong to cause anomalies by coupling into the interconnecting cable in the original configuration as flown.

5.3.4. The electronics subsystem, including cabling, was subjected to laboratory tests with the following results:

(a) The unit was tested in very large electric fields at 10 kHz, which is effectively DC as far as the unit is concerned. It required a field of greater than 100 kV m^{-1} across the subsystems boxes and cables to produce anomalies. This is much larger than any field expected in the vicinity of the flight unit during substorms, so large DC fields can be ruled out as causitive agents in anomalous events.

(b) The magnetic field tests showed that near DC fields of up to 50 G, and certainly much greater than 5 G, were required to cause anomalies. Again, we see no plausible sources for such very large magnetic fields near the flight unit.

5.3.5. It is our conclusion that the anomalies in behavior of the electronics subsystems were due to secondary arc discharges with primary cause the differential charging during substorms of spacecraft surfaces and components. The very high frequency, large amplitude electric fields in radiated pulses were picked up in the subsystems cabling, producing the anomaly.

5.3.6. It is also our conclusion that long term thermal degradation of the aluminized mylar thermal blanket material could occur due to arc discharges removing aluminum from a sizable fraction of the total area of the blanket.

5.4. RECOMMENDATIONS FOR SPACECRAFT DESIGN PRACTICES

Based on the models for spacecraft interaction with unfavorable plasma environments, and on our test results, the following recommendations bearing on spacecraft systems are presented:

(a) High intensity, high frequency (arc discharge) sources should be identified by careful examination of the proposed structures and how they would respond to a high energy electron environment, using charging models as they become available, especially from in-flight data from orbiting, instrumented spacecraft.

(b) Once this has been done, these sources should be included in the standard EMI analysis, such as computer simulation programs, to ensure system and subsystem immunity to the pulses from these sources.

(c) Adequate shielding design of boxes, cables and connectors must be employed. Twisted-pair and common mode rejection techniques may be required.

(d) Grounding techniques should be reviewed. All boxes around electronics subsystems, sensors, etc., should be bonded to their ground-plane platforms. Cabling should be shielded, with the shields bonded to local ground at every point where such a ground is possible. Single-point grounding techniques are inappropriate for protection against large amplitude, high frequency pulses from arc discharge sources. Solid grounds of all metallized layers in thermal blankets should be provided. The aluminum honeycomb material in solar cell substrates should be replaced by dielectric material wherever possible, and closure of 'open-ended' spacecraft should be considered.

(e) Circuit design should take into account the possibility of the high-frequency, large amplitude pulses from arcs. Each interbox wire should be 'grounded' at each box for frequencies higher than those in signals carried by that wire. Circuits should be designed to minimize the required bandwidths (or maximize required rise-times) on interbox wiring. Protection should be provided against burnout of any semiconductor components (such as MOSFETS, FETS, diodes and transistors) by the large voltage pulses from arcs.

(f) Spacecraft design practice should be to minimize the surface areas of dielectrics or insulated conductors exposed to bombardment by high energy (substorm-associated or otherwise) electrons. Transparent conductive coatings on solar cell covers and mylar or teflon surface of thermal structures should be considered.

(g) In addition, an electron emitter device to neutralize the spacecraft charge could provide a 'clamp' to force spacecraft-plasma potential differences to safe, small values.

Acknowledgements

The work reported here was supported by the U.S. Air Force Space and Missile Systems Organization (SAMSO) under Contract Number AF-F04701-69-C-0091. The authors are indebted particularly to Dr C. E. McIlwain and Dr Sherman DeForest of the University of California, San Diego, for illuminating discussions of spacecraft charging in general, of their ATS-5 observations, and for permission to include their ATS-5 data for November 26, 1972 prior to publication. The authors are also endebted to Dr N. L. Sanders and Dr G. T. Inouye, of the TRW Space Sciences Dept., who carried out the laboratory tests described in Section 4.

References

DeForest, S. E. and McIlwain, C. E.: 1971, *J. Geophys. Res.* **76**, 3587.
DeForest, S. E.: 1972, *J. Geophys. Res.* **77**, 651.

DeForest, S. E.: 1973, this volume, p. 263.
Grard, R. J. L.: 1972, ESTEC Internal Working Paper No. 663.
Knott, K.: 1972, *Planetary Space Sci.* **20**, 1137.
Rostoker, G.: 1972, *Rev. Geophys. Space Phys.* **10**, 157.

DISCUSSION

Polychronopulos: What sort of pressures were used for the breakdown tests of the satellite skin structure to justify the use of the term 'arcs'.

Fredricks: Sometimes as low as 10^{-6} A. The 'arcs' were 'vacuum' arcs due to surface tracking.

DeForest: Charging presents two problems: one is disturbance of the environment, the other is hazard to the spacecraft. If the outer layer is a conductor, the same potential might be achieved as with an insulator. However, the hazard is greater with an ungrounded conductor than with a floating insulator, since larger amounts of energy can be released in shorter periods of time in a discharge.

Fredricks: A conductive coating on a spacecraft might not be sufficient for protection if its resistivity is very high and its grounding not very good.

Rosenbauer: Since the typical particle currents are rather low ($\approx 10^{-3}$ A for the whole spacecraft), even moderately conductive connections of the surface to the spacecraft main frame should inhibit differential charging to voltages so high that breakdowns can occur.

A SYSTEM FOR MEASURING AND CONTROLLING
THE SURFACE POTENTIAL OF ROCKETS FLOWN
IN THE IONOSPHERE

B. POLYCHRONOPULOS and C. V. GOODALL

Dept. of Space Research, University of Birmingham,
P.O. Box 363, Birmingham B15, 2TT, England

Abstract. A short account is given of the effects on rocket potential of drawing excessive electron currents to Langmuir probes mounted on a rocket. A system is described whereby the rocket to plasma potential can be regulated by means of ejecting electrons into the plasma medium with an electron gun which is servo-controlled. The system was developed and tested in a laboratory plasma, simulating ionospheric conditions, and was consequently incorporated in a rocket payload P119H which reached an apogee of about 140 km. The results presented show the extent to which rocket potential can change as a consequence of the various currents drawn to the probes when no regulation is applied and also the high degree of regulation attained when the control system is brought into operation. Certain improvements, which can be made to the system in order to avoid gross contact potential difference effects, are suggested.

1. Introduction

Since the advent of rockets, Langmuir probes have been widely used for in situ measurements of ionospheric plasma parameters such as electron density and electron temperature. Most of these measurements have been made with single Langmuir probes as opposed to double or multiple probe systems, because of the relative simplicity of construction and ease of data analysis (Sayers, 1970). However, a major problem is to ensure that the rocket potential, which is used as a reference, remains constant with respect to plasma potential since, in general, fluctuations in the rocket potential will occur as a result of insufficient surface area to collect the return current for the probe system (Boyd, 1968). In experiments where the probe bias does not exceed plasma potential the current return area requirements are modest and can be relatively easily provided by a medium size rocket but when the probe is biased positive with respect to plasma potential it may collect considerable electron current and the return area requirements can become stringent. Thus, even the surface area of large rockets may not be adequate for ensuring constancy of the reference electrode potential and steps must be taken for avoiding the large errors which will result if this effect is ignored. Similar problems exist for probe experiments carried on satellites due to the much smaller area usually available.

In order to provide some quantitative basis for assessing the magnitude of this effect, consider a simple probe system immersed in a static Maxwellian plasma where the probe is of area A_p and the reference electrode is of area A_r $(A_r > A_p)$ and between which a voltage, V, may be applied. Assume the absence of any contact potential differences (cpd's) and that the electrodes, although not necessarily the same shape and size, adopt identical floating potentials. When the probe is positive with respect

R. J. L. Grard (ed.), Photon and Particle Interactions with Surfaces in Space, 309–320. All Rights Reserved
Copyright © 1973 by D. Reidel Publishing Company, Dordrecht-Holland

to the reference electrode $(V>0)$ the probe collects a net negative (electron) current and, in order that the net current drawn from the system is zero, the reference electrode collects a net positive current which is effected by a suitable negative shift in potential. Since the electron current to the reference electrode varies exponentially (Mott-Smith and Langmuir, 1926), such a shift can be readily accommodated provided that the electron current drawn to the probe, which can be assumed to greatly exceed the electron current flowing to the probe when at floating potential, is less than or equal to the electron (or ion) current collected by the reference electrode at floating potential.

In the following derivations it is assumed that the positive ion current to the reference electrode does not change appreciably. Let primes and double primes refer to the currents drawn for $V=0$ and $V>0$ respectively, the e and i subscripts refer to electron and ion components of the current and the p and r subscripts refer to the probe and reference electrode. The net current, i'', flowing in the circuit is given by

$$- i'' = i''_{er} + i''_{ir}. \tag{1}$$

Since, as stated above, $i'' \approx i_{ep}$ and $i_{ir} \approx -i'_{er}$, then from (1)

$$- i''_{ep} = i''_{er} - i'_{er}. \tag{2}$$

But the ratio of electron currents to the reference electrode for $V=0$ and $V>0$ is given by (Mott-Smith and Langmuir (1926))

$$\frac{i'_{er}}{i''_{er}} = \exp\left[\frac{e}{kT_e}(U'_r - U''_r)\right] \tag{3}$$

since U'_r and U''_r are the negative potentials of the reference electrode with respect to plasma potential. In terms of normalized potentials, $\eta = eU/kT_e$, Equation (3) can be expressed as

$$- \Delta\eta = \eta'_r - \eta''_r = \ln\left(\frac{i'_{er}}{i''_{er}}\right). \tag{4}$$

Hence from Equations (2) and (4)

$$\Delta\eta = - \ln\left(1 - \frac{i''_{ep}}{i'_{er}}\right). \tag{5}$$

Assuming that the probe is spherical and that it is operated under orbital limited conditions (Mott-Smith and Langmuir, 1926), then direct substitution for i''_{ep} and i'_{er} in the above equation yields

$$\Delta\eta = - \ln\left[1 - \frac{A_p(1 + \eta''_p)}{A_r \exp(\eta'_r)}\right]. \tag{6}$$

But the normalized floating potential of the reference electrode is given approximately by (Swift and Schwar, 1970)

$$\eta'_r \simeq - \tfrac{1}{2} \ln\left[\frac{m_i}{m_e}\right]$$

which, with Equation (6), gives

$$\Delta \eta = - \ln \left[1 - \frac{A_p}{A_r} (1 + \eta_p'') \left(\frac{m_i}{m_e} \right)^{1/2} \right]. \tag{7}$$

It should be noted that the approximate Equation (7) will give a finite value of $\Delta \eta$ only if

$$\frac{A_r}{A_p} > (1 + \eta_p'') \left(\frac{m_i}{m_e} \right)^{1/2}. \tag{8}$$

If this condition is satisfied, a first order approximation can be obtained by expanding (7) in series and retaining only the first term. Hence

$$\Delta \eta \sim \frac{A_p}{A_r} (1 + \eta_p'') \left(\frac{m_i}{m_e} \right)^{1/2}. \tag{9}$$

If condition (8) is not satisfied, the physical implication is that $i_{er}' < i_{ep}''$ and therefore the electron current collected by the probe cannot be balanced by a decrease in the electron current flowing to the reference electrode alone and an increase in the ion current collected by the latter is required. Since, in general, the accelerated ion current to a probe of large radius to Debye length ratio, r_p/λ_D, is only a weak function of voltage (Laframboise, 1966), such an increase would necessitate a considerable negative excursion of the reference electrode potential, this being of the order of several normalized voltage units. Equation (8) represents the most stringent case, since a spherical probe operated in this mode has the steepest current-voltage characteristics, particularly for $r_p/\lambda_D < 1$, as can be seen from Laframboise curves (Laframboise, 1966). Therefore for other probe geometries and for higher ratios of r_p/λ_D, condition (8) can be relaxed accordingly. For example, for a cylindrical probe operated under orbital limited conditions Equation (9) becomes

$$\Delta \eta \sim \frac{A_p}{A_r} \frac{2}{\sqrt{\pi}} (1 + \eta_p'')^{1/2} \left(\frac{m_i}{m_e} \right)^{1/2}. \tag{10}$$

In practice, in order to satisfy condition (8), certain compromises have to be made. The simplest solution is to decrease the probe area but this can lead to technical difficulties. Another possibility is to increase the available conducting surface of the reference electrode which can be effected by painting it with Aquadag (colloidal graphite) to form a stable conducting surface or by deploying additional surfaces (Huang, 1970) which could also be 'aquadaged'. It is also possible to increase the effective value of i_{er}' (see Equation (5)) by incorporating an electron gun which emits a constant electron current, part of which is drawn into the plasma. This method, which involves the use of heated filaments and a constant accelerating potential to a grid connected to the reference electrode, was used for the FR1 satellite (Blanquart and Ramond, 1966). A similar technique was adopted in the present investigations although the inclusion of an electronic feedback system enabled not only accurate control of the current emitted from the filament which permitted excellent voltage regulation, but also allowed measurements to be made to check the performance of the system.

2. Principle of Operation

The mode of operation of the system is to maintain the potential difference between the rocket body and the plasma constant. Use is made of the fact that under normal operating conditions the floating potential of an electrode is fixed with respect to plasma potential and so it is the potential of a small floating probe which is used to reference the potential of the rocket. When the rocket potential becomes negative as a result of excess electron current drawn to a probe, a positive voltage is developed at the input of an inverting differential amplifier. The negative voltage output is applied to the filament of an electron gun thus increasing the potential difference which accelerates the electrons and hence increases the electron current dumped into the plasma. This increase tends to shift the rocket potential in a positive sense thus offsetting the initial disturbance.

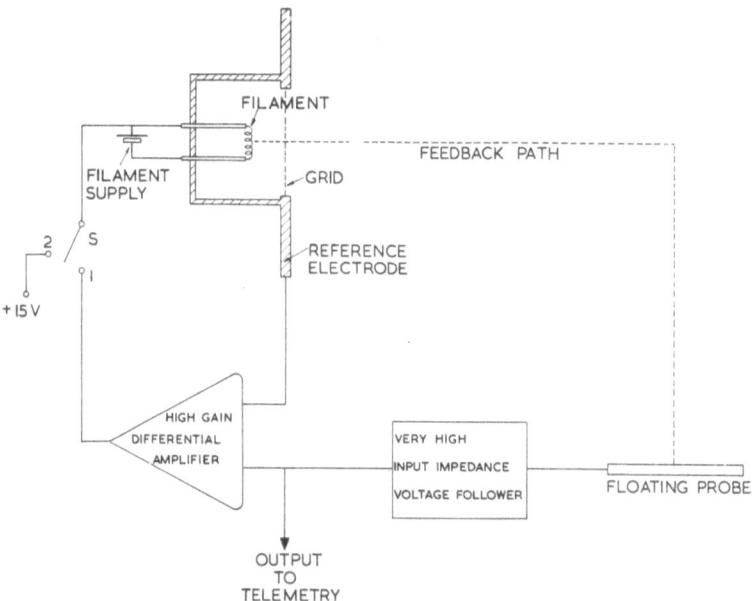

Fig. 1.   Schematic diagram of the basic electron gun system.

The voltage follower shown in the block diagram of Figure 1 senses the potential difference between the floating probe and rocket body. The very high input impedance ($\sim 10^{12}$ Ω) of the voltage follower ensures that negligible net current is collected by the probe and hence its potential remains unchanged. The voltage follower acts as a unity gain buffer amplifier with low output impedance so that it can drive a monitoring network enabling floating potential to be telemetered. The high gain differential amplifier (gain $\sim 10^5$) forms the basis of the servo system which controls the rocket potential by adjusting the voltage applied to the filament of the electron gun arrangement. The magnitude of residual deviation in the servo system is inversely proportional to the open loop gain of the differential amplifier (DeBarr, 1962), hence the requirement for selecting an amplifier with very high gain. To avoid instabilities which could easily arise in such a high gain system, the bandwidth of the differential amplifier was reduced to the absolute minimum of about 500 Hz, dictated by the highest sampling frequency of the telemetered data. A switch, S, enables the feedback control system to be switched in and out of action thus permitting detailed comparisons to be made of the effect of the electron gun system on rocket potential. When the feedback system is switched off, a large positive voltage is applied to the filament of the electron gun, thus inhibiting any electrons from reaching the ambient plasma. This obviates the need for switching the filament which is a relatively slow process.

The system as described is unable to regulate rocket potential under circumstances where the rocket potential would normally be positive with respect to floating probe potential. This could arise, although unlikely, if very high negative voltages are applied to probes of large collecting areas compared with that of the rocket. More important, it could also result if a cpd of the required polarity existed between the rocket and floating probe. The servo-loop mechanism would not then come into operation until sufficient current had been drawn by one of the probes to drive the rocket potential negative by an amount at least equal to the cpd. An improved system not dependent on cpd polarity and magnitude is proposed and discussed in a later section.

3. Design Considerations

A diagram of the components of the electron gun arrangement is shown in Figure 2. Attention was paid to a simple mechanical design requiring the use of readily available parts. The filament of the electron gun was that of a 6 V, 0.5 A commercially available bulb with the glass envelope removed and it was powered by a rechargeable battery of 7 V on load voltage, 300 mA h^{-1} capacity. A switch (not shown in Figure 1) is incorporated to turn the filament on after nose-cone separation, so that it is not burned accidentally when the ambient pressure is high. The filament is overrun slightly in

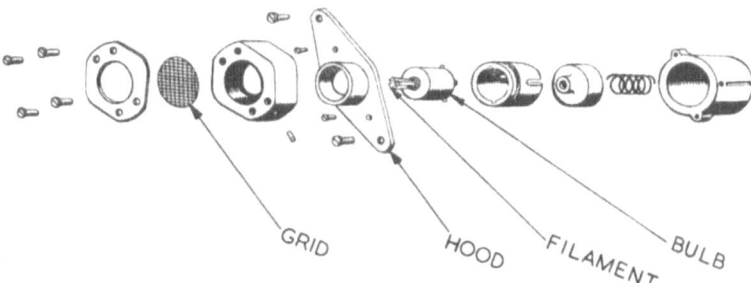

Fig. 2. Diagram of the electron gun.

order to ensure copious thermionic emission. The accelerating grid is a fine, high transparency, stainless-steel grid manufactured by a photographic etching process. Its
function is to control the electron current dumped into the ambient plasma. In order
to avoid the complications associated with a separate high tension source and its
control circuitry, relatively small voltages (± 10 V) are applied between filament and
grid. Thus the electron gun operates under space-charge limited conditions and the
smallest practical filament to grid spacing (1 to 2 mm) must be employed for maximum
electron beam intensity. Laboratory tests have shown that the gun is capable of ejecting about 0.2 mA of electron current with an accelerating bias of 10 V at a background pressure of about 10^{-4} torr. This performance figure was adequate for the
payload for which the system was designed (P119) since the maximum electron current
drawn to the Langmuir probes was not expected to exceed 0.1 mA. The positioning
of the electron gun in the payload relative to the various probes in their deployed
position is critical since the electrons of the beam emerging from the gun must be
prevented from being collected by any of the probes as this could lead to errors in
measurement. In order to implement this it was found necessary in the P119 payload
to reduce the angular width of the beam to about 80° by an insulating hood (see
Figure 2).

In designing the floating probe for the feedback system two important points must
be considered. First, its surface must be very clean and stable since the voltage employed
for the regulation of rocket potential is that sensed by the floating probe. Any changes
in the contact potential difference (cpd) between this probe and plasma will therefore
be reflected on the regulated value of rocket potential. For this reason the surface of
the floating probe should be treated with the same care as the surface of the Langmuir
probes. The second point is that the floating probe must be placed as far away from
the payload surface as possible, so that it samples a virtually undisturbed plasma.
Unfortunately technical problems did not permit this condition to be satisfied in the
P119 payload, the distance between the floating probe and the nearest conducting
payload surface being only about 1 cm. The effects of this are discussed in a later
section.

4. Laboratory Tests

Before incorporating the system into a rocket payload, performance tests were carried
out in a plasma tank. An argon plasma whose electron density and electron temperature could be varied from about 10^5–10^7 cm^{-3} and 10^3–10^5 K respectively was produced at a background pressure of about 10^{-3} torr in a stainless steel tank approximately 2 m long and 1 m diam (Polychronopulos, 1970). A schematic diagram of the
experimental arrangement is shown in Figure 3. The dimensions of the electron gun
were nominally the same as those used on the flight payload but the dimensions of
the probe and the reference surface were suitably scaled, the respective surface areas
being approximately 1 cm^2 and 500 cm^2. As can be seen from Figure 3, the complete
system is electrically isolated from ground, high impedance devices being used to
measure the potentials of the reference electrode and floating probe with respect to

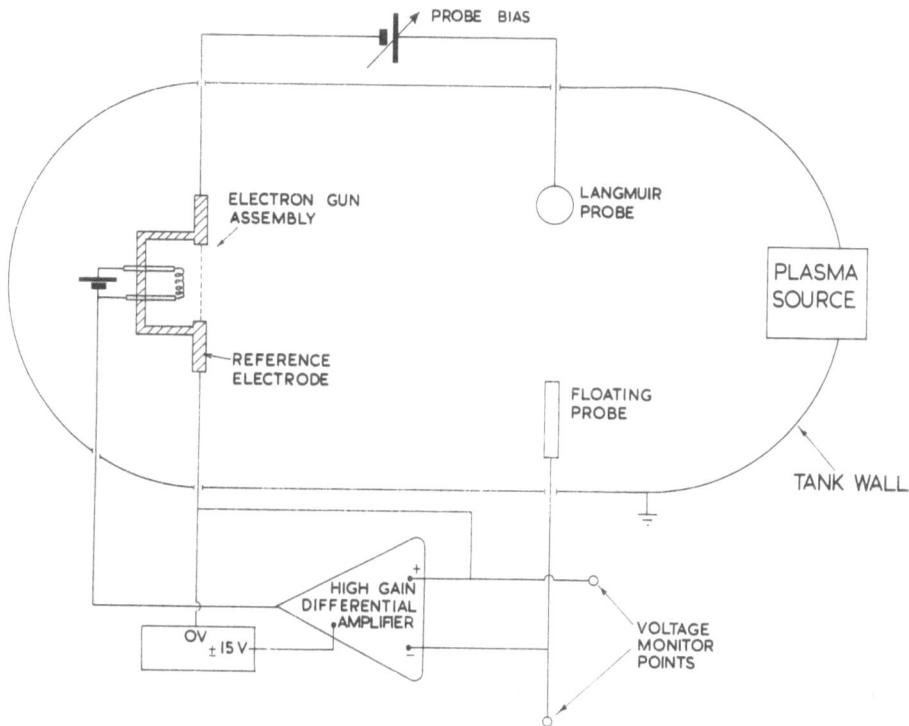

Fig. 3.   Schematic diagram of the experimental arrangement for laboratory tests.

the tank wall. The plasma is at some fixed potential with respect to the grounded tank
wall and is independent of the currents drawn to the various probes. However, as the
current to the Langmuir probe is varied, in order to maintain the net current to the
probe zero, the potential of the reference electrode plate will adjust unless the servo
loop controlled by the floating probe is operative.

To test the effectiveness of the electron gun the potential of the reference electrode
was monitored while the Langmuir probe was swept from −6 to +6 V initially with
the gun off and then on. This test was carried out under a number of plasma conditions,
the figures given below being those appropriate to a plasma electron density of about
10^6 cm^{-3}. The potentials, with respect to the tank wall, of the reference electrode and
floating probe when electrically isolated from each other were −2.7 V and −1.3 V
respectively indicating a cpd of 1.4 V. When the Langmuir probe was voltage swept
as described, the potential of the reference electrode changed from −2.7 V to −3.8 V,
whereas with the gun operative the reference electrode potential changed to a level of
approximately −1.3 V and remained constant to within ±10 mV. It can be seen from
these figures that even for a probe to reference electrode surface area ratio of about
1:500, the potential of the plate changed by several kT_e/e to compensate for the im-
balance in current drawn to the Langmuir probe. Furthermore it should be noted
that when the servo-loop was operative, the potential of the reference electrode was

maintained constant and at the same potential as that of the floating probe, that is an initial change in reference electrode potential equal to the cpd between this electrode and the floating probe had to be effected before compensation could be made for the excess current drawn to the Langmuir probe. This underlines the limitation of the system described in an earlier section and was demonstrated in the plasma tank when the cpd rendered the plate potential positive with respect to the floating probe. To offset this cpd an additional bias between the plate and floating probe was required.

5. In-flight Results

The rocket payload P119 which was flown from South Uist on December 10th, 1971 to an altitude of about 140 km was designed to study the behaviour of Langmuir probes operated in the electron accelerating region of the current-voltage characteristics under conditions where the plasma conditions are reasonable well known and understood. The payload consisted of a guarded cylindrical probe, a guarded spherical probe and a double-grid probe system, the latter to measure electron temperature (Tyler, 1972). The meaningful operation of the two former probes requires either a knowledge of the rocket potential with respect to plasma potential or that rocket potential remains constant. To achieve the second of the conditions an electron gun and floating probe system was incorporated in the payload. The three probes were approximately symmetrically deployed on flexible steel tapes to a distance of about 2 m and the floating probe deployed to a distance of about 20 cm from the rocket axis. The floating probe in its stowed position can be seen in Figure 4 which also shows the electron gun flush mounted on the side of the upper platform. The respective current collecting areas of the floating, spherical, cylindrical and gridded probes were approximately 1.0, 3.1, 36.7 and 130 cm^2 and the effective collecting area of the Petrel rocket was calculated to be about 1.2×10^4 cm^2.

The switching sequence within the experiment can be inferred from Figure 5 which shows a section of the relevant current and voltage waveforms though it should be noted that the sequence shown is not a complete operational cycle. In brief, the electron temperature experiment operated for a period of six, 0.25 s sawtooth waveform voltage sweeps and was then inhibited for an identical period. The magnitude of the voltage sweep was -1 to $+3$ V. The electron gun was operated for the period only when the electron temperature experiment was off. The spherical and cylindrical probes were operated continuously but switched between the inputs of two amplifiers each of which supplied a different range of probe voltages, these being approximately -1 to $+2$ V and -1 to $+4$ V; the switching occurred at the end of each triangular waveform voltage sweep period of 0.5 s. In this way the effects of the operations of each probe on one another and on the rocket potential could be examined.

Figure 5 shows a section of the in-flight data obtained near apogee over a total duration of two seconds before and after the instant in time when the electron temperature experiment is switched out of its inhibited state. It is clearly seen from the data that when the electron gun is operative, the potential of the rocket is maintained

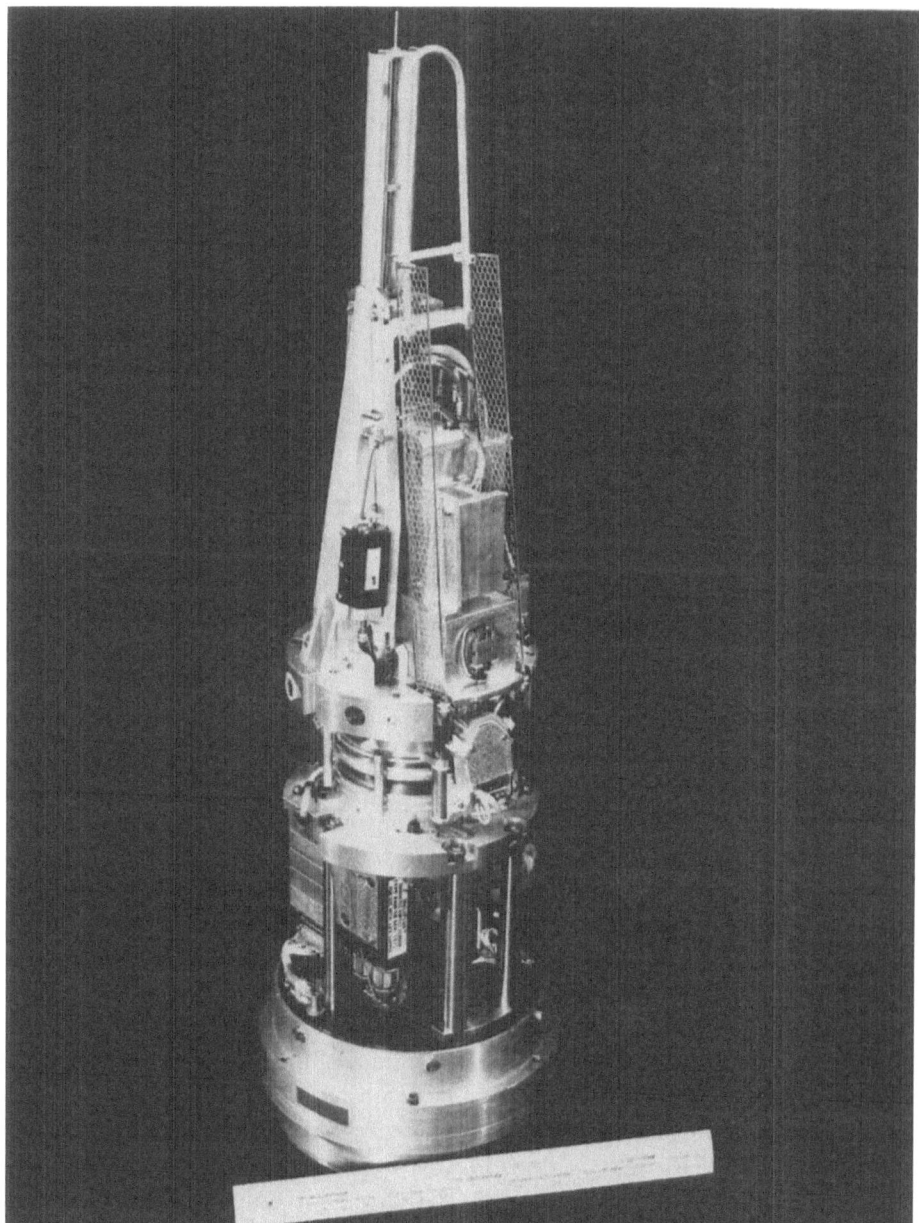

Fig. 4. Photograph of the P119 payload.

constant except for small random variations due to telemetry errors. However, when the electron dumper is not on, marked changes in rocket potential occur, especially as a consequence of the currents drawn to the gridded probe and the cylindrical probe. This is confirmed by examining the distortions in the current-voltage characteristics

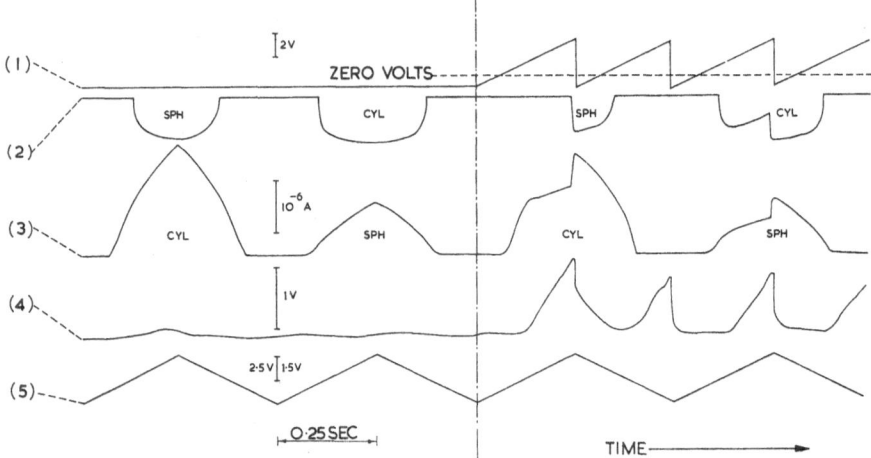

Fig. 5. Display of telemetry channels: (1) voltage sweep applied to the gridded probes; (2) logarithmic amplifier output for the Langmuir probe currents; (3) linear amplifier output for the Langmuir probe currents; (4) floating probe voltage; (5) voltage sweep applied to the Langmuir probes.

of the spherical and cylindrical probes resulting not only from the current drawn to the gridded probe which is readily seen at the instant when the potential of the latter is switched from its positive limit to its negative limit, but also resulting from the current drawn to the probes themselves and which can be seen by referencing the characteristics to those obtained when the electron gun was on and the electron temperature experiment inhibited. By superimposing the various characteristics, changes in rocket potential can be independently determined. A comparison of the two types of results shows them to be in disagreement in two ways. Firstly the magnitude of the changes in rocket potential as measured by the floating probe system is approximately a factor 1.5 smaller than that inferred from comparing current-voltage characteristics of both the spherical and cylindrical probes. Secondly, it would appear from an examination of the floating probe data that the effect on the rocket potential of the current drawn to the gridded probe is about the same as that resulting from the current drawn to the cylindrical probe whereas the magnitude of these currents differ by about a factor of three. This is not substantiated by comparing current-voltage characteristics which show the magnitude of the change in rocket potential to vary approximately as the magnitude of the current drawn to the respective probes, as can be seen in Figure 6. The reason for this discrepancy is thought to be due to the fact that the floating probe is mounted too close to surfaces at rocket potential and is being influenced by the space charge sheaths surrounding these surfaces. Hence it is not acting as a true reference probe. Although it is not possible to make detailed calculations because of the complex geometric configuration to give any quantitative support, the potential of the floating probe could be significantly influenced by rocket potential since the Debye length of the ionospheric plasma at this altitude is about 1 cm and in general terms the thickness of the space charge sheath near a surface at a normalised potential

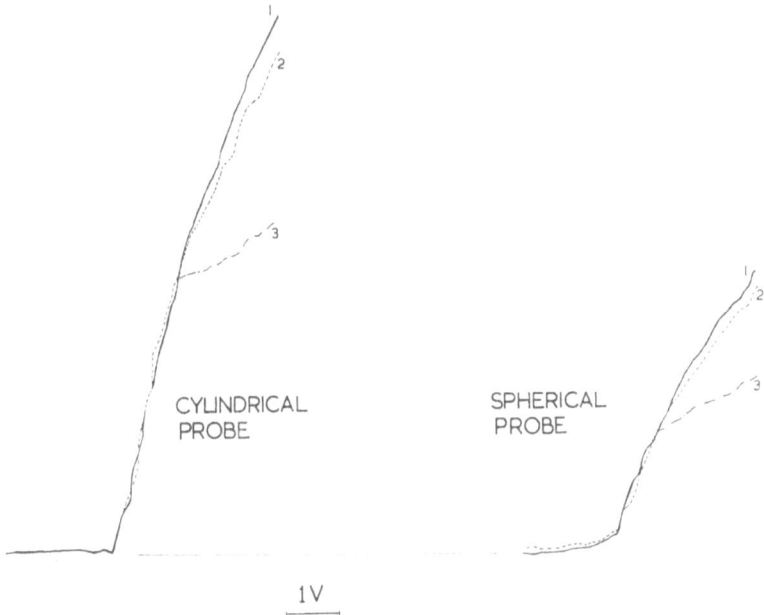

Fig. 6.   Superimposed Langmuir probe characteristics for the conditions: (1) rocket potential regulated (solid line); (2) rocket potential not regulated but grids negative with respect to plasma potential (dotted line); (3) rocket potential not regulated and grid voltage swept (broken line).

of a few kT_e/e is of the order of several Debye lengths. Such an effect could certainly reduce the magnitude of the changes in rocket potential measured by the floating probe. However it is possible that the floating probe could still effectively control the rocket potential since it is expected that there is a unique rocket potential for which there is zero potential difference between the rocket and floating probe. In a modified system the floating probe is deployed to much greater distances at the end of a cable whose outer surface is not at rocket potential.

6. Conclusions

The preliminary results from this initial flight indicate that changes in rocket potential of up to about 2 V with respect to plasma potential can result from currents up to about 6×10^{-6} A being drawn to a probe system where the ratio of rocket to probe areas is of the order of 100:1. This figure is in agreement with theory, in that it is greater than the kT_e/e change which would be expected if the change in electron current to the rocket body at floating potential was alone sufficient to compensate the electron current collected by the probe. The results imply that a shift of about 2 V in rocket potential was required to increase the positive ion current to the required magnitude. Changes in rocket potential were effectively overcome by means of the electron gun although it is thought that the floating probe controlling the gun was situated too close to the payload structure and not in the undisturbed plasma. Modifications in the payload design have been completed enabling the floating probe to be deployed to a distance of about 2 m.

As mentioned earlier, the main limitation of the present system is its failure to operate under conditions where cpd's render the rocket potential positive with respect to that of the floating probe. The consequences of this limitation can be reduced by painting the surfaces with Aquadag to minimize cpd's (Trendelenburg *et al.*, 1970; Goodall, 1971) but this solution is not entirely satisfactory. A more effective solution which would also provide for greater versatility could be achieved by detecting the magnitude and polarity of the cpd by a separate circuit, storing it and compensating for it by applying it to the non-inverting input of the high gain differential amplifier (see Figure 1). Such a system is under development at the present time and it is planned to incorporate it in future payloads.

Acknowledgements

The authors would like to express their gratitude to Prof. Sayers for the interest shown in this work and also to Prof. Willmore for helpful discussions. They would also like to thank Mr M. J. G. Clark for his technical assistance.

References

Boyd, R. L. F.: 1968, in Lochte-Holtgreven (ed.), *Plasma Diagnostics*, Chapt. 12, North Holland Pub. Co., Amsterdam.
Blanquart, P. and Ramond, J.: 1966, *L'Echo des Recherches*, No. 45. (1966)
De Barr, A. E.: 1962, *Automatic Control*, Chapt. 2, Reinhold Pub. Co.
Goodall, C. V.: 1971, *Planetary Space Sci.* **19**, 827.
Huang, F. T.: 1970, Ph.D. Thesis, Maryland University, U.S.A. (1970)
Laframboise, J. G.: 1966, Univ. of Toronto Inst. for Aerospace Stud. Report No. 100, Canada.
Mott-Smith, H. M. and Langmuir, I.: 1926, *Phys. Rev.* **28**, 727.
Polychronopulos, B.: 1970, M.Sc. Thesis, Birmingham University, U.K.
Sayers, J.: 1970, *J. Atmospheric Terrest. Phys.* **32**, 663.
Swift, J. D. and Schwar, M. J. R.: 1970, *Electrical Probes for Plasma Diagnostics*, Chapt. 1, Iliffe.
Trendelenburg, E. A., Fitton, B., Page, D. E., and Pedersen, A.: 1970, *Eldo-Cecles / Esro-Cers Sci. Tech. Rev.* **2**, 1.
Tyler, A. F.: 1972, *Radio Electr. Eng.* **42**, 309.

2.6. SPACECRAFT SURFACE MATERIALS

WORK FUNCTION VARIATION ACROSS THE SURFACE
OF TUNGSTEN AND VITREOUS CARBON*

M. BUJOR

Groupe de Recherches Ionosphériques 4 Avenue de Neptune
94 – St. Maur-des-Fossés, France

Abstract. The Kelvin Zisman method was used in ultra high vacuum conditions to compare the uniformity in work function of a tungsten monocrystal and of a vitreous carbon sample for Langmuir probe material selection. The resolution, measured on a tungsten polycrystal, was found to be twice the diameter of the probe. Rates of contamination were also compared in residual atmosphere on the clean and oxygen-covered materials. Work function values of 4.42 eV for the clean vitreous carbon surface and 4,47 eV for the oxygen covered surface were obtained. Corresponding values for the tungsten were 4.62 eV and 6.29 eV respectively. A similar uniformity in work function and a much lower rate of contamination can be associated with other practical advantages of vitreous carbon over tungsten as a suitable material for Langmuir probe construction.

1. Introduction

Uniformity of the work function of the surface of a Langmuir probe is a necessary condition for obtaining satisfactory results when ionic and electronic currents are measured in a plasma as functions of the probe potential.

Vitreous carbon was first reported by ESTEC research workers (Fitton *et al.*, 1971) to be a suitable material for Langmuir probes. In order to check its work function uniformity, it was compared with a tungsten monocrystal whose work function is known to be absolutely constant. Both samples were studied by the Kelvin-Zisman method (Zisman, 1932) under the same ultra-high vacuum conditions in order to minimize work function variations caused by residual atmosphere contamination.

Both clean and oxygen covered materials were compared and the results are presented in this paper.

2. Experimental Procedure

In the Kelvin-Zisman method the work function difference:

$$\Delta W = W_E - W_R$$

between the sample E and the reference R is measured when R is vibrating in front of E.

The apparatus consists of two main parts shown in Figure 1:

– The vacuum chamber can reach an ultimate low pressure of 1×10^{-10} torr after baking at 250°C. An adjustable leak valve allows gas introduction in the 10^{-5} to 10^{-10} range. A quadrupole gas analyser gives the residual atmosphere composition.

– The electronic equipment, which puts the reference electrode into vibration, also amplifies and detects the resulting signal at the same frequency by means of a lock-in amplifier.

* This work was performed at the Commisariat à l'Energie Atomique (DTCE/SEEN), and sponsored by the G.R.I.

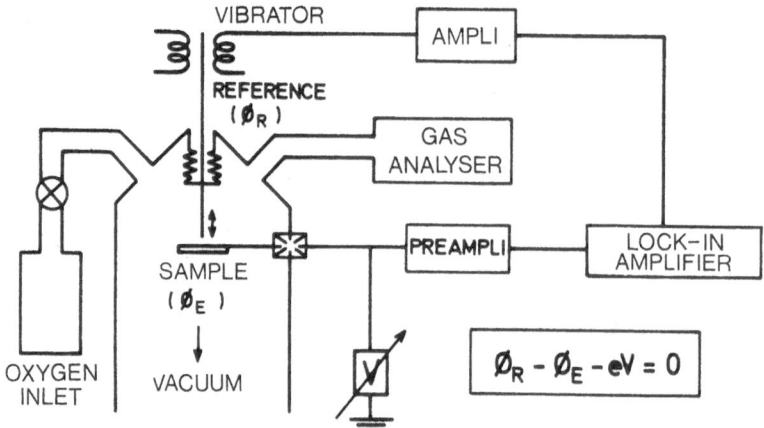

Fig. 1.   Experimental apparatus.

The sample is a disk 6 mm wide and 2 mm thick; the reference electrode is a recrys-
talised tungsten rod 1 mm wide, which has been reduced to only 0.3 mm at the end
facing the sample. This improves the resolution at the surface without introducing
transverse vibrations.

This reduction was obtained by electrolysis in a 20% NaOH solution, in which the
tungsten rod was the anode and a surrounding stainless steel ring was the cathode.

The sample can be moved horizontally in a direction perpendicular to the vertical
reference electrode. The motion is obtained by a finely threaded rod and a stainless
steel bellows. Displacements of 0.01 mm are measurable but are without effect, due
to the size of the reference electrode.

The true spatial resolution of the method was determined with a polycrystalline
tungsten sample formed of very large grains. When the sharp boundary between two
grains passed in front of the reference electrode a continuous work function variation

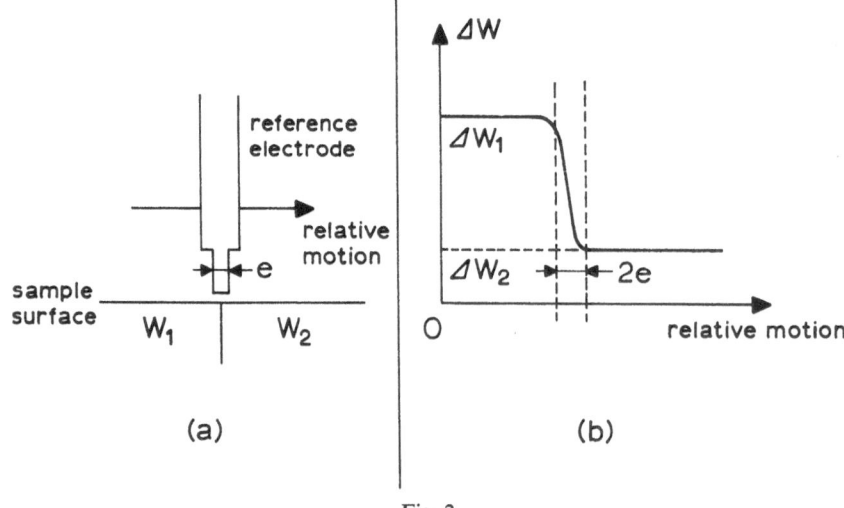

Fig. 2.

was detectable over a distance of 0.7 mm which is a little more than twice the diameter of the reference (see Figure 2).

The reference electrode can be cleaned by electron bombardment heating to 2500 °C for 15 s. The same treatment can also be used to clean the tungsten sample. In the case of vitreous carbon, however, its low conductivity does not permit it to be heated uniformly at such temperatures, where it is damaged locally by the beam coming from the filament of the electron gun situated 1 mm below. Desorption of the adsorbed gases from the carbon surface is obtained by electron bombardment heating at 1000 °C for 30 s.

3. Contamination Rates in Residual Atmosphere

After cleaning both electrodes, the work function variation was recorded at a given point of the sample surface, as a function of time, for different pressures. This gives only an indication of the contamination rate, as both electrodes are adsorbing gases simultaneously at different rates. The results shown on Figure 3 for the tungsten are, however, sufficiently explicit to prove the absolute necessity of ultra-high vacuum conditions in order to be able to make any significant measurements of work function variations across the sample surface. The pressure indicated for each curve is the maximum obtained during the cleaning process. In order to minimize the rise of pressure during outgassing, and to keep it in the low 10^{-10} torr range, a cryogenic panel was mounted inside the vacuum chamber.

The absolute contamination of tungsten and vitreous carbon was determined with the cleaned reference after one night in a 10^{-10} torr vacuum. Comparing with the values measured 2 minutes after cleaning both electrodes (reference and sample),

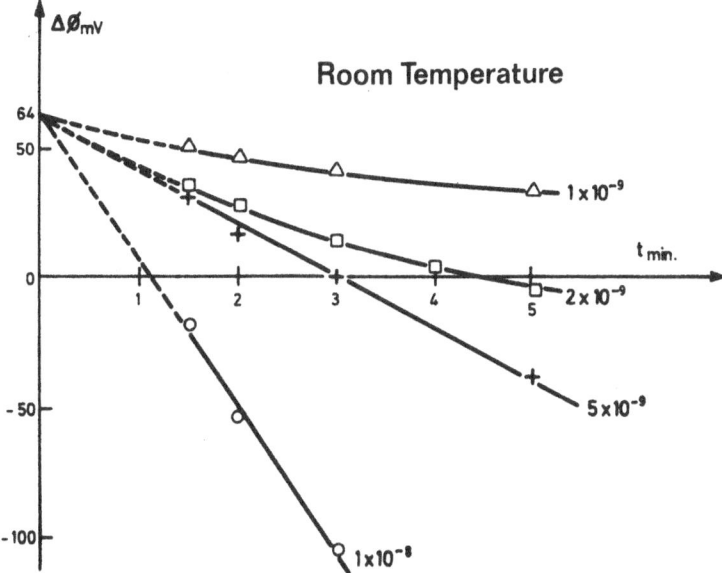

Fig. 3.   Work function variation of tungsten vs time for different pressures.

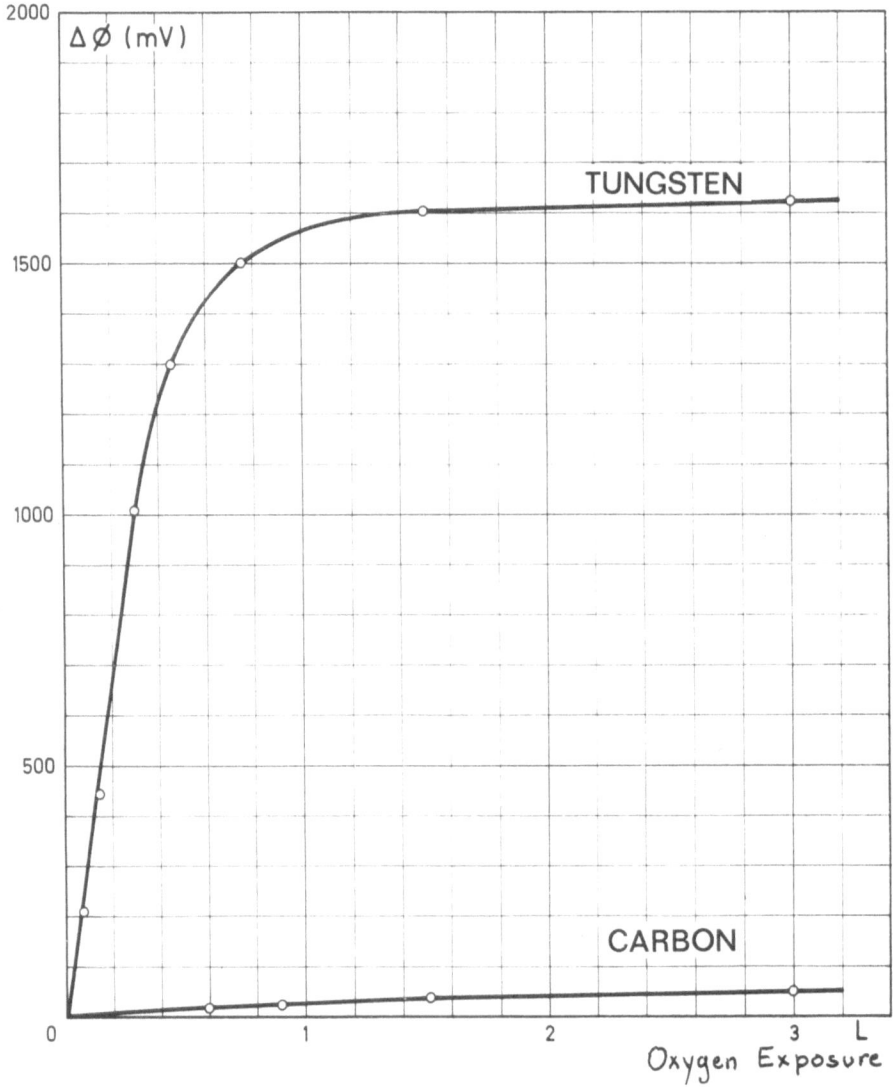

Fig. 4.   Work function variations of tungsten and carbon vs oxygen exposure.

variations of 600 mV for the carbon were recorded, thus showing that the later is 20 times less contaminated than the former under the same conditions.

The difference in contamination is also illustrated by the way in which oxygen is adsorbed on both materials, thus changing their work function. Figure 4 shows the work function variations as a function of oxygen exposure measured in Langmuirs (1 Langmuir $= 10^{-6}$ torr $\times$ second). As the reference was cleaned before each measurement, this is a measure of the contamination of the sample itself. Again, the advantage of vitreous carbon as compared with tungsten is clearly visible.

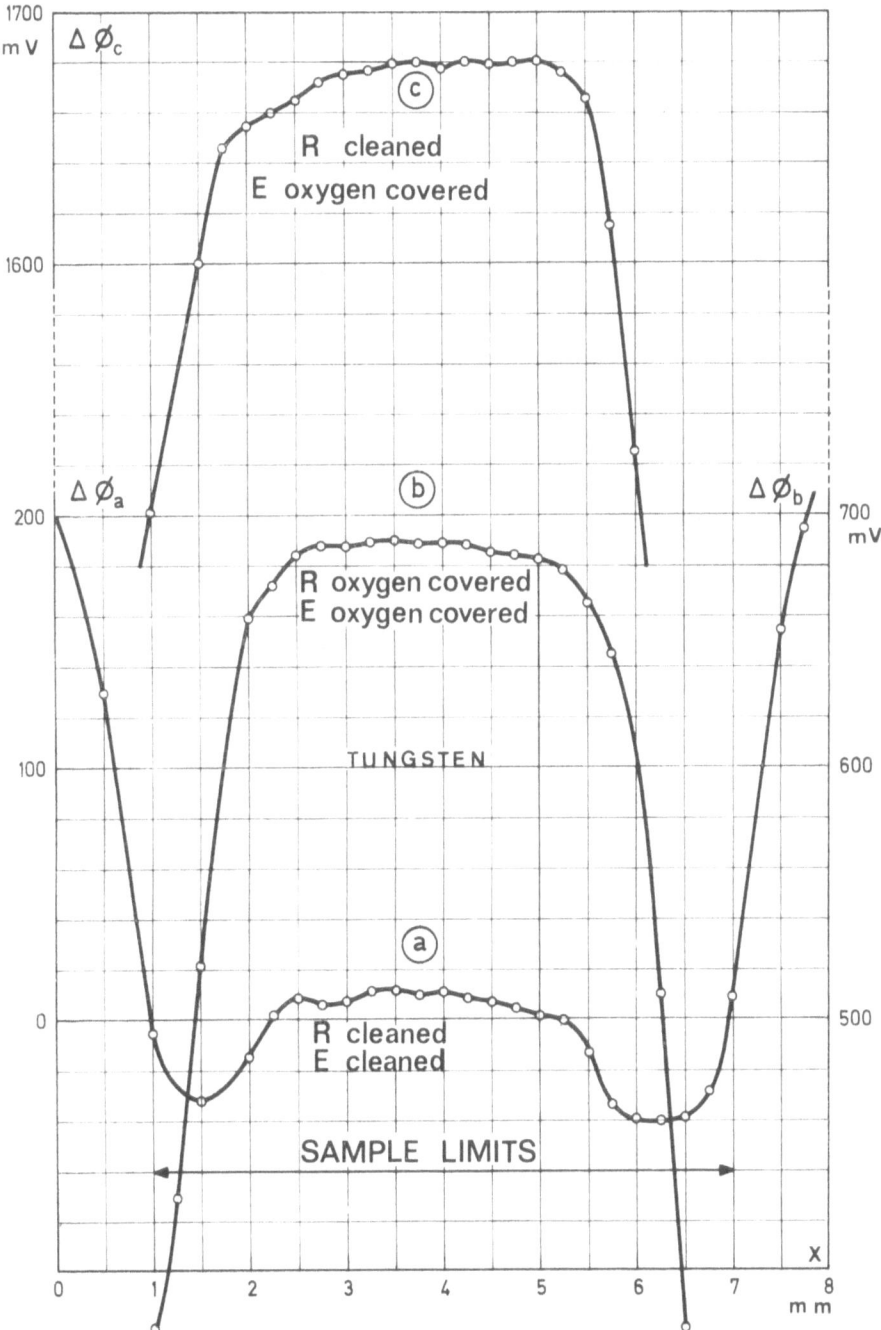

Fig. 5. Work function variation across a diameter of the tungsten sample.

4. Work Function Variations Across the Surface

Figure 5 shows these variations along a diameter of the tungsten sample in three cases:
  (a) E and R cleaned;
  (b) E and R covered with an oxygen monolayer;
  (c) E covered with an oxygen monolayer and R cleaned;
  X represents the relative displacement of the sample with respect to the reference. From the values $\Delta\phi_a = 10$ mV and $\Delta\phi_c = 1680$ mV (and utilizing the value $\phi_0 = 4.62$ eV for the clean tungsten) one can obtain $\phi_1 = 6.29$ eV for the oxygen covered tungsten.

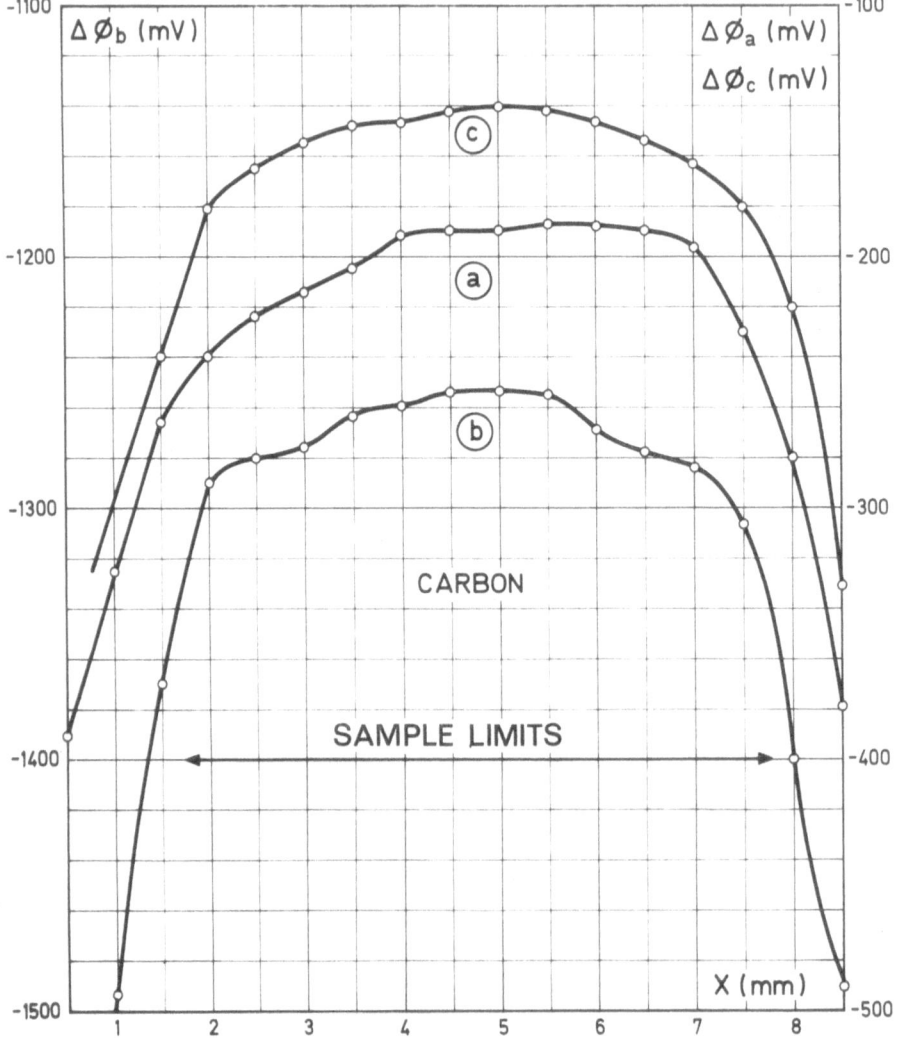

Fig. 6.   Work function variation across a diameter of the carbon sample. (a) E and R cleaned; (b) E and R covered with an oxygen monolayer; (c) E covered with an oxygen monolayer and R cleaned.

This compares quite well with measurements previously made in the same laboratory (Bacal *et al.*, 1972).

Figure 6 shows the corresponding results for the carbon sample. From the values $\Delta\phi_a = -190$ mV, $\Delta\phi_c = -140$ mV and $\phi_R = 4.61$ eV, one can obtain $\phi_0 = 4.42$ eV for the clean vitreous carbon surface and $\phi_1 = 4.47$ eV for the oxygen-covered surface.

As for the tungsten, the repeatability of the measurements is of the order of 10 mV. The comparison between Figures 5 and 6 shows that the work function uniformity across the surface is almost as good for the vitreous carbon as it is for the tungsten monocrystal.

5. Conclusion

Vitreous carbon presents the degree of work function uniformity across the surface which is believed to be necessary for Langmuir probe measurements. Moreover its low density and low contamination rate are in its favour when it is compared with tungsten. Finally, spherical probes are easier to construct with this material than with tungsten, which must be deposited from the vapour phase onto a molybdenium substrate in order to ensure the necessary uniformity in work function.

All these reasons contribute to make vitreous carbon a suitable material for the construction of Langmuir probes.

Acknowledgements

This work forms part of a program of research and development on diagnostic methods for space plasma physics, carried out at the Groupe de Recherches Ionospheriques (G.R.I.) of the Centre National de la Recherche Scientifique, and sponsored by the Centre National d'Études Spatiales (C.N.E.S.). Related work at the G.R.I. is being performed by the group directed by Dr J. J. Berthelier, to whom we are grateful for first calling our attention to the interest of vitreous carbon as a material for making electrostatic probes.

References

Bacal, M., Desplat, J. L., and Alleau, T.: 1972, *J. Vac. Sci. Techn.* **9**, (2), 851–856.
Fitton, B., Page, D. E., Pedersen, E., and Trendelenburg, E. A.: 1971, *ELDO/ESRO Sci. Techn. Rev.* **3**, 321–426.
Zisman, W. A.: 1932, *Rev. Sci. Instr.* **3**, 367.

Discussion

Wiesemann: Is it possible to improve the spatial resolution by using a reference electrode with a guard ring to compensate for fringe fields?

Bujor: From the literature I know that people using guarded electrodes have the same result as we, i.e. the resolution is twice the diameter of the reference. The resolution can be increased by using an oscillating electron beam of a few μm diameter, which we will be able to do in some weeks.

Cauffman: Could you comment on the effect of the roughness of the surface of the sample?

Bujor: We have not yet compared the same material with different surface roughnesses, but we intend to do so. Our vitreous carbon sample was not polished.

Laude: The work function you measured of 4.42 eV for vitreous carbon specimens manufactured by Carbone Lorraine differs considerably from the value I obtained of 5.1 eV for the vitreous carbon from Plessey Co. This difference may be due to the higher impurity contents of the material prepared by Carbone Lorraine. Other intermediate values for the work function have been obtained. We shall be carrying out a full study of the differences in work function for vitreous carbon from various sources.

Fitton: I just wish to point out that our earlier data on vitreous carbon were obtained on samples obtained from both Carbon Lorraine and from the Plessey Co. Our measurements showed that the Plessey material was superior in surface finish and in work function uniformity. I think it would be very interesting if you could carry out an intercomparison in more detail.

Bujor: A piece of a vitreous carbon sample manufactured by the Plessey Co. for ESTEC was given to us by Dr Laude and was tested in our apparatus. A work function value of 4.60 eV was measured by the Kelvin-Zisman method instead of the 5.10 eV value which was measured by the photoemission method. Such a difference between the results of the two methods is not unusual and has been reported previously in the case of graphite.

However, we found that, when it was heated, the Carbone Lorraine sample released continuously an amount of hydrogen which was increasing with the temperature, while the Plessey Co. sample did not release any amount of gas, even when heated at 2000°C. This is obviously in favour of the latter for Langmuir probe construction.

EXPERIMENTAL INVESTIGATION OF PHOTOEMISSION FROM SATELLITE SURFACE MATERIALS

M. ANDEREGG, B. FEUERBACHER, and B. FITTON

Surface Physics Division, European Space Research Organisation, Noordwijk, The Netherlands

Summary. Calculations on the electron sheath around illuminated satellites suffer from a lack of experimental data on the photoelectron characteristics of surface materials commonly used in space. The majority of photoemission data available today have been taken in ultrahigh vacuum under carefully controlled conditions in order to ensure clean surfaces, since the primary interest has been the aspects of solid state physics. These data cannot be used for the calculation of photoemission from the type of materials actually used in space, due to the extreme sensitivity of photoemission properties to adsorbed gases and surface layers down to fractions of a monolayer thickness.

In this paper data are presented on photoelectric threshold, photoelectric yield, as a function of photon energy, photoelectron energy distribution, for a set of photon energies, and optical reflectance in an energy range up to 25 eV, for the materials gold, aluminium, stainless steel, vitreous carbon, Aquadag, graphite, silica and semiconducting Indium oxide. The surfaces were treated such that the results are most likely to be characteristic of materials actually used in space.

By combining these data with the solar flux energy spectrum we present the flux of photoelectrons under solar irradiation for the various materials. A method is proposed of quenching the photoelectron flux in space by overcoating with a thin layer of LiF.

The present paper is a combination of two articles (Feuerbacher and Fitton, 1972; Trendelenburg *et al.*, 1970) published recently, and the reader will find detailed information in these references.

References

Feuerbacher, B. and Fitton, B.: 1972, *J. Appl. Phys.* **43**, No. 4, 1563–1572.
Trendelenburg, E. A., Fitton, B., Page, D. E., and Pedersen, A.: 1970, *ELDO-ESRO Scient. Tech. Rev.* **2**, 53.

DISCUSSION

Norman: Your apparatus was used to obtain the photoelectric yield from a specimen at near to normal incidence. Have you measured the dependence of the photoelectric yield on the angle of incidence of the illuminating beam?

Anderegg: No, but the apparatus can be used to measure the yield for a range of incident angles. The yield becomes greater for increasing angles of incidence.

Wrenn: We have heard about the stability of work function of vitreous carbon, but three values have been given: 4.42 eV, 5.1 eV and now 4.75 eV. Can you explain this?

Anderegg: These three work function values might be attributed to production differences.

PRESENT STATE OF THE ART IN
CONDUCTIVE COATING TECHNOLOGY

H. KÖSTLIN

Philips Forschungslaboratorium Aachen GmbH, 5100 Aachen,
Weisshausstrasse, Germany

and

A. ATZEI

Energy Conversion Division, European Space Research and Technology Centre,
Noordwijk, Holland

Abstract. The existence of an electrically conductive coating ensures potential uniformity on the surface of a solar array. The development and preparation of this coating are described.

The conductive layer consists of a very thin Sn doped In_2O_3 film which reduces the solar flux received at the cell's surface by only 1–2%; it has a surface resistance of the order of 10 kΩ $\square^{-1}$ and does not affect the equilibrium temperature of the cells. Results of space qualification tests indicate face that their properties remain unchanged under space environment condition.

1. Requirements of a Conductive Coating

The purpose of conductive coating technology is to provide a homogeneous electrical potential all over the surface of a space craft (Köstlin *et al.*, 1970). The metallic parts of the craft automatically reach a homogeneous potential whereas the other part, that is especially the solar generator, does not. This is due to the fact that here in the area of the solar generator the surface potential can be influenced by two facts:

(1) The photovoltages of the solar cells themselves which are additionally partly connected in series.

(2) The local charging of the isolating solar cell coverglasses by ions and electrons coming from the space.

Both effects can be avoided if one coats the coverglasses with an electrically conductive layer and connects this to the metallic part of the surface. This arrangement is the so called conductive coating, short CC. The layer must meet the condition that it does not effect the solar cells efficiency, i.e.

(1) high transparency for sunlight,

(2) no attenuation of the thermal emissivity of the coverglasses.

A measure for the electrical effectiveness of the required coating is its surface resistance. A surface resistance is defined as the resistance between the opposite sides of a square of the layer. This value must be low enough to ensure a good shielding and a sufficient leakage. Resistances of 10 kΩ $\square^{-1}$ or less would be about two orders of magnitude on the right side at current densities of charged particles between 10^{-8} and 10^{-7} A cm^{-2} and a maximum potential difference of 0.5 V (requirement first stated for AEROS). Thus an upper limit for CC resistance should be about 10 kΩ $\square^{-1}$.

All these properties of the CC have of course to be 'stable' under space environment conditions.

R. J. L. Grard (ed.), Photon and Particle Interactions with Surfaces in Space, 333–341. All Rights Reserved

2. Development of the Conductive Coating

The material from which CCs are made has to combine both electrical conductivity and high solar transmittance. This would be the case for semiconductors with a high bandgap like SnO_2 or In_2O_3. Both materials have their band edge in the UV and show no absorption in the visible and near *IR*, they can be simply doped to give quite good *n*-conductors. There exists a lot of experience about both systems at Philips from the application as electrooptical device and as heat radiation reflecting filter for Na-low pressure discharge lamps.

The interesting spectral region for the space craft's solar generator CC is defined by the solar radiation, the solar cell sensitivity and the normal coverglass transparency. This extends from 0.4 to 1.1 μm, the maximum contribution in this rather broad spectral band lying at $\lambda_{max} \approx 0.7$ μm.

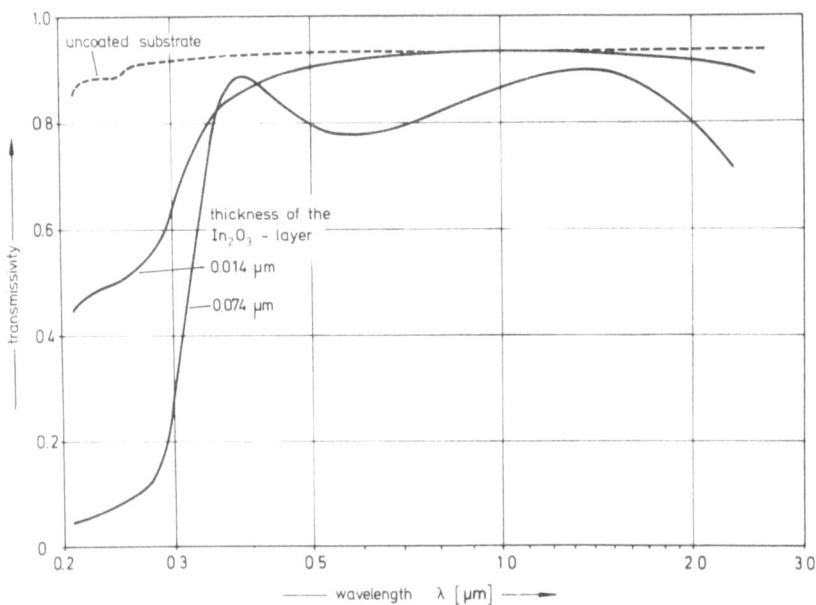

Fig. 1.   Spectral transmission of two In₂O₃ layers of different thickness
and of an uncoated vitreous quartz substrate.

As mentioned the CC materials are absorptionless in this region however they can cause a transmittance loss by reflection. This is because they have a rather high index of refraction, which is about $n=2$. In Figure 1 the transmission spectra of two samples with In_2O_3-layers and of an uncoated substrate are shown. At wavelengths below 0.3 μm the spectrum is controlled by the intrinsic absorption of the material. At longer wavelengths reflection determines the spectrum, which in the case of thin layers is strongly modified by interference effects. Especially the somewhat thicker sample shows a distinct interference structure and demonstrates how large the transparency loss caused by reflection and enhanced by interference can rise. As one needs high transparency all over the whole spectral band from 0.4 to 1.1 μm, not much is gained, if one adapts an interference transmission maximum to the wavelength of maximal sensitivity at 0.7 μm by using a suitable thickness of the layer. In the best case the integral transmittance loss would still be more than 5%.

The size of this loss can be kept low much more efficiently, if the CC is prepared with such a small thickness, that the layer is hardly optically effective and no interference extrema can be seen as for example at the thinner sample in Figure 1. This will be the case if the thickness d follows the relation

$$d \ll \frac{\lambda_{max}}{4n}.$$

At the last mentioned sample $d = 0.014$ μm the loss compared to an uncoated substrate is only hardly 2%. For samples with $d < 0.03$ μm this loss is almost proportional to the square of the thickness.

To achieve a sufficient conductivity in such a very thin layer, it is necessary to dope up to 1 at.%. Resistance and thickness are inversely proportional to one another for more than 0.01 μm thick layers, below this the resistance rises more rapidly and spreads rather widly. The minimal thickness, for which 10 kΩ $\square^{-1}$ can be realized, is about 0.007 μm. From an optical point of view this is the optimal case and the integral transparency loss is only 0.5%.

With increasing conductivity the films become more and more reflective in the infrared. This is caused by the high density of the doped free charge carriers. The immediate consequence of this is that the ability of the glass substrate to emit approximately black body radiation in the IR at some 100 K is suppressed by the presence of this metallic like coating and, therefore, the radiative cooling of the solarcells by their coverglasses is no longer effective.

In Figure 2 the thermal emissivity of such samples is compared to an uncoated quartz slide. The measurements were taken at 100 °C and correspond closely to the theoretical curve. To a first approximation this effect, the reduction of the emissivity, is only determined by the total number of free charge carriers per unit area, as samples with the same surface resistance but with different thickness and dope, show roughly the same emissivity. It should be noted that resistances above 1 kΩ $\square^{-1}$ reduce the emissivity by less than 2%. Therefore optimum conductive coating layers should have relatively small thicknesses (not more than 0.02 μm, but never less than 0.007 μm) and additionally the doping should be chosen such that the surface resistance lies between 1 and 10 kΩ $\square^{-1}$.

3. Preparation of Conductive Coating

There are various methods of preparing such layers. Examples are reactive evaporation or pyrolytic deposition. At Philips Aachen the last method to prepare the

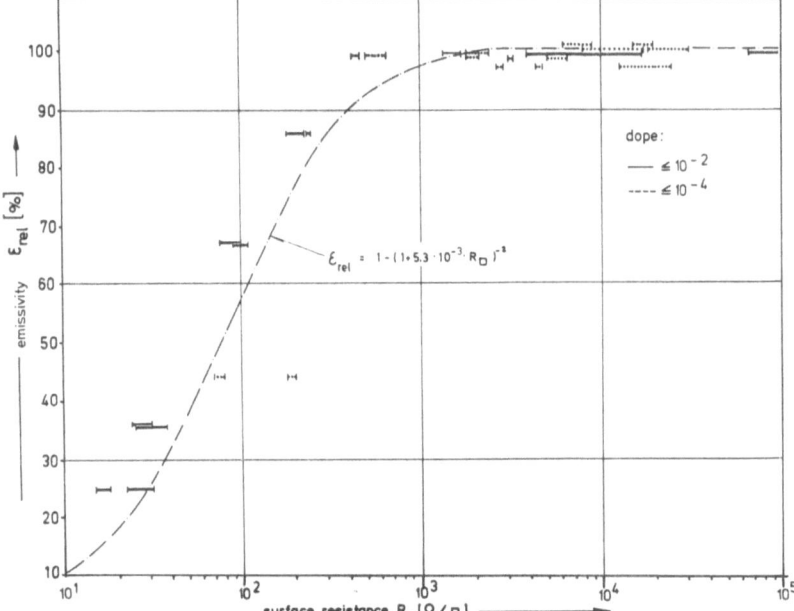

Fig. 2.   Thermal emissivity of In_2O_3 coated quartz samples (relative to an uncoated sample) plotted against the surface resistance of the layer. The layers have different thicknesses and dope. The dashed-dotted line is calculated following (Brügel, 1961).

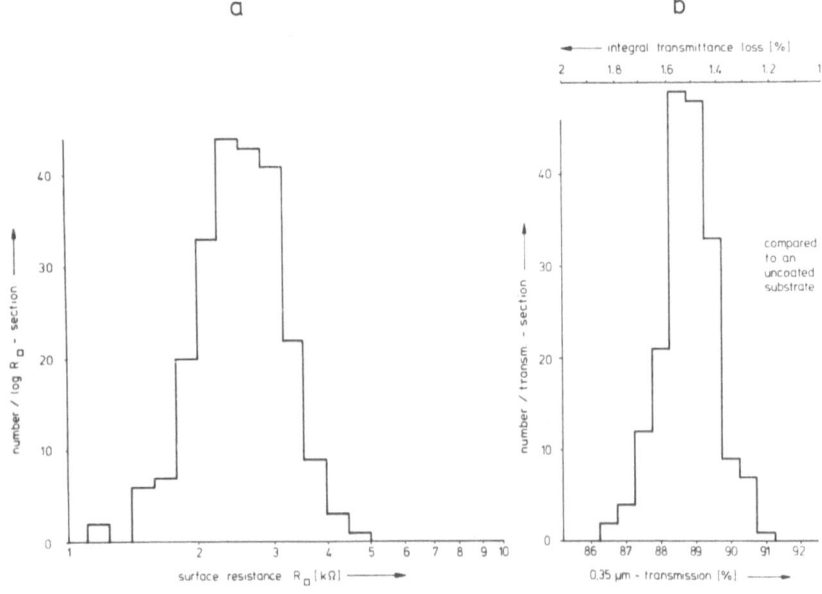

Fig. 3.   Histograms of the surface resistance (a) and of the transmission at 0.35 μm (b) relative to an uncoated sample for a batch of 200. For given material properties the whole transmission spectrum of a layer can be determined by measuring the transmission at one wavelength. This was experimentally confirmed for the CC. Thus the measurement at 0.35 μm is correlated to the integral transmittance loss (upper scale).

conductive coatings is used. The films are formed on a quartz substrate, heated up to about 450 °C, from an aerosol which consists of atomized solutions of $InCl_3$ and – as dope – $SnCl_4$ in butyl acetate.

The reproducibility of this preparation technique is shown from the histogram in Figure 3. In a batch of about 200 successively prepared samples the surface resistance has a mean scattering of $\pm 30\%$ around a required value of about 2.5 kΩ $\square^{-1}$. The transmittance, which can be checked best at short wavelengths, shows at 0.35 μm 89% compared to an uncoated quartz substrate and a mean scattering less than $\pm 1\%$. This corresponds to a total solar transparency loss (upper scale) of only $1.5 \pm 0.1\%$.

4. Space Qualification

As stated before, the parameters defined as critical are: transmittance, emissivity and resistance. The first two parameters directly affect the solar cells output, i.e. the available power, which is proportional to the light intensity and roughly inversely proportional to the cell temperature. The initial performance of the conductive coating should be retained during the all satellite mission. Special environmental tests, which simulate typical space conditions, were performed, and the behaviour of the optical, thermal and electrical properties of the conductive coating and the solar cell were observed. A complete flow chart of these tests done at ESTEC for the GEOS project is shown in Figure 4. The method adopted allowed single, partial and cumulative effects to be examined.

The test conditions were:

(1) humidity: 90%, 40 °C, 96 h

(2) UV irradiation: 1150 ESH (2500 W Xe-lamps with 10 cm water filter), 40 °C, 10^{-6} torr

(3) electron irradiation: 2×10^{14} cm^{-2} 1 MeV equivalent electrons, ambient condition

(4) thermal cycling: $+100$ °C to -102 °C, 101 cycles of 52 min each

(5) thermal vacuum: $+60$ °C, 10^{-5} torr, 5 days.

The results of these different tests always gave the same picture: Nearly no change of transmission ($<0.5\%$), no measurable change of emissivity. Only the surface resistance can be affected, mainly by vacuum and by humidity, but not more than maximum some 10%, which moreover is reversible (Table I).

The adherence of the CC-layer to the substrate is very strong, as diamond needle scratches showed.

The photoelectric emission is of interest for the whole satellite charging and was measured as a function of the photon energy. Conventional coverglasses and coverglasses with CC showed comparable photoyield (emitted electrons/incoming photon). The threshold energy is located at about 5 eV, while above 15 eV a saturation with a yield of about 0.1 is attained.

If one compares these CC-coverglasses which are made of quartz slides and have a conventional *UV*-reflection filter on one side, with normal microsheet coverglasses,

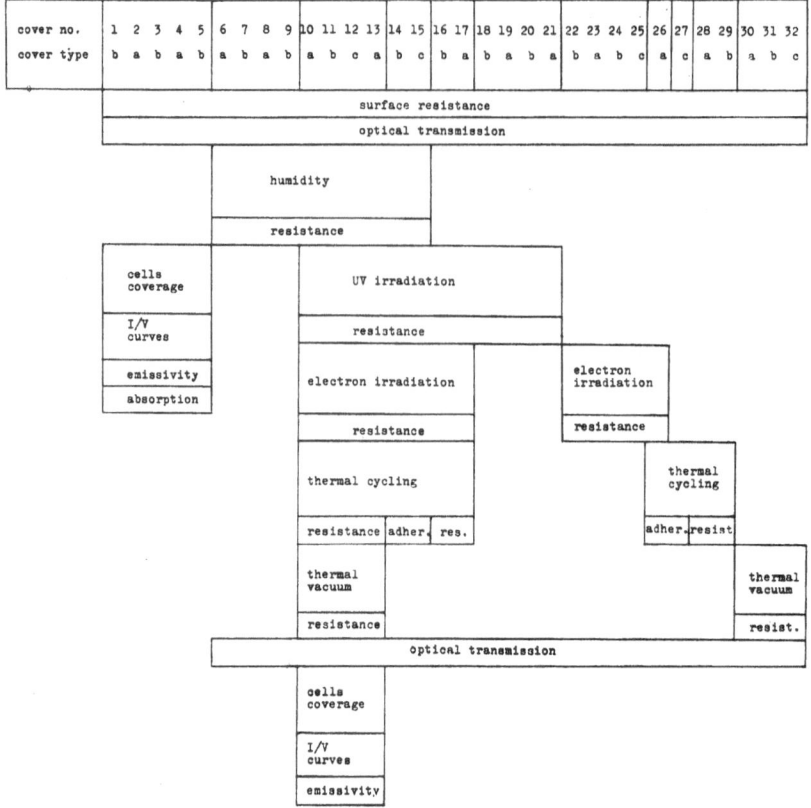

Fig. 4. Flow chart of the test schedule. Cover types: quartz slide with conductive coating (a), type a) with an *UV*-filter at the other side (b), type b) with Cr–Ni/Ag contacts on the CC (c). 'adher.' denotes adherence.

one finds no great differences: The output of a solar cell will have a loss of only 1 to 2%.

The ratio of solar absorptance α and thermal emissivity ε, i.e. α/ε, is 1.02 for solar cells with microsheet coverglasses ($\alpha=0.87$, $\varepsilon=0.85$); for those with CC coverglasses a value of 1.00 was determined ($\alpha=0.81$, $\varepsilon=0.81$). Therefore it can be concluded that, despite the fact that cells covered with conductive coating possess lower emissivity than conventionally covered cells their absorptance properties are such that the cell equilibrium temperature will not increase, on the contrary it decreases a little.

5. Interconnection Possibilities

The CC can be contacted in many different ways. Very strong contacts up to 1 kp rupture strength can be achieved by soldering with In. The thermal properties meet the requirements up to 100 °C. Another successful method developed by Philips is ultrasonic welding of Al-wires. The strength is a minimum of 200 p, this depends

TABLE I

Surface resistance (% of initial value)

Cover No.	Before test	After humidity	After UV irradiation	After electron irradiation	After thermal cycling	After thermal vacuum
1	100					
2	100					
3	100					
4	100					
5	100					
6	100	100				
7	100	95				
8	100	89				
9	100	106				
10	100	113	100	89	89	95
11	100	117	79	75	64	92
12	100	163	68	61	61	86
13	100	138	119	89	89	106
14	100	112	50	50		
15	100	216	47	53		
16	100		40	40	49	
17	100		92	105	105	
18	100		50			
19	100		70			
20	100		45			
21	100		34			
22	100			91		
23	100			50		
24	100			62		
25	100			93		
26	100					
27	100					
28	100				87	
29	100				72	
30	100					79
31	100					86
32	100					88

mainly on the rather thin wire. Very often thermal compressive bonding with thin gold wire (strength 50 to 100 p) is used in the laboratory. All these techniques can be applied directly to the CC layer. With the use of an additionally deposited layer at the interconnection point for instance of Cr–Ni/Ag or Mo/Ag also other soldering (Serre, 1972) or welding (Gochermann, 1971) techniques are appropriate.

A complete module made by SAT will demonstrate this CC-technique (Figure 5).

6. Conclusion

It has become common practice in satellite manufacture to ensure magnetic cleanliness and today the conductive coating described here enables us to fulfill the requirements of a uniform surface potential on the space craft's solar array. Its use will give

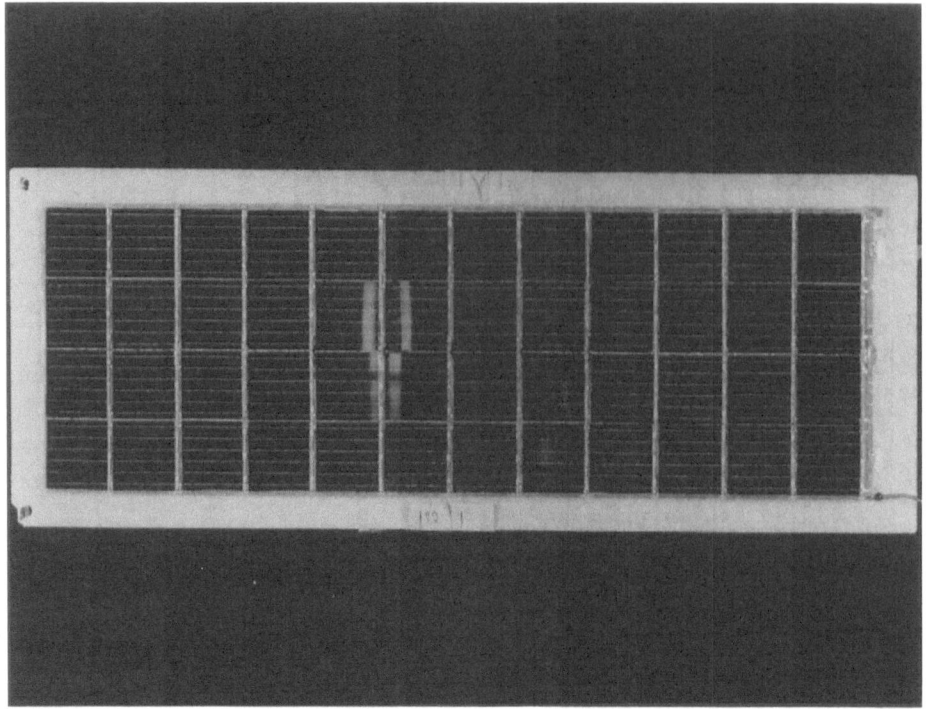

Fig. 5.   Module of a solar cell generator built by SAT with conductive coating (Serre, 1972). Every cell cover CC is soldered at three points to the interconnector which runs along each row of 4 parallel cells. This module passed the thermal, acoustic and vibrational tests according the GEOS launcher's specifications.

only a very little loss in the available electrical power, while its performance is very reproducible and is retained after the standard tests required for space qualification.

References

Brügel, W.: 1961, *Physik und Technik der Ultrarotstrahlung*, Vincenz, Hannover, p. 57.
Gochermann, H.: 1971, Research report (BMBW) RV 1-07/16/70 Z.
Köstlin, H. and Andrich, E.: 1970, Endbericht zum GfW-Kontrakt: RV II/2-3/69-QB-03-07-13.
Serre, F.: 1972, Rapport final, Contrat d'Estec No. 1213/70.

DISCUSSION

Kaiser: Will Dr Köstlin comment upon the technical problems of ensuring that the proportion of non-conductive surface is sufficiently low to satisfy the experimentor's requirements.

Köstlin: For the area of the solar generator this problem is solved quite well, manufacturers today guarantee a protection with CC up to 98%. This also holds for areas with second surface mirrors (SSM, OSR), as there is no difference in adapting the CC-technology to glass sheets which act as coverglasses or as mirrors. Mixed solar cell SSM panels can also be manufactured. – Problems still exist in those areas which are used for thermal control, as the coating applied there today do not satisfy both the optical and the electrical requirements.

Dauphin: Regarding the possible use of conductive coating for non-solar cell covered surfaces in GEOS spacecraft, it may be mentioned that some very preliminary tests made in Materials Section in ESTEC tend to show that it is possible to base such coatings on a flexible SSM type or a simple anodized aluminium surface, having a very thin cromium flash on the front face.

DeForest: Since the total dosage used in the radiation damage test is a small fraction of a year's dosage at synchronon altitude, I suggest a separate test with longer exposure.

Köstlin: Of course a longer irradiation dose would improve our estimation of CC's behaviour in space. But as all irradiation test up to now did not show any real change of the transmission or of the surface resistance I do not expect a remarkable effect in such a test. As regards the possibility of sputtering away the thin CC, a short calculation shows that a year's dose of van Allen belt protons would diminish the CC thickness only by less than 1% (protons have a low sputtering efficiency, $\sim 10^{-3}$). On the other hand the very small thickness makes the coating insensitive to optical degradation by irradiation, because the radiation can only create a small number of defects when passing through the short distance in the layer.

3. INTERACTIONS WITH CELESTIAL OBJECTS

3.1. ELECTRIC PROPERTIES OF THE MOON SURFACE AND ENVIRONMENT

PLASMA AND POTENTIAL AT THE LUNAR SURFACE

ROBERT H. MANKA

Space Science Dept., Rice University, Houston, Tex. 77001, U.S.A.

Abstract. The various plasma environments of the Moon are described for the lunar orbit through the solar wind and geomagnetic tail. The sources of lunar surface charge, including plasma, photo, and secondary currents, are compraed for their expected range of values. The electric potential is calculated from probe theory as a function of local position on the lunar surface; the theory includes plasma environments which are both stationary and flowing. In agreement with several other calculations, the potential at the subsolar point is likely to be a few volts positive. However, for the case when the Moon is in the solar wind, the potential is calculated to be a few tens of volts negative at the terminator, and is expected to go to larger negative values on the dark side. If the Moon traverses a significant plasma sheet in the geomagnetic tail, then the dark side potential tends toward several kilovolts negative but may be strongly limited by the secondary electron current. Associated surface electric fields are estimated.

1. Discussion

The electrostatic potential of the Moon and its effect on the lunar surface environment have been discussed by a number of authors. The problem is complex due to the fact that there are the competing but interrelated charging mechanisms of positive and negative plasma flux to the surface, production of photo electrons, and production of secondary electrons, the magnitude of these not all being well known for the lunar surface. A dominant photoelectric effect was considered by Öpik and Singer (1960) to obtain potentials of $+20$ and $+30$ V resulting in significant loss processes in the lunar atmosphere while Bernstein *et al.* (1963) estimated that the solar wind electron flux may exceed the photoelectron flux resulting in a neutral or negative surface potential and less effective ejection of ions from the lunar atmosphere. Grobman and Blank (1969) did a calculation of potential as a function of position over a portion of the sunlit side of the Moon. The lunar surface electric field was estimated for the possible plasma environments by Manka and Anderson (1968, 1969) and Anderson and Manka (1968, 1970).

2. Theory-Probe Equations

The current density per unit solid angle at the probe due to a positive or negative species of the plasma is proportional to the radial component of the velocity times the distribution function, evaluated at the probe, integrated over all velocity space; the current to the probe is given by an integral of the current density over all solid angle.

For an isotropic, thermal plasma, the ambient plasma distribution function at infinity is given by the Maxwell-Boltzmann distribution and at finite distances from the probe in the presence of a spherically symmetric electrostatic potential $\phi(\bar{r})$, the distribution function is

$$f(\bar{r}, \bar{v}) = n \left(\frac{m}{2\pi kT}\right)^{3/2} \exp\left[\frac{-1}{2}\frac{mv^2}{kT} - \frac{e\phi(\bar{r})}{kT}\right], \tag{1}$$

R. J. L. Grard (ed.), Photon and Particle Interactions with Surfaces in Space, 347–361. All Rights Reserved
Copyright © 1973 by D. Reidel Publishing Company, Dordrecht-Holland

where n is the ambient plasma electron density, m is the species mass, k is the Boltz-mann constant, T is the species temperature, and $\bar{v}$ is the species velocity. For a Maxwellian plasma with a superimposed flow velocity $\bar{V}$, the distribution at infinity of the ambient plasma is

$$f(\bar{r}, \bar{v}) = n \left(\frac{m}{2\pi kT}\right)^{3/2} \exp\left[\frac{-m(\bar{v} - \bar{V})^2}{2kT}\right]. \tag{2}$$

For a stationary plasma, we use the basic probe equations assuming spherical symmetry first derived by Langmuir and Mott-Smith (1926) and summarized by Whipple (1965) and Fahleson (1967); in general, these will be a good approximation. The solution is readily obtained for the isotropic plasma surrounding a spherical probe; however in the case of a flowing solar wind, a special expression for the plasma current must be used. This expression reduces to the stationary plasma solution when flow goes to zero (Whipple, 1959). The sheath can be assumed to be thin in comparison to the lunar radius for most of the cases considered (at least on the sunward side) and in addition, the DC conductivity at the lunar surface can be assumed to be very small so that the potential of the local lunar surface can be calculated as if the equilibrium condition at that point existed over the entire surface; i.e., the solution at any point is independent of what occurs elsewhere.

The local lunar surface will reach a potential such that the net current to it is zero,

$$I_e + I_i + I_p + I_s = 0, \tag{3}$$

where I_e = electron current, I_i = ion current, I_p = photoelectric current, and I_s = secondary electron current. Since these currents depend on the surface potential, the equation can be solved for the equilibrium potential. The form of the expressions for current as a function of potential depends on whether the species is attracted or repelled.

The form of the current equations depends on whether the potential is positive or negative. The complete set of current equations for both stationary and flowing plasma and either positive or negative potentials is given in the Appendix. The equations for a couple of cases are discussed below.

2.1. POSITIVE POTENTIAL, $\phi > 0$

The current density due to repelled thermal plasma ions is

$$\begin{aligned} I_i &= ne \sqrt{\frac{kT_i}{2\pi m_i}} \exp\left(\frac{-e\phi}{kT_i}\right) \\ &= I_{i0} \exp\left(\frac{-e\phi}{kT_i}\right). \end{aligned} \tag{4}$$

The current density due to attracted thermal plasma electrons is

$$I_e = - ne \sqrt{\frac{kT_e}{2m_e}} \frac{a^2}{r^2} \left[1 - \frac{(a-r^2)}{a^2} \exp\left(\frac{-r^2}{a^2-r^2} \frac{e\phi}{kT_e} \right) \right]$$

$$\simeq - ne \sqrt{\frac{kT_e}{2m_e}} \left(1 + \frac{2t}{r} \right), \quad \text{thin sheath}$$

$$\simeq - ne \sqrt{\frac{kT_e}{2m_e}}, \quad \text{very thin sheath} \tag{5}$$

$$= I_{e0}$$

in which

$$a = r + t, \, t = \lambda \, 0.83 \left(\frac{r}{\lambda} \right)^{1/3} \left(\frac{e\phi}{kT_e} \right) \tag{6}$$

with r the body radius, a the sheath radius, t the sheath thickness, and λ is the Debye length given by

$$\lambda = 6.9 \times 10^{-2} \sqrt{\frac{T}{n}} \, (m), \tag{7}$$

where T is in degrees Kelvin and n is per cm^3. For a probe large compared to the sheath thickness, the current is weakly dependent on potential since the area collecting the current is hardly increased by a large potential. I_{i0} and I_{e0} are the flux currents when the lunar potential is at the plasma potential (i.e., zero).

The current density of photoelectrons can approximately be written

$$I_p = i_p \cos \theta \, \exp\left(\frac{-e\phi}{kT_p} \right), \tag{8}$$

where i_p is the photo current density from an area of the lunar surface at the plasma potential with normally incident sunlight and θ is the polar angle of the local surface with respect to the subsolar point. This expression assumes that for purposes of calculation the photoelectric distribution function can be taken to be Maxwellian with an equivalent temperature, T_p.

The secondary electron current is proportional to the flux of primary particles and is given by

$$I_s = (I_e \delta_e + I_i \delta_i) \exp\left(\frac{-e\phi}{kT_s} \right), \tag{9}$$

where δ_e and δ_i are the secondary production coefficients for primary electrons and protons respectively and depend in a complicated fashion on the primary energy, angle of incidence and lunar surface material. In general the coefficients for protons is assumed to be small, and only for primary electrons with one hundred to several hundred electron volts energy is the electron coefficient likely to be significant.

For the following plasma, the thermal contribution to the current (such as given in Equation (5)) is interrelated with the flow contribution, and for example the electron current becomes

$$I_e = \frac{nev_{me}}{2\sqrt{\pi}} [e^{-U^2_e} + \sqrt{\pi} U_e (1 + \mathrm{erf}(U_e))], \tag{10}$$

where

$$U = \frac{V \cos \theta}{v_m} \tag{11}$$

and v_m is the species' mean speed. We note that when $V \to 0$, (10) reduces to (5); likewise, if the thermal velocity, v, goes to zero, the equation reduces to the pure flow current. Thus the equations in the Appendix give the *complete* description of the plasma currents to the sunlit hemisphere of the lunar surface.

2.2. NEGATIVE POTENTIAL, $\phi < 0$

When the thermal electron current exceeds the ejected electron currents the potential becomes increasingly negative until the remaining repelled electron current reaching the surface equals the positive current to the surface.

The repelled electron current density is given by (for a stationary plasma)

$$I_e = -ne \sqrt{\frac{kT_e}{2\pi m_e}} \exp\left[\frac{e\phi}{kT_e}\right] \tag{12}$$

while the attracted ion current has the same form as the attracted electron current in the previous section, and in the limit of a very large body

$$I_i = ne \sqrt{\frac{kT_i}{2\pi m_i}}. \tag{13}$$

The emitted photoelectron current density will be

$$I_p = i_p \cos \theta \tag{14}$$

and the secondary current density is

$$I_s = I_e \delta_e + I_i \delta_i. \tag{15}$$

3. The Lunar Environment

3.1. LUNAR PROPERTIES AND ORBIT

The orbit of the Moon takes it through several environments: the solar wind, the Earth's magnetosheath, and the geomagnetic tail. In addition to the general environment of the Moon, the charging of the local surface will depend on the direction of the photon and plasma fluxes and on the physical properties of the lunar surface. Also, it is important to note that the lunar orbital plane has an inclination with respect to the ecliptic plane of about 5 deg and a relatively long precession period of 18.6 yr; thus when the Moon is in the geomagnetic tail, its position can vary as much as $\pm 5 R_E$ with respect to the ecliptic plane, depending upon the time of the year.

3.2. SOLAR ELECTROMAGNETIC RADIATION AND LUNAR PHOTOCURRENTS

The solar spectrum and resulting photoemission has been discussed by several authors and is reviewed by Whipple (1965), who gives an integrated photocurrent for metals of $I_p = 2 \times 10^{-9}$ to 8×10^{-9} A cm^{-2} while Hinteregger et al. (1959) gave 4×10^{-9} A cm^{-2}. If we assume an effective emissivity ε_0 for metals corresponding to a photo current of 5×10^{-9} A cm^{-2}; then an emissivity $\varepsilon = 10^{-1} \varepsilon_0$ corresponds to a photo-current of 5×10^{-10} A cm^{-2}, etc.

The energy distribution of the emitted photoelectrons is a complicated function of the energy of the photon and the characteristics of the emitter (see, for example, McDaniel, 1964); typical photoelectron energies are one to several electron volts. Fahleson (1967) quotes equivalent temperature, T_p, of 1 eV while Whipple found an equivalent temperature of about 2 eV from satellite photocurrents (private communication).

3.3. SECONDARY ELECTRON CURRENTS

It is difficult to estimate the role of secondaries in determining the lunar potential, however, in a few cases they may be important. If the secondary spectrum is assumed to be quasi-Maxwellian, then an equivalent temperature is about 2 eV (Whipple, 1965). The secondary coefficient, δ, is largest when the incident primaries are electrons striking at oblique angles with energies of 200 to 1000 eV and δ may be as large as 4 to 6 for some insulators (McDaniel, 1964). One case where the secondary electrons may be important is that of the dark side of the Moon in the plasma sheet where electron temperatures are higher than in the solar wind; here the secondary current, like a weak photocurrent, may act to limit the large negative potentials which are expected as shown later in this chapter. The value of δ is difficult to predict since it depends on the material and the condition of the surface; however, by choosing values of 0.1, 0.8, and 2, it is possible to get a complete picture of the effect of the secondary current on the potential.

3.4. THE PLASMA SHEET

Measurements made on Vela satellites at 15 to 20 R_E have been reported by Bame et al. (1967) and by Bame (1968) who found a plasma sheet extending across the tail with a thickness of 4 to 6 R_E. The omnidirectional electron flux is typically 10^8 to 10^9 cm^{-2} s^{-1} which corresponds to an electron current range of 2×10^{-11} to 2×10^{-10} A cm^{-2} while the average electron energy is 0.6 keV but varies from 200 eV to 12 keV. The electron density varies from 0.1 to 1 cm^{-3} and has an average of about 0.5 cm^{-3}. It is important to note for our calculations that the spectrum is quasi-Maxwellian so that it can be characterized with a temperature. The plasma pressure in the region of the neutral sheet is sufficient to balance the magnetic pressure outside of the plasma sheet; thus even though there are large fluctuations of plasma energy and number density, the plasma pressure and flux are expected to stay more constant.

Similar results are reported by Vasyliunas (1968) of measurements made at distances of 20 to 24 R_E in the geomagnetic tail. The electron density is found to vary from 0.3 to 30 cm^{-3} and the average electron energy from 50 to 1600 eV.

Observations to greater distances in the geomagnetic tail were made during the outbound passage of Pioneer 7 (Lazarus, et al., 1968) which stayed mostly in the plasma sheet until it passed through the side of the magnetosphere at an anti-solar distance of 40 R_E. Typical electron fluxes are $\Phi_e = 1.25 \times 10^8$ cm^{-2} s^{-1} so $I_{e0} = = 2 \times 10^{-11}$ A cm^{-2}; the number density and thermal speed stay fairly constant throughout the flight.

Very recently, measurements on Explorer 35 in lunar orbit reported by Nishida and Lyon (1972) indicate a plasma sheet at lunar distances which is several Earth radii thick with electron flux, density, and energy of the order of 10^9 cm^{-2} s^{-1}, 1 cm^{-3}, and 0.6 keV. The Charged Particle Lunar Environment Experiment (CPLEE) on the lunar surface occasionally sees fluxes as high as 3×10^8 cm^{-2} s^{-1}, but more often the fluxes are of the order of a few times 10^7 cm^{-2} s^{-1}. (Burke and Reasoner, 1972; Reasoner, Burke, and Rich, personal communication).

3.5. THE LOW-DENSITY REGIONS

From the data reported it is clear that the tail regions above and below the plasma sheet contain a plasma with only relatively low number and energy densities and that the magnetic field provides most of the pressure; we will refer to these regions as the low density or high latitude regions. The Vela detectors were able to see some high latitude plasma and Bame (1968) shows a sample spectra. The average electron energy is about 70 eV and the proton energy appears to be somewhat higher and an electron density of 0.04 cm^{-3} was calculated. If we take $n \leqslant 0.05$ cm^{-3} and $\bar{E}_e = 50$ eV, then $\Phi_e \leqslant 6 \times 10^6$ cm^{-2} s^{-1} and $I_{e0} \leqslant 10^{-12}$ A cm^{-2}. Since this is a magnetosheath like plasma, if we take the same number density for ions and an average energy $\bar{E}_i = 500$ eV, then $\Phi_i = 4.5 \times 10^5$ cm^{-2} s^{-1} and $I_{i0} = 7 \times 10^{-14}$ A cm^{-2}.

Further measurements will give more exact information on the plasma parameters at 60 R_E, whether there is a net flow toward or away from the Earth and the thickness of the sheet. Because of the shift of the location of the plasma sheet with the tilt of the geomagnetic poles, along with the ± 5 R_E latitude variation in the lunar orbit, it is possible that the Moon will not be in the plasma sheet on all passes but will instead be in the low density region.

3.6. PLASMA FLUXES IN THE SOLAR WIND

The average properties of the solar wind at 1 AU have been reviewed by several authors (Dessler, 1967; Axord, 1968). Some average properties which we will use for our calculations are a flow velocity $V_{sw} = 430$ km s^{-1}, a proton flow energy $E_{swi} =$

TABLE I

Estimated properties of the geomagnetic tail plasma at 60 R_E

Question marks indicate uncertainty in the value(s)

	Flux Φ(cm^{-2} s^{-1})	Current I_0(A cm^{-2})	Average energy $\bar{E}$(keV)	Energy range (keV)	Average density n(cm^{-3})	Average density range (cm^{-3})
Plasma sheet						
Electrons	10^8 to 10^9	2×10^{-11} to 2×10^{-10}	0.6	0.3 to 1	0.5	0.1 to 1
Protons	10^7 to 10^8	10^{-12} to 10^{-11}	5.0	3 to 10	0.5	0.1 to 1
High latitude magnetotail						
Electrons	$\leqslant 6\times10^6$	10^{-12}	0.05 (?)	(?)	0.05 (?)	(?)
Protons	$\leqslant 5\times10^5$	$\leqslant 7\times10^{-14}$	0.5 (?)	0.2 to 1 (?)	0.05 (?)	(?)
Magnetosheath						
Electrons	5×10^8	10^{-10}	0.025	(?)	10	5 to 20 (?)
Protons (thermal)	10^7	2×10^{-12}	0.25	(?)	10	5 to 20 (?)
Solar wind						
Electrons (thermal and flow)	8×10^8	1.3×10^{-10}	0.010	0.006 to 0.020	10	4 to 15
Protons (flow)	4×10^8	7×10^{-11}	1.0	0.5 to 1.5	10	4 to 15

$=1$ keV, superimposed thermal temperatures $T_i \approx T_e \approx 10$ eV, particle densities $n_i = n_e = 10$ cm^{-3} with a range of 5 to 15 cm^{-3}.

On the dark side of the Moon, while there is generally a plasma void, we expect that there will be a few high temperature particles from the solar wind which reach the lunar surface by traveling along magnetic field lines. In the absence of the inter-planetary magnetic field, the condition for a particle flowing in the solar wind to reach a given point on the lunar surface is that its transverse thermal velocity be great enough that the particle velocity vector in the lunar system be at least tangent to the lunar surface at that point. The consequence is that increasingly higher temperatures are required to reach the surface at locations approaching the antisolar point. However, we will not calculate the plasma fluxes quantitatively as a function of position since two factors strongly modify this picture: one is that the presence of the interplanetary magnetic field will introduce a large asymmetry in the fluxes; the other factor is the apparently significant affect on particle motion by the flow structure at the edge of the lunar wake. However, it is apparent that ion energies of the order of the flow energy are required to reach some positions on the dark side.

A summary of some possible, average plasma properties at 60 R_E are given in Table I.

4. The Lunar Surface Potential

There are three quite different plasma environments for the Moon: the solar wind, possibly the plasma sheet of the geomagnetic tail, and the low density plasma in the

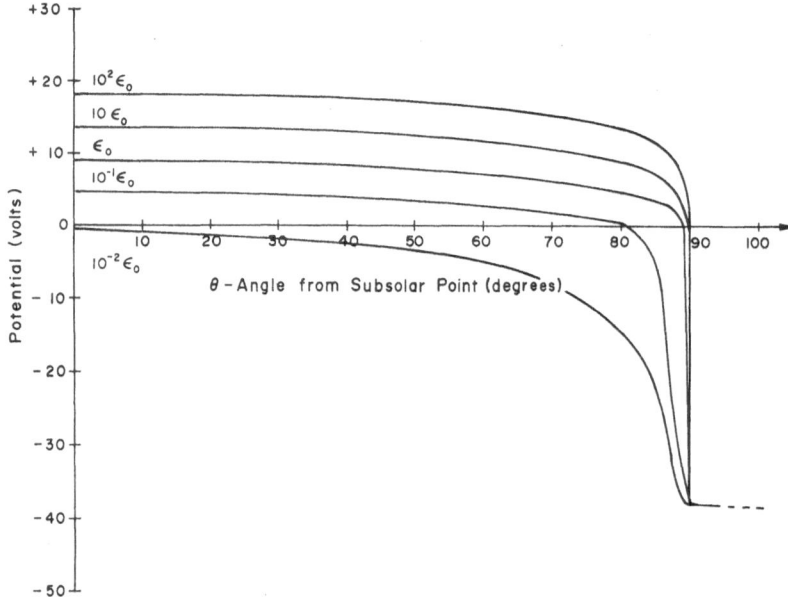

Fig. 1.   Plot of the lunar surface potential as a function of angle from the subsolar point. Potentials are calculated for five effective photoelectron emissivities.

high latitude regions of the tail; the magnetosheath is not considered in detail as it is similar to the solar wind. The potential is found by solving Equation (3) using the appropriate expressions for the currents depending on whether the ambient positive or negative current dominates.

4.1. MOON IN SOLAR WIND

The most likely answers to the question of the sign of the potential of the lunar surface in the solar wind can now be formulated. In the absence of an external magnetic field, the deciding parameter on the sunlit surface is the photoelectric emissivity; it appears that for most of the values in the expected range of emissivities that the photocurrent will exceed the plasma electron current and the central area of the sunlit

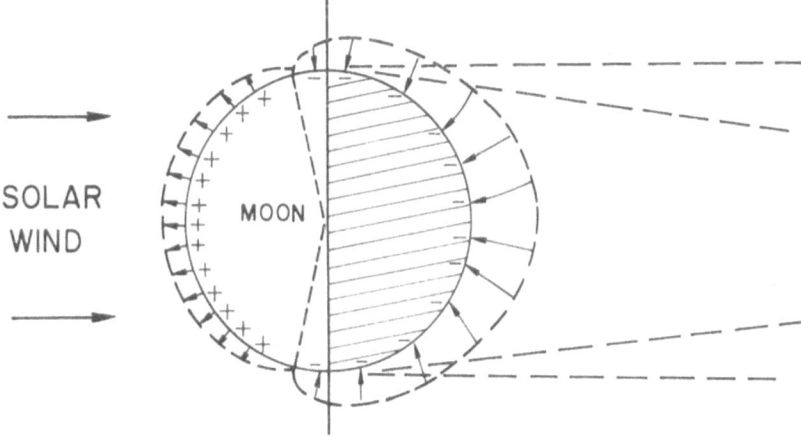

Fig. 2. Sketch of the expected potential and field distribution when the Moon is in the solar wind. The interplanetary magnetic field is neglected.

side will be positive. The potential for different values of ε as a function of angle from the subsolar point is shown in Figure 1 and sketched in Figure 2. Note that the thermal electron current in the solar wind dominates as the angle approaches 90° and that the potential goes to -38 V at the solar wind terminator.

4.2. MOON IN GEOMAGNETIC TAIL

4.2.1. *Plasma Sheet*

Using the equations for a stationary plasma and the currents from the lower end of the expected range in Table I, we calculate that at the subsolar point the potential varies from $+1.8$ to $+20$ V and is equal to $+11$ V for $\varepsilon = \varepsilon_0$.

The potential at the terminator and on the dark side will tend to be negative and since the plasma temperature is high, large negative potentials could exist. However, the secondary current will tend to limit these potentials and could even make the surface potential positive. For the plasma currents $I_{e0} = 2 \times 10^{-11}$ A cm^{-2} and

$I_{i0} = 10^{-12}$ A cm^{-2} then $I_{i0} = 0.05\ I_{e0}$ and $\delta_{ec} \simeq 0.95$ is the critical value for crossover from negative to positive potential. Assuming that the above current ratios occur uniformly from the terminator over the entire dark side, and using the approximation that the sheath is very thin, then the potential is constant and is given by Table II.

TABLE II

Potential at terminator and over the dark side when the Moon is in the plasma sheet, as a function of secondary electron coefficient

Secondary coefficient δ_e	Potential at terminator and on dark side (V)
0	− 1800
0.1	− 1730
0.8	− 830
$\delta_{ec} = 0.95$	~ 0
2	+ 1.5

Since a reasonable estimate for the lunar secondary coefficient could be either less than or greater than the critical value, the range of secondary coefficients from 0.1 to 2 should be considered. The interesting effect on the potential is that if the secondary coefficient is greater than δ_{ec}, then the potential will be several hundred volts negative and will vary with the plasma sheet electron temperature.

4.2.2. High Latitude Regions

As was pointed out when the plasma fluxes were discussed, at certain times conditions are favorable for the lunar orbit through the tail to be out of the plasma sheet and in the high latitude region. From Table I the plasma currents are quite low, $I_{e0} = 10^{-12}$ A cm^{-2} and $I_{i0} = 7 \times 10^{-14}$ A cm^{-2}, so that the photocurrent will diminate over most of the sunlit surface. The potential at the subsolar point varies from $+8.0$ to $+26.4$ V for $\varepsilon = 10^{-2}\ \varepsilon_0$ and we get $+17.2$ V for $\varepsilon = \varepsilon_0$. At the terminator the potential goes to -135 V which is simply the probe potential corresponding to the given ion and electron temperatures.

It is important to *note* that the recent measurements of lunar surface photoelectrons by CPLEE (O'Brien and Reasoner, 1971) could increase these subsolar potentials. They measure the high energy tail (>40 eV) of the spectrum and find the expected power law rather than Maxwellian response. In all calculations in this paper we have used a Maxwellian spectrum and in the other regions when the plasma fluxes are greater, then the quasi-Maxwellian peak of the photoelectron spectrum determines the potential. However, in the low density regions, the plasma electron flux is so low that it is sufficiently balanced by the high energy tail of the photoelectron spectrum whose power law distribution may give larger positive potentials when Equation (3) is solved. It should also be noted that lower plasma fluxes in the plasma sheet, as measured by CPLEE, would also tend to increase the calculated positive potential there due to both the fact that the power law tail of the photoelectron spectrum becomes important, and that just having a lower external electron flux raises the potential.

5. Surface Electric Field

We can now immediately calculate the surface electric field in the plasma sheath. When the currents determining the potential are Maxwellian then the surface field decays exponentially above the surface, and the field at the surface is given by

$$E \cong \frac{\phi}{\lambda},$$
(16)

where λ is the Debye length which is determined by the density and temperature of the *attracted* species of the plasma. All of the relevant parameters and the resulting fields are summarized in Table III.

TABLE III

Electric potential and field at the lunar surface for different plasma environments

Position of Moon	Location on surface	T (eV)		λ (m)	Φ (V)	E (V m^{-1})
Solar wind	Subsolar	$T_p =$	2	0.41	9.0	22
Solar wind	Terminator	$T_i =$	10	7.4	$-$ 38	$-$ 5.1
High latitude tail	Subsolar	$T_p =$	2	0.41	17.2	42
High latitude tail	Terminator and dark side	$T_i =$	500	738	$-$ 135	$-$ 0.18
Plasma sheet	Subsolar	$T_p =$	2	0.41	11.0	27
Plasma sheet	Terminator and dark side $(\delta = 0)$	$T_i =$	5×10^3	744	-1800	$-$ 2.4
Plasma sheet	Terminator and dark side $(\delta = 2)$	$T_s =$	2	4.6	1.5	0.33

6. Conclusions

The lunar surface potential, and associated electric field are calculated using the concept of an effective photoemissivity where emission throughout the solar spectrum is related to total photo currents measured by satellite experiments. For the case of the Moon in the solar wind, the potential is calculated using plasma flux equations containing both thermal and flow terms. In agreement with other calculations, the surface at the subsolar point is expected to be a few volts positive; however, we find that the potential should go to tens of volts negative at the terminator, and possibly more negative on the dark side.

Other results are that in the geomagnetic tail the sunlit potential is again expected m be positive; however, in the high latitude regions the potential over the dark side toay be about 100 V negative. It is striking that when the Moon is in the plasma sheet,

the terminator and dark side potential could be about 1800 V negative (due simply to
the hot plasma); however, if the secondary electron emissivity is greater than about
unity, this large negative potential could be reduced to a small positive one.

An advantage of the formulation of the problem as described here is that as more
experimental data on the lunar plasma environment and photoemissivity become
available, then these more accurate parameters can readily be incorporated into the
equations. Accurate calculations of the surface potential are important, both because
the potential is part at the lunar environment, and because of the effect of the potential
on charged particles measured by lunar surface experiments.

Acknowledgements

The author would like to thank Drs E. C. Whipple, F. S. Mozer, D. L. Reasoner,
W. J. Burke, and F. J. Rich for helpful discussions, and especially to thank Dr H. R.
Anderson for many helpful suggestions. Much of this work was done when the author
was at the NASA Manned Spacecraft Center; it has been supported at Rice under
Contract NAS 9-5911.

Appendix: Plasma Equations for Potential Calculation

This appendix summarizes the probe equations for currents to a surface. The equations
and approximations are the appropriate ones for calculating the potential of a body
the size of the Moon in the type of plasma environments encountered in the lunar
orbit. The physical interpretation of the equations is Section 2 of the text.

The equations for a body in a stationary thermal plasma follow the work of Whipple
(1965) and Fahleson (1967) but with added terms for secondary electron emission.

The equations for a body in a flowing as well as thermal plasma are based on the
plasma equation developed by Whipple (1959). The flow and thermal contributions to
the current are interrelated but the equations reduce to the static plasma equations
when the flow velocity goes to zero and to the pure flow equations when the thermal
velocity goes to zero. When the potential is positive, the Repelled Species is the plasma
ion and the Attracted Species is the plasma electron.

In the equations, ϕ is the local potential of the Moon with respect to the adjacent
plasma potential, and I_e, I_i, I_p and I_s are the local currents *to* the lunar surface from
the plasma electrons and ions and from the surface photo- and secondary currents
respectively. In the stationary plasma equations, k is the Boltzmann constant, m_e and
m_i are the electron and ion masses, T_e and T_i are the electron and ion temperatures, t
is the plasma sheath thickness and a is the radius of the Moon, λ is the Debye length,
ε is the effective photoemissivity of the surface, n is the external plasma electron
(or ion) density, e is the magnitude of the electron charge, T_p and T_s are the photo- and
secondary electron temperatures, and δ_e and δ_i are the secondary electron coefficients
for primary electrons and ions respectively.

In the equations for a flowing plasma, V is the flow velocity, v_{mi} and v_{me} are the most probable ion and electron thermal velocities, v_{ti} and v_{te} are the average ion and electron thermal velocities, θ is the angle from the subsolar point, and the functions U and X are defined in the equations as functions of the other variables.

POTENTIAL DETERMINED BY BALANCE OF CURRENT DENSITY

$$I_e + I_i + I_p + I_s = 0.$$

A. *For Stationary Plasma, Positive Potential* $(\phi>0)$

Repelled: $\quad I_i = ne\sqrt{\dfrac{kT_i}{2\pi m_i}}\exp\left(\dfrac{-e\phi}{kT_i}\right)$

attracted: $\quad I_e = -ne\sqrt{\dfrac{kT_e}{2\pi m_e}\dfrac{a^2}{r^2}}\left(1 - \dfrac{a^2 - r^2}{a^2}\exp\left[\dfrac{-r^2}{a^2 - r^2}\dfrac{e\phi}{kT_e}\right]\right)$

$\qquad\qquad\;\; \cong -ne\sqrt{\dfrac{kT_e}{2\pi m_e}}\left[1 + \dfrac{2t}{r}\right],\quad$ thin sheath

$\qquad\qquad\;\; \cong -ne\sqrt{\dfrac{kT_e}{2\pi m_e}},\quad$ very thin sheath

where $\quad a = r + t, t = \lambda\, 0.83\left(\dfrac{r}{\lambda}\right)^{1/3}\left(\dfrac{e\phi}{kT_e}\right)^{1/2},\quad \lambda = 6.9\times10^{-2}\sqrt{\dfrac{T}{n}}\, m$

emitted: $\quad I_p = i_p(\varepsilon)\cos\theta\,\exp\left(\dfrac{-e\phi}{kT_p}\right)$

$\qquad\qquad I_s = (I_e\delta_e + I_i\delta_i)\exp\left(\dfrac{-e\phi}{kT_s}\right).$

B. *For Stationary Plasma, Negative Potential,* $(\phi<0)$

Repelled: $\quad I_e = -ne\sqrt{\dfrac{kT_e}{2\pi m_e}}\exp\left(\dfrac{e\phi}{kT_e}\right)$

attracted: $\quad I_i \cong ne\sqrt{\dfrac{kT_i}{2\pi m_i}} = \dfrac{nev_{ti}}{4} = \dfrac{nev_{mi}}{2\sqrt{\pi}}$

emitted: $\quad I_p = i_p\cos\theta$

$\qquad\qquad I_s = I_e\delta_e + I_i\delta_i.$

C. *For Flowing Plasma, Positive Potential* $(\phi>0)$

Repelled: $\quad I_i = \dfrac{neV\cos\theta}{2}\left[1 + \mathrm{erf}(X) + \dfrac{v_{mi}}{V\cos\theta\sqrt{\pi}}e^{-X^2}\right]$

$\qquad\qquad = \dfrac{nev_{mi}}{2\sqrt{\pi}}\left[e^{-X^2} + \dfrac{V\cos\theta\sqrt{\pi}}{v_{mi}}(1 + \mathrm{erf}(X))\right]$

$\qquad\qquad \sim neV\cos\theta,$

where $\qquad X = \dfrac{V\cos\theta}{v_m} - \sqrt{\dfrac{e\phi}{kT}} = U - \sqrt{\dfrac{e\phi}{kT}}$

attracted: $\qquad I_e = \dfrac{nev_{me}}{2\sqrt{\pi}}\left[e^{-U_e{}^2} + \sqrt{\pi}\,U_e(1 + \mathrm{erf}(U_e))\right]$

$\qquad\qquad \cong \dfrac{nev_{me}}{2\sqrt{\pi}}\left[1 + \sqrt{\pi}\,U_e\right]$

emitted: $\qquad I_p = i_p \cos\theta \exp\left(\dfrac{-e\phi}{kT_p}\right), \qquad I_s = I_e\delta_e \exp\left(\dfrac{-e\phi}{kT_s}\right).$

D. *For Flowing Plasma, Negative Potential* ($\phi < 0$)

Repelled: $\qquad I_e = \dfrac{-nev_{me}}{2\sqrt{\pi}}\left[e^{-X^2} + \dfrac{V\cos\theta\sqrt{\pi}}{v_{me}}(1 + \mathrm{erf}(X))\right]$

$\qquad\qquad \sim \dfrac{-nev_{me}}{2\sqrt{\pi}}\,e^{-e|\phi|/kT_e},$

where $\qquad X_e = \dfrac{V\cos\theta}{v_{me}} - \sqrt{\dfrac{e|\phi|}{kT}} = U_e - \sqrt{\dfrac{e|\phi|}{kT}}$

attracted: $\qquad I_i = \dfrac{neV\cos\theta}{2}\left[1 + \mathrm{erf}(U) + \dfrac{1}{U\sqrt{\pi}}\,e^{-U^2}\right]$

$\qquad\qquad \sim neV\cos\theta$

emitted: $\qquad I_p = i_p \cos\theta, \qquad I_s = I_e\delta_e.$

References

Anderson, H. R. and Manka, R. H.: 1968, 'Electric fields at the Lunar Surface – Sources and Methods of Measurement', paper presented at the *NASA Ames Conference on Electromagnetic Exploration of the Moon*, Moffett Field, California.

Anderson, H. R. and Manka, R. H.: 1970, in W. I. Linlor (ed.), *Electromagnetic Exploratoni of the Moon*, Mono, Baltimore, p. 117.

Axford, W. I.: 1968, *Space Sci. Rev.* **8**, 331.

Bame, S. J., Asbridge, J. R., Felthauser, H. E., Hones, E. W., and Strong, I. B.: 1967, *J. Geophys. Res.* **72**, 113.

Bame, S. J.: 1968, in B. M. McCormac (ed.), *Particles and Fields*, Rinehold, New York, p. 373.

Bernstein, W., Fredricks, R. W., Vogl, J. L., and Fowler, W. A.: 1963, *Icarus* **2**, 233.

Burke, W. J. and Reasoner, D. L.: 1972, *Planetary Space Sci.* **20**, 429.

Dessler, A. J.: 1967, *Rev. Geophys.* **51**, 1.

Fahleson, U.: 1967, *Space Sci. Rev.* **7**, 238.

Grobman, W. D. and Blank, J. L.: 1969, *J. Geophys. Res.* **74**, 3943.

Hinteregger, H. E., Damon, K. R., and Hall, L. A.: 1959, *J. Geophys. Res.* **64**, 961.

Langmuir, I. and Mott-Smith, H. M.: 1926, *Phys. Rev.* **28**, 727.

Lazarus, A. J., Siscoe, G. L., and Ness, N. F.: 1968, *J. Geophys. Res.* **73**, 2399.

Manka, R. H. and Anderson, H. R.: 1968, *Trans. Am. Geophys. Union* **49**, 227.

Manka, R. H. and Anderson, H. R.: 1969, *Trans. Am. Geophys. Union* **50**, 217.

McDaniel, E. W.: 1964, *Collision Phenomena in Ionized Gases*, John Wiley and Sons, New York, 775 pp.

Nishida, A. and Lyon, E. F.: 1972, *J. Geophys. Res.* **77**, 4086.

O'Brien, B. J. and Reasoner, D. L.: 1971, in *Apollo 14 Preliminary Science Report*, NASA SP-272, 193.

Öpik, E. J. and Singer, S. F.: 1960, *J. Geophys. Res.* **65**, 3065.
Vasyliunas, V. M.: 1968, *J. Geophys. Res.* **73**, 2839.
Whipple, E. C., Jr.: 1959, *Proc IRE*, 2023.
Whipple, E. C., Jr.: 1965, Thesis, George Washington University, Washington, D.C.

DISCUSSION

Fredricks: MAGCONS have been found by the Apollo 15 subsatellite, and fields ~100 γ have been seen by hand-carried instruments by astronauts on the surface of the Moon. Have you attempted to account for these relatively large field values in your probe theory?

Manka: We have considered the effect of a surface magnetic field, though not in this paper. The effect of a surface magnetic field existing above the surface might be similar to the effect of an external magnetic field, such as the interplanetary magnetic field, that I discussed. Here, in some cases the magnetic field prevents photoelectrons from leaving, also, it restricts some external plasma fluxes from reaching the surface and in general complicates the problem. The surface field also deviates incoming ions, thus possibly selecting certain ions to be seen by a detector.

THE ELECTRIC POTENTIAL OF THE MOON
IN THE SOLAR WIND

J. W. FREEMAN, Jr., M. A. FENNER, and H. K. HILLS

Dept. of Space Science, Rice University, Houston, Tex. 77001, U.S.A.

Abstract. Acceleration and detection of the lunar thermal ionosphere in the presence of the lunar electric field yields a value of approximately $+10$ V for the lunar electric potential for solar zenith angles between $20°$ and $45°$ and in the magnetosheath or solar wind. The ion number density of the thermal ionosphere observed is compatible with a surface neutral number density of about 10^5 atoms cm^{-3}.

1. Introduction

Theoretical studies of the sunlit lunar surface electric potential have predicted values ranging from $+20$ V down to a few volts positive. Originally, Öpik and Singer (1960), and Öpik (1962) proposed a value of about $+20$ V. More recently Grobman and Blank (1969) revised this estimate downward by about an order of magnitude and Manka (1972) estimates a value near $+10$ V. Reasoner and Burke (1972) have reported potentials as high as $+200$ V in the magnetospheric tail.

In this paper we discuss evidence for a sunlit potential in the magnetosheath and solar wind of about $+10$ V. This evidence comes from the analysis of the energy spectra of lunar ionosphere thermal ions accelerated toward the Moon by an artificial electric field. The data are provided by the Apollo Lunar Surface Experiment Package (ALSEP) Suprathermal Ion Detector Experiment (SIDE) deployed at the Apollo 14 and 15 sites.

2. The SIDE

The Suprathermal Ion Detector Experiment (SIDE) is designed to measure positive ions down to a few tenths of an electron-volt. To accomplish this in the face of possible lunar surface potentials of the order of several tens of volts, the instrument is equipped with a ground plane electrode whose potential with respect to a wire grid above the ion entrance apertures (also instrument ground) is stepped through a series of 24 voltages (see Table I). Some of these voltages match the pass-band center energies for the two SIDE ion detectors, the total ion detector (TID) and the mass analyzer (MA). Both the TID and MA possess energy per unit charge filters for the entering ions.

The ground plane electrode is a circular wire grid 65 cm in diameter which lies on the lunar surface beneath the SIDE. Close-up photographs show that the grid makes contact with the surface at many points, so good electrical contact will be assumed. The SIDE is DC isolated from the ALSEP central station and other experiments.

These features of the SIDE allow the determination of the electric potential of the lunar surface under certain conditions. This is accomplished by examination of the energy spectra of thermal ions accelerated into the SIDE by the ground plane voltage in the presence of the electric field due to the surface charge of the Moon.

R. J. L. Grard (ed.), Photon and Particle Interactions with Surfaces in Space, 363–368. All Rights Reserved

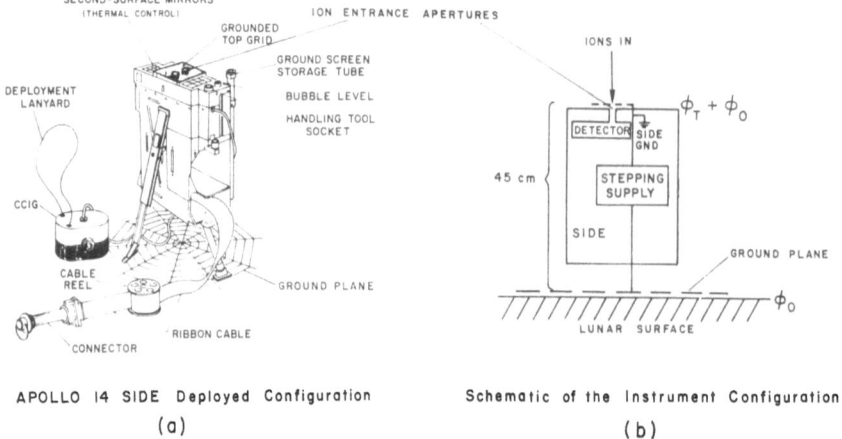

APOLLO 14 SIDE Deployed Configuration   Schematic of the Instrument Configuration

(a)   (b)

Fig. 1.   The Suprathermal Ion Detector Experiment (a) as deployed on the lunar surface, and (b) showing the ground plane and top wire grid configuration schematically. Thermal ions are attracted to or repelled from the top of the instrument by the top grid. All potentials are referenced to infinity. ϕ_T is the voltage produced by the stepping supply. The ground plane is in actual contact with the lunar surface.

TABLE I

Voltages in the ground plane
stepping sequence

0.0	0.0
0.6	− 0.6
1.2	− 1.2
1.8	− 1.8
2.4	− 2.4
3.6	− 3.6
5.4	− 5.4
7.8	− 7.8
10.2	− 10.2
16.2	− 16.2
19.8	− 19.8
27.6	− 27.6

Figure 1 shows the instrument and the ground plane in the deployed configuration. Further details of the experiment can be found in Freeman *et al.* (1972).

3. Instrument Function

Assuming good electrical coupling between the ground plane and the lunar surface the instrument would be expected to function as follows:

(a) In the case of a near-zero lunar surface potential a negative ground plane voltage accelerates thermal positive ions into the instrument with an energy approximately that of the stepper voltage.

(b) When the lunar surface is substantially positive and the stepper voltage negative, but larger in absolute value, the energy of the detected ions is less than the stepper voltage by an amount equal to the lunar surface potential. When the stepper voltage exactly matches but is of opposite sign from the lunar surface potential, the ions are seen unaccelerated.

(c) When the lunar surface is negative the ions may be repelled by a large enough positive stepper voltage.

We can assume that the ions seen are principally ionized neutral gas in thermal equilibrium with the lunar surface $(T < 400\,\mathrm{K})$ and further that most of the ions appear to have come from infinity so far as the energy acquired from the electric fields is concerned. Under these conditions, the foregoing can be made more explicit by expressing the ion energy, E, seen at the detector by

$$E = E_i - (\varphi_t + \varphi_0)\, q, \tag{1}$$

where E_i is the initial ion energy (the neutral thermal energy), φ_t the potential of the top wire grid of the SIDE relative to the ground plane and hence the lunar surface, φ_0 the lunar surface potential, and q the ion charge (assumed to be $+1$). We will be concerned with cases in which φ_t is negative and $|\varphi_t + \varphi_0| \gg E_i$ so that (1) becomes

$$E \cong - (\varphi_t + \varphi_0)\, q. \tag{2}$$

4. Observations

A feature of the data when the solar zenith angle is between approximately $20°$ and $45°$ is the frequent appearance of narrow peaked low-energy ion flux spectra which show a correlation with the ground plane stepper voltage. That is, certain peak energies recur with specific ground plane stepper voltages. There is often a complete absence of ions except at the resonant ground plane stepper voltage.

This phenomenona has been seen with the Apollo 14 and 15 SIDEs and exclusively in the solar wind or magnetosheath on either side of the magnetospheric tail. Figure 2 shows the location in lunar orbit of these observations.

Figure 3 shows an extended set of data from the Apollo 14 SIDE. Here the TID count rates from two low energy channels have been grouped according to the ground plane voltage. The energies that stand out near the center of the figure are 7 eV at ground plane voltage -16.2 and 17 eV at ground plane voltage -27.6.

Similar results have been obtained from the SIDE mass analyzer detectors where the passband energies extend down to 0.2 eV. In this case the enhanced fluxes are seen when the ground plane stepper voltage is -10.2.

Equation (2) can be solved for these examples to yield a lunar surface potential, φ_0, of approximately $+10$ V. A similar potential is found from the SIDE data at the Apollo 15 site $26.4°$ north of the equator. Also, the potential does not change detectably at the equatorial site (Apollo 14) over the solar zenith angle range $20°$ to $45°$.

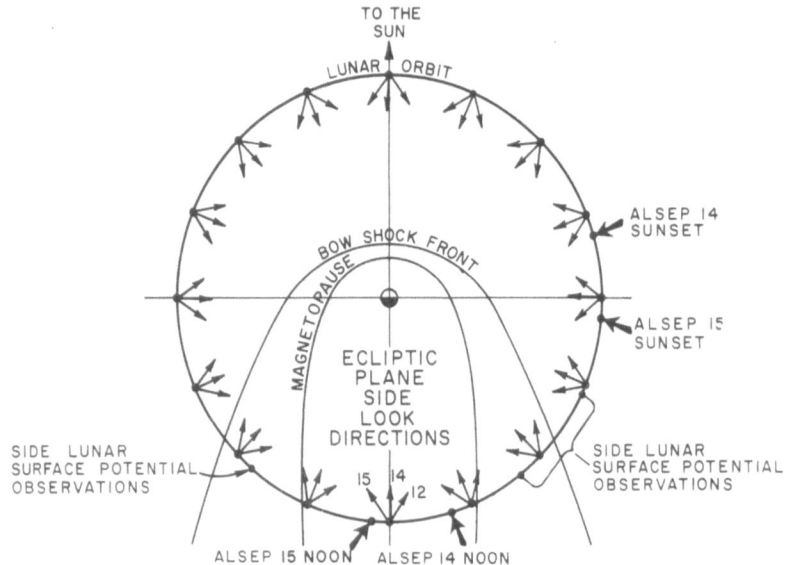

Fig. 2.   Location in lunar orbit of the lunar surface potential observations.

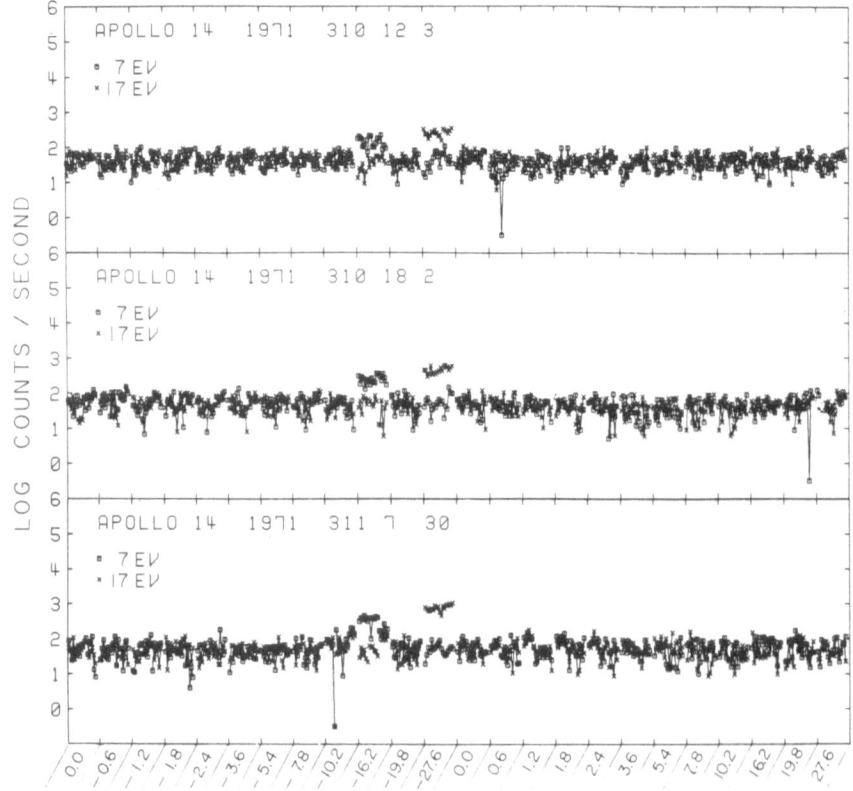

Fig. 3.   SIDE total ion detector count rates in two energy channels for three successive time interval grouped by ground plane stepper voltage. The time given is the start time of the interval. Each interval is approximately three hours long.

5. The Thermal Lunar Ionosphere

The foregoing assumes that the thermal lunar ionosphere more than one screening length from the SIDE is being accelerated into the instrument in the presence of a lunar surface electric field. We can now use the observed ion flux to calculate the requisite ion number density and hence the neutral number density in the lunar atmosphere.

Referring to Figure 3 we take 300 counts s^{-1} as typical of the 17 eV ions and assuming the ions to be isotropic over 1 sr and absent elsewhere we calculate a flux of 3×10^6 ions $cm^{-2} s^{-1}$. Johnson (1971) has shown that the lunar atmosphere is expected to consist largely of neon. Using this as the dominant gas we obtain an ion number density of approximately 2 ions cm^{-3}. Using Johnson's number of 10^{-5} (Johnson, 1971) for the degree of ionization of neon in the lunar atmosphere we find a neutral number density of 2×10^5 atoms cm^{-3}. This is exactly the sunset number density measured by the ALSEP cold cathode ionization gauge experiment (Johnson *et al*, 1972). On the basis of this we have confidence that the SIDE is measuring the thermal lunar ionosphere. (Here thermal refers to temperature equilibrium with the lunar surface and not the solar wind.)

6. Discussion and Summary

Acceleration of the lunar thermal ionosphere with known voltages has lead to the value of about $+10$ V for the dayside lunar surface potential. It should be emphasized this value holds only in the solar wind or magnetosheath plasma. Reasoner and Burke (1972) have reported evidence for potentials as high as $+200$ V in the greater vacuum of the magnetospheric tail.

These results suggest that the sunlit, solar-wind lunar surface potential is limited to 10 V by the large sea of solar wind electrons of mean thermal energy ~ 10eV. Empirically, φ_0, the sunlit lunar surface potential in the solar wind, can be well approximated by

$$\varphi_0 \cong -\frac{kT_e}{e}, \tag{3}$$

where k is Boltzmann's constant, T_e the solar wind electron temperature, and e the electron charge.

This $+10$ V value appears to hold at least between solar zenith angles of 20° and 45°. Elsewhere the potential may change so that the ion energy lies between the discrete energy bands of our detector for the discrete ground plane voltages available to us. More complete analysis of the data may provide data on the change of this potential with solar zenith angle and changing plasma conditions.

Acknowledgement

We have profited from discussions with several Rice University scientists; particularly Dr Richard Vondrak and Dr Robert Manka. One of us, John W. Freeman, was also a Visiting Scientist at the Lunar Science Institute during preparation of this paper. This research was supported by NASA contract NAS 9-5911.

References

Freeman, J. W., Jr., Fenner, M. A., Hills, H. K., Lindeman, R. A., Medrano, R., and Meister, J.: 1972, *Icarus*, **16**, 328.

Grobman, W. D. and Blank, J. L.: 1969, *J. Geophys. Res.* **74**, 3943.

Johnson, F. S.: 1971, *Rev. Geophys. Space Phys.* **9**, 813.

Johnson, F. S., Evans, D. E., and Carroll, J.: 1972, paper presented at the 3rd *Lunar Sci. Conf.*, January 10–13, Houston.

Manka, Robert Hall: 1972, Ph. D. Thesis, Rice University.

Öpik, E. J.: 1962, *Planetary Space Sci.* **9**, 221.

Öpik, E. J. and Singer, S. F.: 1960, *J. Geophys. Res.* **65**, 3065.

Reasoner, David L. and Burke, William J.: 1972, this volume, p. 369.

DISCUSSION

Willis: What is the neutral atom concentration, the major component, and how are your measurements affected by polution of the lunar atmosphere due to the spacecraft?

Freeman: The measurements suggest that N_e of mass 20 to be the most abundant element. Based on this mass, we obtain an ion density equivalent to density of neutrals of the order of 10^5 atoms per cm^3. This is in very good agreement with the lunar cold cathode measurements and estimations from radiofrequency data. The number density does vary due to spacecraft polution effects, but the above value is thought to be due to the Moon's intrinsic atmosphere, measurements being made over a long period of time (>1 month).

Feuerbacher: Do you have any evidence that excludes a potential difference between your instrument and the lunar surface due to photoemission from your instrument box?

Freeman: We have no direct evidence that excludes such a potential difference, however, based on conductivity calculations and the fact that we have direct contact with the surface many places, it is hard to imagine that such a potential could be very large. It is true that a different potential will be established on the sides of the instrument (which are covered with insulating paint), however, the electric field from such a potential should not influence the incoming ion energy due to the large grounded grid on the top surface where the ions enter the instrument.

Manka: I think that rather than relate your possible observation of a $+10$ V surface potential to the ~ 10 eV temperature of the solar wind electrons, that it is more correct to relate it to the temperature of the photoelectrons emitted from the lunar surface. Since the measurement indicates a positive surface potential, this implies that the emitted photocurrent exceeds the current from the incoming thermal electrons. Thus, in equilibrium, the surface potential goes positive to that value which allows only the hotter photoelectrons to escape such that their loss is replaced by the solar wind electrons coming to the surface; i.e., the photocurrent escaping is restricted to equal the total current of incoming solar wind electrons. The resulting surface potential then depends on the photoelectron temperature; the solar wind electron temperature is of significance only in that together with the solar wind number density, it determines the total solar wind electron current to the surface.

MEASUREMENT OF THE LUNAR PHOTOELECTRON LAYER IN THE GEOMAGNETIC TAIL

DAVID L. REASONER and WILLIAM J. BURKE

Dept. of Space Science, Rice University, Houston, Tex. 77001, U.S.A.

Abstract. The Charged Particle Lunar Environment Experiment (CPLEE), a part of the Apollo 14 ALSEP, is an ion-electron spectrometer capable of measuring ions and electrons with energies between 40 eV and 50 keV. The instrument, with apertures 26 cm above the surface, has detected a photo-electron gas layer above the sunlit lunar surface, with energies ranging up to 200 eV. Experimental data for periods while the Moon was in the Earth's magnetotail for electrons with energies $40 \text{ eV} \leqslant E \leqslant 200 \text{ eV}$ follow a power law spectrum $j(E) = j_0(E/E_0)^{-\mu}$ with $3.5 \leqslant \mu \leqslant 4$. In the absence of photoelectrons with $E > 200$ eV, we assume that the surface potential is at least 200 V. The modulation of this potential in the presence of intense plasma sheet fluxes has been observed.

Numerical solutions for the variation of electron density and potential as functions of height above the lunar surface were obtained. The solar photon spectrum $I(h\nu)$, obtained from various experi-mental sources, and the photoelectron yield function of the surface materials, $Y(h\nu)$, are two para-meters of the solution. Energy spectra at the height of the measurements for various values of $Y(h\nu)$ were computed until a fit to experimental data was obtained. Using a functional form $Y(h\nu) = = [Y_0(h\nu - W)]/(W/2)$ for $6 \text{ eV} \lesssim h\nu \lesssim 9 \text{ eV}$ and $Y(h\nu) = Y_0$ for $(h\nu) > 9$ eV where W, the lunar surface work function, was set at 6 eV, we calculated a value of $Y_0 = 0.1$ electrons photon^{-1}. The solution also showed that the photoelectron density falls by 5 orders of magnitude within 10 m of the surface, but the layer actually terminates several hundred meters above this height.

1. Introduction

The general problems of photoelectron emission by an isolated body in a vacuum and in a plasma have been the objects of several investigations. For example, Medved (1968) has treated electron sheath formations about bodies of typical satellite dimen-sions. Guernsey and Fu (1970) have considered the properties of an infinite, photo-emitting plate immersed in a dilute plasma. Grobman and Blank (1969) obtained expressions for the lunar surface potential due to photoelectron emission while the moon is in the solar wind. Walbridge (1970) developed a set of equations for obtaining the density of photoelectrons as well as the electrostatic potential as functions of height above the surface of the Moon while the Moon is in the solar wind. By assuming a simplified form of the solar photon emission spectrum he could provide analytic expressions for these quantities.

In this paper we report on observations of stable photoelectron fluxes, with energies between 40 and 200 eV by the Apollo XIV Charged Particle Lunar Environment Experiment (CPLEE). These observations, made in the magnetotail under near vacuum conditions, are compared with numerically calculated photoemission spectra to determine the approximate potential difference between ground and CPLEE's apertures (26 cm). Numerically calculated density and potential distributions, when compared with our measured values, help us estimate the photoelectron yield function of the dust layer covering the Moon.

R. J. L. Grard (ed.), Photon and Particle Interactions with Surfaces in Space, 369–387. All Rights Reserved

2. The Instrument

Complete descriptions of the CPLEE instrument has been given by O'Brien and Reasoner (1971) and Burke and Reasoner (1972). The instrument contains two identical charged-particle analyzers, hereafter referred to as analyzers A and B. Analyzer A looks toward the local lunar vertical, and analyzer B looks 60° from vertical toward lunar west.

The particle analyzers contain a set of electrostatic deflection plates to separate particles according to energy and charge type, and an array of 6 channel electron multipliers for particle detection. For a fixed voltage on the deflection plates, a five band measurement of the spectrum of particles of one charge sign and a single-band measurement of particles of the opposite charge sign are made. The deflection plate voltage is stepped through a sequence of 3 voltages at both polarities, plus background and calibration levels with zero voltage on the plates. A complete measurement of the spectrum of ions and electrons with energies between 40 eV and 50 keV is made every 19.2 s. Of particular relevance to this study are the lowest electron energy passbands. With a deflection voltage of −35 V, the instrument measures electrons in five ranges centered at 40, 50, 65, 90 and 200 eV. With +35 V on the deflection plates, electrons in a single energy range between 50 and 150 eV are measured. At the next higher deflection voltages of ±350 V, the energy passbands given above are scaled upward by approximately a factor of 10.

3. Observations

In this section we present data from the February 1971 passage of the moon through the magnetotail. Because these are so typical, the display of data from subsequent months would be redundant. At approximately 0300 UT on February 8 CPLEE passed from the dusk side magnetosheath into the tail. The five minute averaged counting rates for analyzer A, channel 1, at −35 V measuring 40 eV electrons are plotted for this day in Figure 1. Almost identical count rates are observed in analyzers A and B during this period of observation. As CPLEE moves across the magnetopause the counting rate drops from ~ 200 cycle^{-1} to the magnetotail photoelectron background of ~ 35 cycle^{-1} (1 cycle = 1.2 s). Enhancements at ~ 0530 h and at ~ 0930 h correspond to plasma events associated with substorms on Earth (Burke and Reasoner, 1972). There is a data gap from 1000 to 1200 h. With the exception of the short lived ($\leqslant 1$ h) enhancements the detector shows a stable counting rate when the Moon is in the magnetotail.

Our contention is that these stable fluxes observed in the magnetotail during periods of low magnetic activity are photoelectrons generated by ultraviolet radiation from the Sun striking the surface of the Moon. In support of this thesis we have reproduced the counting rates observed in the same detector on February 10 when the Moon was near the center of the tail (Figure 2). First, we note that the stable count level is the same at the center as it was when CPLEE first entered the tail. Secondly, from about 0500 to 1000 h the Moon was in eclipse. During this time we observe the

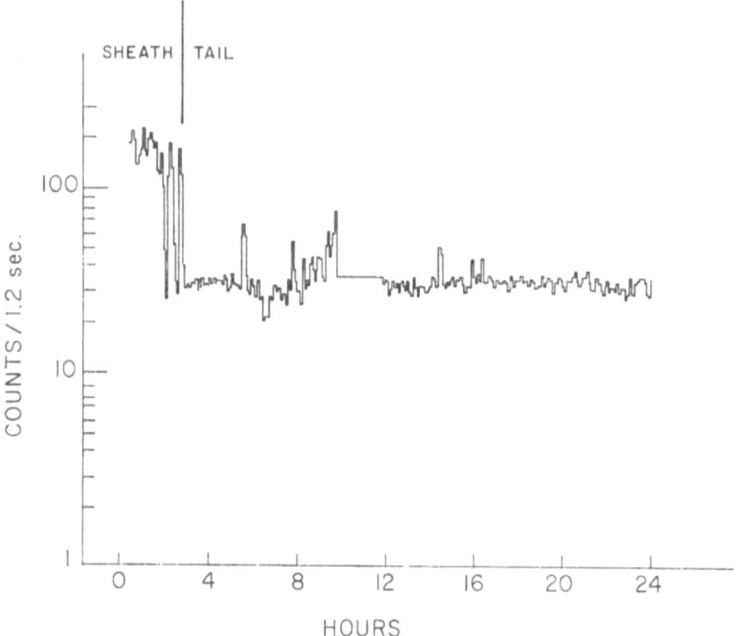

Fig. 1. 5-min averaged counting rates for CPLEE, analyzer A, channel 1 at −35 V, measuring 40 eV electrons on February 8, 1971. After 0300 UT counting rates fell from high magnetosheath to stable photoelectron levels.

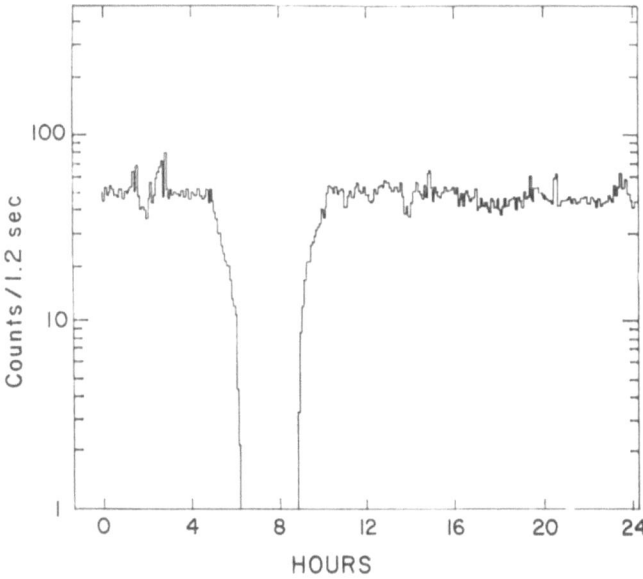

Fig. 2. 5-min averaged deep tail counting rates of 40 eV electrons on February 10, 1971. The lunar eclipse (0500–0900 UT) is marked by vanishing photoelectron counting rates.

counting rates go to zero. As the moon emerges from the Earth's shadow, the counting rates return to their pre-eclipse levels. If the stable low energy electrons were part of an ambient plasma, rather than photoelectrons, the counting rates would not be so radically altered as the moon moved across the Earth's shadow.

It could be argued that the observed counting rates were due to photons scattering within the detectors themselves and not due to external photoelectrons. This however is not the case. Preflight calibrations with a laboratory ultraviolet source showed enhanced counting rates only when the angle between the look direction of the detector and the source was less than 10°. Given the 60° separation between the look directions of analyzers A and B, it would be impossible for the Sun, essentially a point source, to produce identical counting rates in both analyzers simultaneously.

A typical spectrum of photoelectrons shown in Figure 3 was observed by analyzer

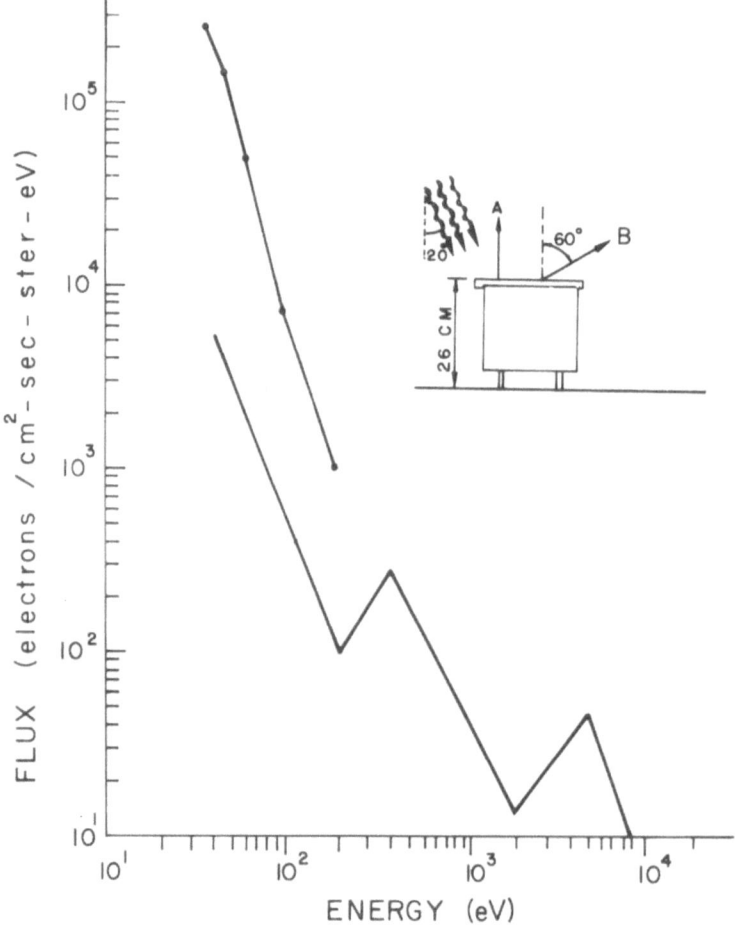

Fig. 3.   Typical photoelectron spectrum observed by CPLEE at the lunar surface in the high latitude magnetotail.

A at ~ 0400 h on February 10, shortly before the Moon entered penumbral eclipse. The dark line marks the differential flux equivalent to a background count of one per cycle in each channel. (Channels 1–5.) For all five channels, with the deflection plates at -35 V, the differential flux is well above this background level. During geomagnetically quiet times no statistically significant counts are observed when the deflection plates are at -350 or -3500 V corresponding to electrons with $E > 500$ eV (Burke and Reasoner, 1972).

With the exception of periods of ultraviolet contamination in analyzer A, we always observe nearly the same counting rate due to photoelectrons in analyzers A and B. For all purposes, we can say that the spectrum displayed in Figure 3 is just as typical as for analyzer B. We have found no case of anisotropy in the photoelectron fluxes. In all cases too, we found that the photoelectron spectra observed in both analyzers were close to a power law dependence on energy. If we write the differential flux in the form $j(E) = j_0 (E/E_0)^{-\mu}$, μ is between 3.5 and 4. In the following section the details of this spectrum are more carefully studied.

Also in Figure 3 we display a schematic cross section of our instrument as it is deployed on the surface of the Moon. The apertures of both analyzers are elevated 26 cm from ground. Their geometry is such that they observe only electrons with a component of velocity in the downward direction. Since we continually observe photoelectrons with energies up to ~ 200 eV, we must assume that the lunar surface potential is at least 200 V during these times. This measurement will seem high to those familiar with the work of Walbridge (1970) and Grobman and Blank (1969), who calculate a surface potential that is at least an order of magnitude lower. The difference is that their models deal with photoemissions from the surface of the Moon in the presence of the solar wind. Our measurements in the magnetotail are made under near-vacuum conditions. After further analysis of the problem we return to considerations of the surface potential.

To summarize: During geomagnetically quiet times, when the Moon is in the magnetotail and not in eclipse, stable photoelectron fluxes with energies between 40 and 200 eV are observed. These fluxes are isotropic and obey a power law, $E^{-\mu}$, where μ is between 3.5 and 4. From the fact that CPLEE is observing downward moving electrons we conclude that in the high-latitude magnetotail the lunar surface potential is on the order of 200 V.

4. Numerical Analysis

4.1. GENERAL THEORY

The variations of photoelectron density and electrostatic potential above the surface of the Moon can be calculated numerically. We approximate the lunar surface by an infinite plane, with the x direction normal to the surface, and assume spatial variations of physical quantities only with the height.

At a height x above the surface the electron density is $\int f(\mathbf{v}, x)\, d^3v$. $f(\mathbf{v}, x)$ is the electron distribution function. If we assume an isotropic flux at the surface, the

Liouville Theorem can be used to show that the distribution function is independent of angles at all heights. Writing

$$d^3v = \sqrt{\frac{2E}{m^3}}\, dE\, d\Omega$$

and integrating over solid angles, the density is

$$n(x) = 4\pi \int_0^\infty \sqrt{\frac{2E}{m^3}}\, f(E, x)\, dE. \tag{1}$$

Since the distribution function is a constant along particle trajectories, $f(E, x) = f(E_0, x=0)$, where $E = E_0 - q[\varphi_0 - \varphi(x)]$. By changing the variable of integration from E to E_0 Equation (1) can be expressed

$$n(x) = 4\pi \int_{q[\varphi_0 - \varphi(x)]}^\infty \sqrt{2m(E_0 - q[\varphi_0 - \varphi(x)])}\, f(E_0, x=0)\, dE_0. \tag{2}$$

To calculate the distribution function of photoelectrons at the surface consider the quantity

$$j(E_0)\, dE_0 = \left[\int I(hv)\, Y(hv)\, \varrho(E_0, hv)\, dhv \right] dE_0 \tag{3}$$

the upward moving flux of photoelectrons emitted from the surface with energies between E_0 and $E_0 + dE_0$. $I(hv)\, d(hv)$ is the flux of photons reaching the lunar surface with energies between hv and $hv + d(hv)$. $Y(hv)$, the quantum yield function, gives the number of electrons emitted by the surface per incident photon with energy hv. $\varrho(E_0, hv)\, dE_0$ is the probability that an electron emitted from the surface, due to a photon with energy hv, will have a kinetic energy between E_0 and $E_0 + dE_0$. $\varrho(E_0, hv)$ is normalized so that

$$\int_0^\infty \varrho(E_0, hv)\, dE_0 = 1.$$

The total upward moving flux at the surface is $S_\uparrow(x=0) = \int_0^\infty j(E_0)\, dE_0$. But

$$S_\uparrow(x=0) = \int_0^{2\pi} \int_0^{\pi/2} \int_0^\infty v_0 f(E_0, \theta, \varphi, 0)\, v_0^2\, dv_0\, \sin\theta\, d\theta\, d\varphi.$$

Since $v_0 = v_0[\mathbf{i} \cos\theta + \mathbf{j} \sin\theta \cos\varphi + \mathbf{k} \sin\theta \sin\varphi]$ and f is independent of angle,

$$S(x=0) = \pi \int_0^\infty \frac{2E_0}{m^2} f(E_0, x=0)\, dE_0. \tag{4}$$

Thus

$$f(E_0, x = 0) = \frac{m^2 j(E_0)}{2\pi E_0} \tag{5}$$

and

$$n(x) = 2 \int_{q[\varphi_0 - \varphi(x)]}^{\infty} \sqrt{2m(E_0 - q[\varphi_0 - \varphi(x)])} \frac{j(E_0)}{E_0} dE_0. \tag{6}$$

The potential as a function of height is evaluated by multiplying Poisson's equation, $\partial^2\varphi/\partial x^2 = -4\pi q n(x)$, by $\partial\varphi/\partial x$ and integrating in from $x = \infty$ to get

$$\left(\frac{\partial\varphi}{\partial x}\right)^2 = -8\pi q \int_{\varphi(x)}^{0} n(\varphi') d\varphi', \tag{7}$$

where we have written

$$\int_{x}^{\infty} n(x') \frac{\partial\varphi}{\partial x'} dx' = \int_{\varphi(x)}^{0} n(\varphi') d\varphi'.$$

A further integration out from the surface, gives us the potential at a point x.

4.2. COMPUTATIONAL METHODS AND RESULTS

To determine the upward moving differential flux at the surface, upon the knowledge of which the distribution function, number density and potential depend, we must first calculate the integral in Equation (3). The solar photon differential flux at 1 AU, $I(hv)$, is taken from Friedman (1963) for the range 2000 to 1800 Å and from Hinteregger (1965) from the range 1775–1 Å and is plotted in Figure 4. Following the suggestion of Walbridge (1970) we have:

(1) Adopted a work function W of lunar material of 6 eV.

(2) Assumed a photoelectron yield function of the form

$$Y(hv) = \begin{cases} 0 & hv < 6 \ eV \\ Y_0 \dfrac{hv - 6}{3} & 6 \leqslant hv < 9 \ eV \\ Y_0 & hv \geqslant 9 \ eV, \end{cases} \tag{8}$$

where Y_0 is a free parameter of our calculation.

(3) Chosen a probability function

$$\varrho(E, hv) = \begin{cases} 6E(E_1 - E)/E_1 & 0 \leqslant E \leqslant E_1 \\ 0 & E > E_1, \end{cases} \tag{9}$$

where

$$E_1 = \begin{cases} hv - W & hv \geqslant W \\ 0 & hv < W. \end{cases}$$

In general the probability function is a complicated function depending on the nature of the photoemission material. However, Grobman and Blank (1969) have shown that for the purpose of calculating Equation (3) any broad function with zeros at $E=0$ and $E=E_1$ and a width $\Delta E \sim h\nu$ will suffice. A plot of $\varrho(E, h\nu)$ is shown in Figure 5 for various value of E_1.

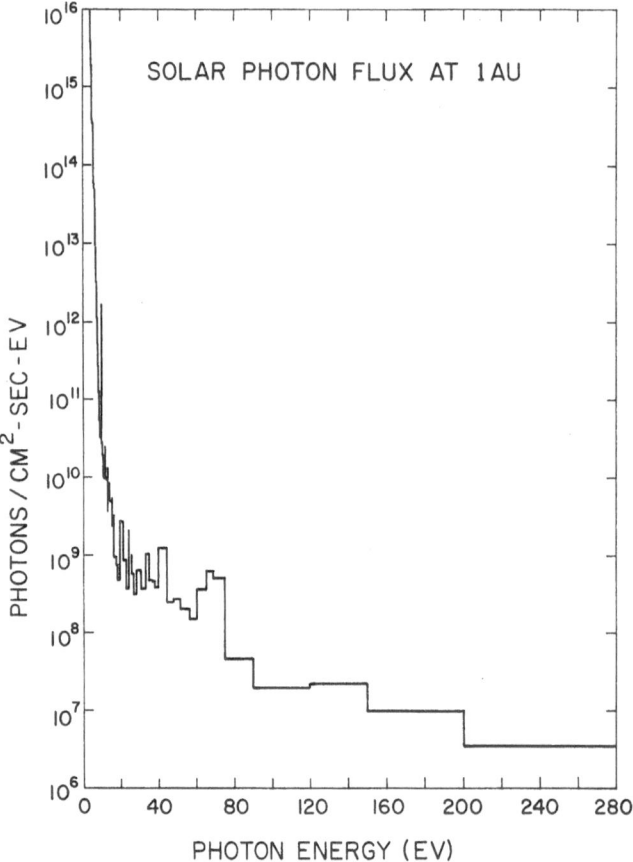

Fig. 4.   Solar photon energy spectrum at 1 AU from 2000 to 1 Å.

The upward directed differential flux in electrons $\mathrm{cm}^{-2}\,\mathrm{s}^{-1}\,\mathrm{sr}^{-1}\,\mathrm{eV}^{-1}$ for the values $Y_0 = 1$, 0.1, 0.01 were numerically computed and have been plotted in Figure 6. We have also inserted the photoelectron differential flux observed by CPLEE at 26 cm. The Liouville theorem allows us to set a lower bound on Y_0 of 0.1. That is if there were no potential difference between the surface and 26 cm the yield function would be 0.1 electrons photon^{-1}. After estimating the potential difference between 26 cm and the surface we can also determine an upper bound on Y_0.

Solving the integro-differential Equation (7) for $\varphi(x)$ involves an integration from the surface outward, with an assumed value of φ_0. However the expression for $\partial\varphi/\partial x$ involves an integral from infinity in to x, or equivalently from $\varphi=0$ to $\varphi(x)$. By the expedient of dividing the integral into pieces in E_0 space and using an analytic approximation to the function $j(E_0)$ in each of these intervals, a solution was effected.

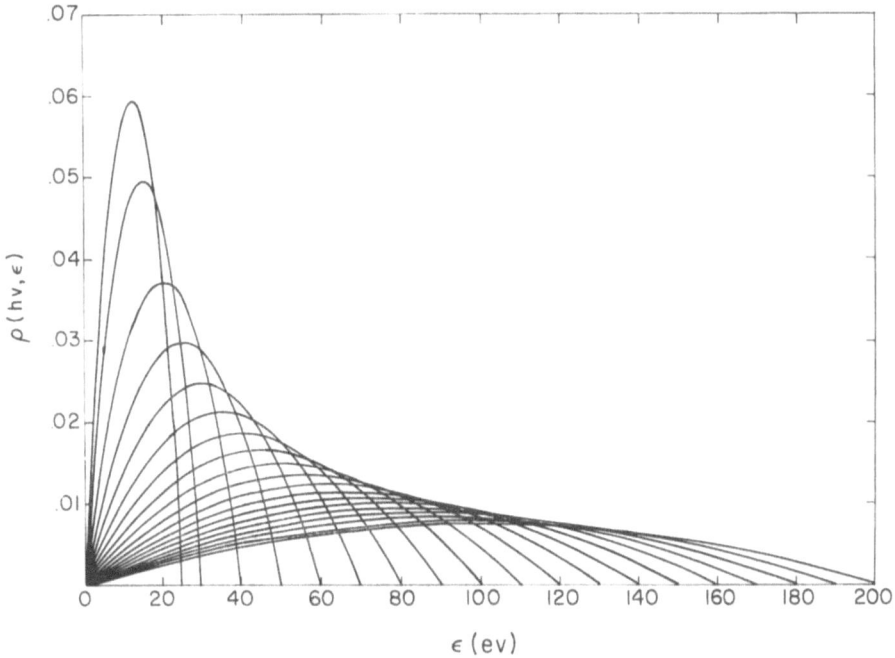

Fig. 5. Probability function that a photon of energy hv will cause the lunar surface material to emit a photoelectron of energy E with different values of $E_1 = hv - W$.

In this way it was only necessary to know the values of $\varphi=\varphi(x)$ and $\varphi=0$ at the end points of the interval, and the solution would proceed. In Figure 7 we show families of solutions for $\varphi(x)$ with several values of the parameter Y_0.

The value of Y_0 calculated by assuming no potential difference between the surface and $x=26$ cm was 0.1. Figure 7 shows that for $Y_0=0.1$, the potential difference $\Phi(x=0)-\Phi(x-26$ cm$)$ is only 3 V. Obviously, we could now use an iterative procedure, modifying our spectral measurement at 26 cm to obtain the surface spectrum according to the Equation $f(E, x)=f(E_0, 0)$ and hence obtain a new estimate of Y_0. However, the procedure is hardly justified considering the small potential difference (~ 3 V) and the energy range of the measured photoelectrons (40–200 eV). Hence we conclude from our measured photoelectron fluxes and numerical analysis a lunar surface potential of at least 200 V and a value of the average photoelectron yield of $Y_0=0.1$ electrons photon^{-1}.

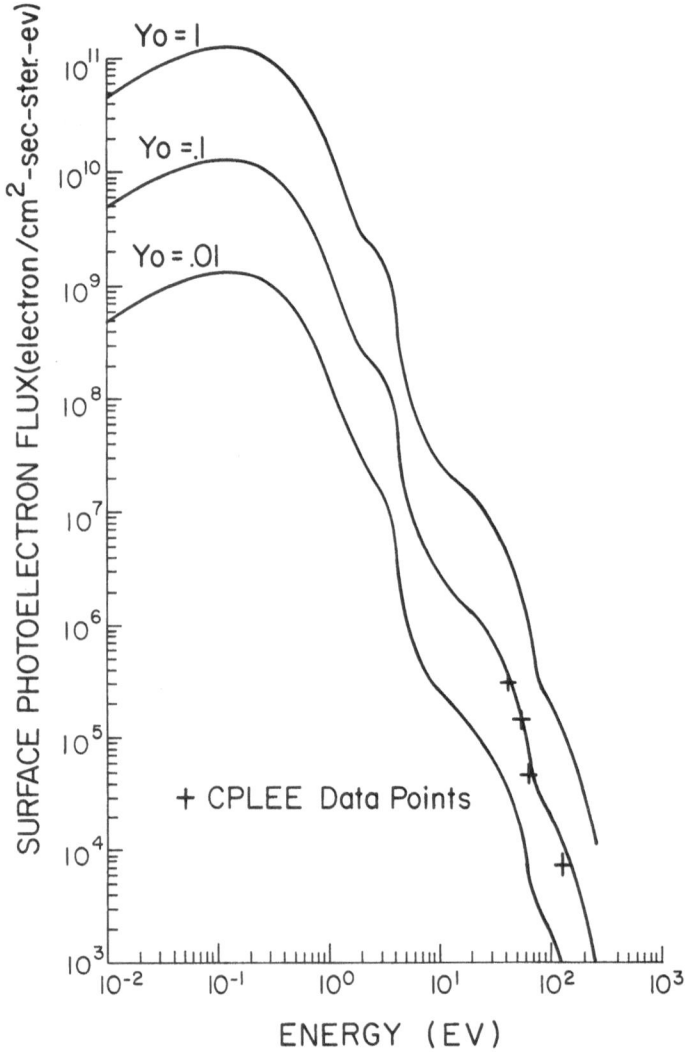

Fig. 6. Numerically computed photoelectron spectra emitted for the yield functions $Y_0 = 1$, 0.1 and 0.01 electrons per photon. The photoelectron spectrum measured by CPLEE is found to fall close to the $Y_0 = 0.1$ line.

For the sake of completeness, Figure 8 shows the variation of potential and photoelectron density and pressure with height obtained from the numerical solution discussed above. For this solution, a value of $Y_0 = 0.1$ and $\varphi_0 = 200$ V were chosen. It may be tempting to define a Debye length for the photoelectron gas according to the formula

$$\lambda = \left(\frac{P_0}{2\pi n^2 q^2} \right)^{1/2} = 4 \text{ cm} \quad \text{at} \quad x = 1 \text{ cm} .$$

However, this characteristic length is by no means the distance over which the entire potential drop is developed. It is important to keep in mind that the photoelectron layer is not a plasma, but rather is a one-component gas and hence the concept of a Debye length as a potential shielding distance is not applicable.

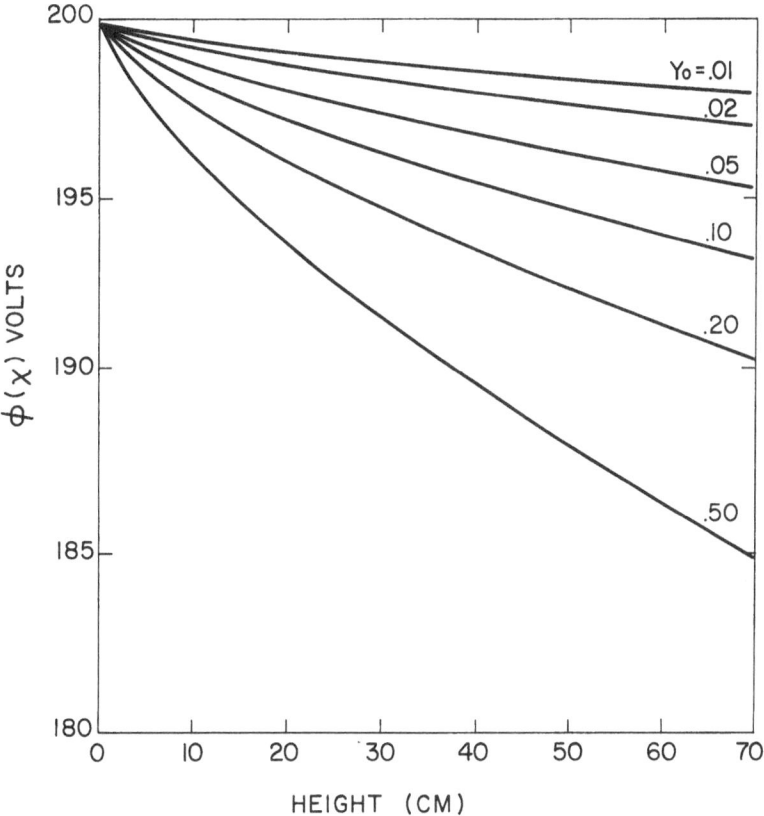

Fig. 7. Numerically computed potential distribution above the lunar surface for several values of the yield function Y_0. For $Y_0 = 0.1$ the potential difference between ground and 26 cm is about 3 V.

5. The Lunar Surface Potential φ_0

The experimental measurements of photoelectrons at 200 eV but no significant fluxes in the next highest energy channel at 500 eV lead us to conclude that the lunar surface potential is at least 200 V. The data of Hinteregger *et al.* (1965) shows significant solar photon fluxes up to 400 eV, and presumably the lunar surface potential under vacuum conditions could be 400 V. However, we detected no photoelectrons with $E > 400$ eV, and in fact the extrapolation of our measured spectrum (Figure 3) to 400 eV is below the instrument background. For this reason therefore we have adopted a conservative value of 200 V as the lunar surface potential for purposes of the calculations in the preceding and following sections.

The lunar surface potential can be decreased however by the presence of a hot ambient plasma which furnishes an electron return current which partially balances the emitted photoelectron current. In effect, the highest energy photoelectrons can escape from the potential well, since electrons from the ambient plasma furnish the return

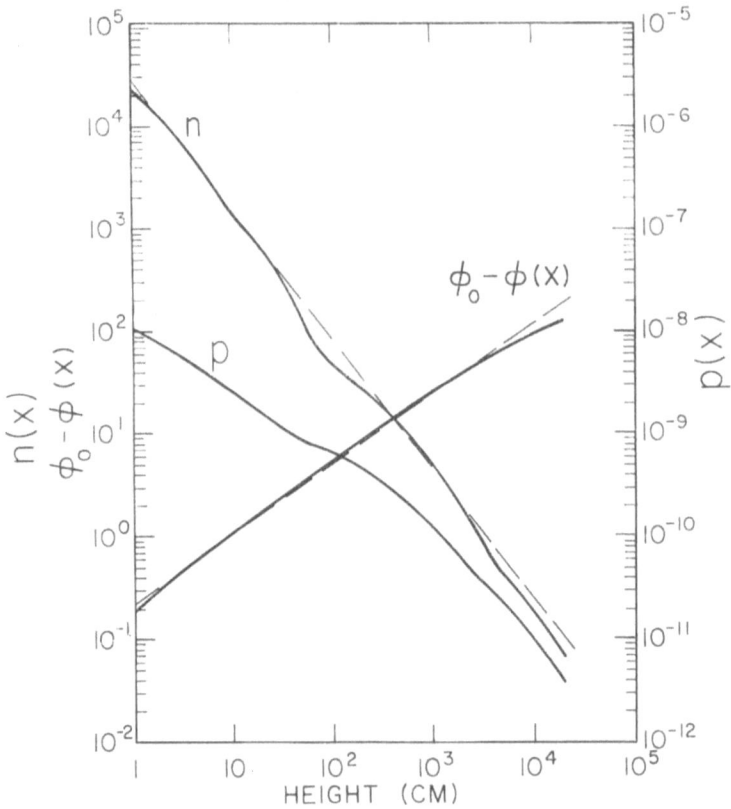

Fig. 8.   Numerically computed values of electron density, potential difference and pressure as a function of distance from the lunar surface.

current to balance these escaping photoelectrons. Quantitatively, if F_s is the net negative flux to the lunar surface from the ambient plasma, and $j(E_0)$ is the emitted photoelectron energy spectrum in units of electrons cm^{-2} s^{-1} eV^{-1}, then:

$$F_s = \int_{E_{\parallel} > q\varphi_0}^{\infty} j(E_0)\, dE_0 \tag{10}$$

and this equation can be solved for φ_0, the lunar surface potential. The results of this calculation for $30\ V < \Phi_0 < 200\ V$ are shown in Figure 9. The curve was computed for $Y_0 = 1.0$, but can be scaled for other values of Y_0.

Our measurements of photoelectrons were taken during periods in the magnetotail when all of the channels of the instrument except the lowest-energy electron channels were at background levels. Thus we can establish an upper limit to the electron flux from the ambient plasma for electrons with $40 \text{ eV} < E < 50 \text{ keV}$. Figure 3 shows the 'background spectrum', calculated by converting the background counting rate of

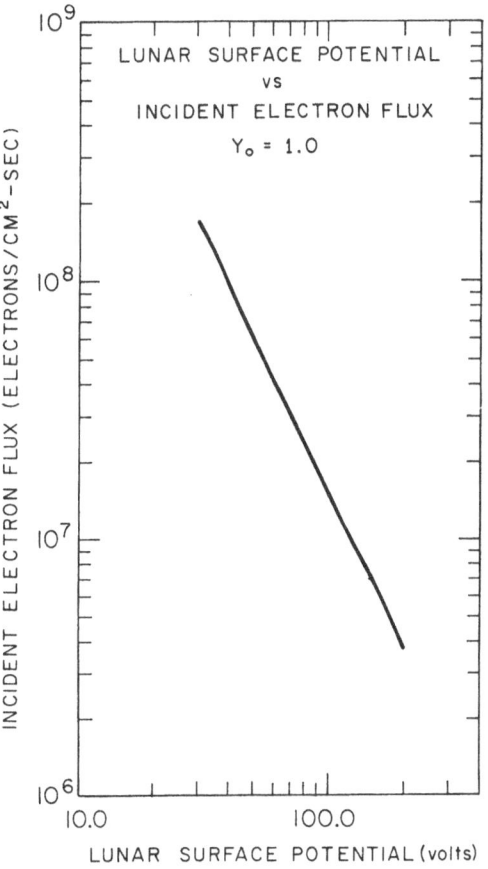

Fig. 9. Computed curve, from the emitted photoelectron spectra shown in Figure 6, of the lunar surface potential versus incident electron flux for $Y_0 = 1.0$. The curve can be scaled directly for any other value of Y_0.

~ 1 count s^{-1} to equivalent flux in each of the energy channels. Integrating over this spectrum and converting to flux over the hemisphere gives $F_s \leqslant 3.4 \times 10^6$ electrons $\text{cm}^{-2} \text{ s}^{-1}$. We feel that this is a valid upper limit, as the range of measurement in energy includes the peak energy of the plasma sheet spectrum (~ 1 keV).

We note that Vasyliunas (1968) obtained an upper limit to the electron concentration for locations outside of the plasma sheet based on OGO-3 data. The relation expressed was $NE_0^{1/2} < 10^{-2} \text{ cm}^{-3} \text{ keV}^{1/2}$ where N is the electron density and E_0 is the energy at the peak of the spectrum. For an isotropic plasma where the bulk motion can be neglected relative to the thermal motion, the electron flux to a probe is given by $F_s = N\bar{v}/2\pi^{1/2}$. Applying the appropriate conversion of factors, the expression of Vasyliunas results in an upper limit to the electron flux of $F_s < 5.6 \times 10^7$ electrons $cm^{-2} s^{-1}$.

TABLE I

Electron flux $(cm^{-2} s^{-1})$	Y_0	$\Phi_0(V)$
3.4×10^6	1.0	181
3.4×10^6	0.1	114
3.4×10^6	0.01	44
5.6×10^7	1.0	96
5.6×10^7	0.1	36
5.6×10^7	0.01	8

In Table I we show results of surface potential computations for the two electron flux upper limits given above and for values of Y_0 of 1.0, 0.1, and 0.01.

The lower half-height of the channel 5 energy passband is 160 eV. Hence the surface potential could be as low as 160 V and still result in particle fluxes in channel 5. This estimate of the potential is seen to be not inconsistent with a value of $Y_0 = 0.1$, $F_s \leqslant 3.4 \times 10^6$ resulting in a surface potential (Table I) of 114 V.

One rather obvious prediction of our arguments about the surface potential is that when the electron flux reaching the surface of the Moon is sufficiently high, surface generated photoelectrons with energies in the range of our detector should vanish. This happens in the solar wind and magnetosheath, but it is impossible for CPLEE to provide conclusive observational evidence in the presence of contamination by solar wind and magnetosheath electrons that the $E > 40$ eV photo electrons are not returning to the surface and hence entering the detectors. Electron densities > 1 cm^{-3} and temperatures ~ 15 eV such as those commonly encountered in the solar wind and magnetosheath provide much higher fluxes in the 40–200 eV range than photoelectron fluxes observed in high latitude magnetotail.

On April 9, 1971 a world-wide magnetic storm was observed by CPLEE (Burke *et al.*, 1972) in which the magnetosheath moved in to $Y_{SE} = 15 R_E$. As the magnetopause moved out past the Moon, intense plasma sheet fluxes were observed. At this time the fluxes observed in the 40, 50 and 70 eV channels fell below photoelectron levels. In Figure 10 we show the count rates for the 40 eV and 500 eV electron channels from 11:00 to 12:00 on April 9. A heavy line has been drawn at the 40 eV photoelectron level. As the 500 eV count rate rises the 40 eV count rate falls below this line. From 11:28 to 11:35 when the 500 eV count rate dropped the 40 eV channel returned to the photoelectron level. In Figure 11 we have plotted the electron spectra at 11:26:40 (photoelectron) and at 11:27:19 (plasma sheet). The total incident fluxes, calculated by subtracting the photoelectron contribution and by assuming isotropy over the upper hemisphere are indicated for each of the two spectra. The data show that the lunar surface photoelectron yield in the range 40 eV $< E < 200$ eV is of sufficient

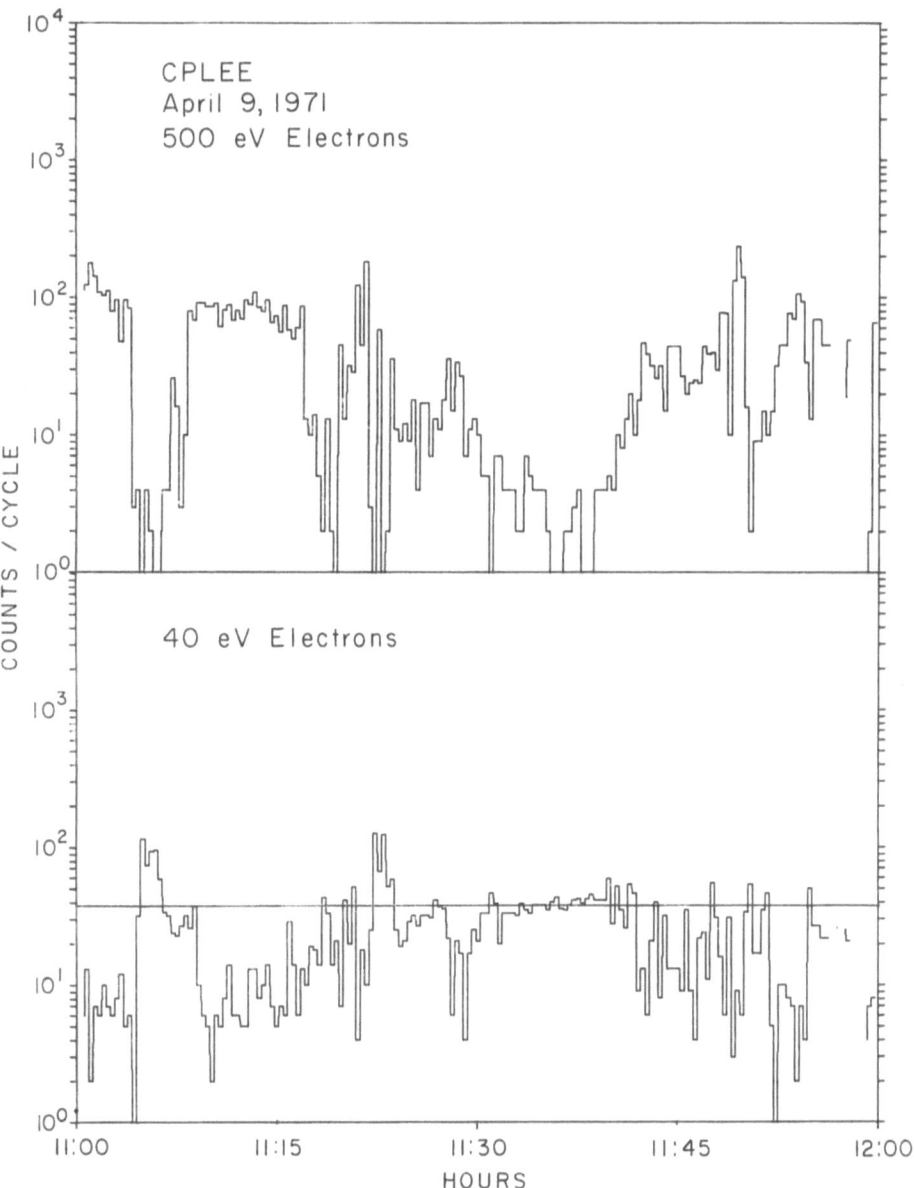

Fig. 10. Comparison of the counting rates due to 40 eV and 500 eV electrons for the period 11:00 to 12:00 on April 9, 1971. The horizontal line on the 40 eV plot is the normal photoelectron level. Note the anti-correlation between photoelectron flux and higher-energy electron flux, indicating modulation of the lunar surface potential by plasma sheet electron fluxes.

magnitude to maintain the potential at 40 V for an incident flux of 2.5×10^7 electrons cm^{-2} s^{-1}. This can be used with the curve of surface potential vs incident electron flux (Figure 9) to compute a value of $Y_0 = 0.3$. The discrepancy between this value and the value calculated previously ($Y_0 = 0.1$) is probably due to an error associated with the assumption of isotropy of the incident flux. If the incident flux were smaller at large zenith angles, the total flux and hence the value of Y_0 computed by this method would be correspondingly less.

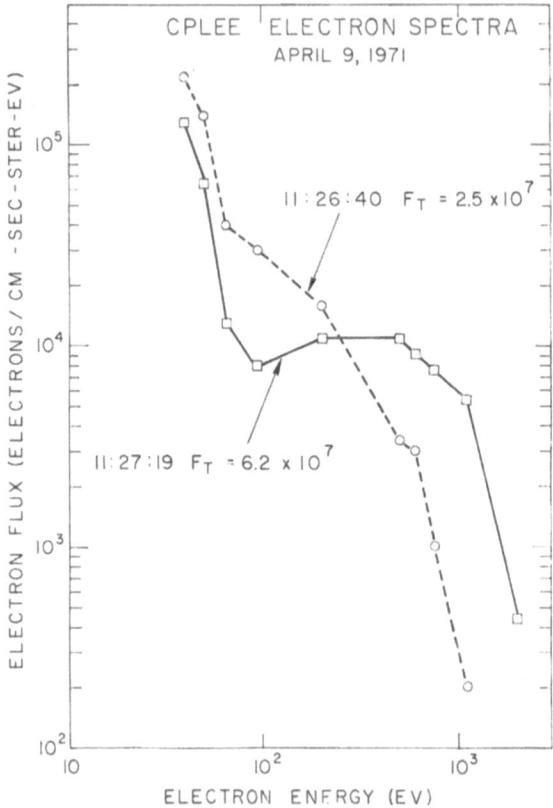

Fig. 11. Electron spectra just prior to and after the photoelectron decrease at 11:27 shown in Figure 10. These spectra show in detail the modulation of the lunar surface potential by an incident electron flux impinging upon the lunar surface. The total fluxes, F_t, were calculated by first subtracting the photoelectron contribution and then assuming that the higher energy electron flux was isotropic over the upper hemisphere.

6. Discussion

Feuerbacher *et al.* (1972) have measured the photoelectron yield of a lunar fine sample. In the photon energy range $5 < E < 20$ eV they found a yield function which reaches a maximum value of ~ 0.08 at 15 eV then drops to 0.01 at 20 eV. B. Fitton (private communication, 1971) has suggested that our value of 0.1 is more a measurement of the CPLEE instrument case yield function than that of the lunar surface. We find it difficult to understand how this could be the case.

First, were CPLEE an electrically isolated package then the requirement that the net current to the instrument be zero would result in a measured yield function that is representative of the case material. Relative to the lunar surface the case would bear a positive potential in order to maintain an enhanced photoelectron density in its immediate vicinity.

The fact is, however, that CPLEE is not electrically isolated. CPLEE is connected to the central ALSEP station. Further, at any given time, only the top surfaces and one or two sides of ALSEP packages are illuminated by the Sun, while the remaining area is shadowed. These unilluminated surfaces provide receptor areas for return current from the photoelectron gas. Thus no large potential difference can develop between CPLEE and the lunar surface, and photoelectrons emitted at the lunar surface and at the CPLEE case are indistinguishable. Geometrical considerations of electron trajectories would lead us to expect that the bulk of the photoelectrons measured by CPLEE were emitted at the lunar surface at least several meters from CPLEE. If the ALSEP instrument cases had a photoelectron yield much larger or much smaller than the yield of the lunar surface, then one would expect a perturbation of the photo-electron flux in the vicinity of ALSEP. This perturbation would depend not only on the ratio of photoelectron yield, but more importantly on the ratio of the area of the ALSEP instruments to the area encompassed by the trajectories of $E > 40$ eV photo-electrons from a point source. This last area is on the order of the square of the scale height of photoelectrons with $E > 40$ eV (~ 10 m, Figure 11). Since the ALSEP area is ~ 3 m^2, the ratio of areas is on the order of 2%. Thus, even if the yield of the instrument cases was a factor of 10 greater or smaller than the surface yield, the flux perturbation would only be 20%.

Second, were CPLEE measuring its own photoelectrons one would expect to observe changes in the relative fluxes observed in the two analyzers with solar zenith angle as the electron cloud surrounding the instrument adjusts to changing illumination conditions. Specifically, the ratio of the flux observed analyzer B (looking 60° west of vertical) to the flux observed in analyzer A (looking to the vertical) should be larger after than before lunar noon. Our data shows isotropic photoelectron fluxes across the entire magnetotail.

7. Summary and Conclusions

In this paper we have reported the observation of stable, isotropic photoelectron fluxes 26 cm above the lunar surface. In the energy range $40 \leqslant E \leqslant 200$ the flux obeys a power law of the form $j(E) = j_0 (E/E_0)^{-\mu}$ where μ is between 3.5 and 4. Because these fluxes were moving down we conclude that in the near vacuum conditions of the high latitude magnetotail the lunar surface potential is at least 200 V. The modulation of the surface potential in the presence of intense plasma sheet fluxes has also been observed. It was shown that these electrons can be explained in terms of the measured solar photon spectrum producing an isotropic flux of photoelectrons at the surface. A photoelectron yield function of $Y_0 = 0.1$ electron photon^{-1} was calculated.

Acknowledgements

We gratefully acknowledge the assistance of Wayne Vogel and Patricia Moore who developed the computer programs used in the analysis.

This work was supported by National Aeronautics and Space Administration Contract No. NAS 9-5884.

References

Burke, W. J. and Reasoner, D. L.: 1972, *Planetary Space Sci.* **20**, 429.
Burke, William J., Rich, Frederick J., Reasoner, D. L., Colburn, D. S., Goldstein, B. E., and Lazarus, A. J.: 1972, 'Effects on the Geomagnetic Tail at 60 R_e of the World-Wide Storm of April 9, 1972', Rice University Preprint, to be published.
Feuerbacher, B., Anderegg, M., Fitton, B., Laude, L. D., and Willis, R. F.: 1972, in C. Watkins (ed.), *Lunar Science III*, Lunar Science Institute No. 88, p. 253.
Friedman, H.: 1963, in D. P. LeGallet (ed.), *Space Science*, John Wiley & Sons, New York, p. 549.
Grobman, W. D. and Blank, J. L.: 1969, *J. Geophys. Res.* **74**, 3943.
Guernsey, R. L. and Fu, J. H. M.: 1970, *J. Geophys. Res.* **75**, 3193.
Hinteregger, H. E., Hall, L. A., and Schmidtke, G.: 1965, in D. K. King-Hele, P. Muller, and G. Righini (eds.), *Space Research V*, North Holland Publishing Co., Amsterdam, p. 1175.
Medved, D. B.: 1968, in S. F. Singer (ed.), *Interaction of Space Vehicles with an Ionized Atmosphere*, Pergamon Press, Oxford, p. 305.
O'Brien, B. J. and Reasoner, D. L.: 1971, *Apollo 14 Preliminary Science Report*, NASA SP-272, 193.
Vasyliunas, V. M.: 1968, *J. Geophys. Res.* **73**, 2839.
Walbridge, E.: 1970, 'The Lunar Photoelectron Layer. 1. The Steady State', preprint.

DISCUSSION

Feuerbacher: Did you take into account the photoemission from your box?

Reasoner: The resolution of this question is a matter of the contact between the instrument cases and the local surface potential. It is certainly true that if the instrument case were truly isolated, then $\Delta J = 0$ would immediately imply that the measured photoelectrons were indicative of the instrument case yield. However, there is more area in electrical contact with the instrument that is not illuminated than is illuminated, and therefore these 'cold' areas can act as receptor areas to collect current from the photoelectron gas, and thus prevent large potential differences from developing between the instrument and the lunar surface. Granting this to be the case, then, and recalling that photoelectrons from both a vertical direction and a direction 60° from vertical were measured with equal intensities at all Sun angles, we therefore conclude that the measurement is a true representation of the lunar surface photoelectron flux.

Walker: Let me get this clear: what is the basic discrepancy between Reasoner's measurement and the laboratory measurements of the ESTEC group?

Feuerbacher: I do not think that there is a basic discrepancy between our results and those of Dr Reasoner. Our surface potentials have been calculated for solar wind conditions, while the CPLEE data refer to measurements in the high latitude magnetotail, where the incoming plasma flux is negligible. Our experimental data cannot possibly be extrapolated to electron energies of 200 eV, since the highest photon energy used was 23 eV.

Fitton: I would like to refer to Figure 1. Could you indicate the likely effect of the cover sheet, on the right hand side of the figure, so far as the properties of the local photoelectron sheath measured by your instrument is concerned?

Reasoner: The photograph of Figure 1 shows the CPLEE, the ALSEP central station and a portion of the thermal blanket of the passive seismic experiment (PSE). This thermal blanket is to the best of my recollection composed of many layers of aluminized mylar and is located approximately 1.5 m from CPLEE. The top surface of the blanket is, however, not in electrical contact with the ALSEP system. Furthermore, photoelectrons emitted from this surface would produce highly anisotropic fluxes at the instrument, and this is not observed.

PHOTOEMISSION AND SECONDARY ELECTRON EMISSION FROM LUNAR SURFACE MATERIAL

R. F. WILLIS, M. ANDEREGG, B. FEUERBACHER, and B. FITTON

Surface Physics Division, European Space Research Organisation, Noordwijk, The Netherlands

Abstract. The photoelectric yield, secondary electron yield and energy distribution of low energy electron emission from Apollo 14 and 15 dust samples are reported for photon energies, 4 to 21 eV, and incident electron energies, 50 to 2500 eV. The differential photoelectron flux and energy distribution spectrum due to solar irradiation are derived, and the results are used to estimate the electrical characteristics of the lunar photoelectron sheath. Auger electron spectroscopic analysis of the elemental composition of the lunar dust samples is described, and a comparison is made with the results obtained by the S-161 X-ray fluorescence experiment on-board the Apollo 15 Command Service Module, while in lunar orbit. The results are consistent with those expected for a finegrain, predominantly insulating, powder.

1. Introduction

Calculations have shown that the steady-state electrostatic potential and charge distribution above the sunlit lunar surface is determined primarily by the photoemissive properties of the lunar surface materials and the impacting solar wind flux. Only a fraction of the emitted photoelectrons can escape as the sunlit lunar surface attains a small positive electrostatic potential, the magnitude of which is determined by the number and energy distribution of the photoemitted electrons and the incident solar wind particle energy distribution and density (Öpik and Singer, 1960, 1962). Secondary electron emission (SEE), due to the impact of solar wind particles, had previously been assumed to be negligible as a consequence of the low energies of solar wind electrons (<10 eV) protons (~ 1 keV) (Grobman and Blank, 1969; Brandt, 1970). However the measurements by the charged particle lunar environment experiment (CPLEE) (Reasoner and Burke, 1972) placed on the lunar surface during the Apollo 14 mission revealed the incidence of electrons in the energy range from 300 eV to several keV respectively, during the Moon's passage through the magnetosphere. These higher energy electrons will give rise to secondary electron emission at the lunar surface, which in turn, could well effect the lunar surface charge and potential significantly, depending on the SEE-characteristics of lunar surface materials. The photoelectron density will be many orders of magnitude larger than the density of secondary electrons over the sunlit surface. It is conceivable however that SEE may play a significant role on the dark-side of the Moon where magnetospheric particles impact the surface in the shadow region (Knott, 1973).

Quantitative estimates of the parameters which define the lunar-surface electron sheath have been frustrated to some extent by a lack of data on the photoelectric and secondary electron emission properties of the lunar surface. In this paper, measurements of the photoelectric yield, secondary electron yield and energy distribution of low energy electron emission from Apollo 14 and 15 dust samples are reported for photon energies, 4 to 21 eV, and incident electron energies, 50 to 2500 eV. The

R. J. L. Grard (ed.), Photon and Particle Interactions with Surfaces in Space, 389–401. All Rights Reserved
Copyright © 1973 by D. Reidel Publishing Company, Dordrecht-Holland

differential photoelectron flux spectrum and energy distribution due to solar irradiation is derived, and the results are used to estimate the electrical characteristics of the lunar photoelectron sheath. Auger electron spectroscopic (AES) analysis of the elemental composition of the lunar dust samples is also described and compared with the results obtained by the S-161 X-ray fluorescence experiment on-board the Apollo 15 Command Service Module while in lunar orbit (Adler *et al.*, 1972). The results are consistent with those expected for an insulating dust layer, the average elemental composition of which may vary in a systematic manner between the highlands and maria regions of the lunar terrain (Adler *et al.*, 1972).

2. Experimental

The samples were handled in a dry nitrogen atmosphere, and the measurements were made in a vacuum between 10^{-7} and 10^{-9} torr, using a hemispherical collector system and retarding-field, voltage-modulation energy analysis, details of which have been described (Feuerbacher *et al.*, 1972; Willis *et al.*, 1972). Bakeout of the ultrahigh vacuum system and sample was carried out at 150°C for 1 week in order to simulate lunar thermal conditions and to reduce contamination by adsorbed gases. In situ sputter cleaning by Ar ion bombardment was not used in view of chemical reduction effects, which are known to occur (Hapke *et al.*, 1970).

3. Photoemission Properties

3.1. Photoelectric yield

The differential photoelectric yield, $Y(\omega)$, of samples from the Apollo 14 and 15 missions is shown in Figure 1. The number of emitted electrons per incoming photon is shown as a function of wavelength for the range, 500 Å to 2500 Å ($\simeq 5$ to 21 eV). The photoelectric threshold (or work function) was measured (Feuerbacher *et al.*, 1972) and found to be 5 eV (2480 Å), which is rather high compared with values of the order of 4.5 eV, characteristic of many common materials. A maximum yield value of about $9 \pm 3\%$ is observed for both samples at 900 Å wavelength. This is low compared with values observed for bulk metals (Au$\simeq 14\%$) and insulators (Al$_2$O$_3 \simeq 30\%$). The yield curve from threshold to about 500 Å (25 eV) is shown in Figure 1. The differences between samples from the Apollo 14 and 15 mission, indicated by the error bars, are of the same size as variations in the yield for identical samples due to different compactness in the sample holder. An interesting feature of the yield curves (Figure 1) is the decreasing yield values observed for wavelengths shorter than the peak value at 900 Å (14 eV). This would indicate that the lunar dust has a maximum yield value of about 10% at this wavelength falling to $\lesssim 1\%$ for wavelengths shorter than 584 Å (21.2 eV). In contrast, bulk insulators exhibit yield curves which increase continuously over this range, 900 Å to 584 Å.

  The differential photoelectron flux under solar irradiation is obtained by multiplying the observed differential photoelectric yield, Figure 1, and the solar energy flux

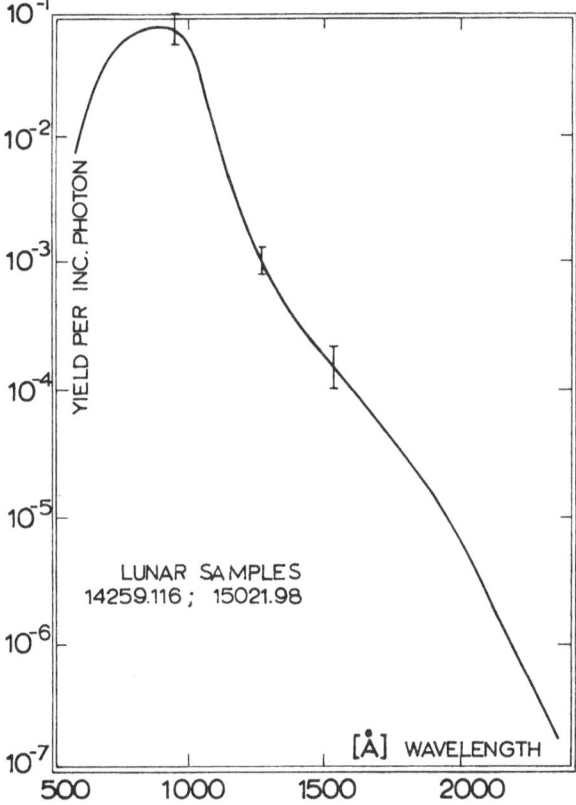

Fig. 1. Differential photoelectric yield, $Y(\omega)$, of lunar dust samples 14259,116 and 15021,98.

spectrum (Hinteregger *et al.*, 1964; Tousey, 1966):

$$H(\omega) = S(\omega) \cdot Y(\omega). \tag{1}$$

The histogram of Figure 2 representing $H(\omega)$ was determined from the yield values $Y(\omega)$, obtained for the Apollo 14 sample and shows the contributions of the various regions of the solar spectrum to the total photoelectron flux. Two spectral regions dominate the curve. One peaks around 6 eV, as a result of the rapid increase of the solar photon flux towards lower energies (Hinteregger *et al.*, 1964; Tousey, 1966) combined with the fall-off in yield towards the photoelectric cutoff, $W = 5$ eV (Feuerbacher *et al.*, 1972). The second region centred about 14 eV arises from the maximum in the differential photoelectric yield at 900 Å, Figure 1. Integration of the differential photoelectron flux curve, Figure 2, with respect to photon energy gives the saturation photoelectron flux under solar irradiation, I_s:

$$eI_s = e \int_0^\infty H(\omega) \, d\omega. \tag{2}$$

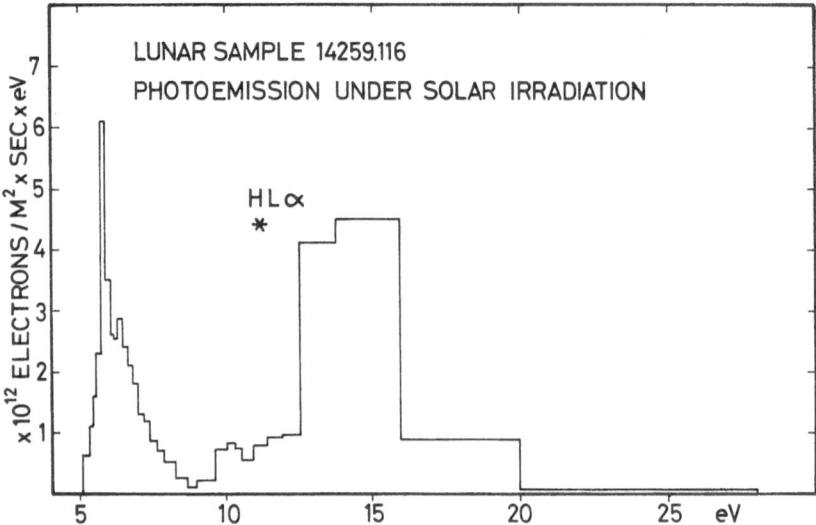

Fig. 2. Differential photoelectron flux from lunar sample 14259,116 due to solar irradiation. The asterisk indicates the position of a column 1 eV wide equivalent to the hydrogen Lα line flux of $4,5 \times 10^{12}$ electrons m^{-2} s^{-1}.

Numerical evaluation gives a value of 4.5 μA m^{-2} (or 2.8×10^{13} electrons m^{-2} s^{-1}), which is low compared with values for metal-coated surfaces and insulating materials (Feuerbacher and Fitton, 1972).

3.2. PHOTOELECTRON ENERGY DISTRIBUTION

The energy or velocity distribution of photoemitted electrons, $F(\psi, \omega)$, after escape from the surface has been measured as a function of electron energy, $e\psi$, for a series of photon energies, $\hbar\omega$, between 4 and 21 eV (Feuerbacher et al., 1972). The photo-electron energy distribution under solar irradiation, $p(\psi)$, is obtained by multiplying the energy distribution spectrum for monochromatic light, $F(\psi, \omega)$ by the differential photoelectron flux, $H(\omega)$, and integrating with respect to photon energy:

$$p(\psi) = \frac{1}{I_s} \int_0^\infty F(\psi, \omega) H(\omega) \, d\omega. \tag{3}$$

The result for the Apollo 14 sample is shown in Figure 3 (full curve), the curve being normalized such that the area under it is unity. The bulk of the photoelectrons are seen to be emitted with energies between 1 and 4 eV, with a mean kinetic energy:

$$e\psi_e = (8me)^{1/2} \frac{I_s}{N_0} \int_0^\infty \sqrt{\psi} \, p(\psi) \, d\psi \tag{4}$$

which yield a value of $e\psi_e \simeq 2.2$ eV for lunar dust. This value is higher than those values determined for a range of common materials, $e\psi_e \simeq 1.2$–1.5 eV (Grard, 1972). The tail of electrons with energies greater than 6 eV is seen to be very weak (Figure 3), as expected in view of the decreasing yield observed for higher photon energies (Figure 1).

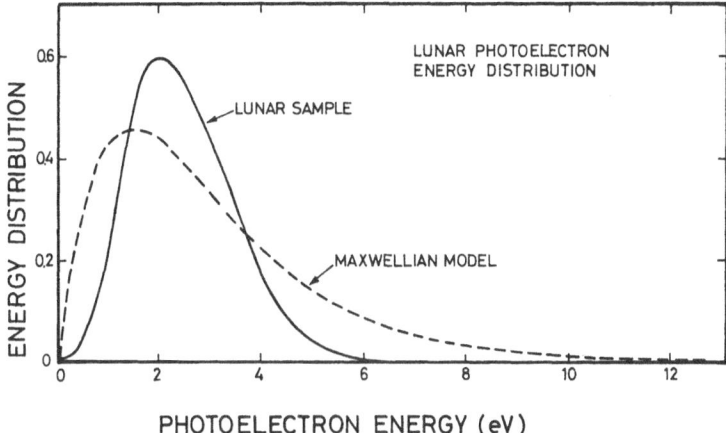

Fig. 3. Energy distribution of lunar photoelectrons due to solar irradiation. The area under the curve is normalized to unity. The dashed curve shows the Mawellian approximation used to calculate the photoelectric sheath characteristics.

4. The Electrical Characteristics of the Lunar Photoelectron Sheath

The energy distribution curve, Figure 3, and the saturation photocurrent, I_s, provide all necessary information to calculate the properties of the photoelectron sheath, if proper assumptions are made about the angular distribution of the emitted electrons. In the present work, however, the energy distribution (Equation (3)) curve has been approximated by a Maxwellian distribution

$$p(\psi) = \frac{\psi}{\psi_p} e^{-\psi/\psi_p}, \qquad (5)$$

where $e\psi_p$ is the most probable electron energy. This curve is shown by the dashed line in Figure 3. For high electron energies it obviously is not a good approximation. However it is expected that the sheath properties close to the surface are described rather well (Grard, 1972).

With this model, the parameters that describe the lunar photoelectron sheath are derived from very simple formulae. The decay length for the electric field close to the surface is given by

$$\lambda_0 = \left(\frac{2}{3} \frac{\varepsilon_0}{N_0} \frac{\psi_e}{e} \right)^{1/2}, \qquad (6)$$

where ε_0 is the dielectric constant of the electron sheath with density N_0 at the surface. The mean kinetic energy ψ_e is related to the most probable energy by $\psi_p = \frac{2}{3}\psi_e$ in this model. The electron density at the curface is given by

$$N_0 = \left(\frac{2\pi e}{m\psi_p}\right)^{1/2} I_s,$$ (7)

the electric field E_0 and the surface charge σ_0

$$E_0 = \frac{2\sqrt{2}}{3}\frac{\psi_e}{\lambda_0},$$ (8)

$$\sigma_0 = \sqrt{2} N_0 \lambda_0$$ (9)

The surface potential Φ_0 will be determined by the saturation photoelectron current and the incident plasma flux I_a

$$\Phi_0 = \frac{2}{3}\psi_e \ln \frac{I_a}{I_s}.$$ (10)

The parameters calculated for the lunar surface, using the Maxwellian approximation to the measured data, are listed in Table I. The solar wind electron flux was taken to be 2×10^{12} electrons s^{-1} m^{-2}, corresponding to a solar wind density of 5 electrons cm^{-3} and a flow speed of 320 km s^{-1}. In Table I, the values for lunar dust are compared to those expected from a typical smooth insulating surface, Al_2O_3, which should represent an upper limit to the expected electron densities. It is seen from the table, that the surface potential is hardly affected, though Al_2O_3 gives a much denser electron sheath. The variation of the electron density and electric field with distance from the lunar surface are plotted in Figure 4 as calculated from the Maxwellian approximation.

The values listed in Table I for lunar dust are in reasonable agreement with the prediction of Grobman and Blank (1969), while Manka's values (Manka, 1973) are slightly higher, in agreement with experimental data of Freeman et al. (1973), as derived from an analysis of the energy spectra of lunar ionosphere thermal electrons accelerated toward the Moon's surface. Reasoner and Burke (1973) have obtained photoelectron data in the 35 to 200 eV energy range from the CPLEE experiment. Those measurements were made during the Moon's passage through the magnetospheric tail of the Earth, where surface charges of 200 V have been observed due to the lack of an incident plasma flux. Extrapolation of their data to lower photon energies, using a photoemission model of Walbridge (1971), yields electron densities at the surface in the order of 10^4 cm^{-3} (the surface potential cannot be compared because of vanishing ambient flux). This has to be compared to a value of $N_0 = 130$ cm^{-3} derived from the present measurements. It is very likely that this discrepancy is due to the choice of the yield curve $Y(\omega)$ in the calculations of Walbridge (1971) and Reasoner and Burke (1973). A linear increase of the yield curve between the threshold and 9 eV greatly overemphasizes the number of low energy electrons when the solar spectrum

TABLE I

Electrical characteristics of lunar surface material under solar wind ambient conditions

Material	Saturation photocurrent density $eI_s (\mu A\ m^{-2})$	Volume electron density at surface $N_0 (electrons\ m^{-3})$	Surface charge density $\sigma_0 (electrons\ m^{-2})$	Surface electric field $E_0 (V\ m^{-1})$	Surface potential $\phi_0 (V)$	Shielding distance $\lambda_0 (m)$
Lunar dust	4.5	130×10^6	140×10^6	2.65	$+4.3$	0.78
Al$_2$O$_3$ (Smooth surface)	42.0	1670×10^6	410×10^6	7.40	$+4.4$	0.17

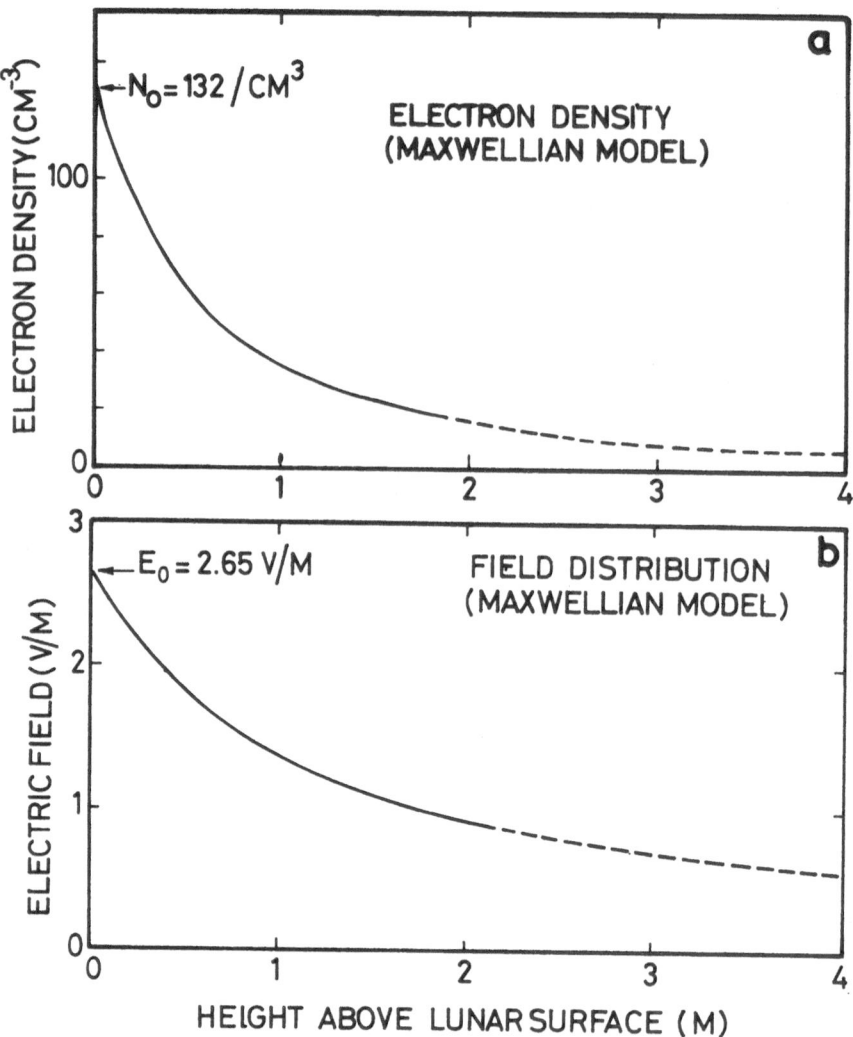

Fig. 4.   Electron density (a) and electric field distribution (b) in the lunar photoelectron sheath, calculated using a Maxwellian distribution approximation.

is folded into it. An extrapolation from high energy electrons will therefore give an unrealistic high electron density.

5. Secondary Electron Emission

The secondary electron emission (SEE) yield curves for the Apollo 14 and 15 samples, Figure 5, show a maximum of 1.5 ± 0.1 in the energy region 300–700 eV. Values of the SEE yield greater than unity cause the dust to charge positively i.e., for incident electron flux energies in the range from about 100 eV to 2000 eV. It is significant perhaps that the increased electron flux density observed by the CPLEE at energies of 300 to 500 eV (O'Brien and Reasoner, 1972), during the Moon's passage through the geomagnetic tail of the magnetosphere, is also the energy at which the lunar dust samples possess maximum secondary yield. Incident electron energies below 100 eV, for which the secondary yield will be negligible, and above 2000 eV, at which energies magnetospheric electron fluxes have also been detected (O'Brien and Reasoner, 1972),

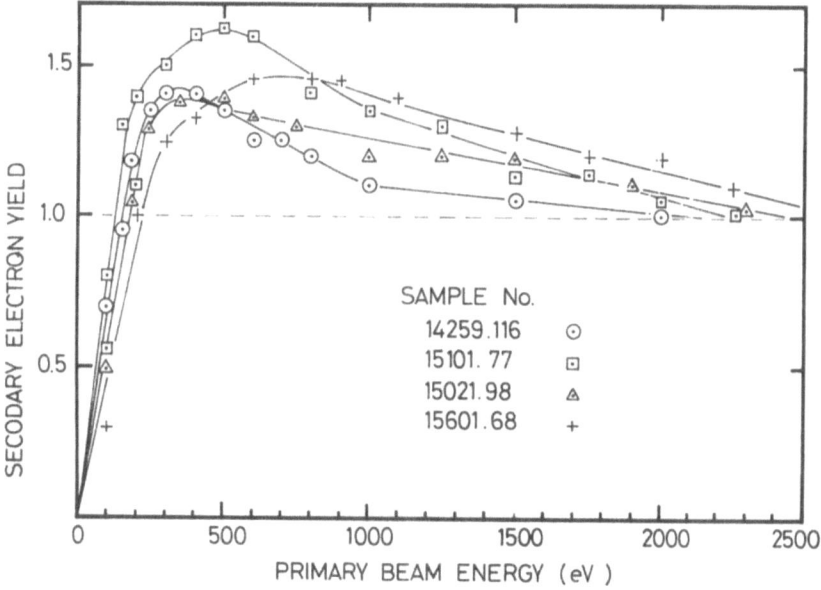

Fig. 5. Secondary electron yield from Apollo 14 and 15 lunar dust samples for incident primary energies of 50 to 2500 eV.

will cause the dust to charge negatively. The consequences of such SEE charging effects as the Moon enters the Earth's magnetosphere have been discussed fully by Manka (1973). Some of the curves, notably those from samples 14259,116 and 15101,77 exhibit a pronounced 'kink' about 1000 eV, which is consistent with the behaviour expected from a mixture of components with maximum yields close to unity and 1.5 respectively. All curves are typical of those expected for a highly particulate, predominantly insulating, sample (Gibbons, 1966).

6. Auger Electron Spectroscopic Analysis

Auger electron spectra from an Apollo 15 sample, 15601,68, and an Apollo 14 sample, 14259,116, are compared in Figure 6a and 6b respectively. Spectra from all of the Apollo 15 samples were very similar to that of sample 15601,68. The following elements were identified: Al, Si, K, Ca, O and Fe were the major constituents, small amounts of

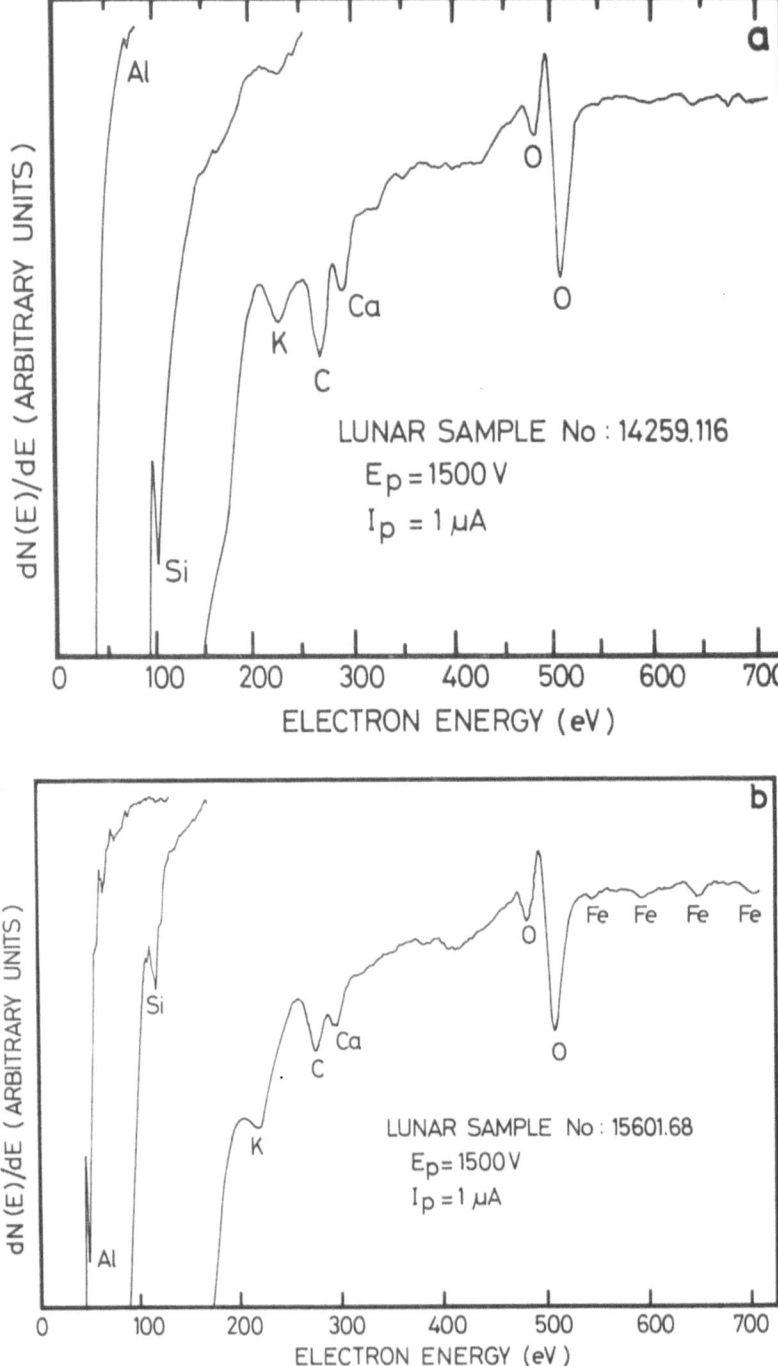

Fig. 6.   Auger electron spectra from (a) lunar sample 14259,116 and (b) lunar sample 15601,68.
The relative gain for the three curves (left to right in each figure) is approximately × 1, × 10 and × 40,
corresponding to modulation-voltage amplitudes of 0.5, 5.0 and 20.0 V peak-to-peak.

C, S and Cl appearing as probable contaminants, in agreement with an Apollo 12 sample analysis reported previously (Connel *et al.*, 1971). The results are consistent with a chemical analysis which shows the samples to consist mainly of: SiO_2 ($\simeq 48\%$), Al_2O_3 ($\simeq 18\%$), CaO ($\simeq 11\%$), FeO ($\simeq 10\%$), TiO_2 ($\simeq 2\%$) and K_2O ($\simeq 0.5\%$). A major difference between the Apollo 14 and 15 samples, however, was the more pronounced Fe peaks at 560, 600, 650 and 700 eV and the high Al/Si abundance in the latter (Figure 6). This would appear to endorse similar compositional differences reported for the two lunar regions from X-ray fluorescent analysis of the lunar terrain (Adler, *et al.*, 1972). The latter investigation indicated that the albedo difference between the highlands and maria regions of the lunar surface may be related to distinct chemical and mineralogical differences, in particular, the relative abundance of elemental Al/Si.

7. Concluding Remarks

Photoemission and secondary electron emission measurements have been made on lunar dust samples returned from the Apollo 14 and 15 missions, and the results appear to be representative of the lunar surface. However, since no ultra-high vacuum samples were available, the measurements had to be performed with samples handled in an inert atmosphere of less than 30 ppm impurity. Since low energy electron emission is strictly a surface property, absorbed gaseous impurities may change the properties drastically. The Auger electron spectroscopic analysis, however, suggests that any contamination present was limited to less than 10% of one monolayer.

With these limitations in mind, the present work indicates that the lunar photo-electron sheath is tenuous, the electron density at the surface being of the order of 130 electrons cm^{-3} and shielding distance extending out to 0.78 m under solar wind ambient conditions. Both the differential photoelectric and secondary electron emission yield values are consistent with that expected of a fine-grain, insulating dust mixture. The surface charge and potential of the sunlit lunar surface will be dominated by the denser photoelectron continuum and secondary electron emission will be negligible in comparison. However, any surface charging on the dark side of the moon and along the Moon's terminator may be significantly affected by secondary electron emission, particularly during the Moon's transit through the magnetosphere (Knott, 1973; Manka, 1973). Future calculations should take such considerations into account in view of their pertinence to related phenomena, such as lunar soil charging effects (Gold, 1964). The relationship between lunar surface composition and albedo variation requires further analysis on future Apollo samples before these preliminary results can be clarified.

Acknowledgements

We thank Dr R. J. L. Grard for his advice on the analytical aspects of the determination of the electrical parameters which characterise the lunar photo-electron sheath. The encouragement of Dr E. A. Trendelenburg at all times is gratefully acknowledged.

References

Adler, I., Gerard, J., Trombka, J., Schmadebeck, R., Lowman, P., Blodgett, H., Yin, L., Eller, E., Lamothe, R., Gorenstein, P., Bjorkholm, P., Harris, B., and Gursky, H.: 1972, *Proc. 3rd Lunar Sci. Conf., Geochim. Cosmochim. Acta Suppl.* 3, 2157.

Brandt, J. C.: 1970, *Introduction to the Solar Wind*, W. H. Freeman & Co., p. 119.

Connel, G. L., Schreidmiller, F. R., Kraatz, P., and Gupta, Y. P.: 1971, *Proc. 2nd Lunar Sci. Conf., Geochim. Cosmochim. Acta Suppl.* 3, 2083.

Feuerbacher, B. and Fitton, B.: 1972, *J. Appl. Phys.* 43, 1563.

Feuerbacher, B., Anderegg, M., Fitton, B., Laude, L. D., Willis, R. F., and Grard, R. J. L.: 1972, *Proc. 3rd Lunar Sci. Conf., Geochim. Cosmochim. Acta Suppl.* 3, 2655.

Freeman, J. W., Jr., Fenner, M. A., and Hills, H. K.: 1973, this volume, p. 365.

Gibbons, D. J.: 1966, in A. H. Beck (ed.), *Handbook of Vacuum Physics*, Vol. 2, Part 3, Pergamon Press.

Gold, T.: 1964, in J. W. Salisbury and P. E. Glaser (eds.), *The Lunar Surface Layer*, Academic Press, p. 345.

Grard, R. J. L.: 1972, ESTEC/ESRO Internal Working Paper No. 663.

Grobman, W. D. and Blank, J. L.: 1969, *J. Geophys. Res.* 74, 3943.

Hapke, B. W., Cohen, A. J., Cassidy, W. A., and Wells, E. N.: 1970, *Proc. Apollo 11 Lunar Sci. Conf., Geochim. Cosmochim. Acta Suppl. 1* 3, 2199.

Hinteregger, H. E., Hall, L. A., and Schmidtke, G.: 1964, *Space Research V*, North Holland, p. 1175.

Knott, K.: 1973, ESTEC Int. Rep., to be published.

Manka, R. H.: 1973, this volume, p. 347.

Öpik, E. J. and Singer, S. F.: 1960, *J. Geophysical Res.* 65, 3065.

Öpik, E. J. and Singer, S. F.: 1962, *Planet. Space Sci.* 9, 221.

Reasoner, D. L. and Burke, W. J.: 1972, *Proc. 3rd Lunar Sci. Conf., Geochim. Cosmochim. Acta Suppl.* 3, 2639.

Reasoner, D. L. and Burke, W. J.: 1973, this volume, p. 369.

Tousey, R.: 1966, in A. E. S. Green (ed.), *The Middle Ultraviolet*, John Wiley, p. 1.

Walbridge, E.: 1971, Internal Report, High Altitude Observatory, Boulder, Colorado.

Willis, R. F., Anderegg, M., Feuerbacher, B., and Fitton, B.: 1972, in J. W. Chamberlain and C. Watkins (eds.), *The Apollo 15 Lunar Samples*, Lunar Science Institute, Houston.

DISCUSSION

Gold: Secondary emission and photoemission act very differently on a microscopic scale. Secondary emission tends to an instability making the surface charge very uneven; this arises because the incoming electron has its velocity affected by the potential of the surface it is hitting, and that in turn affects the secondary emission ratio, and therefore the potential in turn.

Manka: I would like to comment that your measured secondary electron coefficients that are greater than unity for primary energies of about 200–2000 eV, imply that in relation to my calculations of the dark side potential when the Moon is in the plasma sheet of the geomagnetic tail, then the very large negative potentials (− 1800 V) associated with a low secondary emissivity will not occur, and rather the potential will be about zero or a few volts positive. That is, the secondary electron emission on the dark side has an effect similar to the photoemission on the front side, in which it prevents large negative potentials which would result from the hot plasma environment alone.

Reasoner: The secondary electrons emitted from the lunar surface will remain trapped and make no contribution to the net current unless the lunar surface potential falls below ∼40 V.

Willis: I agree that secondary electron emission will be negligible on the sunlit lunar surface, the photoelectron flux being many orders of magnitude greater. Where secondary electron emission may be important is in dynamic effects, along the Moon's terminator in particular, when the Moon passes through the plasmasheet of the magnetosphere. Also, such perturbations may be extremely local. Thus, I agree that the overall secondary electron current may well be negligible compared to the saturation photoelectron current, but dynamic perturbations may occur in the lunar sheath due to secondary charging.

Criswell: Are the lunar dust samples likely to be contaminated with gas adsorption and could this increase the SEE yield considerably?

Willis: The lunar samples appeared to be relatively free of gas adsorption since Auger spectra are sensitive to contamination effects. Hydrogen up to one monolayer could be present since this has no spectral lines. Higher concentrations would tend to obscure the observed Auger spectral lines of the lunar material. H_2O, N_2 and other likely gases in concentrations above 1/100th of monolayer would have been detected and were not. Whether or not the lunar surface contains a high degree of adsorbed gases is not known. The yield of dust samples containing such heavily adsorbed gas is also unknown as far as I can recollect with any certainty.

Criswell: What are the expected yield and differential energy distribution of high energy photo-electrons (photon energy > 100 eV) likely to be for highly insulating ($> 10^{17} \Omega$ cm), pure lunar material?

Willis: The X-ray photoemission yield in the 75–300 Å range for metals has been measured to be approx. 8% and for insulators approx. $25 \pm 5\%$. Upon grinding the bulk solids up into a powder, the yield drops by a factor of 10. This would indicate a figure of $< 3\%$ for lunar dust material, which is less than our figure of 7%, measured at approximately 750 Å. Our measurements indicate a decrease in yield below 600 Å, which would suggest that a photoyield of 7% is an *upper* limit. It is hoped to measure the differential energy distribution for photon energies > 100 eV on our sample in the near future to verify this. Our sample is not contaminated by adsorbed gases > 0.1 monolayers since well resolved substrate Auger spectra were obtained.

3.2. INFLUENCE OF SURFACE EMISSION ON THE PROPERTIES OF OTHER CELESTIAL OBJECTS

SHEATH ACCELERATION OF PHOTOELECTRONS
BY JUPITER'S SATELLITE IO

STANLEY D. SHAWHAN, RICHARD F. HUBBARD,
GLENN JOYCE, and DONALD A. GURNETT

Dept. of Physics and Astronomy, The University of Iowa, Iowa City, Iowa 52242, U.S.A.

Abstract. We are investigating a model of the influence of Jupiter's moon Io on Jovian decametric radiation due to plasma sheaths formed around Io's surface. With Io, we are dealing with a large, partially-conducting, and photoemitting body with a large $\mathbf{V} \times \mathbf{B}$ potential (~ 700000 Volts) across its diameter. A well-known model by Goldreich and Lyden-Bell (1969) assumes Jupiter's field lines are frozen to Io, while we are proposing instead that sheaths form around Io and electrons are accelerated across these sheaths.

Two types of sheaths are considered. A Debye sheath forms around regions of Io's surface which are negative with respect to the plasma potential while a photoelectron sheath forms where the surface potential is positive. The Debye sheath (of area A_p) accelerates emitted photoelectrons away from the surface (with current density J_p) while the photoelectron sheath (of area A_e) collects an ambient electron current ($J_e A_e$). The current balance is $J_p A_p = J_e A_e$. The boundary between the two regions has zero potential.

We estimate J_e and J_p to be 3×10^{-7} A m^{-2}, but both may vary considerably. The emitted particle spectrum is critically dependent on the ratio J_e/J_p. Estimates of total power available in the accelerated photoelectrons are 10^{10}–10^{13} W, well above the 10^7 W contained in a typical decametric burst. We have also studied the effect of Io's orbital position on our model since decametric bursts are strongly coupled to Io's position.

Although $\mathbf{E}$ is probably radial through the sheath, we believe that the electron gyroradius is small compared with sheath dimensions so that the particles emerge almost parallel to Jupiter's field lines. Using an 'oblique' version of Child's Law, we estimate the typical sheath thickness as 10–50 km. High energy (up to several hundred keV) electrons thus travel along $\mathbf{B}$ field lines and eventually produce the observed radio noise.

1. Introduction

Bigg (1964) demonstrated that the orbital position of Io, the innermost Galilean moon of Jupiter, was related to the modulation of the Jovian decametric wavelength radio bursts. Since that time a number of models to explain these Io related bursts have been published. Gledhill (1967) proposed that Io stimulates plasma waves from a plasma discus surrounding Jupiter. Piddington and Drake (1968), Goldreich and Lynden-Bell (1969), Dermott (1970), and Piddington (1972) suggested that the Io interaction for a sufficiently high electrical conductivity could be due to Io dragging its associated flux tube through the Jovian magnetosphere and ionosphere. Goldreich and Lynden-Bell suggested that the emission is due to Doppler-shifted coherent cyclotron radiation from energetic electrons streaming along the magnetic field lines. The radio emission is explained by Piddington as due to plasma oscillations from thin sheets with densities $\sim 10^7$ cm^{-3} in the lower ionosphere.

Duncan (1970) and Schatten and Ness (1971) consider the tilted dipole magnetic field geometry of Jupiter to explain the observed combinations of Io orbit position and longitude for which the noise storms occur. These storms occur as a result of emission from perturbed trapped electrons as Io moves through various magnetospher-

R. J. L. Grard (ed.), Photon and Particle Interactions with Surfaces in Space, 405–413. All Rights Reserved

ic boundaries according to Duncan. Schatten and Ness suggest that a Moon-like (non-magnetic, non-conducting) body could scatter radiation belt particles to produce the bursts by synchrotron radiation. Mozer and Bogott (1972) propose that the photoelectric emission from Io in sunlight produces sufficient cold plasma locally to cause growing electrostatic waves which perturb the trapped energetic electrons. Gurnett (1972) concludes that a plasma sheath may form around Io allowing it to slip through the magnetospheric magnetic field. Bursts are emitted by photoelectrons accelerated to several hundred kilovolts by the motional emf of Io developing across this sheath.

The paper reports the progress on further calculations related to the plasma sheath model of Gurnett. In particular a range of values are estimated from a number of different models for the parameters assumed for the model: the photoelectron and thermal electron current densities, the Jovian ionospheric and Io surface conductivities. Numerically integrating an oblique version of Child's Law the plasma sheath dimensions are determined. Also, the particle gyroradius is shown to be small compared to the sheath dimension. Consequently the photoelectrons from Io are accelerated predominantly along the magnetic field line to several hundred kilovolts of energy. An estimate of the resulting particle maximum energy, current, and power as a function of Io's orbital position is given.

2. Review of Basic Model

The basic model to be discussed is that due to Gurnett (1972). An illustration of the model geometry with the important parameters indicated is shown in Figure 1.

Io's orbital velocity is different from the rotational velocity of Jupiter. Io, therefore, moves through the Jovian magnetic field producing a motional electric field. This field is directed toward Jupiter and results in a potential of up to ~ 700 kV across Io's diameter. Sheaths form around Io as a consequence of the current balance between photoelectrons emitted from the surface and thermal electrons collected from the magnetosphere. On the face of Io away from Jupiter, the surface potential is positive with respect to the plasma potential where a photoelectron sheath is formed. On the face toward Jupiter the surface potential is negative which creates a Debye sheath of thermal electrons.

Because of the large surface potentials compared to the potentials of the thermal or photoelectrons, the current I_p emitted in the Debye sheath region is predominantly from photoelectrons and the current I_e collected in the photoelectron sheath region is due to thermal electrons. The transition between the sheath regions (where $\Phi = 0$) is determined by the current balance condition $I_e = I_p$ where $I = JA$; and J and A are the current density and the area of each region. The conductivity of Io σ_I is assumed sufficiently large that there is not a significant potential drop due to this current through Io.

These sheaths around Io can limit the current to Io and prevent the plasma from being frozen to the magnetic flux tube passing through Io. This frozen plasma is assumed by the models of Piddington and Drake (1968), Goldreich and Lynden-Bell (1969) and Dermott (1970). The plasma is unfrozen from the motion of Io if the height integrated Pedersen Σ_p ionospheric conductivity is sufficiently high that currents flowing up and down the field lines are shorted across the field lines in the ionosphere.

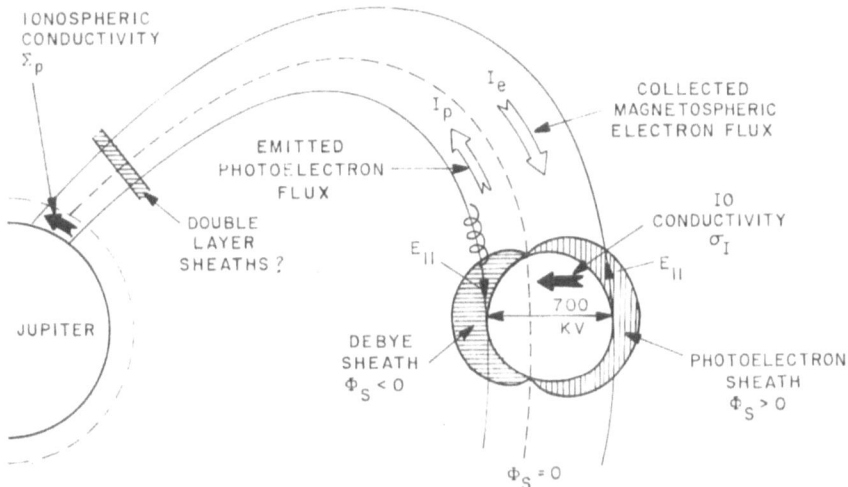

Fig. 1.   A sketch of the basic Io sheath and current model with the important parameters indicated. Because of Io's motion through the Jovian magnetic field, a potential of $\sim 700\,\mathrm{kV}$ is developed across Io. The position of the zero potential line ($\Phi_s = 0$) is determined by the current balance between the emitted photoelectrons and the collected magnetospheric electrons $I_e = I_p$. Where $\Phi_s < 0$ a Debye sheath forms and where $\Phi_s > 0$ a photoelectron sheath forms. If Σ_p the ionospheric height integrated Pedersen conductivity and the Io conductivity σ_I are sufficiently large, then the total potential drop will be across the sheaths. The emitted photoelectrons are accelerated through the Debye sheath along the magnetic field lines with energies of several hundred kilovolts.

The critical ionospheric conductivity for the plasma freezing is given by

$$\Sigma_c' = \frac{J_p l_p}{E_{\mathrm{Io}}} \left(2 \cos \theta_{\mathrm{colat}}\right)^{-1}, \tag{1}$$

where l_p is the linear dimension in the ionosphere corresponding to the Debye sheath dimension at Io, J_p the photoelectron current density, E_{Io} the motional electric field across Io, and the $2 \cos \theta_{\mathrm{colat}}$ accounts for the field line convergence. If $\Sigma_p > \Sigma_c'$ then the motional emf is developed across the sheath at Io. Photoelectrons emitted from the surface in the Debye sheath region are accelerated to hundreds of keV. These energetic particles may cause the observed decametric radio bursts related to Io.

3. Estimation of Model Parameters

In order to further assess the plausibility of this sheath model, working values for the key parameters have been re-evaluated based on a combination of calculations

TABLE I

Range and working value of assumed parameters

Parameter	Gurnett (1972)	Range	Working value
Photoelectron current density, J_p (A m^{-2})	4.5×10^{-6}	1×10^{-7}–3×10^{-6}	3×10^{-7}
Thermal electron current density, J_e (A m^{-2})	4.0×10^{-7}	1×10^{-7}–1×10^{-6}	3×10^{-7}
Critical ionospheric conductivity, Σ'_c (mhos)	4.8	0.2–3.2	0.6
Expected ionospheric conductivity, Σ_p (mhos)		0.33–20.	> 0.6
Critical Io conductivity, σ_c (Ω^{-1} cm^{-1})	2×10^{-8}	5×10^{-11}–2×10^{-8}	5×10^{-11}
Expected Io conductivity, σ_I (Ω^{-1} cm^{-1})		10^{-31}–10^{-1}	$\simeq 0$ darkside $\simeq \infty$ lightside

and available literature. These parameters as indicated on Figure 1 and listed in Table I are the photoelectron current density, the thermal electron current density, the Jovian ionospheric conductivity and the conductivity of Io.

3.1. Photoelectron Current Density

A realistic estimate of the photoelectron current density from Io must await more detailed knowledge of the state and composition of the surface material. An empirical determination of the photoelectron current density has been made, however, by scaling results from the Charged Particle Lunar Environment Experiment of Apollo 14. Typical measurements of photoelectron fluxes between 40 and 200 eV have been fit by curves which extrapolate the fluxes to lower energy (Reasoner and Burke, 1971, Figure 7 for $V=0.1$). We have integrated this curve to cutoff energies of 3 eV and 0.01 eV and scaled the value by 1/27 to account for the decrease in solar photon flux at Io. The resulting photoelectron current densities range between 1×10^{-7} and 3×10^{-6} A m^{-2}. We find that cutting off the spectrum near 3 eV is consistent with the photoelectron energy density value of 7.4 eV cm^{-3} which is 1/27 of 200 eV cm^{-3} determined for the Moon by Criswell (1971). We, therefore, choose a working value of $J_p \sim 3 \times 10^{-7}$ A m^{-2}. Gurnett (1972) had used 4.5×10^{-6} A m^{-2}, as indicated in Table I, but considered that this value might be an order of magnitude high.

3.2. Thermal Electron Current Density

The thermal electron current density depends on the ambient electron number density and temperature. A magnetospheric model by Ioannidis and Brice (1971, Figure 11) gives values of ~ 3 electrons cm^{-3} and 10^5 K in the vicinity of Io. This value of number density is consistent with the upper limit < 10 electrons cm^{-3} placed on the magnetospheric number density from Faraday rotation measurements by Warwick and Dulk

(1964). Using these numbers we obtain a value for the thermal electron current density of $J_e = 3 \times 10^{-7}$ A m^{-2}. This value is consistent with Gurnett (1972). We consider that the value could easily range between 1×10^{-7} and 1×10^{-6} A m^{-2} as indicated in Table I due to Io's orbital inclination to the 'donut' of plasma density in the magnetic equatorial plane.

3.3. IONOSPHERIC CONDUCTIVITY

In order for the plasma to be unfrozen from the flux tube passing through Io at any instant, the ionospheric (height integrated Pedersen) conductivity must exceed a certain value. From Gurnett (1972) this value is given by Equation (1). Using the range of possible values for the photoelectron and thermal electron current densities we find the critical conductivity can range between 0.22 and 3.2. Using the working values for J_p and J_e, $\Sigma_c' = 0.6$ mhos. This value is lower than the 4.8 mhos of Gurnett primarily because of the lower photoelectron current density value.

Goldreich and Lynden-Bell (1969) calculate a height integrated Pedersen conductivity of 0.57 mhos but suggest that this value could be low by more than a factor of five. Brice and Ioannidis have estimated the conductivity to be 2–20 mhos. Using the Gross and Rasool (1964) and the Prasad and Capone (1971) Jovian upper atmosphere models, we obtain values of the conductivity ranging from 0.3 to 1.0 mhos. As pointed out by Gurnett (1972) these calculated values are valid only for the undisturbed ionosphere. Since our model predicts precipitated particles of up to a few hundred kilovolts energy, a significant amount of impact ionization would be expected which would increase the conductivity. Therefore it seems from a number of estimates that Io may be effectively decoupled from its field lines.

3.4. IO CONDUCTIVITY

In our model the conductivity of Io is important as it relates to the voltage drop across the Moon for the total current passing through it. Goldreich and Lynden-Bell (1969) estimate a critical conductivity for Io of $\sigma_c \approx 2 \times 10^{-8}$ Ω^{-1} cm^{-1}. Dermott (1970) using a different model for the temperature dependent conductivity, obtains $\sigma_c \approx 5 \times 10^{-11}$ Ω^{-1} cm^{-1} at 100K. This conductivity corresponds to a resistance through Io of 0.2 Ω. For 3×10^{-7} A m^{-2} collected over 25% of Io's area this resistance would drop the motional emf to approximately two thirds its value. We, therefore, take $\sigma_c \approx 5 \times 10^{-11}$ Ω^{-1} cm^{-1} as being the critical Io conductivity.

Possible values for Io's conductivity cover a very large range. Goldreich and Lynden-Bell (1969) guess the conductivity might be like the Earth's upper mantle and use $\sigma_I \sim 10^{-5}$ Ω^{-1} cm^{-1}. If Io is composed of olivines, in particular fayalite, then Dermott (1970) finds it conductivity would be $\sigma_I \ll 10^{-10}$ Ω^{-1} cm^{-1} and perhaps as low as 10^{-31} Ω^{-1} cm^{-1} at 100K. An outer layer of (water) ice on Io would give a conductivity at 100K of $\sigma_I \approx 10^{-25}$ Ω^{-1} cm^{-1}. Lewis (1971) discusses a steady state thermal model for icy satellites including Io. He estimates that the inner Galilean satellites may have a surface coating of aqueous NH$_3$ solution with a conductivity as high as 10^{-1} Ω^{-1} cm^{-1}, but cautions that this value may not apply directly to Io.

In order to present the least complicated model, we assume at present that $\sigma_1 \sim \infty$ on the dayside and $\sigma_1 \sim 0$ on the nightside compared to $5 \times 10^{-11} \, \Omega^{-1} \, cm^{-1}$.

4. Energy Gain Across Sheath

A comparison of estimated conductivity values to critical values allows the possibility that the large potentials develop across the sheaths surrounding Io. Gurnett (1972) has implicitly assumed that the emitted photoelectrons will gain energy along the magnetic field lines equivalent to this potential. We have investigated this assumption in more detail using an oblique version of Child's law.

In Figure 2 is shown in geometry and coordinate system for a particle emitted from Io in the Debye sheath region. Two extreme possibilities exist for the emitted particle. If the gyroradius for the particle is larger than the sheath dimension or if the v_D is much larger than the velocity along the magnetic field $v_\parallel$, the particle would follow the path ab and would be recollected on the surface. At least it would not gain a significant amount of parallel energy. Path ac could be the case for $v_D \ll v_\parallel$ and for the gyroradius small compared to the sheath dimension.

Taking z along the magnetic field direction and x perpendicular to the magnetic field and the electric field $\mathbf{E}$ initially radial, Poisson's equation can be written

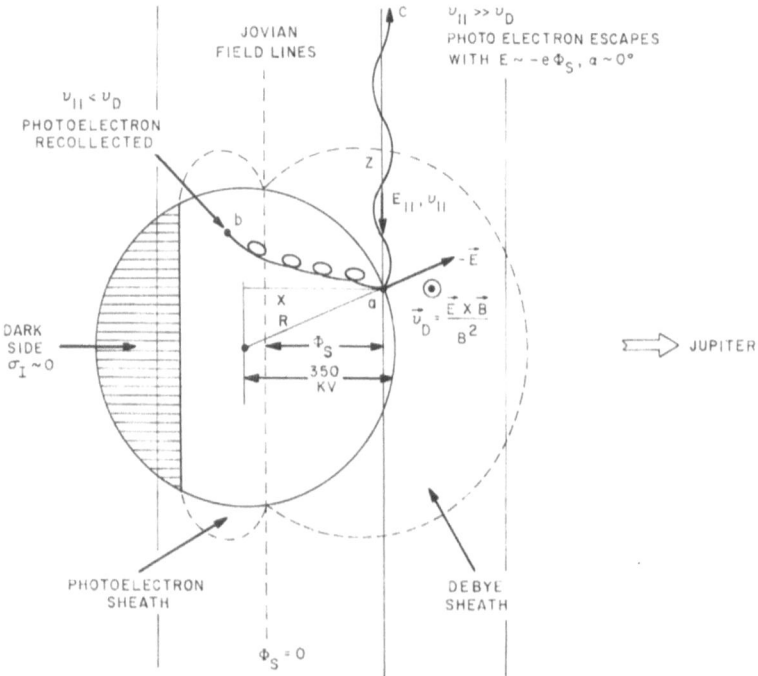

Fig. 2. Geometry for evaluating Child's Law where $\mathbf{E}$ is not parallel to the magnetic field direction. For the electron gyro-radius large compared to the sheath dimension or $v_D \gg v_\parallel$ an emitted photo-electron would follow path ab. However, the sheath dimension is tens of kilometers and $v_\parallel > 10 \, v_D$. Photoelectrons gain nearly the surface potential at the emission point along the path ac.

$$\frac{d^2\Phi}{dz^2} = \frac{J}{\varepsilon_0}\left(\frac{m}{2e}\right)^{1/2}\Phi^{-1/2}\left(\frac{z^2}{x^2+z^2}\right),\tag{2}$$

where Φ is the potential and J the current density. We have solved this equation numerically for a photoelectron emitted at 45° latitude. The resulting sheath thickness is approximately 40 km $(=L_D)$. For a 300 keV electron at Io the gyroradius would be approximately 200 m which is significantly less than the sheath thickness. Over the sheath region the ratio of drift to parallel velocity ranges from 10^{-1} to 10^{-2}. Consequently the assumption that photoelectrons follow path ac is well justified. Emitted photoelectrons therefore gain the large sheath potential in energy and these leave the sheath region with pitch angles peaked about 0°. A significant fraction of these particles can penetrate to the atmosphere of Jupiter.

5. Variation with Io's Orbital Position

From the proposed sheath model of Io's interaction with Jupiter, the main characteristics we can predict are those related to the accelerated particles which may be the cause of the observed decametric radio bursts. It is, therefore, of interest to see how the particle characteristics change with Io's orbital position. Our present model assumes that no photoelectrons are emitted and no magnetospheric electrons are collected on the dark side. The particles gain the surface potential in energy from the point that they are emitted. As Io moves around Jupiter, it is the position of the $\Phi_s=0$ V line and the photoelectron emitting area that changes. We have taken $J_p=J_e=3\times10^{-7}$ A m^{-2} so that the current balance condition is reduced to equating the photoelectron emitting and magnetospheric electron collecting areas $A_p=A_e$.

TABLE II

Range of model dependent quantities with Io's orbit

Quantity	Range	Io angle at maximum
Particle maximum energy	156–396 keV	120°, 240°
Particle current	8.4 × 10⁵ A	
Particle power	4.4 × 10¹¹–0 (occultation) W	90°, 270°
X-ray flux	5 × 10²⁴ photons s⁻¹	

In Figure 3 the maximum potential obtainable by a photoelectron is plotted as a function of the Io orbital position. Toward the Earth (and Sun) is defined as the 180° position. At 0°, 90° and 240° the illumination and $\Phi_s=0$ line are illustrated. This maximum potential is seen to vary from 156 to 396 kV with peaks at 120 and 240°. Maximum power is imparted to the particles at 90° and 270°. This power is approximately 10⁴ times that in a typical Jovian decametric burst. Other quantities of interest are recorded in Table II.

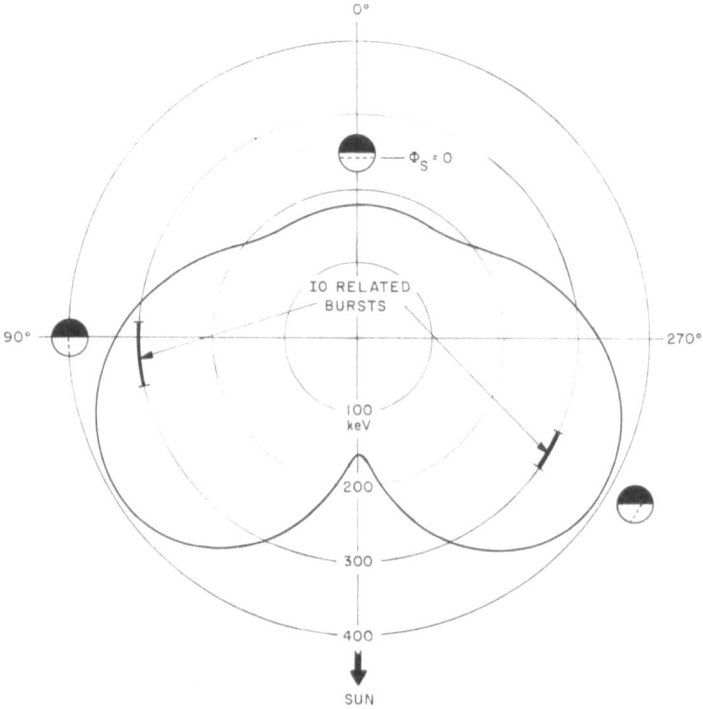

Fig. 3. The maximum energy gained by photoelectrons as a function of the Io orbit position. 180°
is toward the Earth and Sun. The regions of Io related bursts are also indicated.

6. Discussion

By considering a number of points in detail, we feel that the sheath model proposed for Io and the acceleration of charged particles is plausible. Further development of the model should include detailed calculations of the particle energy and pitch angle distributions so that various emission mechanisms for radio noise can be tested. The effect of the tilted and possibly offset Jovian magnet dipole must be included. For instance, because of the 'donut' density distribution of thermal plasma, the thermal current density will change significantly with Io's orbital position. Also consideration should be given to the possibility of other sheaths detached from Io where particles might be accelerated or turbulent regions where anomolous resistivity could limit the current and modify the proposed model. Block (1972) discusses the possibility of double layer sheaths forming at density minima from field aligned currents flowing in the Earth's ionosphere. Several of these layers could be formed (indicated on Figure 1). In the laboratory these layers have been observed to appear and disappear at a 10–100 kHz rate. Such a time scale is suggestive of the Jovian S-burst duration.

Acknowledgements

The authors wish to thank Dr Lars P. Block for his helpful comments and suggestions concerning this model.

This research was supported by the Atmospheric Sciences Section, National Science Foundation, NSF Grant GA31676.

References

Bigg, E. K.: 1964, *Nature* **203**, 1008.
Block, Lars P.: 1972, submitted to *Cosmic Electrodyn.* **3**.
Criswell, David R.: 1971, *EOS* **52**, 855.
Dermott, S. F.: 1970, *Monthly Notices Roy. Astron. Soc.* **149**, 35.
Duncan, R. A.: 1970, *Planetary Space Sci.* **18**, 217.
Gledhill, J. A.: 1967, *Nature* **214**, 155.
Goldreich, P. and Lynden-Bell, D.: 1969, *Astrophys. J.* **156**, 56.
Gross, R. H. and Rasool, S. I.: 1964, *Icarus* **3**, 311.
Gurnett, Donald A.: 1972, *Astrophys. J.* **175**, 525.
Ioannidis, G. and Brice, N.: 1971, *Icarus* **14**, 360.
Lewis, John S.: 1971, *Icarus* **15**, 174.
Mozer, F. S. and Bogott, F. H.: 1972, *Astrophys. J.* **174**, 153.
Piddington, J. H. and Drake, J. F.: 1968, *Nature* **217**, 935.
Piddington, J. H.: 1972, *Cosmic Electrodyn.* **3**, 240.
Prasad, S. S. and Capone, L. A.: 1971, *Icarus* **15**, 45.
Reasoner, David L. and Burke, William J.: 1971, 'Direct Observation of the Lunar Photoelectron Layer', preprint Dept. of Space Science, Rice University.
Schatten, K. H. and Ness, N. F.: 1971, *Astrophys. J.* **165**, 621.
Warwick, J. W. and Dulk, G. A.: 1964, *Science* **145**, 380.

DISCUSSION

Thomas: The decametric radiation has a well-known frequency versus time structure. Do you attribute this structure to variations at the source or to changes in group-travel times of the various frequency components as they travel through Jupiter's ionosphere, as is done for example by Wu and co-workers (Univ. of Md.)?

Shawhan: I feel that the frequency time characteristics are due both to the emission mechanism and the propagation characteristics of the emitted radio waves.

Srnka: Your values of electron current densities mean that the drift velocities of the electrons are much larger than the ion sound speed, and nearly equals the electron thermal speed. This means the current-carrying plasma should be turbulent, and exhibit an anomalous plasma resistivity. Have you considered this?

Shawhan: We have considered anomalous plasma conductivity, but we have not yet made detailed calculations. At the moment, we cannot say whether a highly resistive plasma path for the current would be important in the problem.

Fredricks: Brice's model of the distribution of thermal plasma around Io's orbit, as I remember, is a pancake-like distribution confined quite closely to the equatorial plane. If this model is correct, then the strong field-aligned currents due to your runaway electrons would not lead to plasma turbulence and anomalous resistivity along field lines at higher latitudes. Thus, the absence of plasma along the field lines away from the equator means no background necessary to carry the plasma instability due to beam-plasma or streaming interactions, and thus it may be that a very high conductivity exists over most of the field line. Then, the important plasma wave stimulation would mainly occur when your runaway electron currents enter the low altitude Jovian ionosphere at the base of the field line.

ELECTROSTATIC CHARGING AND FORMATION
OF COMPOSITE INTERSTELLAR GRAINS

B. FEUERBACHER, R. F. WILLIS, and B. FITTON

*Surface Physics Division, European Space Research and Technology Centre,
Noordwijk, The Netherlands*

Abstract. The equilibrium potential is calculated for interstellar grains subjected to typical UV radiation fields and plasma parameters, using two different grain materials. For an environment typical of an H I region, the results show that the various components of interstellar grains in the multicomponent model may charge to potentials of opposite sign, thus favouring the formation of mixed composite grains due to Coulombic attraction. In H II regions, a three-shell model which defines the grain charge is presented, consisting of a sphere of positively charged grains around the central star bounded by a transition region, in which the two types of grains are oppositely charged. Here grain growth and the formation of mixed composite grains may take place. In the outer shell all grains are charged negative. Estimated radii of these zones are given for an exciting O5 type star.

1. Introduction

Interstellar matter is an important constituent of the universe, contributing about half of the total mass density of our Galaxy (van Oort, 1932). This interstellar matter consists primarily of hydrogen gas with an average density of about 1 atom cm^{-3}. Only about 1% of the mass of the interstellar matter is found in the form of solid grains. It is this part however which is reponsible for the obscuration of distant stars. Measurements of the interstellar extinction show that the absorption of starlight increases towards shorter wavelengths, following roughly a $1/\lambda$ law. Mie scattering theory allows conclusions on the possible size and shape of the grains, and leads to a rough indication of the type of grain material in order to fit the observed extinction curves. The size estimations range from 150 Å (Gilra, 1972) to 3000 Å (van de Hulst, 1946). Present views of the nature of the grain material assume a mixture of different constituents, mainly silicate, graphite, iron, quartz, and silicon carbide (Hoyle *et al.*, 1969; Gilra, 1971; Gilman, 1969).

The physical condition of the interstellar gas varies for different regions in the Galaxy. Close to a hot early-type star, where a large density of high energy photons is present, the interstellar hydrogen will be fully ionized and therefore unable to absorb photons. As the radiation flux density decreases away from the star, recombination becomes more and more important. Neutral hydrogen absorbs light with energies above the hydrogen ionization limit at 13.6 eV. One therefore expects a relatively sharp boundary between regions of fully ionized hydrogen and neutral hydrogen called the *Strömgren* radius. The region within the Strömgren radius, containing ionized hydrogen in a hard radiation field around a hot star is termed a *H II region*, while the region outside it, where hydrogen is ionized only to a degree of 1 in 10^3 and the radiation field is shielded for energies above 13.6 eV, is called *H I region*. Typical kinetic plasma temperatures are 100 K in *H I* regions and 10^5 K in *H II* regions.

R. J. L. Grard (ed.), Photon and Particle Interactions with Surfaces in Space, 415–426. All Rights Reserved
Copyright © 1973 by D. Reidel Publishing Company, Dordrecht-Holland

Interstellar grains are expected to be charged as the result of two competing processes. Photoelectric emission will tend to remove electrons from the grains, leaving
them at a positive potential. On the other hand, charge accretion from the ambient
plasma simultaneously takes place. In thermal equilibrium, electrons move much
faster than ions, therefore electron encounters on the grain will be much more frequent
than ion encounters. If the sticking coefficient for ions and electrons is not too different,
the grain will tend to charge negatively under the impacts from the ambient plasma.
The balance between the two competing processes, photo-emission and charged
particle capture, will determine the grain equilibrium potential.

In the present paper, an attempt is made to calculate the grain potential by for-
mulating the current balance at the grain. Current-voltage characteristics are set up
independently for the case of photoelectric emission and charged particle capture.
If these characteristics are expressed in terms of outgoing electrons or incoming net
negative unit charge respectively, then the crossing point of the two curves represents
the equilibrium potential of the grains. This method is illustrated in Figure 1, which
shows a photoelectron emission characteristic PE and three charged particle characteristics CP. At the crossing points of these curves, the in- and outgoing currents
balance. In order to obtain the equilibrium potential of a grain as a function of the
electron density in the ambient plasma, the curves for charged particle capture are
scaled by the electron density as shown in Figure 1 for three factors. Using this

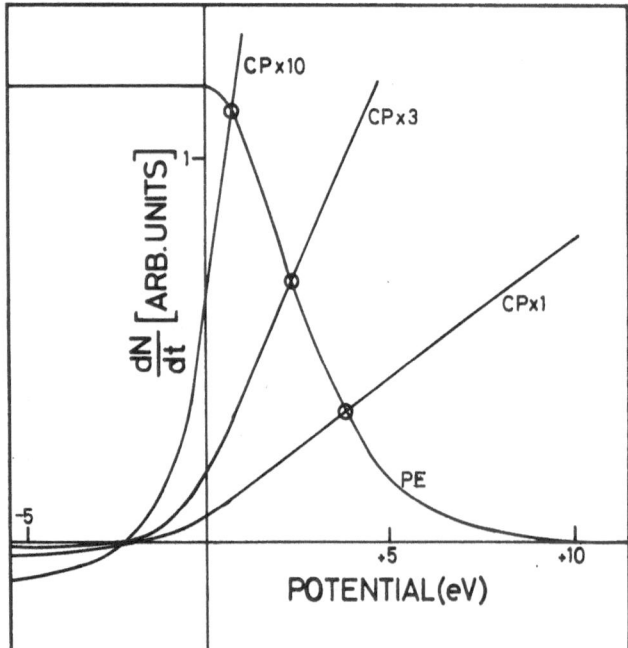

Fig. 1.   Schematic illustration of the calculation of the grain equilibrium potential as a function of
electron density by crossing a photoelectron current-voltage characteristic (PE) with a charged
particle characteristic (CP) multiplied by different factors.

method, one can now plot the equilibrium potential as a function of electron density.

2. Photoelectric Emission

The current-voltage characteristics for photoelectric emission are obtained from laboratory data (Anderegg *et al.*, 1973). The photoelectric yield, measured as a function of incoming photon energy, is multiplied by a photon spectrum typical for the

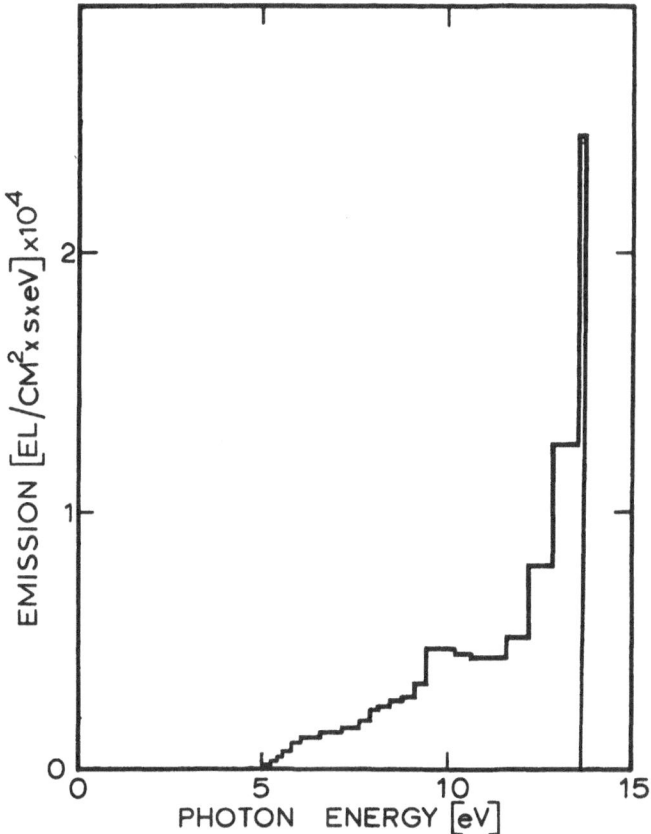

Fig. 2.   Photoelectron flux from a graphite grain in an H I region.

interstellar region of interest. The result, calculated for graphite in a radiation field theoretically derived for an H I region (Habing, 1968), is shown in Figure 2. The photon spectrum essentially drops to zero for energies above the hydrogen cutoff, so that significant photoemission is limited to a range between the photoelectric threshold and 13.6 eV. The next step is to scale photoelectron energy-distribution spectra measured at photon energies $\hbar\omega_0$ such that the integral under the spectrum is equal to

the photoelectron number read from Figure 2 at energy E_0. These scaled curved are
then integrated with respect to the photon energy. The result is shown in **Figure 3**,
which gives the energy distribution of the photoelectrons emitted from a graphite
grain subjected to a radiation field typical for an H I region. Most electrons are seen
to be emitted with energies below 4 eV. It is obvious that a Maxwellian velocity
distribution is not a good approximation in this case.

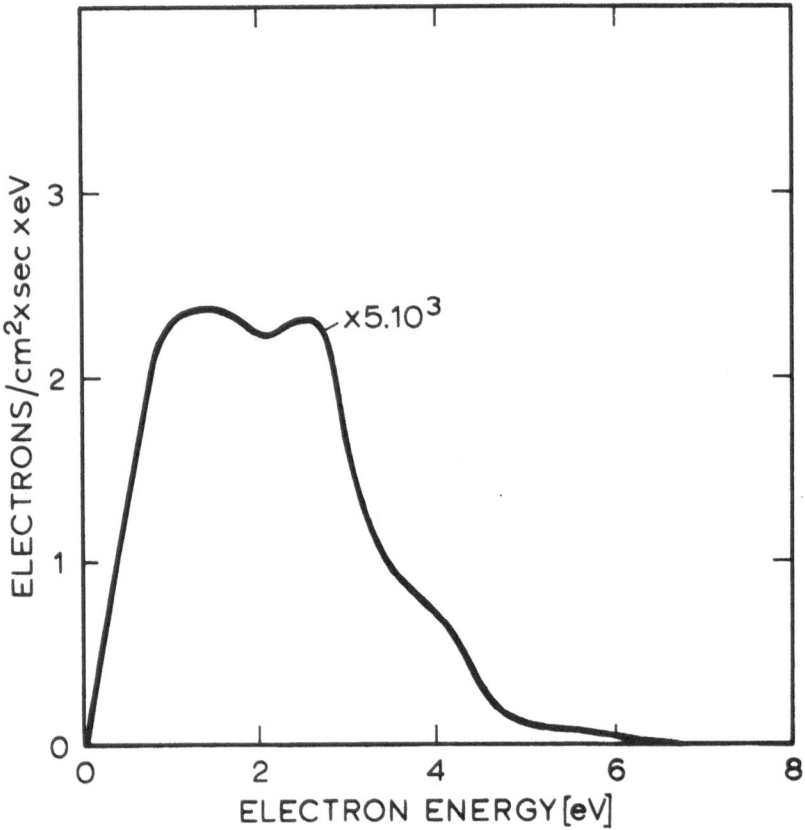

Fig. 3.   Energy distribution of photoelectrons emitted from a graphite grain in an H I region.

If the grain is at a certain potential U, all electrons with kinetic energy $E_0 > U$, will
be able to leave the grain. The number of these electrons is given by the integral over
the energy distribution curve, Figure 3, taken from E_0 to infinity. Consequently,
the current-voltage characteristic of the grain is obtained by performing this integration for various values of E_0 and plotting the result as a function of E_0, which may be
identified with the grain potential U. The current-voltage characteristic obtained in
this way for graphite grains in a H I region is shown in Figure 4. A constant saturation
current is drawn for negative potentials, since all electrons can leave the grain.

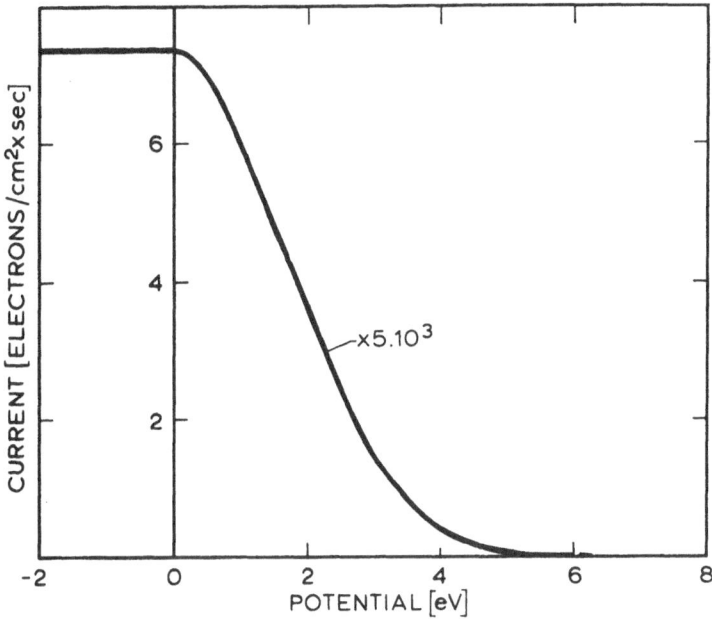

Fig. 4. Current-voltage characteristic of a graphite grain for photoelectric emission in an H I region.

3. Charged Particle Capture

The treatment of the current-voltage characteristic due to charged particle capture follows mainly the early work of Spitzer (1941, 1948). The change of the number of electrons on the grain, dN/dt, is given by

$$\frac{dN}{dt} = neA\left(s_e\sigma_e v_e - s_I\sigma_I v_I\right),$$

where n is the electron density, equal to the ion density for plasma neutrality, e is the electron charge, and A is the geometrical cross section of the grain. The sticking coefficients for electrons and ions of velocity v_e and v_I are given by s_e and s_I. The collision cross sections σ_e and σ_I for electrons and ions are different from the geometrical cross section A due to Coulomb interaction. They are given by Spitzer (1948).

$$\sigma_e = 1 + \frac{2U}{m_e v_e^2} \qquad \sigma_I = 1 - \frac{2U}{m_I v_I^2},$$

where U is the grain potential. Introducing a Maxwellian velocity distribution $f(v)$ for electrons and ions, one may write

$$\left.\frac{dN}{dt}\right|_{U<0} = nseA\left(\int_{v_0}^{\infty} \sigma_e v_e f(v_e)\, dv_e - \int_0^{\infty} \sigma_I v_I f(v_I)\, dv_I\right).$$

Here the sticking coefficient for electrons and ions has been assumed to be equal and independent of velocity. For negative potentials, ions of any energy may reach the grain, while some electrons will be repelled. The lower integration limit for electrons is therefore taken at the point, where the cross section σ_e vanishes. For positive grain potentials, the same considerations apply on the integration limits for ions.

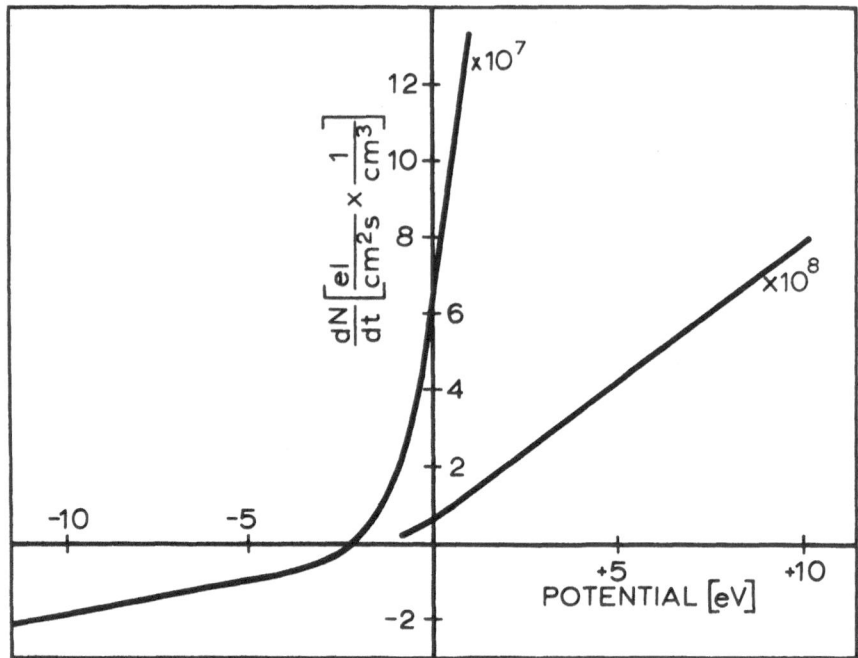

Fig. 5. Current-voltage characteristic of an interstellar grain for charged particle capture in a plasma consisting of electrons and protons with a kinetic temperature of 10000 K.

A typical current-voltage characteristic for charged particle capture is shown in Figure 5 for the case of a grain in a plasma consisting of electrons and protons with a kinetic temperature of 10000 K (H II region). For negative potentials, mainly ions are captured, and the slope of the curve is small. As more electrons reach the grain, the capture rate depends strongly on the grain potential. For the case of negligible photoelectric emission the zero crossing point represents the equilibrium potential, which is seen to be slightly negative as expected.

4. Results

The present calculations of the equilibrium potential of interstellar grains have been performed for two different materials, which are taken to represent two components of the constitutents of interstellar grains with extreme properties as regards the photo-electric emission. The material chosen to represent a low yield emitter was graphite.

It is an obvious choice since graphite is known to emit relatively few photoelectrons (Anderegg *et al.*, 1973) and is also very likely to form a major constituent of interstellar grains (Hoyle, 1969). The choice of a high yield photoemitter was difficult because photoemission data on relevant insulating materials are very limited and difficult to obtain due to charging problems. The problem was overcome by selecting aluminium oxide since not only does it have a high photoelectric yield but also it can be prepared as a thin layer on an aluminium substrate so as to reduce the charge build-up problem. The results calculated for aluminium oxide are typical of those expected for silicon oxide, which is believed to be a second major component of interstellar grain clouds (Gilra, 1971).

Grain potential calculations are presented for grains in both H I and H II regions. For H I regions, a radiation field has been taken which was theoretically derived by Habing (1968) and agrees with experimental data (Hayakawa *et al.*, 1969) in the region of overlap. Oxygen was assumed to be the most abundant ionic species for this region. For the case of an H II region, a very hot exciting star of type O5 was chosen, with an ultraviolet radiation temperature of 50000 K.

4.1. H I REGIONS

The results for the equilibrium potentials of interstellar grains in an H I region is given in Figure 6. The potential is given as a function of the electron density N_e. The upper diagram shows the results as calculated for the case of a low yield photoemitter (graphite), while the lower diagram was calculated for a high yield emitter (aluminium oxide). The curves are shown for three different values of the kinetic temperature of the ambient plasma. The shaded band represents a range of a factor of two greater or less than the 'standard' value of 100 K.

For the case of negligible photoelectric emission, represented by the low yield material in a high ambient electron density (upper diagram of Figure 6, right hand side), a negative potential of several tenths of an eV is obtained. For the other extreme, dominating photoemission (lower diagram of Figure 6, left hand side), the potential is positive. Its value is determined by the upper limit for the photon energy, 13.6 eV, minus the photoelectric threshold of the material. The arrows indicate estimated or measured electron densities for typical H I regions. BS marks an upper and lower limit given by Bergeron and Souffrin (1971), SS and B mark values given by Spitzer and Scott (1969) and Ball *et al.* (1970).

From the curves, it can be seen that a low yield grain in a plasma at 100 K kinetic temperature changes from positive to negative potential as the electron density exceeds 6×10^{-3} cm^{-3}. The same behaviour is observed for a high yield grain, but at a much higher electron density, namely 2×10^{-1} cm^{-3}. As a consequence, over the whole range of most likely electron densities (Bergeron and Souffrin, 1971; Spitzer and Scott, 1969; Ball *et al.*, 1970), grains composed of these two different materials will be charged to potentials of opposite sign. This, in turn, leads to the conclusion that, in the multicomponent model of interstellar grains, it is possible that grains composed of different materials will charge to potentials of opposite sign under those conditions expected for H I regions. Grain charging therefore will give rise to attractive Coulomb forces between the grains, which may lead to the formation of mixed composite particles having extinction properties quite different from those expected for a cloud composed of a mixture of individual grains of different nature.

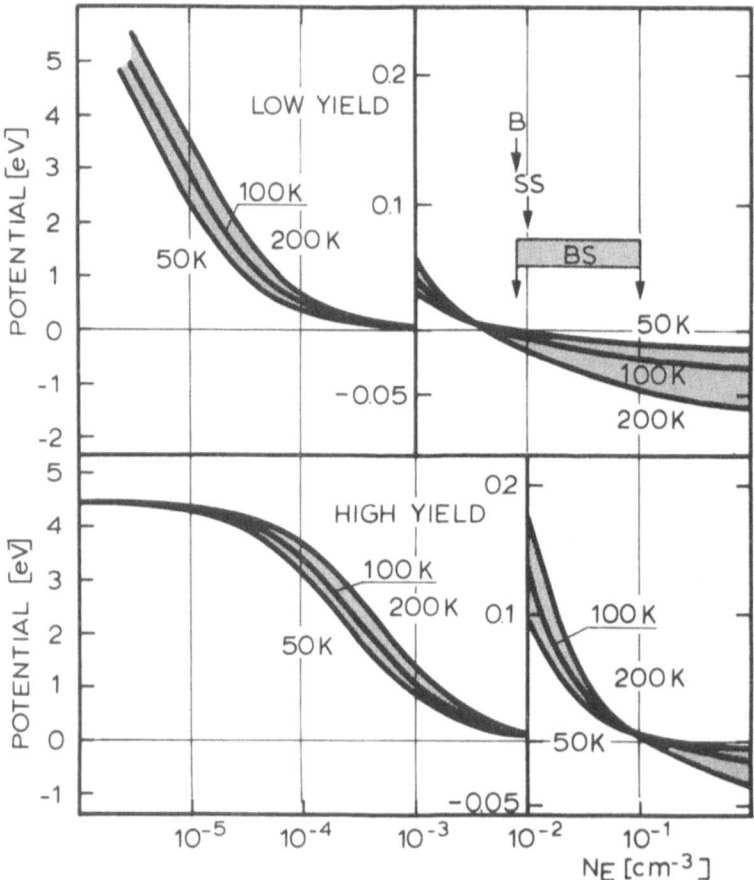

Fig. 6. Equilibrium potential of interstellar grains in an H I region for three different kinetic plasma temperatures. The upper diagram is calculated for a material of low photoelectric yield (graphite), the lower for a high yield material (aluminium oxide). The arrows mark observed or estimated values for the electron density.

The rate of formation of composite grains will depend on their random velocity and the mean number of grains per unit volume. Estimates based on a grain velocity equal to the Brownian motion corresponding to the ambient gas temperature show that densities well above the average interstellar value are required for the electrostatic forces to be effective (Purcell, 1973). The proposed grain growth mechanism may however constitute a significant process in relatively dense dust clouds.

The present results may also have consequences for the process of molecule formation, which is known to take place on the surface of grains. Many of the most abundant atoms in H I regions will be in the form of ions, for example carbon and silicon. If the grain potential should exceed 0.1 eV positive, the majority of ions in an ambient plasma of 100 K kinetic temperature will be unable to impact with the grain. Graphite grains, on the other hand, will possess a negative charge over the pertinent range of electron densities and, as a consequence, will be most effective in molecule formation. In contrast, in the interior of dense clouds the radiation field will be shielded (Greenberg, 1971), and this will shift the curves in Figure 6 to the left, resulting in more negative grain potentials for the high photoyield materials also. This will favour molecule formation due to ion-charged grain impact in the interior of dense dust clouds for a wide range of grain materials, which is in agreement with observations.

4.2. H II REGIONS

Figure 7 shows the grain potentials as determined for the above materials for an environment typical of an H II region. The radiation field has been calculated for an exciting O5 star with an ultraviolet radiation temperature of 50000 K at a distance of 10 pars. The calculation was performed for three different radiation temperatures, again a factor of 2 deviation from a 'standard' value of 10000 K. The arrows mark measured values of the electron density as given by O'Dell (1965) (D) or from a list of 40 observations collected by Johnson (1968) (J). A 'standard' value S is taken at the lower limit of the observed range because of observational selection.

At first glance it seems surprising that an equilibrium potential is obtained, which is close to the values expected for vanishing photoelectron emission, since the photon flux is higher close to a hot star. However, due to the complete ionization of hydrogen the electron density is much higher in an H II region, thereby compensating for the higher photoelectron flux at an arbitrary distance of 10 pars. In fact Figure 7 may be used also to find the grain charge as a function of the distance from the exciting star. Assuming that the photon spectrum varies only by quantity and not by quality, i.e., the spectral distribution remains constant as the distance is varied, the horizontal scale may be given in terms of the distance from the central star. Taking a constant electron density of 10 cm^{-3}, the scale on the top of Figure 7 gives the distance from the star in pars. Close to the exciting star, photoelectron emission dominates, so all grains will be charged positive. In an intermediate range, both positive and negative grain charges are possible, while further out all grains will be negatively charged. The results of the present investigation may be interpreted, therefore, in the form of a three shell model based on the sign of the grain charge. For a 50000 K exciting star and an electron density of 10 cm^{-3}, all grains will be charged positive, up to a distance of 1.5 pars. In a shell-region, between $r = 1.5$ pc and $r = 6.5$ pc, opposite sign potentials are possible on grains having different photoelectric properties. Here grain impact and growth may take place by Coulombic attraction until finally the grains are blown out of this region by radiation pressure. Further out, up to the Strömgren radius, the potential on all grains will be negative. An illustration of a possible consequence of the above model is given in Figure 8, which shows a photograph of the Rosette Nebula, NGC 2237. This emission nebula has an obvious dark hole in its center. According to Mathews (1967), this central hole is due to the fact that the Coulombic drag essentially 'freezes'

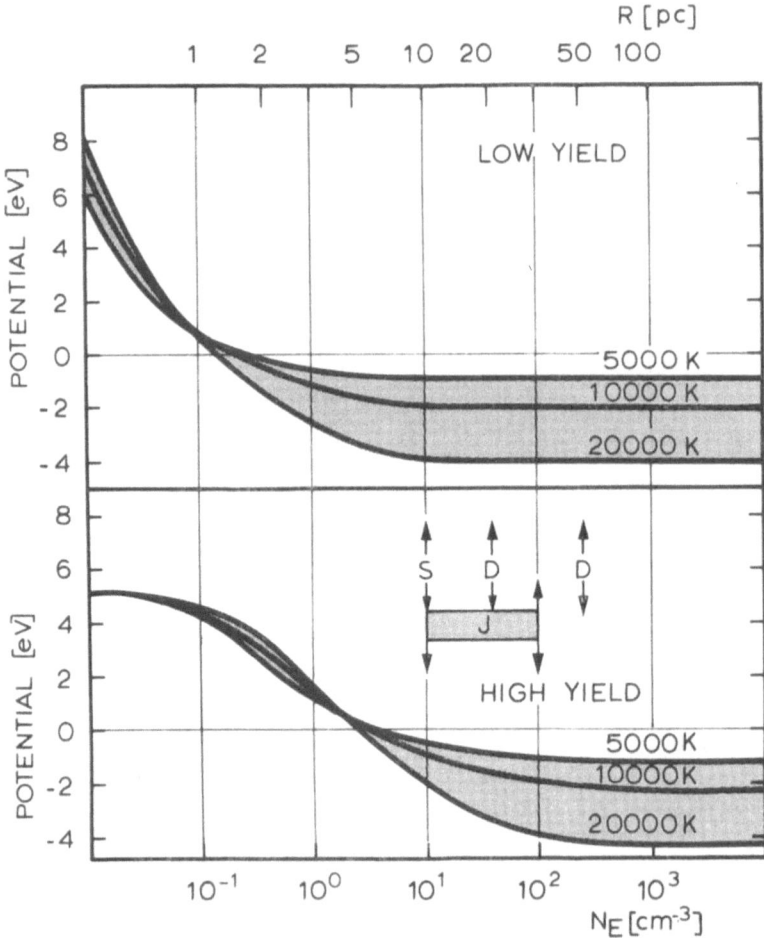

Fig. 7. Equilibrium potential of interstellar grains in an H II region of an O5 star at 10 pc distance of the exciting star. Upper diagram, low yield material (graphite); lower diagram, high yield material (aluminium oxide). The plasma kinetic temperature is used as a parameter. The scale on the top indicates the potential as a function of the distance from the exciting star for a constant electron density of 10 cm^{-3}.

the grains with a high positive charge near the center into the ambient plasma. As the grains are pushed away from the center by radiation pressure, they drag the plasma along, leaving a central hole without gas to emit light. The radius of the central hole is found to be about 4 pc, in good general agreement with the radii given above for positive grain charges around an O5 star.

A final remark remains to be made concerning the possibility of negative drag effects, such as those discussed in the paper of Walker (1973). Here, it was mentioned that negative drag effects could arise for the case when ions are elastically reflected after being discharged on the surface, provided the potential of the body is large and negative and the shape is assymmetric. While our knowledge to date about the way

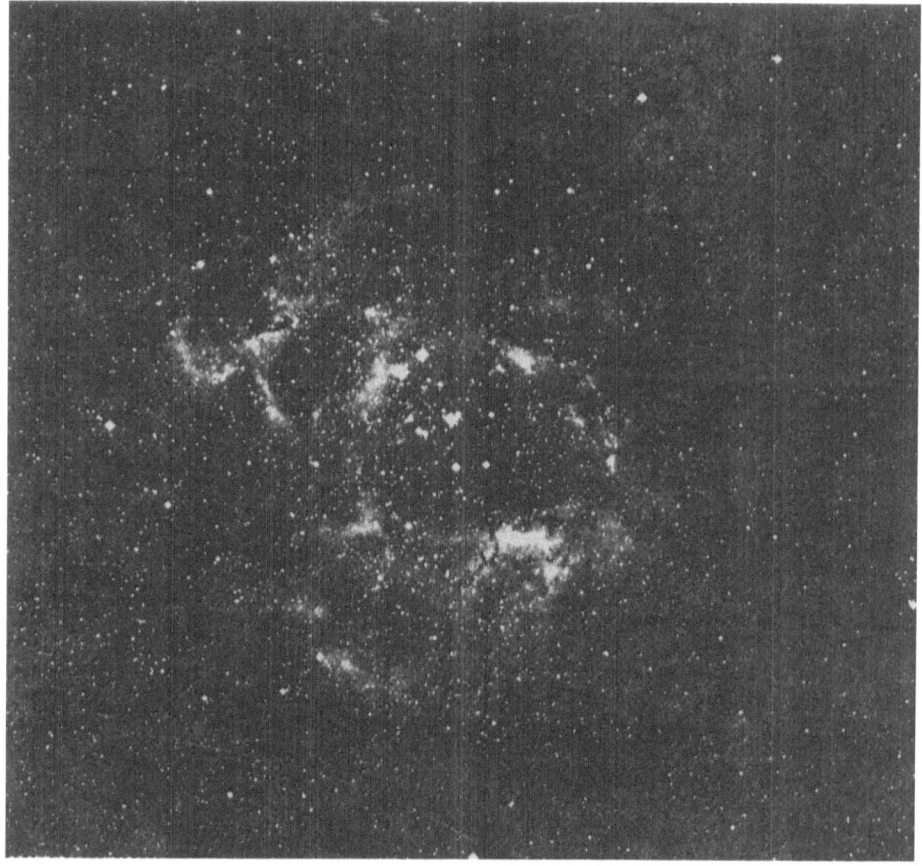

Fig. 8. The Rosette Nebula, NGC 2237, photographed with the 48-in, telescope at Palomar. The radius of the central hole is about 4 pc. (Photograph courtesy of The Hale Observatories).

low energy ions interact with interstellar grains is sparse, it is widely believed that the grains are indeed asymmetric in shape, as shown by the polarization effects arising from interstellar absorption. As the present results indicate, negative potentials on the grains may be maintained in the outer shell of an H II region where the high temperature plasma is able to keep the grains at potentials several volts negative. This would indicate that the only region where negative drag effects might be important in the mechanics of interstellar grains therefore is in the outer shell zone of an H II region.

References

Anderegg, M., Feuerbacher, B., and Fitton, B.: 1973, this volume, p. 331.

Ball, J. A., Cesarski, D., Dupree, A. K., Goldberg, L., and Lilley, A. E.: 1970, *Astrophys. J.* **162**, L25.

Bergeron, J. and Souffrin, S.: 1971, *Astron. Astrophys.* **11**, 40.

Gilman, R. C.: 1969, *Astrophys. J.* **155**, L185.

Gilra, D. P.: 1971, *Nature* **229**, 237.

Gilra, D. P.: 1972, in J. M. Greenberg and H. C. van de Hulst (eds.), 'Interstellar Dust and Related Topics', *IAU Symp.* **52**, in preparation.

Greenberg, M. J.: 1971, *Astron. Astrophys.* **12**, 240.

Habing, H. J.: 1968, *Bull. Astron. Inst. Netherlands* **19**, 421.

Hayakawa, S., Yamashita, K., and Yoshioka, S.: 1969, *Astrophys. Space Sci.* **5**, 493.

Hoyle, F. and Wickramasinghe, N. C.: 1969, *Nature* **223**, 459.

Johnson, H. M.: 1968, *Stars and Stellar Systems* **7**, 65.

Mathews, W. G.: 1967, *Astrophys. J.* **147**, 865.

O'Dell, C. R.: 1965, *Astrophys. J.* **143**, 168.

Purcell, E. M.: 1973, private communication.

Spitzer, L.: 1941, *Astrophys. J.* **93**, 369.

Spitzer, L.: 1948, *Astrophys. J.* **107**, 6.

Spitzer, L. and Scott, E. H.: 1969, *Astrophys. J.* **158**, 161.

Van Oort, J. H.: 1932, *Bull. Astron. Inst. Netherlands* **6**, 249.

Van de Hulst, H. C.: 1946, *Rech. Astron. Obs. Utrecht* **11**, part 1.

Walker, E. H.: 1973, this volume, p. 73.

DISCUSSION

Wickramasinghe: Have you made an estimate of time scales involved, e.g. in reaching steady state potential, and infusion of oppositely charged grains leading to composite grains?

Feuerbacher: We haven't so far, the main problem being the magnitude of the turbulent velocities of the grains in a cloud. We have estimated the screening lengths and found them in the same order of magnitude as the average intergrain distance, so Coulomb interaction is possible even with very small random velocities. The time scales involved in fact are very large. For example, photoelectric emission from the grains takes place with a frequency of about one photoelectron per year.

Gold: The plasma forces you mentioned may be more important for causing interstellar grains to spin rather than to accelerate their motion.

Feuerbacher: I fully agree. All I was trying to say is that the only region where such processes are possible are the rims of H ii regions.

3.3. SOLAR WIND INTERACTIONS WITH CELESTIAL OBJECTS

LUNAR ION FLUX AND ENERGY

R. H. MANKA and F. C. MICHEL

Space Science Dept., Rice University, Houston, Tex. 77001, U.S.A.

Abstract. The dynamics of the lunar ionosphere and resulting flux of lunar ions to the lunar surface are reviewed. In the Lunar rest frame, ions formed from the neutral lunar atmosphere are accelerated by the interplanetary electric and magnetic fields. The trajectories of heavier ions are primarily along the electric field; the ion flux is in a direction perpendicular to the solar wind flow, correlated to the orientation of the interplanetary magnetic field, and impacts the surface with energies of tens of electron volts to a few keV. Thus we predict a relatively energetic (compared to thermal energies), highly directional, lunar ion flux but with the possibility that light ions such as hydrogen can execute orbits that return them to portions of the lunar surface not directly exposed to the solar wind. The effects of surface electric and magnetic fields are discussed, as is the ion energy spectrum. We calculate the trapping of atmospheric ions which impact the surface and from this the density of neutral Ar^{40} in the lunar atmosphere. We show that using the acceleration model, the lunar atmosphere total neutral number density can be calculated from ion detector data.

1. Introduction

The purpose of this paper is to very briefly review the properties, for the dynamics of the lunar ionosphere, which have been presented in a series of preceding talks and publications. It has been shown that the principal acceleration of lunar ions is by electric and magnetic fields in the solar wind and that these fields control much of the dynamics of the lunar ionosphere (Manka and Michel, 1970a, b). Furthermore, for those ions whose trajectories carry them to the lunar surface, some impact with sufficient energy to be trapped in the surface and this is the likely source of the anomalously high concentration of Ar^{40} found in the surface of lunar sample grains. (Manka and Michel, 1970c; this idea was also suggested by T. Gold at the 1*st Lunar Sci. Conf.*, see Heymann and Yaniv, 1970) As we shall see, the acceleration model for lunar ions also predicts the flux, energy, and direction of incoming ions measured by ion detectors at the lunar surface. Furthermore, if measured ion fluxes are selected which have the proper direction, flux, and spectrum to indicate that they represent the lunar ionosphere, rather than some external source, then the density and pressure of the neutral lunar ionosphere can be calculated (Manka *et al.*, 1972b). The study of the dynamics of the lunar ionosphere is related to a number of other measurements of the lunar environment, such as radio measurements of the lunar ionosphere.

The lunar atmosphere has been reviewed by Bernstein *et al.* (1963), Michel (1964), and Hinton and Taeusch (1964) who discuss sources such as solar wind accretion and volcanism, and losses such as ionization and interaction with the solar wind, chemical reactions, and gravitational escape; mean lifetimes against ionization are 10^6 to 10^7 s. Using the ion trajectories described in this paper, it will be possible to calculate more accurately the largest loss mechanism, interaction with fiels in the solar wind. Some species generally expected in the lunar atmosphere are solar wind radiogenic gases such as H, He, Ne, Ar, Kr and Xe, and possibly some volcanized gases such as H_2O or CO and their dissociative products.

Atmospheric species except hydrogen and helium are gravitationally bound (have lifetimes against gravitational escape much greater than against ionization) and thus form part of the equilibrium lunar atmosphere. The density of each species decreases approximately exponentially with height

$$n(r) \cong n_0 e^{-r/h}, \tag{1}$$

where h is the scale height for the species given by

$$h = \frac{kT}{mg} \tag{2}$$

and n_0 is the density per cm^3 at the surface, k is the Boltzmann constant, T and m are species temperature and mass and g is the lunar gravitational acceleration. The Ar^{40} scale height is about 50 km at the subsolar point.

The properties of ions which are reviewed in this paper should also apply to ions measured by detectors in the ALSEP package. The Suprathermal Ion Detector (Freeman *et al.*, 1971) detects 'cloudlike' sporadic bursts of ions and also ions associated with Apollo vehicle impacts. The Solar Wind Spectrometer (Snyder *et al.*, 1971) has detected ions associated with vehicle impact events and obtained the directional spectrum of the ions. The relationship of these measurements to the ion trajectories will be briefly discussed later.

2. Trajectories of Atmospheric Ions

When an ion is formed in the lunar atmosphere, whether it escapes the Moon or is accreted to the surface will depend on the interplanetary electric field, and in some cases on electric and magnetic fields at the lunar surface. In a frame of reference at rest with respect to the Moon the interplanetary electric field is given by

$$\bar{E}_{sw} = - \bar{V}_{sw} \times \bar{B}_{sw}, \tag{3}$$

where $\bar{V}_{sw}$, $\bar{B}_{sw}$, and $\bar{E}_{sw}$ are, respectively, the solar wind velocity, magnetic field, and electric field. For an ion formed at rest in the lunar frame and accelerated by $\bar{E}_{sw}$ and $\bar{B}_{sw}$, the trajectory is in a plane perpendicular to $\bar{B}_{sw}$ and containing $\bar{E}_{sw}$ and $\bar{V}_D$, where $\bar{V}_D$ is the drift velocity given by

$$\bar{V}_D = \frac{\bar{E}_{sw} \times \bar{B}_{sw}}{B_{sw}^2}. \tag{4}$$

The plane of $\bar{V}_D$ can be thought of as defining a 'magnetic' longitude in analogy to the 'electric' latitude which will be introduced shortly. In the simplifying case where $\bar{B}_{sw}$ is perpendicular to $\bar{V}_{sw}$, then $\bar{V}_D$ is along $\bar{V}_{sw}$ and the ion drift plane passing through the center of the moon is just the noon-midnight plane.

For an ion formed at rest at $x=y=0$, then its orbit in the crossed electric and magnetic fields is an ordinary cycloid and is given by

$$x = \frac{-V_D}{\omega_c} \sin \omega_c t + V_{sw}t \tag{5}$$

and

$$y = \frac{V_D}{\omega_c} (1 - \cos \omega_c t), \tag{6}$$

where ω_c is the angular cyclotron frequency. The electrostatic force far exceeds the gravitational force so the initial motion of an ion is along $\bar{E}_{sw}$ and as the ion gains energy the magnetic force curves the ion in the direction of the solar wind flow with

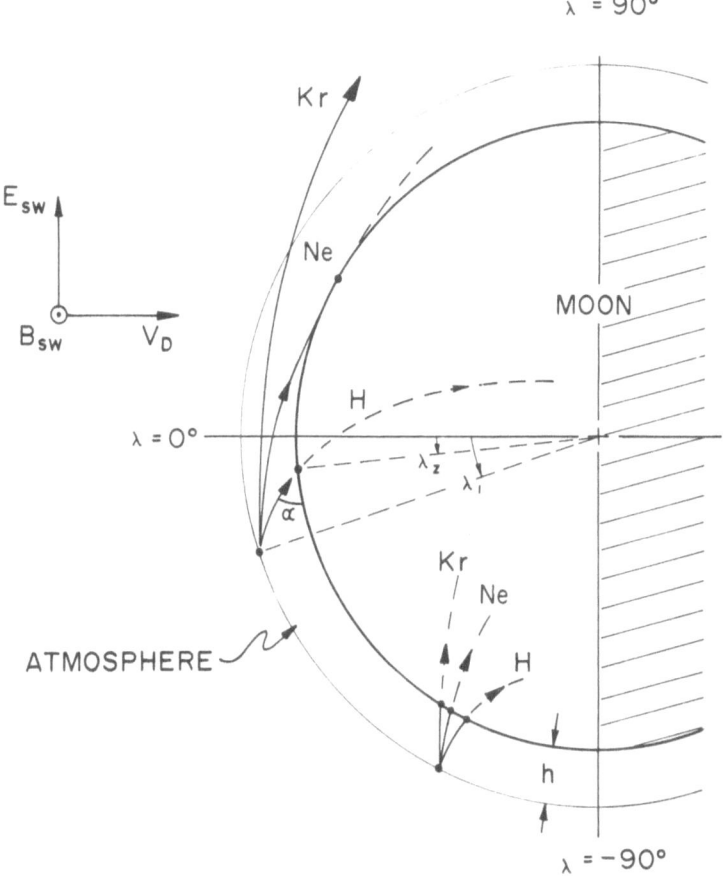

Fig. 1. Sketch of orbits of ions formed outside the plasma sheath and at a distance h from the surface. (Actually krypton and oxygen have different scale heights and hydrogen is not bound.) The sketch illustrates that ions heavier than hydrogen will strike primarily the hemisphere into which they are driven by the interplanetary electric field; when the polarities of $\bar{B}_{sw}$ and $\bar{E}_{sw}$ reverse, the other hemisphere is struck. The trajectories are shown in a plane passing through the center of the Moon and containing the drift velocity $\bar{V}_D$ as well as $\bar{E}_{sw}$.

a resulting cycloidal orbit. The height of the cycloid, twice the cyclotron radius for the ion in the lunar rest frame, is much greater than lunar dimensions. Thus the ion's trajectory from formation to impact is the initial part of a cycloid, and the motion in nearly parallel to $\bar{E}_{sw}$ and most of the flux of lunar ions to the surface is in a direction perpendicular to the solar wind flow. In general, ions formed in the lower sunlit atmosphere are driven up (with respect to $\bar{E}_{sw}$) into the Moon while ions formed at the equator and in the upper hemisphere escape, as illustrated in Figure 1. Depending upon the direction of $\bar{B}_{sw}$, the interplanetary electric field is generally upward or downward out of the solar ecliptic plane; and, when the direction of $\bar{B}_{sw}$ reverses several times during each solar rotation (as is common due to the sector structure of the interplanetary magnetic field), the direction of $\bar{E}_{sw}$ and the ion flux also reverse.

The energy of the ion at impact is just the energy gain along the interplanetary electric field

$$\mathscr{E} = eE_{sw}y_i, \tag{7}$$

where e is the ion charge (we assume single ionization) and y_i is the y-coordinate at impact. The coordinates and other parameters of impact are found by solving the equations for the cycloid with the equation for the locus of the lunar surface. As shown in Figure 1, an ion formed at some height, say one scale height, above the surface and at an 'electric' latitude λ_1 impacts at λ_2 with an impact angle α and an energy $\mathscr{E}$. (The latitude is chosen with respect to the direction to $\bar{E}_{sw}$; thus if $\bar{E}_{sw}$ reverses direction, the $\lambda = +90°$ and $-90°$ poles also exchange positions with respect to the fixed selenographic coordinates.) There is a critical starting latitude for which an ion just grazes the surface, and ions formed at latitude nearer the equator, and in the upper hemisphere, escape. The sketch of Figure 1 illustrates, with some exaggeration, the trajectories of three ions of quite different mass. Actually, hydrogen is not gravitationally bound, scale height does not have a meaning in the usual sense, and most hydrogen ions are formed at much greater heights than for bound species. Tables IA and IB show the impact angle, latitude, and energy for ions formed at altitude h and latitude λ_1. The average interplanetary magnetic field varies considerably about 6γ (gamma) and is oriented at about 45 deg to the flow direction. The characteristic features of the trajectories can be obtained by calculating the trajectory in a plane containing $\bar{V}_D$, perpendicular to $\bar{B}_{sw}$, and through the center of the Moon; the orbit parameters are calculated numerically for both $B_{sw} = 5\gamma$ and 10γ (which give $E_{sw} = 2$ V km^{-1} and 4 V km^{-1} respectively) where 5γ is probably closer to the actual average conditions. Examination of the impact coordinates shows relatively small differences for the two cases; however, the impact energies for the 10γ case are about twice those for the 5γ case since the energies depend on E_{sw} as shown in Equation (7). Physically we can see that since only the initial part of the cycloid is involved, the ion trajectory is primarily along $\bar{E}_{sw}$ and thus variations in the magnitude of B_{sw} do not greatly alter the impact coordinates; however, variations in the magnitude of $\bar{B}_{sw}$ do proportionally change the magnitude of $\bar{E}_{sw}$ and thus the value of $\mathscr{E}$. The impact energies are higher near the equator than at the poles so atmospheric ions should be

TABLE IA

Orbit parameters for ions starting at rest one scale height from the lunar surface. Solar wind parameters are $V_D = 400$ km s^{-1} and $B_{sw} = 5\gamma$; calculations are for the lunar noon-midnight meridian plane

λ_1 (deg)	λ_2 (deg)	α (deg)	$\mathscr{E}$ gain (eV)
Ne^{20} ($h = 100$ km)			
− 90	− 90.1	93.3	200
− 60	− 58.3	61.6	228
− 30	− 24.5	28.9	397
−14.9	0	6.8	951
− 13.7	8.2	0	1363
Ar^{40} ($h = 50$ km)			
− 90	− 90.0	91.6	100
− 60	− 59.1	60.8	115
− 30	− 27.2	29.4	199
− 11.1	0	4.2	687
− 10.4	4.7	0	935
Kr^{84} ($h = 23.8$ km)			
− 90	− 90.0	90.8	48
− 60	− 59.6	60.4	55
− 30	− 28.7	29.7	95
− 8.0	0	2.4	487
− 7.6	2.6	0	627

TABLE IB

Orbit parameters for ions starting at rest one scale height from the lunar surface. Solar wind conditions are $V_D = 400$ km s^{-1} and $B_{sw} = 10\gamma$; calculations are for the noon-midnight meridian plane

λ_1 (deg)	λ_2 (deg)	α (deg)	$\mathscr{E}$ gain (eV)
Ne^{20} ($h = 100$ km)			
− 90	− 90.2	94.6	400
− 60	− 58.3	63.1	451
− 30	− 24.8	30.9	763
− 13.7	0	9.3	1747
− 11.3	11.7	0	2857
Ar^{40} ($h = 50$ km)			
− 90	− 90.0	92.3	200
− 60	− 59.1	61.5	228
− 30	− 27.3	30.4	392
− 10.3	0	5.6	1286
− 9.1	6.8	0	1960
Kr^{84} ($h = 23.8$ km)			
− 90	− 90.0	91.1	96
− 60	− 59.6	60.7	110
− 30	− 28.7	30.2	189
− 7.5	0	3.4	912
− 6.9	3.7	0	1289

more easily trapped near the equator. The impact energy is also approximately inversely proportional to the mass of the ion.

In the geometry of the trajectories discussed so far, the plane of the trajectory is determined by the direction of $\bar{E}_{sw}$ which in turn is orthogonal to $\bar{B}_{sw}$. While $\bar{B}_{sw}$ lies on the average in the vicinity of the plane of the ecliptic, it actually fluctuates widely

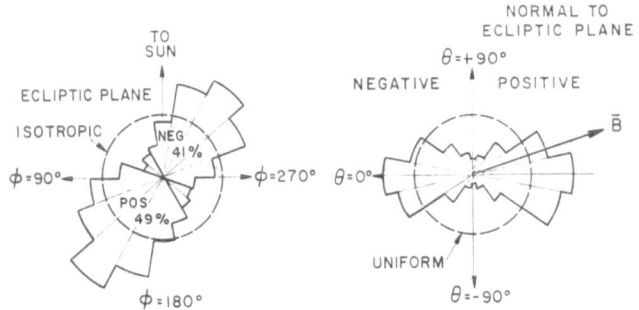

REPRESENTATIVE INTERPLANETARY MAGNETIC FIELD DIRECTIONS
(Burlaga and Ness, 1968)

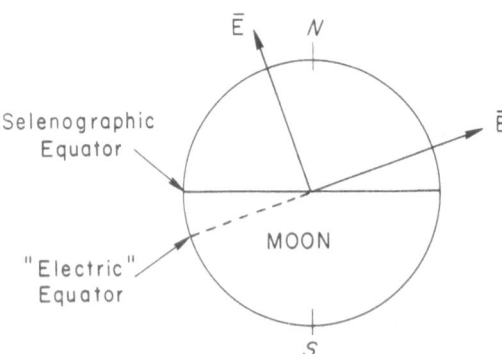

"ELECTRIC" COORDINATES – PROJECTED ON MOON

Fig. 2. Directional histograms of interplanetary magnetic fields parallel and perpendicular to the ecliptic as measured by Pioneers 6 and 7 (Burlaga and Ness, 1968). Also shown is an example of a particular orientation of $\bar{B}_{sw}$, and corresponding $\bar{E}_{sw}$, projected onto the Moon.

in time and may even be at right angles to the ecliptic. In this latter case, the electric field lies in the ecliptic as does the plane of the ion trajectory. Thus although the ion flux is generally in the sunlit portion of the upper or lower hemisphere, this distribution is smeared considerably owing to fluctuations of $\bar{B}_{sw}$. Figure 2 contains data of Burlaga and Ness (1968) showing the distribution of magnetic field directions in the plane of, and perpendicular to, the ecliptic; also shown is an example of one magnetic field and corresponding electric field orientation projected on the Moon as seen looking along the solar wind flow direction.

As mentioned earlier, light species such as hydrogen and helium are not gravitationally bound, at least not on time scales comparable to the ionization time, and thus in the process of escape may be ionized at heights much greater than the scale heights discussed so far. Figure 3 shows the trajectory to scale, for hydrogen ions formed at a height of one lunar radius. It is especially interesting to note that these ions can execute trajectories which carry them over the lunar poles and onto the dark side.

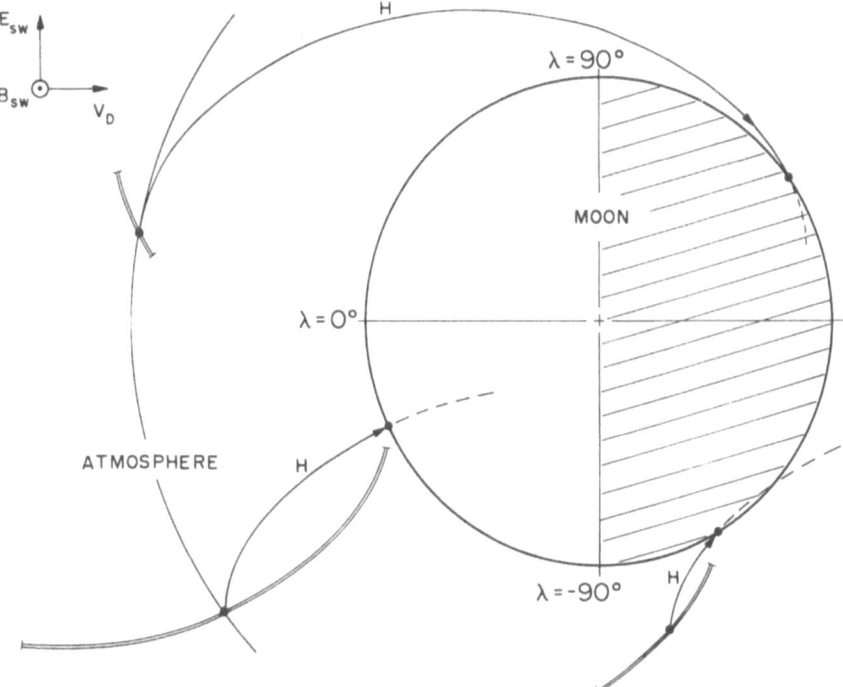

Fig. 3. Trajectories, to scale, of hydrogen ions starting about one lunar radius above the lunar surface. Ions starting along a part of the double lines will all impact at the same point on the surface.

3. Application to Lunar Samples

The analysis of lunar samples continues to exhibit unexpected compositions in some gaseous elements. One of the most pronounced anomalies is the excess of surface correlated Ar^{40}, compared to solar wind implanted Ar^{36}, which was discussed by several researchers, including Heymann *et al.* (1970) who proposed a lunar source for the Ar^{40}, being produced by K^{40} decay in the Moon. In a paper showing that the lunar atmosphere is the likely source of the surface Ar^{40}, Manka and Michel (1970c) suggested that the atmosphere would also be the source of other unexpected surface elements, or unexpected concentrations of elements, including solar wind elements which have been cycled through the atmosphere and reimplanted in a manner characteristic of atmospheric ions. However, the atmospheric contribution to these other elements is not as obvious as in the case of Ar^{40}; the nonradiogenic noble gas isotopes are not as readily attributable to lunar origin as is Ar^{40} and the gases directly implanted from the solar wind tend to mask those cycled through the atmosphere. In addition, gases other than the noble gases can react chemically when they impact the surface, and are then difficult to distinguish from trace elements in the sample minerals.

For an ion whose trajectory causes it to impact the Moon, whether the ion is firmly implanted in the surface or is quickly released will depend on both the impact energy and the trapping proability $\eta_t(\mathscr{E})$ for the lunar material. Studies at Berne on ion implantation in aluminum foils with their associated oxide layer can be used to estimate the trapping characteristics for silicate grains. The trapping probability for helium (e.g. Bühler *et al.*, 1966) rises nearly linearly for energies to 1 keV and then levels off to unity at about 3 keV; the data for other elements is similar. This data was previously approximated by a three part fit (Manka and Michel, 1970c)

$$\eta_t(\mathscr{E}) = \begin{cases} 0.6\mathscr{E}, & 0 \leqslant \mathscr{E} < 1 \text{ keV} \\ 1 - 1.8 \exp(-1.5\mathscr{E}), & 1 \leqslant \mathscr{E} < 3 \text{ keV} \\ 1, & \mathscr{E} \leqslant 3 \text{ keV} \end{cases} \tag{8}$$

which is awkward to integrate but is useful because it clearly indicates the dependence of trapping on energy.

For more recent calculations we use the equally good fit obtained with a single than function; for this helium data it is

$$\eta_t(\mathscr{E}) = \tanh(0.75\mathscr{E}). \tag{9}$$

From the energies calculated in Table I, we see that it is this initial part of the trapping curve, up to a few keV, that will determine the implantation efficiency of the lunar ionosphere. An important feature of the trapping curve is whether there is a finite, and perhaps not very small, cutoff energy at which the trapping probability goes to zero (Meister, 1969); recently this cutoff has been included in the trapping caculations (Manka and Michel, 1971). The total trapping of ions is calculated by integrating the trapping probability over the incident ion energy spectrum

$$N_t = \int_0^\infty \eta_t(\mathscr{E}) n(\mathscr{E}) d\mathscr{E}, \tag{10}$$

where $n(\mathscr{E})$ is determined from the exponentially decreasing number density in the atmosphere and the resulting trajectory along the interplanetary electric field into the Moon.

There is the possibility of some mass fractionation of solar wind isotopes which have been released into the lunar atmosphere and then retrapped via the acceleration mechanism described here (see Manka and Michel, 1971 for a discussion of the experimental data and the calculations).

4. Application to Lunar Ion Detectors

Clearly the considerable fluctuations of $\bar{B}_{sw}$ out of the ecliptic plane will rotate the planes of lunar trajectories. While the interplanetary electric field may be the principal energy source of the ions, the lunar surface electric and magnetic fields could play important roles in changing an ion's direction and modifying its energy as it approaches the surface. The Suprathermal Ion Detector (Freeman *et al.*, 1971) is oriented so that it looks in the ecliptic plane and sees sporadic 'clouds' of ions arriving at the lunar surface. The fluctuations in the direction of $\bar{E}_{sw}$, may combine with the surface fields to allow the detection of these bunches of ions. The measured energies of the ions, of tens to hundreds of electron volts, are reasonable for $\bar{E} \times \bar{B}$ acceleration. The Solar Wind Spectrometer (Snyder *et al.*, 1971) looks both vertical and horizontal to the local surface and the majority of the ion flux detected after the Apollo 13 S-IVB impact event was directed horizontal to the local surface and from north to northeast. That is, the trajectories lie in a plane generally perpendicular to the ecliptic plane and which is qualitatively in agreement with the predictions of Manka and Michel (1970a, c) for the case when $\bar{B}_{sw}$ is generally in the ecliptic plane.

From Figure 2, we see that in order for lunar ions to enter one of the SIDE experiments, which point in the ecliptic with a narrow look angle, the interplanetary magnetic field must be strongly out of the ecliptic and in a direction such that the interplanetary electric field, $\bar{E}_{sw} = -\bar{V}_{sw} \times \bar{B}_{sw}$, is pointed toward the detector. A further immediate consequence is that since the SIDE's generally look within 15° east/west of the local vertical, and since $\bar{E}_{sw}$ lies in a plane perpendicular to $\bar{V}_{sw}$, then ions from the lunar ionosphere should be observed mostly near local sunrise and sunset at the detector. This apparently is the case, with even the ions having energies higher than the ambient lunar ionosphere tending to peak at sunrise/sunset. (Freeman *et al.*, 1972).

Several ion events have been analyzed for correlation between observed ion flux and direction of $\bar{B}_{sw}$. (For a discussion of the Lunar Surface and Explorer 35 Magnetometer data see Dyal and Parkin, 1971 and Colburn *et al.*, 1971.) Two ion events, one at local sunrise and the other at sunset, were found when $\bar{B}_{sw}$ was almost 90° out of the ecliptic plane, an ideal condition for observation according to our model. During this time an intense ion flux was detected by the SIDE, the flux appearing to turn on when $\bar{B}_{sw}$ rotated a few degrees away from this critical angle. These events strongly support the acceleration model. Other events were studied which do not fit the model as well, though $\bar{E}_{sw}$ is generally toward the detector for these events. A basic difference between these two types of events is that the ion events which correlate extremely well with $\bar{B}_{sw}$ have broad, low energy spectra which would be expected for ions accelerated from the atmosphere; on the other hand, the non-correlating events are higher energy, 250 eV to 1 keV, and are usually monoenergetic. This latter type of event is not yet understood, and might consist of ions from the Earth's bow shock or result from an energy-selection effect due to the lunar surface magnetic field.

The energy spectrum for ions from the lunar atmosphere is illustrated in Figure 4

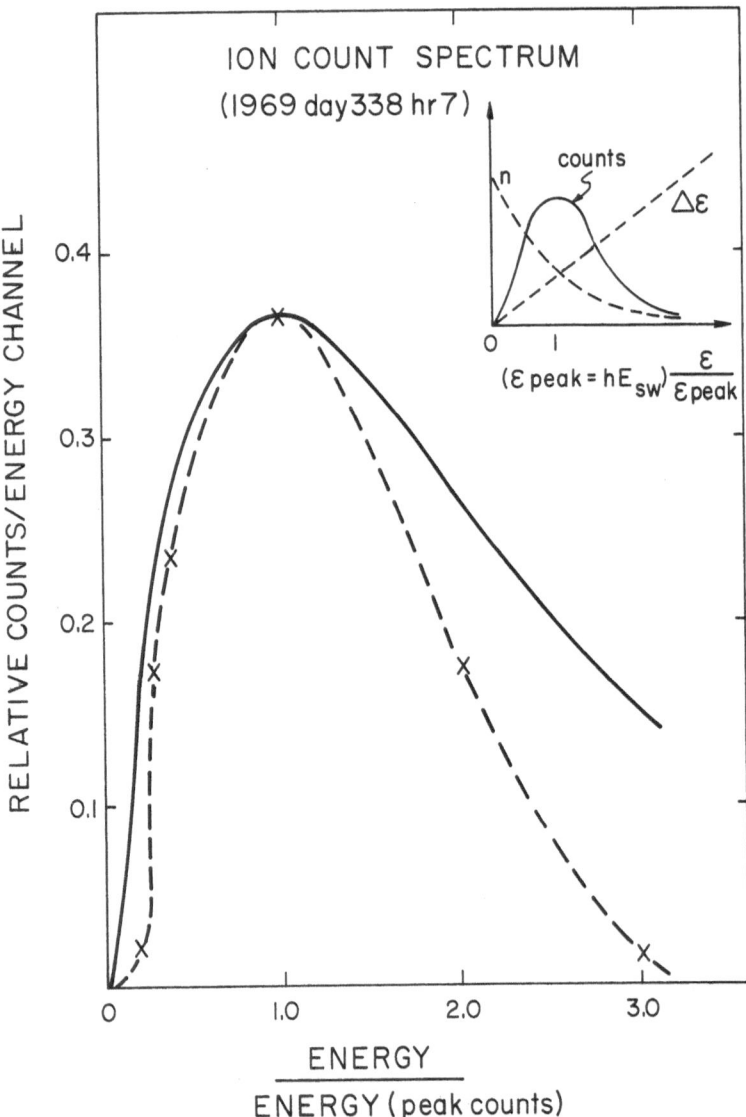

Fig. 4. Plot of the predicted and observed ion count spectrum for Event 7 seen by the Apollo 12 SIDE. The energy is normalized to the energy for which the peak number of counts occurs (250 eV for this event).

(Manka *et al.*, 1972a, b). The solid line is the predicted theoretical spectrum which should be observed by the SIDE, which has an energy band width proportional (∼10%) to each observation energy. This spectrum assumes an exponential density distribution in the atmosphere and that all ions accelerated to the surface by $\bar{E}_{sw}$ are collected. The dashed line is the spectrum observed for one of the events where the ion flux correlates well with the orientation of $\bar{B}_{sw}$. It can be seen that the observed spectrum follows the predicted spectrum fairly closely except for a decreased count rate at higher energies. This decrease could be due to a number of factors such as a density distribution which is not quite Maxwellian, or slight differences in the direction of incoming ions so that not all energies are collected (see Table 1). It should be noted that the peak at the spectrum occurs at an energy

$$\mathscr{E} = hE_{sw},$$
(11)

where h is an effective scale height for the atmospheric species being detected. This provides an additional tool for studying the ionosphere since $\mathscr{E}$ is measured, $\bar{E}_{sw}$ is often known from magnetometer and solar wind measurements, so that h can be calculated and compared with expected species masses.

5. Neutral Lunar Atmosphere Density

Using the acceleration model described in this paper, it is possible to calculate the density of the neutral lunar atmosphere. So far we have made two applications: one is to estimate the partial pressure of Ar^{40} in the lunar atmosphere and the other is to calculate the total neutral density and pressure. These two calculations are based on quite different experimental data; the Ar^{40} density is deduced from the ratios of solar wind and atmosphere noble gases found trapped in lunar samples while the neutral density is calculated more directly from selected ion fluxes observed by the SIDE instruments.

In the lunar samples Ar^{40} is found in approximately a one to one ratio with Ar^{36} which has been trapped from the solar wind. As discussed earlier, this anomalously high concentration of Ar^{40} is probably trapped lunar atmosphere. The flux of solar wind Ar^{36} can be estimated from the ratios of element measured in lunar samples and by the Solar Wind Composition experiment (Geiss *et al.*, 1970a, b) which in turn gives the trapped flux of Ar^{40}. Then by integrating over an assumed exponentially decreasing atmosphere and by folding in the ion acceleration and trapping probability, the number density of neutral Ar^{40} at the lunar surface is calculated to be (Manka and Michel, 1971; Manka, 1972)

$$n_0 (Ar^{40}) \simeq 10^2 \text{ cm}^{-3}.$$
(12)

It should be noted that atmosphere density is based on an average accumulation of Ar^{40} over geologic times and that the concentration in today's lunar atmosphere could be somewhat lower.

The measurement of selected ion fluxes and their interpretation using the acceleration model essentially amounts to the second direct measurement of the neutral atmosphere density, the first being by the Cold Cathode Gauge Experiment (CCGE, Johnson *et al.*, 1972a, b). The only two 'calibration factors', or assumptions, are that the atmosphere density decreases exponentially with height and that some average ionization lifetime must be used. Since neon may be the most common constituent of the lunar atmosphere (e.g. Johnson, 1971), then a lifetime against photo- and charge exchange ionization can be calculated using solar spectrum and solar wind values. The exponential density variation is a fairly good assumption; an exact kinetic calculation for the terminator region is desirable and the calculations thus far have not addressed themselves specifically to the continuous variation with height (e.g. Yeh and Chang, 1972). Using those selected ion events described earlier which ideally fit the acceleration model and have energy spectra of the type expected for the lunar ionosphere, we then calculate that the neutral density at the surface, near the sunrise or sunset terminator, is

$$n_0 \cong 1 \quad \text{to} \quad 3 \times 10^5 \text{ cm}^{-3}$$

in good agreement with the 2×10^5 cm^{-3} measured by the CCGE.

6. Conclusions

The somewhat unusual dynamics of the lunar ionosphere, which result from acceleration by electric and magnetic fields in the solar wind, have been described. In the vicinity of the Moon, the trajectory of lunar atmosphere ions is primarily along $\bar{E}_{sw}$ and therefore approximately at right angles to both $\bar{V}_{sw}$ and $\bar{B}_{sw}$. Those ions whose trajectories intersect the Moon will impact the surface with energies from tens of electron volts to a few keV. As was shown; these energies are sufficient for significant trapping of lunar ions in the surface and this is the likely source for the excess Ar^{40} found in lunar samples; lunar regolith Ar^{40} from potassium decay diffuses into the lunar atmosphere and is implanted in the surface grains by this acceleration mechanism. A similar effect is the possible isotopic fractionation of solar wind noble gas elements which are trapped in the lunar surface, released into the atmosphere, and retrapped by this mechanism.

From the concentration of Ar^{40} in lunar samples, it is possible to calculate the neutral lunar atmosphere density of Ar^{40}, averaged over the geologic time during which the argon was trapped from the atmosphere. After doing the associated calculations, we obtained a neutral density $n_0(\text{Ar}^{40}) \cong 10^2$ cm^{-3}. It is also possible, using the acceleration and atmosphere models, to convert ion fluxes measured by surface ion detectors into neutral atmosphere density. From measurements near local lunar sunrise/sunset we obtain $n_0 \cong 1$ to 3×10^5 cm^{-3} in excellent agreement with measurements by total pressure gauges.

Acknowledgements

We would like to thank Drs Freeman, Hills, and Vondrak of Rice for very helpful discussions and for assistance in interpretation of the SIDE data. We would also like to thank Drs Colburn, Dyal, Parkin, and Sonett for making available data from the Ames Explorer 35 and Lunar Surface Magnetometers, and Dr Snyder (JPL) for data from the Solar Wind Spectrometer. We also thank Drs Geiss of Bern and Meister (Rice) for data and discussions regarding the trapping of ions, and Dr Heymann (Rice) for suggestions concerning lunar sample gases. This work has been supported under contracts NAS 9-5911 and GA-28033X (NSF).

References

Bernstein, W., Fredricks, R. W., Vogl, J. L., and Fowler, W. A.: 1963, *Icarus* **2**, 233.

Bühler, F., Geiss, J., Meister, J., Eberhardt, P., Huneke, J. C., and Signer, P.: 1966, *Earth Planetary Sci. Letters* **1**, 249.

Burlaga, L. F. and Ness, N. F.: 1968, *Can. J. Phys.* **46**, S962.

Colburn, D. S., Mihalov, J. D., and Sonett, C. P.: 1971, *J. Geophys. Res.* **76**, 2940.

Dyal, P. and Parkin, C. W.: 1971, *J. Geophys. Res.* **76**, 5947.

Freeman, J. W., Hills, H. K., and Fenner, M. A.: 1971, *Proc. 2nd Lunar Sci., Conf., Geochim, Cosmochim. Acta, Suppl. 2* **3**, 2093.

Freeman, J. W., Fenner, M. A., Hills, H. K., Lindeman, R. A., Medrano, R., and Meister, J.: 1972, *Icarus* **16**, 328.

Geiss, J., Eberhardt, P., Bühler, F., and Meister, J.: 1970a, *J. Geophys. Res.* **75**, 5972.

Geiss, J., Eberhardt, P., Signer, P., Bühler, F., and Meister, J.: 1970b, in *Apollo 12 Preliminary Science Report*, NASA Sp-235, 99.

Heymann, D., Yaniv, A., Adams, J. A. S., and Fryer, G. E.: 1970, *Science* **167**, 555.

Heymann, D. and Yaniv, A.: 1970, *Proc. 2nd Lunar Sci. Conf., Geochim. Cosmochim. Acta, Suppl. 1* **2**, 1261.

Hinton, F. L. and Taeusch, D. R.: 1964, *J. Geophys. Res.* **69**, 1341.

Johnson, F. S.: 1971, *Rev. Geophys. Space Phys.* **9**, 813.

Johnson, F. S., Evans, D. E., and Carroll, J.: 1972a, 'Lunar Atmosphere', presented at *3rd Lunar Sci. Conf.*, Houston.

Johnson, F. S., Evans, D. E., and Carroll, J.: 1972b, in C. Watkins (ed.), *Lunar Science III*, Lunar Sci. Inst. Contrib. No. 88, p. 436.

Manka, R. H.: 1972, Ph. D. Thesis, Rice University, Houston, Texas, 1972.

Manka, R. H. and Michel, F. C.: 1970a, *Trans. Am. Geophys. Union* **51**, 408.

Manka, R. H. and Michel, F. C.: 1970b, *Trans. Am. Geophys. Union* **51**, 344.

Manka, R. H. and Michel, F. C.: 1970c, *Science* **169**, 278.

Manka, R. H. and Michel, F. C.: 1971, *Proc. 2nd Lunar Sci. Conf., Geochim. Cosmochim. Acta Suppl. 2* **2**, 1717.

Manka, R. H., Michel, F. C., Freeman, J. W., Jr., Dyal, P., Parkin, C. W., Colburn, D. S., and Sonett, C. P.: 1972a, in C. Watkins (ed.), *Lunar Science III*, Lunar Sci. Inst. Contrib. No. 88, p. 504.

Manka, R. H., Michel, F. C., Freeman, J. W., and Hills, H. K.: 1972b, *Trans. Am. Geophys. Union* **53**, 439.

Meister, J.: 1969, Ph. D. Thesis, University of Berne.

Michel, F. C.: 1964, *Planetary Space Sci.* **12**, 1075.

Snyder, C. W., Clay, D. R., and Neugebauer, M.: 1971, 'An Impact Generated Plasma Cloud on the Moon', presented at *2nd Lunar Sci. Conf.*, Houston, Texas.

Yeh, T. T. J. and Chang, G. K.: 1972, *J. Geophys. Res.* **77**, 1720.

DISCUSSION

Weil: Can you say something about the variation with height of this ionosphere; does it have a layered structure?

Manka: There will be an asymmetry with respect to the body of the Moon in the ion distribution. On the side where the ions are driven directly into the Moon by the interplanetary electric field, there will be a thin ion layer existing in the atmosphere throughout which the ions are being formed. On the other side, the ions, as they are formed, are swept out to large distances before they curve significantly into the solar wind direction.

Gold: The situation discussed is similar to that which must occur in the formation of the tail of comets. It will be interesting to observe the detail there. The same kind of process is also the one I have proposed as the explanation of the contents of Ar^{40} implanted in the lunar material. There the total amount is greatly in excess of the amount that could have been transported in the solar wind, and we therefore suppose that it is lunar argon that evaporated, got ionized and then driven back at high speed into the Moon.

Manka: Your comment about similar effects on comets is very interesting. It relates to the question by Dr Weil about the resulting ion density distribution around the Moon. On the side of the Moon where the ions are driven in, there would be only a thin ion layer, perhaps a few scale heights ($\sim$ 200 km) thick. On the side where the ions are being accelerated away from the Moon, the ions are swept out along the interplanetary electric field into a large region above that side of the Moon. Thus there would be a large asymmetry in the ion distribution which might be detected in radio signals.

I had intended to mention the trapping of lunar ions in the lunar surface. I understand that at the *1st Lunar Sci. Conf.* you suggested acceleration by the interplanetary fields as the method of implanting Ar^{40} from the lunar atmosphere into the surface grains. We also independently made this suggestion and have done a number of detailed calculations. An interesting effect is that there may be fractimation of solar wind isotopes that have been trapped in the surface, released into the atmosphere, and then re-implanted by the mechanism discussed here. The result will be that the lighter of two isotopes will be re-implanted slightly more efficiently than the heavier isotope, merely due to the lighter isotope being higher in the lunar atmosphere and this accelerated further in the interplanetary electric field, impacting with greater energy, and being better trapped.

Grard: What is the life-time of the lunar ions, and what is a typical electron density in the lunar ionosphere?

Manka: The ions have a life-time of the order of 1 s; there is no recombination in the lunar atmosphere itself. That is, the ions are very quickly either ejected into space, or impact the moon; this changes the previously assumed ratio between neutrals and ions in the lunar atmosphere.

Gold: It could be deduced from radio occulation experiment that the electron density in the vicinity of the Moon is limited to about 5 electrons cm^{-3} in excess of the solar wind density.

PHOTOELECTRONS AND LUNAR LIMB SHOCKS

DAVID R. CRISWELL

Lunar Science Institute, 3303 Nasa Road 1, Houston, Tex. 77058, U.S.A.

Abstract. It is suggested that boundary conditions for solar wind/lunar limb interactions are active. The 'whole-Moon' limb does not evoke a shock cone because warm ($\simeq 13$ eV electron^{-1}) solar wind electrons are replaced by cool (< 2 eV electron^{-1}) photoelectrons that are ejected from the generally smooth areas of the lunar terminator illuminated at glancing angles by the Sun. A localized volume of low thermal pressure is created in the solar wind by these cool photoelectrons. The solar wind expands into this volume without shock production. Directly illuminated highland areas exchange hot photoelectrons (< 20 eV electron^{-1}) for warm solar wind electrons. The hot electrons generate a localized pressure increase (Δp) in the adjacent solar wind flow which evokes a shock streamer in the solar wind. Shock streamers are identifiable by a coincident increase in the magnitude ($\Delta B \sim \Delta p$) of the solar wind magnetic field. Shock occurrence is controlled by lunar topography, solar activity in the hard ultraviolet (> 20 eV), solar wind electron density and thermal velocity, and the density of the solar wind magnetic field.

1. Introduction

Three magnetic features characterize solar wind/Moon interactions (Colburn *et al.*, 1971; Ness *et al.*, 1967). The insert in Figure 1 illustrates variations in magnetic field intensity (B) detectable by a high altitude space probe such as Explorer 35 passing through the plasma-void region downstream from the Moon. When the probe is in the

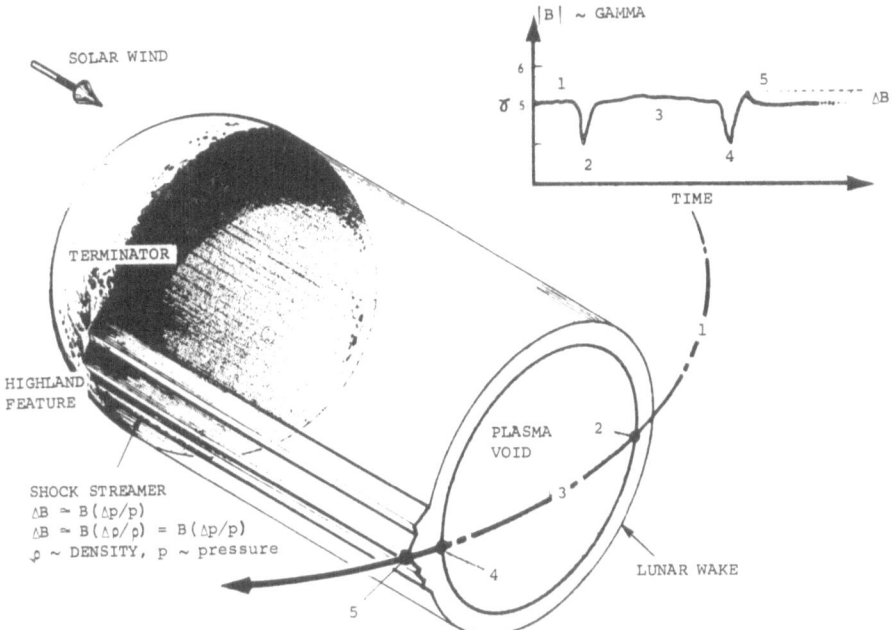

Fig. 1. Schematic representation of the major features of the lunar wake. The insert illustrates the variations in intensity of the magnetic field detectable along the illustrated trajectory which is comparable to that of Explorer-35.

R. J. L. Grard (ed.), Photon and Particle Interactions with Surfaces in Space, 443–451. All Rights Reserved
Copyright © 1973 *by D. Reidel Publishing Company, Dordrecht-Holland*

downstream solar wind, even close to the wake, there is usually no indication of the Moon's presence. In the wake there is a dip in magnetic intensity. Inside the plasma void, the magnetic intensity is usually slightly greater than in the undisturbed solar wind. No indication of a general shock structure in the solar wind due to the presence of the Moon has been observed. Thus, the $+\Delta B$ at the exit point presents a definite dilemma. A similar pattern is observed by the low altitude Apollo subsatellites (Coleman *et al.*, 1972).

Explorer 35 data sporadically display this small increase or exterior enhancement of magnetic intensity in the solar wind at points which are always exterior to the magnetic intensity dip of the wake. Exterior enhancements appear to be shock streamers generated by interaction between the solar wind and sources on the lunar limb. These lunar sources are associated with highland terrain between 30°N and 30°S latitudes and are located primarily on the far-Earth side (Sonett and Mihalov, 1972).

Highlands must evoke a localized pressure (Δp) increase (Equation (1)) in the adjacent solar wind flow in order to deflect the flow 3° to 7° outward from the wake.

$$\Delta p = MNV^2 \sin^2 \theta \simeq (0.4 \text{ to } 1.4) \cdot 10^{-10} \text{ dyn cm}^{-2}, \tag{1}$$

where $M \sim$ proton mass, $N \sim$ number density of solar wind protons, $V \sim$ solar wind bulk velocity, and $\theta \simeq \arctan(V_A/V)$ where $V_A = B/\sqrt{4\pi NM}$ is the local Alfvén velocity. The solar wind magnetic field intensity (B) will increase as the solar plasma is densified by the localized pressure increase and $\Delta B/B \sim \Delta\varrho/\varrho \sim \Delta p/p$. Remanent magnetic fields, induced magnetic fields, and neutral gas releases (Barnes *et al.*, 1971) have been suggested as coupling mechanisms which would induce the local Δp increase. For average solar wind conditions (Hundhausen, 1970) the required pressure increase corresponds to an excess electron thermal energy density of $\Delta\eta \simeq 23$–100 eV cm^{-3} in the coupling region. This corresponds to a total energy per electron the order of $\bar{\varepsilon} \simeq [(5 \text{ to } 20 \text{ eV}) + + 13 \text{ eV}]$ electron$^{-1} = 18$ to 33 eV electron^{-1} assuming $N \simeq 5$ cm^{-3}.

It is the contention of this paper that both the lack of an overall lunar shock and the presence of sporadic limb shocks result from a common mechanism. The mechanism is the exchange of solar wind electrons for lunar photoelectrons as the solar wind flows past the terminator.

Highland photoelectrons will be more energetic, on the average, than solar wind electrons. The extra energy of the highland photoelectrons will provide the excess thermal energy $(\Delta\eta)$ necessary for shock production. Conversely, flat land photoelectrons will be less energetic than the solar wind electrons and decrease the thermal electron energy in the solar wind flow adjacent to the terminator. The deficiency in electron thermal energy constitutes a sink for flow turbulence and suppresses shock production.

2. Schematic Explanation

Figure 2 portrays a totally ad hoc but reasonably simple explanation of the electron-exchange from a highland area. Assume the sun emits photons in only two narrow bands at 10 eV and 80 eV producing the flux levels at earth orbit shown in the left hand graph. Assume in addition that the surface has a photoelectron yield of $Y = 10^{-2}$, a work function $w \simeq 4$ V, and that the slope is sufficiently steep that $\cos \alpha \simeq 1$. Additional, and for illustrative purpose only, assume the 80 eV photons evoke mono-energetic photoelectrons of $\varepsilon_{ph} \simeq 80 - 4 = 76$ eV while the 10 eV photons evoke 2000 times more photoelectrons at 5 eV than at 6 eV $= 10$ eV $- 4$ eV. The input flux of solar wind electrons is taken to be 5×10^{8} cm^{-2} s^{-1} which is characteristic of the average solar wind.

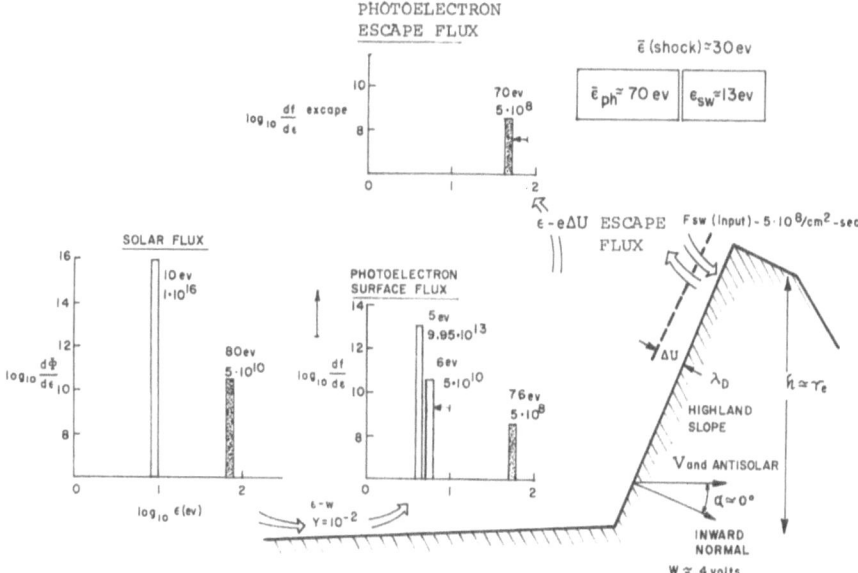

Fig. 2. A schematic representation of the exchange of solar wind and photoelectrons about a highland feature in the terminator region is presented. A photoelectron yield of $Y = 10^{-2}$ and work function of $w = 4$ V are assumed. Solar wind electrons accret from the distance of one electron cyclotron radius ($r_e \simeq 2$ km) above the terminator. A region of high pressure ($\Delta P > 0$) due to the injection of hot photoelectrons is generated just above and behind the mountain. The graphs depict the relative spectra of the solar photons, the evoked surface photoelectrons and finally the escape photoelectrons. The solar wind flow and magnetic field direction are from left to right. The 6 eV photoelectrons can not escape due to the potential drop $\Delta U \simeq 6$ V which adjusts to maintain balance between the inward flux of solar wind electrons and the outward flux of photoelectrons.

Electrical charge neutrality must exit over the scale $\lambda_D \simeq 1000$ cm above the surface where λ_D is the solar wind Debye length. It is natural to expect that all the photo-electrons will attempt to escape but in the process will create a net positive surface charge and a concomitant electric field which will return these lower energy electrons to the surface. There will be an electrical potential drop $\Delta U \simeq 6$ eV on going from the surface to the undisturbed solar wind. 'ΔU' will reduce the energy of the escaping photoelectrons to $\varepsilon_{ph} = 76 - 6 = 70$ eV. Solar wind electrons have $\bar{\varepsilon}_{sw} \simeq 13$ eV. As expected, the escaping photoelectrons have more than sufficient energy to generate a shock ($\varepsilon_{ph} = 70 - 13$ eV $= 57$ eV > 30 eV for a shock).

Figure 3 depicts the situation for a flat land area at sundown. Where $\cos \alpha \simeq 10^{-2}$ for $\alpha \simeq 89\frac{1}{2}°$. The oblique solar illumination drastically reduces the photoelectron output at the surface. However, the solar wind input flux is not significantly decreased. Solar wind electrons possess a random thermal velocity (v_t) approximately 5 times the solar wind bulk velocity which results in an omnidirectional rather than directed flux $F_{sw} \simeq N v_t / 2\pi^{1/2}$. Thus, a photoelectron escape flux of 5×10^8 cm^{-2} s^{-1} is required for

$$\bar{\epsilon}_{ph} = \frac{(1ev)(4.95 \cdot 10^8) + 71(5 \cdot 10^6)}{5 \cdot 10^8} = \boxed{\bar{\epsilon}_p = 1.7 \, ev \ll 13 \, ev = \bar{\epsilon}_{sw}} < \bar{\epsilon} \, (shock) \simeq 30 \, ev$$

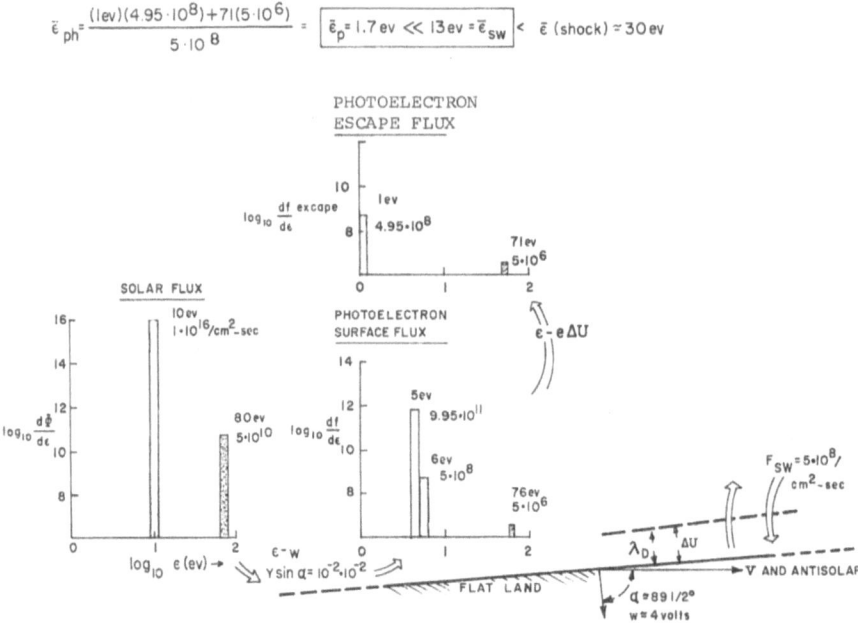

Fig. 3. A schematic representation of the exchange of solar wind and photoelectrons above a flat land area in the terminator zone is presented. The significant features relative to Figure 2 are the decrease in $\cos \alpha$ from 1 to 10^{-2} which reduces the flux of 80 eV photoelectrons and the constancy of the input photoelectron flux. A region of low electron thermal energy density is produced above and downstream of the flat land.

the flat land case just as for the highland case. Flux balance requires that both the 76 eV and 6 eV photoelectrons escape but that the 5 eV electrons be retained. The potential difference ΔU will decrease from 6 V to 5 V to accomodate the photoelectron spectral shift.

The 6 eV photoelectrons lose 5 eV in passing through the photoelectron layer and convey only 1 eV per electron into the solar wind. The average escape energy of 1.7 eV, allowing for the 71 eV photoelectrons, is much less than average energy of the solar wind electrons.

It should be clear from these examples that the required shock pressure (Δp), electron thermal velocity, and intensity of the extreme ultraviolet portion of the solar spectrum are the physically significant parameters.

3. Realistic Models

The schematic model produced the desired results. The numbers were chosen to assure success. A physically reasonable model must incorporate an acceptable model for the solar spectrum (Hinteregger, 1970; Walbridge, 1970; Wende 1972) and experimentally justified models for photon electron yield ($Y(\varepsilon)$) and the photoelectron distribution function ($\varrho(\varepsilon)$). Refer to Criswell (1973) for details of this extended modeling.

Table I presents results based on the photoelectron properties inferred from the Charged Particle Lunar Environment Experiment (CPLEE) which was placed on the Moon during the Apollo-14 mission (Reasoner and Burke, 1972). The left hand section specifies four general solar wind states and the plasma conditions which apply for each state. Column 5 contains the energy per electron necessary to evoke a shock. On the right hand side are displayed the average energies of escape photoelectrons for the two extreme slope values and three degrees of solar activity. The intensity (photons $cm^{-2} s^{-1} eV^{-1}$) of solar flux for $\varepsilon > 10$ eV was modeled as $I = I_0 (\varepsilon/10)^r$ where $r = -1$ (quiet), 0 (average), and $+1$ (active).

These modeling results clearly support the contention that highland regions will be most likely to evoke shocks whereas flat lands will cool the solar wind electrons and suppress shocks. This is readily apparent by comparing the $\bar{\varepsilon}_{sw}$, ε_r and $\bar{\varepsilon}_{ph}$ columns. In addition, the highlands are more likely to evoke shocks as the solar activity increases. The possibility of a 'whole Moon shock' does exist. This will occur if the hard ultraviolet solar radiation increases dramatically during a time of very quiet solar wind flow. A solar limb flare might be adequate.

Table II presents similar calculations based on measurements of the yield and differential energy distribution of lunar fines samples for photon energies between 4 eV and 20 eV (Feuerbacher et al., 1972). The laboratory yield function is approximately 10% of that in the CPLEE model. Its differential energy distribution peaks near the material work function whereas the CPLEE function peaks at $\frac{1}{2}(w+\varepsilon_0)$ where ε_0 is the photon energy. There are two basic differences from Table I. Shock production is much less likely since highlands can evoke a shock only under an active sun and a very quiet solar wind. Disturbed, but non-shocked flow will occur under the conditions which evoked shocks in Table I. Cooling of the solar wind will occur for all other conditions. The second difference is that photo-current limiting occurs in flat land areas prior to local sunset ($\alpha \geqslant 89.5°$). The terminator width for solar wind electrons is the order of $2°$ to $5°$; thus limitations of photocurrent will be important for $\alpha \leqslant 87°$ to $85°$ and will be important for the quiet and average conditions.

Based on the limited literature available concerning the photoelectron properties of insulating materials and the extreme sensitivity of photoelectron yield to the number of conduction band carriers present (Spicer, 1968) it is likely that the 'laboratory' measurements of yield are lower and that the differential energy distributions are skewed more toward w than is actually the case on the lunar surface. Alverez (1972) has argued that changes in the electrical conductivity of very dry rocks on exposure to submonolayer water contamination may be governed by changes in the surface semi-

TABLE I

CPLEE model

Solar wind conditions	Input (average values)		Inferred		$\bar{\varepsilon}_{ph}$ (eV) ~ calculated					
					Solar conditions					
	N(cm^{-3})	$\bar{\varepsilon}_{sw}$(eV)	F_{sw}	$\bar{\varepsilon}$(eV) shock	quiet	average	active	quiet	average	active
Very quiet $T < 2 \times 10^4$K $V \lesssim 300$ km s^{-1}	40	6	1.5×10^8	$\underline{17}$	0.63	3.1	27 [a]	4.9	28 [a]	104 [a]
Quiet $T > 2 \times 10^4$K $300 < V < 400$	20	13	1.1×10^9	$\underline{19}$	0.09	0.43	3.7	6.9	35 [a]	94 [a]
Average $400 < V < 500$	7.7	13	4.3×10^8	$\underline{29}$	0.22	1.1	9.4	8.0	13	112 [a]
Fast-disturbed $V > 500$ km s^{-1}	2	17	1.3×10^8	$\underline{80}$	0.73	3.6	27 [b]	4.5	31 [b]	104 [a]
	$B(\gamma)$		F_{sw}(cm^{-2} s^{-1})		Flat lands $\alpha \simeq 89\frac{1}{2}°$			Highlands $\alpha \simeq 0°$		

[a] Shock will occur.
[b] Turbulent flow but no shock.

TABLE II

Laboratory model

Solar wind conditions	Input (average values)			Inferred	$\bar{\varepsilon}_{ph}$ (eV) ~ calculated					
					Solar conditions					
	$N(\text{cm}^{-3})$		$\bar{\varepsilon}_{sw}(\text{eV})$	$\varepsilon(\text{eV})$ shock	quiet	average	active	quiet	average	active
Very quiet $T < 2 \times 10^4\text{K}$ $V \lesssim 300 \text{ km s}^{-1}$	40	3	6	1.5×10^8	$\underline{17}$	0.28 (87°)[a]	0.98 (87.5°)	6.4[c] (88°)	2.4	6.4	42[b]
Quiet $T > 2 \times 10^4\text{K}$ $300 < V < 400$	20	5	13	1.1×10^9	$\underline{19}$	0.35 (68°)	2.5 (72°)	6.5 (75°)	0.77	3.1	15[c]
Average $400 < V < 500$	7.7	5	13	4.3×10^8	$\underline{29}$	0.34 (80°)	1.1 (82°)	5.6 (85°)	1.8	5.0	22[c]
Fast-disturbed $V > 500 \text{ km s}^{-1}$	2	5	17	1.3×10^8	$\underline{80}$	0.27 (87.5°)	1.1 (87.5°)	5.5 (88.5°)	2.7	5.9	46[c]
$B(\gamma)$				$F_{sw}(\text{cm}^2\,\text{s}^{-1})$	Flat lands $\alpha = (\ )$			Highlands $\alpha = 0°$		

[a] Photocurrent limited for $\alpha > (\ )$.
[b] Shock will occur.
[c] Disturbed flow but no shock.

conductor properties. Several laboratory measurements have revealed a correspondingly extreme sensitivity of the electrical conductivity of lunar material to water
exposure (Chang, 1972; Olhoeft, *et al.*, 1972).

4. Observations

Identification of the shock streamer sources with highland areas has been confirmed by
low altitude magnetic observations from the Apollo-15 subsatellite (Coleman *et al.*,
1972). The subsatellite observations discount the contention by Whang and Ness (1972)
that shock streamers are not highland associated. Careful intercomparison of the
distribution of shock sources with subsatellite magnetometer data, reveals that while
shocks do originate over regions of strong remnant magnetic fields (magcons) such as
Van de Graaff they can also be produced at equal rates over the Levi Civita region
which does not display a large magcon field at 100 km's altitude (Mihalov *et al.*, 1971).

The lunar region centered at 165° E and 22° N produced an 0.3 fractional occurrence
rate of shocks as observed by Explorer-35 at high altitude but no shock activity at the
low altitude of the Apollo-15 subsatellite. Obviously, interaction of a permanent
magcon field and the solar wind is inconsistent with this observational asymmetry.
Other factors, such as hardness of solar spectra, B, $\bar{\varepsilon}_{sw}$, and F_{sw} must influence shock
production. Unfortunately, the proper combination of parameters has not been
analyzed to allow a direct check on the electron exchange theory (Colburn *et al.*, 1971).

Finally, the laser altimeter (Wollenhaupt and Sjogren, 1972) on-board Apollo-15
provide delevation profiles along a single ground track which intersected several shock
production regions. Shock regions typically display larger elevation changes and
larger slopes ($>6°$) than do the non-source regions over 30 km baselines. The elevation
changes in source regions are the order of 2–5 km which covers the expected range of
the cyclotron radius of solar wind electrons. Thus, the mountain faces have sufficient
vertical area to replace the total accretion current to a differential length of lunar
latitude (Criswell, 1973).

5. Conclusion

It has been proposed that the detailed interaction of the solar wind with the lunar limb
is directly controlled by the exchange of solar wind and lunar photoelectrons. If there
is a net flow of electron thermal energy to the lunar surface, then the wake region is
refrigerated and shocks are suppressed. If there is a sufficient flow of electron thermal
energy into the solar wind, then a shock streamer is generated. Further studies of
Moon/solar-wind interactions must take into account this active boundary condition
and adjust for the range of the solar flux and the variations of the solar wind flux,
magnetic field strength, and average electron energy.

Acknowledgement

I am especially grateful to Mr J. Barrios for preparing the computer program utilized in this analysis. Drs B. S. Criswell and R. W. Shorthill provided several comments which aided in clarifying the text. This work was done at the Lunar Science Institute, which is operated by the Universities Space Research Association under Contract No. NSR-09-051-001 with the National Aeronautics and Space Administration. This paper constitutes Lunar Science Institute Contribution No. 112.

References

Alverez, Roman: 1972, Ph.D. dissertation, Engineering Science, Univ. of Calif. at Berkeley.

Barnes, A., Cassen, P., Mihalov, J. D., and Eviatar, A.: 1971, *Science* **172**, 716.

Chung, Dae H.: 1972, *The Moon* **4**, 356.

Colburn, D. S., Mihalov, J. D., and Sonett, C. P.: 1971, *J. Geophys. Res.* **76**, 2940.

Coleman, P. J., Lichtenstein, B. R., Russell, C. T., Sharp, L. R., and Schubert, G.: 1972, *Proc. 3rd Lunar Sci. Conf., Geochim. Cosmochim. Acta. Suppl. 3* **3**, 2271.

Criswell, David R.: 1973, *The Moon* **7**, 202.

Feuerbacher, B., Anderegg, M., Fitton, B., Laude, L. D., Willis, R. F., and Grard, R. J. L.: 1972, *Proc. 3rd Lunar Sci. Conf., Geochim. Cosmochim. Acta. Suppl. 3* **3**, 2655.

Hinteregger, H. E.: 1970, *Ann. Geophys.* **26**, 547.

Hundhausen, A. J.: 1970, *Rev. Geophys. Space Phys.* **8**, 729.

Mihalov, J. D., Sonett, C. P., Binsack, J. H., and Moutsoulas, M. D.: 1971, *Science* **171**, 892.

Ness, N. F., Behannon, K. W., Scearce, C. S., and Cantarano, S. C.: 1967, *J. Geophys. Res.* **72**, 5769.

Olhoeft, G. R., Frisillo, A. L., and Strangway, D. W.: 1972, in J. W. Chamberlain and C. Watkins (eds.), *The Apollo 15 Lunar Samples*, Lunar Science Inst., Houston, p. 477.

Reasoner, David L. and Burke, William T.: 1972, *Proc. 3rd Lunar Sci. Conf., Geochim. Cosmochim. Acta. Suppl. 3* **3**, 2639.

Sonett, C. P. and Mihalov, J. D.: 1972, *J. Geophys. Res.* **77**, 588.

Spicer, W. E.: 1968, in *A Survey of Phenomena in Ionized Gases*, International Atomic Energy Agency, Vienna, p. 271.

Walbridge, Edward: 1970, 'The Photoelectron Layer. 1. The Steady State', High Altitude Observatory, National Center for Atmospheric Research, Boulder, Colo.

Wende, Charles D.: 1972, *Solar Phys.* **22**, 492.

Whang, Y. C. and Ness, N. F.: 1972, *J. Geophys. Res.* **77**, 1109.

Wollenhaupt, W. R. and Sjogren, W. L.: 1972, *The Moon* **4**, 337.

DISCUSSION

Siscoe: Because the solar wind electron thermal speed is so high, they can begin exchanging with the lunar photoelectrons fairly close to the subsolar point, say 15° away. Why then do you look only at the terminator in connection with the maria?

Criswell: One is not sure on theoretical grounds how far along the magnetic field line a plasma electron (< 340 eV) is free to move. The shock region is only 200 km thick 3–5 lunar radii downstream which argues that heat does not flow out of a given 'streaming element' of plasma and thus that the hot electrons do not leave the region quickly. The typical lack of overall front-side plasma disturbances observed by the Apollo subsatellite argues that exchange occurs within 100 km of the surface and that streaming of plasma electrons along B is not significant.

Shawhan: Have you compared your model analysis with respect to solar activity and solar wind properties with actual observations of shocks?

Criswell: I don't have the raw data available to obtain the parameters that I consider necessary, such as the solar EUV spectrum and the average electron energy. The published data are not adequate for making a comparison.

KINETIC THEORY ANALYSIS OF SOLAR WIND INTERACTION WITH PLANETARY OBJECTS

S. T. WU

The University of Alabama in Huntsville, Huntsville, Ala. 35807, U.S.A.

and

MURRAY DRYER

Space Environment Laboratory, NOAA-ERL, Boulder, Colo. 80302, U.S.A.

Abstract. A purely kinetic treatment is proposed for the interaction of the solar wind with any 'small' planetary object. 'Small' refers to those cases when the solar wind proton's thermal gyroradius is arbitrarily taken to be $\geqslant 0.1$ radius of the object under investigation. The 'object' may possibly include an ionosphere or magnetosphere. The collisionless Boltzmann equation, neglecting the magnetic field, is used to calculate steady-state profiles of density and velocity around the obstacle. In order to include the effects of surface reflection from the obstacle, a finite, angularly distributed potential field on the obstacle's surface is also considered. A symmetric potential field is also examined. A low density plasma void in the umbral region and a compression (above the value of the upstream undisturbed density) in the penumbral region are clearly found. The present technique, despite its neglect of the interplanetary magnetic field, is proposed as an alternative zeroth order approach to the continuum, local magnetic anomaly, and guiding center approaches used by others for the particular case of Moon. Some recent, potentially relevant, observations on and in front of the Moon are discussed. Of particular interest is the development of a high density gradient which emanates from the limb and is inclined in the direction of what is known in continuum theory as the 'Mach cone'. This structure leads to the following speculation: if the object's size were to increase (or, alternatively, the proton gyroradius were to decrease) by one or two orders of magnitude, a shock wave would develop first at the limb, gradually building up to a conventional detached bow shock. Thus, gasdynamic theory – as we think of it in Earth's case – can be constructed, as required, on the basis of kinetic theory.

1. Introduction

It is well known that ordinary gasdynamic theory has been used successfully for a zeroth order theoretical continuum approach to the problem of solar wind interaction with Earth, and very possibly, Venus and Mars (Spreiter and Alksne, 1970; Dryer, 1970). The interplanetary magnetic field was considered only in an implicit manner in the early solutions referenced in these papers. Explicit consideration of the field – recently completed by Shen (1972) and Hirsh and Reshotko (1971) – shows that the Lorentz force is effective only in the neighborhood of the subsolar region. The zeroth order approach has also been suggested (in varying degrees of quantitative analysis) for the case of Moon by Lyon *et al.* (1967), Michel (1968), Wolf (1968), Siscoe *et al.* (1969), and Spreiter *et al.* (1970). The boundary condition on the sunlit hemisphere for these continuum studies was modified to permit absorption of all incident particles. An alternative approach for Moon has been used successfully by Whang (1968, 1969, 1970) who used a 'guiding-center' model which reproduced experimental magnetic 'anomalies' in the wake region (Ness *et al.*, 1967; Ness, 1972). This model (also with the complete absorption assumption) uses the microscopic kinetic equation to describe the motion of guiding centers along, but continuum flow perpendicular to, field lines.

R. J. L. Grard (ed.), Photon and Particle Interactions with Surfaces in Space, 453–470. All Rights Reserved
Copyright © 1973 by D. Reidel Publishing Company, Dordrecht-Holland

Whang and Ness (1970) further demonstrated the first experimental observation of standing magneto-acoustic waves which are inclined from the lunar limb in the direction of the 'Mach cone'. In order to interpret the data on a theoretical basis, they returned to the concept of the anisotropic propagation of magnetoacoustic waves as derived not only from the solution of the linearized continuum equations but also from the solution (Whang, 1970) of a guiding center plasma. This approach was formalized by Wang (1971) who used the same absorption boundary condition on the sunlit side of Moon and found similar density profiles in the wake with a three-dimensional kinetic description. In effect, then, Moon might almost be considered to be no more than a point disturbance in the solar wind. Yet, the realities of magnetic and plasma density anomalies discussed earlier strongly suggest that Moon and other 'small' obstacles require further studies which start from first principles. For example, one might conceivably consider a boundary condition whereby particle reflection may be more specular-like, as in the case of an ionopause or magnetopause.

It has been suggested by Dryer (1970) that the obstacle's 'Knudsen number', r_p/d (where r_p is the ambient proton's thermal gyroradius and d is the obstacle's physical or effective diameter) be used as a rough guide in choosing a basic approach to the interaction problem. That is, if r_p/d for an obstacle (a comet, say, at large AU) is large, say $> 10^3$, we would likely use kinetic theory for the study of its interaction. Conversely, if r_p/d were $< 10^{-3}$, we could proceed with great confidence with continuum theory. A 'gray' area lies somewhere between these somewhat arbitrarily-chosen numbers. By analogy with hypersonic flight through the whole gamut of Earth-bound atmospheric and ionospheric levels, this 'gray' area could be labeled as a 'transition' region which conveniently covers our ignorance of the correct procedure for analysis. Alternatively, one might suggest the electron inertial length as the characteristic length of the ambient solar wind. The basis for this suggestion appears to be in its relation to shock wave thicknesses observed in the laboratory and in space. Another candidate is the proton gyroradius, based on the directed velocity, which is the characteristic thickness of the Earth's magnetopause. A related length scale could be the geometric mean of the proton and electron gyroradii. Detailed studies would require consideration of the boundary's nature: magnetic, non-magnetic, or absorbing.

We could, if desired, extend this approach beyond planets, comets, and asteroids (Greenstadt, 1971) to include our heliospheric motion through the interstellar medium. It was in the spirit of reducing the degree of random speculation to a more systematic theoretical approach that this idea was proposed. It is the purpose of this paper to continue in this spirit by providing an alternative approach to our 'shopping list' of theoretical approaches. Thus, we extend the well-known kinetic theory for particle interaction with a sphere (Karamcheti and Sentman, 1965; Sentman and Karamcheti, 1969) to include the effect of an isotropic and anisotropic potential to represent the surfaces of some obstacles. The result may then be diffuse or even specular reflection. We are still interested, regardless of the outcome of the point regarding the boundary condition (on the sunlit side, say, of the Moon) in the limiting case of perfect absorption (c.f., Wang, 1971). This can easily be specified by letting reflection be equal to zero. We will include it, however, in order to maintain generality. Thus the applicability of the boundary condition assumed herein to a specific obstacle (say, Moon, Mercury, etc.) is not our immediate objective. It is our purpose to demonstrate a technique which, in principle, can be adjusted to include other boundary conditions and the interplanetary magnetic field.

2. Lunar Observations

Excluding ionospheric satellites from our discussion, the only potentially 'small' obstacle for which observations exist is the Moon. Some recent experimental observations on and near Moon may be relevant within the context of the fundamental kinetic approach which will be discussed in the next section. Other observations with respect to a possible limb shock (or Mach wave) as well as the magnetic wake anomalies have been extensively discussed in the literature (c.f., Ness, 1972; Sonett and Mihalov, 1972) and, therefore, are not repeated here.

Figure 1 (based on Figure 1 from Freeman, 1972) schematically shows the Moon's position with respect to a nominal position for Earth's shock wave on several dates: 13–15 December 1969 and 27 March 1970. On the latter date, Neugebauer *et al.* (1973) reported solar wind observations obtained with a proton spectrometer which was part

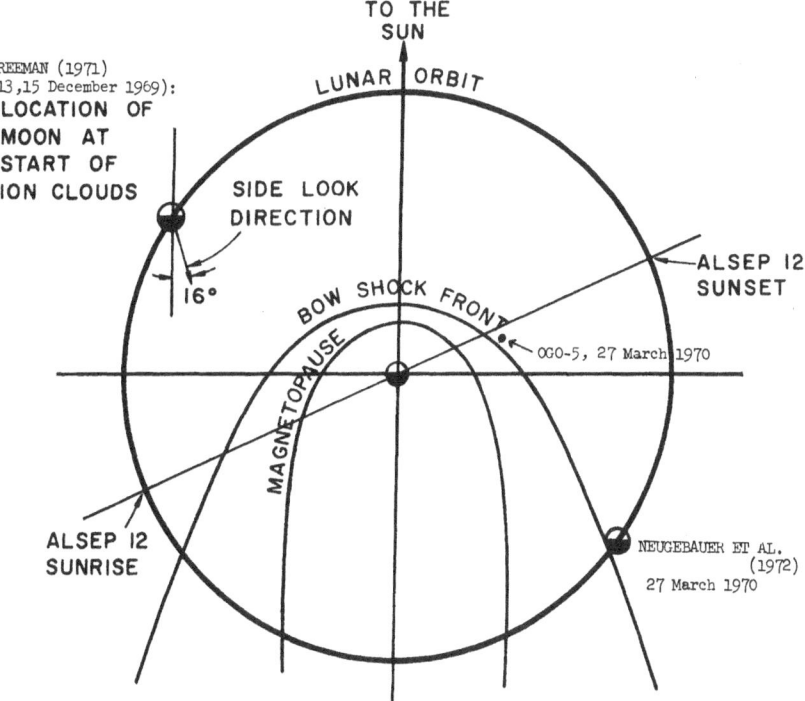

Fig. 1.   Schematic sketch of Moon's position during sunlit (27 March 1970) and night-side (13–15 December 1969) surface observations. (Based on Figure 1 from Freeman, 1972).

of the ALSEP 12 instrument package on the lunar surface. As noted in the figure, the package was on the sunlit side of the Moon (with instrument normal directed about 35° to the east of the sun) and in front of the Earth's shock on 27 March 1970. Figure 2 (from Neugebauer *et al.*, 1973) shows the velocity and density of the solar wind as measured on the surface of Moon.

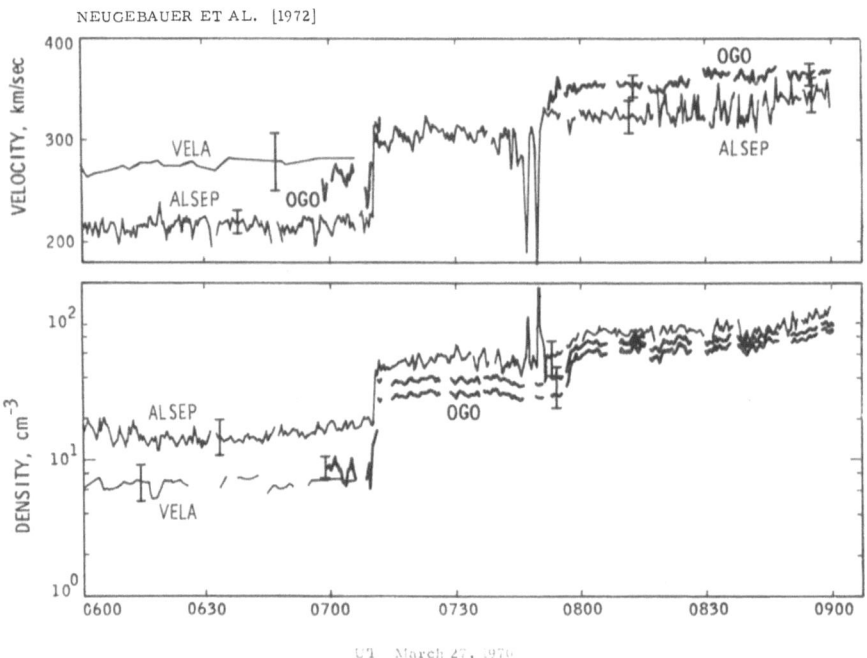

Fig. 2.   Solar wind (proton) velocity and density observations in the free stream (Vela 5 and Ogo 5) compared with lunar surface observations (ALSEP 12). (Neugebauer *et al.*, 1973).

Vela 5 and Ogo 5 were also in front of the bow shock; the approximate position of the latter spacecraft is shown in Figure 1. Their measurements are also shown in Figure 2 in order to present 'freestream' conditions. The time prior to 0711 UT is of interest here. (An interplanetary shock was measured at that time. The data subsequent to 0711 UT are of secondary interest here except for noting the related shock jump conditions at all three instruments. The time scales for these two spacecraft were shifted so as to coincide with ALSEP 12 at the shock wave.) Note that both spacecraft – in the ambient solar wind – measured velocities and densities which were, respectively, higher and lower than those measured on the lunar surface, Neugebauer *et al.* (1973) suggest that this compression effect was caused by local lunar magnetic fields (c.f. Sonett and Mihalov, 1972). It is suggested, however, that the results discussed later in terms of density contours present a physically-plausible alternative explanation.

Several other experimental observations associated with the Moon may also be relevant to the present study. Figure 1 also shows the Moon's approximate position on 13–15 December 1969 when the ALSEP 12 site was deep in the lunar night. Freeman (1972) has reported persistent observations of positive ion bursts, 'probably of solar wind origin', when his suprathermal ion detectors at ALSEP 12 and 14 are located on the night side as indicated for the particular case of ALSEP 12 in Figure 1. The low flux suggested later by density contours and velocity vectors behind a typical spherical object may provide an explanation for these observations.

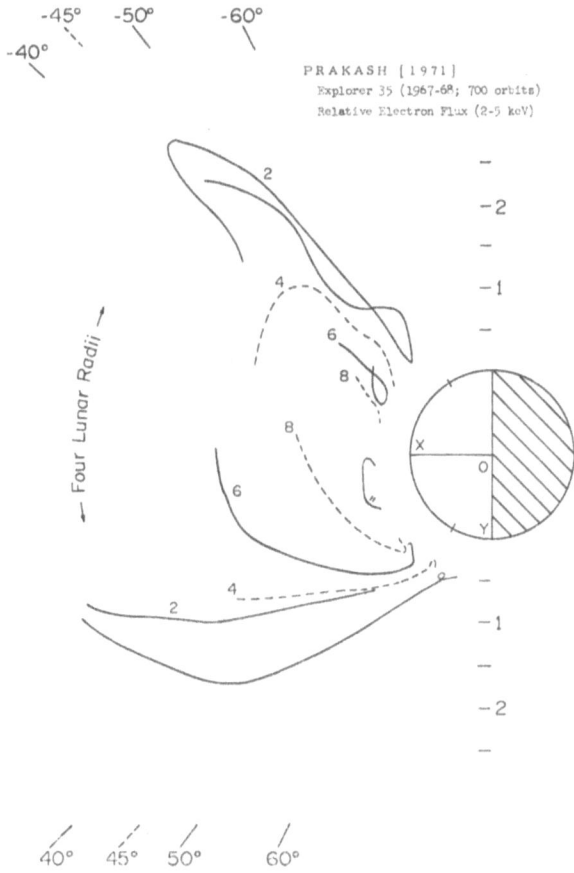

Fig. 3. Relative electron flux contours in front of the Moon as found by Explorer 35 (Prakash, 1970).

Another set of particle observations *in front of the Moon* has been reported by Prakash (1970). Figure 3 shows contours of relative 2–5 keV electron flux on the sunlit side of the Moon as measured by the MIT plasma cup on Explorer 35 during ~700 orbits (outside of the Earth's shock wave). These electrons, of course, represent only the high energy tail of the electron distribution function. Because of charge neutrality which exists in the solar wind, the data shown in Figure 3 must represent only the 'tip of the iceberg' of an equally dense presence of protons in or near the same locations in front of the Moon. The relationship of these preliminary observations to the density contours noted later suggests again that the kinetic approach may be physically plausible for the Moon. If this suggestion is true, the presently-accepted concept of complete solar wind absorption on the sunlit side, with little or no 'sputtering', may require re-examination.

3. Analysis

Let us consider a collisionless neutral plasma (i.e., solar wind) which flows past any 'small' planetary object. On the surface of the object is an assumed angularly distributed positive electrostatic potential field. A schematic diagram of the problem is depicted in Figure 4. This problem is equivalent to that of a free particle flow which interacts

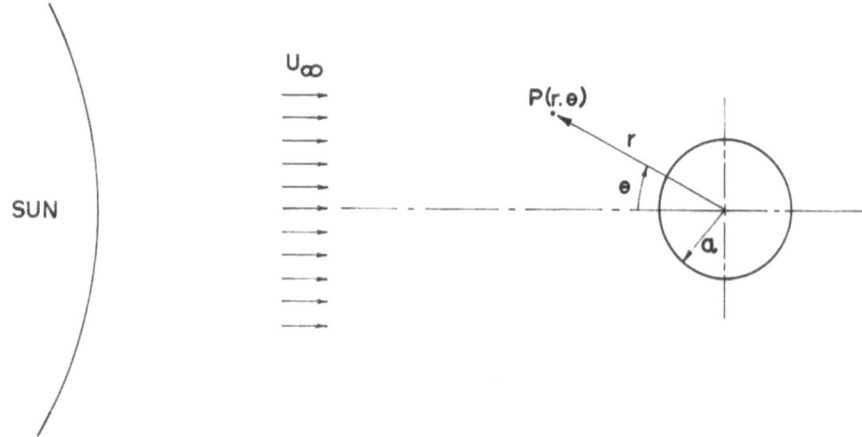

Fig. 4.   Schematic sketch of the interaction of the solar wind with a spherical obstacle. The characteristic radius of the obstacle, a, may include an atmosphere, ionosphere, and/or a magnetosphere. The obstacle is assumed to be 'small' when the characteristic solar wind length (possibly the proton thermal gyroradius) is equal to or greater than, say, $10^{-1}\,a$.

with a solid sphere with a body force due to the electric field on the surface. Hence, the mathematical model for the present problem can be described by the collisionless Boltzmann equation with the body acting as an external force on the gas, thereby eliminating the need for gas-solid surface boundary conditions. For simplicity, we shall consider steady-state ion motion within the solar wind; thus the governing equation for the present problem is

$$\mathbf{v} \cdot \frac{\partial f}{\partial \mathbf{r}} + \frac{\mathbf{F}}{m} \cdot \frac{\partial f}{\partial \mathbf{v}} = 0, \tag{1}$$

where $f = f(\mathbf{r}, \mathbf{v})$ represents the singlet local velocity distribution function of the ions; $\mathbf{v}$, the ion velocity; $\mathbf{F}$, the external force given by

$$\mathbf{F} = - \operatorname{grad} \Phi(\mathbf{r}), \tag{2}$$

where $\Phi(\mathbf{r})$, a scalar function of r and θ, is the potential function due to the properties of the surface of the obstacle under consideration; m, the mass of an ion, finally, $\mathbf{r}$ is the position vector.

The solution of the collisionless Boltzmann equation has been discussed in detail by Karamcheti and Sentman (1965) who direct attention to the advantages that result when solid surfaces are included in the kinetic problem as external forces. Thus, the velocity distribution function $f(\mathbf{r}, \mathbf{v})$ to satisfy Equation (1) for $|\mathbf{r}| \to \infty$ can be written in the following form:

$$f(\mathbf{r}, \mathbf{v}) = A \exp\{- B(\mathbf{v} - U_\infty)^2\} \tag{3}$$

with

$$A = n_\infty (m/2\pi k T_\infty)^{3/2} \tag{3a}$$
$$B = m/2k T_\infty.$$

and where $\mathbf{r}$ and $\mathbf{v}$ are expressed in a spherical coordinate system. U_∞ represents the incoming solar wind velocity, and T_∞, the ambient ion temperature.

The velocity distribution $f(\mathbf{r}, \mathbf{v})$ is calculated numerically for each step. Initially, we assign values of the distribution function to the grid points in $\mathbf{r}-\mathbf{v}$ space. We convect the values of f according to Newton's law:

$$\frac{df}{dt} = 0,$$

$$\frac{d\mathbf{r}}{dt} = \mathbf{v}, \tag{4}$$

$$m \frac{d\mathbf{v}}{dt} = - \operatorname{grad} \Phi(\mathbf{r}).$$

Thus, the distribution function can be calculated successively. It is convenient to express all variables in non-dimensional form by measuring all the lengths, velocities and times in terms of the object's radius, a; the thermal speed of the particle, $c = 1/B^{1/2}$; and the time, $aB^{1/2}$, necessary to travel one 'a' at the particle's thermal speed, respectively. Thus, we calculate the distribution function following the path given by Equation (4). Thus, a non-spherically symmetric potential which represents the body surface can adequately be taken into account. It is important to note that the present solution is the lowest-order approximation to the solution of the Boltzmann equation and consequently, represents a process which ignores inter-particle collisions. Some additional details are given in the Appendix.

On the surface of the body, the velocity distribution function for the particles emitted from the surface is expressed by

$$f_{\text{emit}}(\mathbf{v}, a) = n(\theta_a) \left(\frac{B_w}{\pi}\right)^{3/2} \exp[- B_w v_a^2],$$

where $B_w = (m/2kT_w)$, θ_a and v_a are the constants of integration which will be determined from the trajectory equation. In this study, the accommodation coefficient is assumed to be unity. More details concerning this point can be found from Prager and Rasmussen (1967).

Knowledge of the singlet local velocity distribution enables the calculation of the local density and velocity by taking moments with respect to the local velocity distribution.

4. Method of Solution for Velocity Distribution

Now, the task left us is to calculate the velocity distribution using Equation (4). A detailed discussion of this method is given by Prager and Rasmussen (1967). We shall not repeat it here; however, for completeness, we shall outline the method briefly as follows.

We prescribe a boundary condition for the distribution function at $t = t_0$ in terms of some given function f_1, then,

$$f(\mathbf{r}(t_0), \mathbf{v}(t_0)) = f_1(\mathbf{r}(t_0), \mathbf{v}(t_0)).$$ (5)

Thus, the general solution can be expressed in terms of the function f_1 by

$$f(\mathbf{r}(t), \mathbf{v}(t)) = f_1(\mathbf{r}_0(\mathbf{r}, \mathbf{v}), \mathbf{v}_0(\mathbf{r}, \mathbf{v})),$$ (6)

where

$$\mathbf{v}_0 = \mathbf{v} + \frac{1}{m} \int_{t_0}^{t} \nabla\Phi \, dt$$

$$\mathbf{r}_0 = \mathbf{r} - \int_{t_0}^{t} \mathbf{v} \, dt$$ (7)

which were obtained by integrating Equation (4).

We may note that as $f \to f_0$, the initial value of f is recovered. The vectors $\mathbf{v}_0$ and $\mathbf{r}_0$ represent the six constants of integration required for the equation of motion. Therefore, the general solution is an arbitrary function of the constants of integration of the characteristics of Equation (4). However, this arbitrary function is selected to agree with the boundary condition.

For the present problem, the spherical polar coordinate system for the physical space is chosen, then the solution to the collisionless Boltzmann equation corresponding to the condition Equation (3) is given by

$$f(\mathbf{r}, \mathbf{v}) = A \exp[- B(v_\infty^2(\mathbf{r}, \mathbf{v}) + U_\infty^2 + \\ + 2U_\infty \{v_{r_\infty}(\mathbf{r}, \mathbf{v}) \cos\theta_\infty(\mathbf{r}, \mathbf{v}) + v_{\theta_\infty}(\mathbf{r}, \mathbf{v}) \sin\theta_\infty(\mathbf{r}, \mathbf{v})\})].$$ (8)

Now the task is to determine $v_{r_\infty}(\mathbf{r}, \mathbf{v})$, $v_{\theta_\infty}(\mathbf{r}, \mathbf{v})$, $\cos\theta_\infty(\mathbf{r}, \mathbf{v})$, $\theta_a(\mathbf{r}, \mathbf{v})$ and $v_a(\mathbf{r}, \mathbf{v})$ by solving the equation of motion of a particle in a potential field $\Phi(r, \theta)$.

After some mathematical manipulation, the distribution function for the particles coming from infinity and those emitted from the surface are now

$$f(\mathbf{r}, \mathbf{v}) = A \exp\left[-B\left\{v^2 + \frac{2e}{m}\Phi + U_\infty^2 - 2U_\infty\left(\sqrt{v^2 + \frac{2e}{m}\Phi}\right)\cos\theta_\infty\right\}\right]$$

and

$$f_{\text{emit}}(\mathbf{v}, \mathbf{r}) = n(\theta_a)\left(\frac{m}{2\pi k T_w}\right)^{3/2} \exp\left[-B_w\left(v^2 + \frac{2e}{m}\Phi\right)\right]. \tag{10}$$

The functions $\cos\theta_\infty$ and θ_a again will be determined from the trajectory equation of particles.

For convenience, we shall write the total local density in the form

$$n(r, \theta) = n_i(r, \theta) + n_{\text{emit}}(r, \theta), \tag{11}$$

where $n_i(r, \theta)$ is the net contribution to the number density by the incident particles, i.e., the contribution of those particles that are moving toward the object together with the effect of those particles which impact the object; and $n_{\text{emit}}(r, \theta)$ is the contribution to the number density by the particles emitted from the surface with the temperature of the body surface, T_w. This is the effect of the boundary condition on the surface.

After some mathematical manipulation, we obtain

$$n_i(r, \theta) = \int_{-\infty}^{0} \int_{-\infty}^{\infty} \int_{-\infty}^{\infty} \left(1 + \frac{2e}{m}\frac{\Phi(a, \theta)}{\chi^2}\right)^{1/2} f_i \, d\chi_r \, d\chi_\theta \, d\chi_\psi, \tag{12}$$

with

$$\chi^2 = v^2 + \frac{2e}{m}\Phi(r, \theta),$$

where

$$\chi = (\chi_r, \chi_\theta, \chi_\psi); \qquad \mathbf{v} = (v_r, v_\theta, v_\psi)$$

and

$$f_i = n_\infty\left(\frac{B_\infty}{\pi}\right)^{3/2} \exp\left[-S_\infty\left\{\left(\frac{\chi}{U_\infty}\right)^2 + 1 - 2\chi_r\cos\theta + 2\chi_\theta\sin\theta\right\}\right] \tag{13}$$

with

$$S_\infty = U_\infty / \left(\frac{2k T_\infty}{m}\right)^{1/2},$$

the particle speed ratio.

$$n_{\text{emit}}(r, \theta) = n(\theta_a)\left(\frac{B_w}{\pi}\right)^{3/2} \int_{0}^{\infty} \int_{-\infty}^{\infty} \int_{-\infty}^{\infty} \left(1 + \frac{2e}{m}\frac{\Phi(a, \theta)}{\chi^2}\right)^{1/2} \times$$

$$\times e^{-B_w\chi^2} \, d\chi_r \, d\chi_\theta \, d\chi_\psi, \tag{14}$$

where $n(\theta_a)$ will be determined by the assurance of continuity in the characteristic equation and can be found in terms of n_∞.

The mean velocity field is determined by the first moment of the velocity distribution function,

$$u_j(r, \theta) = \frac{1}{n(r, \theta)} \left[\int_{-\infty}^{0} \int_{-\infty}^{\infty} \int_{-\infty}^{\infty} \chi_j \left(1 + \frac{2e}{m} \frac{\Phi(a, \theta)}{\chi^2}\right)^{1/2} f_i(\mathbf{r}, \mathbf{v}) \, d\chi_r \, d\chi_\theta \, d\chi_\psi + \right.$$

$$\left. + \int_{0}^{\infty} \int_{-\infty}^{\infty} \int_{-\infty}^{\infty} \chi_j \left(1 + \frac{2e}{m} \frac{\Phi(a, \theta)}{\chi^2}\right)^{1/2} f_{\text{emit}} \, d\chi_r \, d\chi_\theta \, d\chi_\psi \right] \tag{15}$$

with
$$j = r, \theta.$$

Finally, the potential on the surface is derived from a charged particle interacting with an angularly-distributed charged particle layer on the surface and is given as follows:

$$\Phi(r, \theta) = \int_{-\pi/2}^{\pi/2} \frac{Q(\cos\theta')^{1/2} \, d\theta'}{\left[1 + \left(\frac{r}{a}\right)^2 - 2\left(\frac{r}{a}\right)\cos(\theta - \theta')\right]^{1/2}} \tag{16}$$

and/or

$$\Phi(r) \sim \frac{Q'}{r^n}, \qquad n = 2 \text{ and } 5 \text{ in the present study}, \tag{17}$$

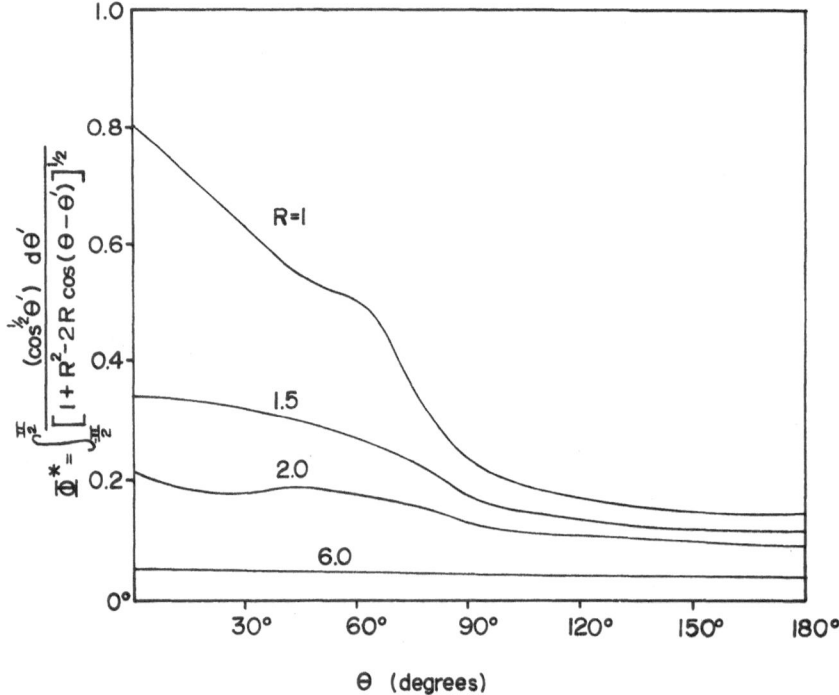

Fig. 5. Dimensionless electrostatic potential, Φ^*, in the vicinity of a spherical obstacle. $R = r/a$. The potential, Φ, is normalized by the quantity Q which includes the surface properties and quantum efficiency of the photoionization process (Walbridge, 1973).

where Q indicates a constant representing the product of quantum efficiency and sub-solar potential and Q' is the usual force constant. A repulsive electric field of 1 V m^{-1} (volt/m) at the subsolar point will be assumed for both asymmetric and spherically symmetric cases. In the latter case, of course, Φ is independent of θ so that the two cases are normalized at the subsolar point.

In the figures which are discussed below, the nondimensional radial distance, r/a, is denoted as R. Figure 5 shows the result of the integration of Equation (16). It is seen that Φ becomes essentially independent of θ within several radii, thus, the main effect (of the potential which represents the force on the fluid) is confined to small values of R where a shock wave could eventually develop. In such a calculation, we have weighted the potential energy by the reflected particle's thermal energy, kT_w, as seen, for example, in Equation (10).

5. Results

A steady solar wind is assumed to be flowing past a 'small' object with the following assumed parameters: $U_\infty = 400$ km s^{-1}, $n_\infty = 5$ cm^{-3}, and $T = 10^5$ K. The particle speed ratio, S, is therefore 9.85. Other typical solar wind parameters will generally give $S > 1$. Sentman and Karamcheti (1969) show that the results for density contours are insensitive to variations in S once $S > 1$. We will assume that the present case, with the electric field, is similar and will not investigate this point any further in the belief that the present case will be sufficiently representative.

Using Equations (11) through (14) together with (4) through (10), the density distribution along the Sun-obstacle axis (i.e., $\theta = 0°$) is first found. Figure 6 shows n/n_∞ plotted versus the non-dimensional radius, R, in the absence of an electric field as well as with the presence of a field as discussed earlier. A sharp rise from the ambient value of density to 2 and 3 times that value occurs within half of the obstacle's radial distance from the surface for the two cases, respectively. The results for the major portion of the affected flow field are shown by contours of constant density (with effective non-central force field) in Figure 7. It is seen that the surface density rapidly decreases as θ moves toward the object's limb. A substantial paucity of particles is also seen in the immediate base region as shown also by Wang (1971) for the alternative case of absorption on the sunlit surface. Of particular interest, however, is the steep density gradient in the immediate vicinity of the limb. This result is, not unexpectedly, similar to that are shown by Sentman and Karmacheti (1969) for the case of no electric field. The density contours for $S \approx 10$ (with and without an electric field) are also given by Dryer (1970). Also, the contours of constant density with effective central force field which are described by Equation (17) are shown in Figures 8 and 9 for $n = 2$ and 5, respectively. Note that there is no substantive difference between the case of an asymmetrical force field (which represents a surface electric field as suggested for Moon by Walbridge (1968, 1973) and Grobman and Blank (1969)) and a spherical inverse force field.

Finally, Equations (16) and (17) provide additional information about the flow field, namely, the streamline pattern. Figure 10 shows the results of the calculation for

non-central force field where the length of the individual velocity vectors is proportional to their magnitude. Note, for reference, the length of the undisturbed velocity vector
at the left side of the figure. The expected deceleration of the flow on the sunlit side of
the obstacle with subsequent acceleration in the expansion phase around the limb and

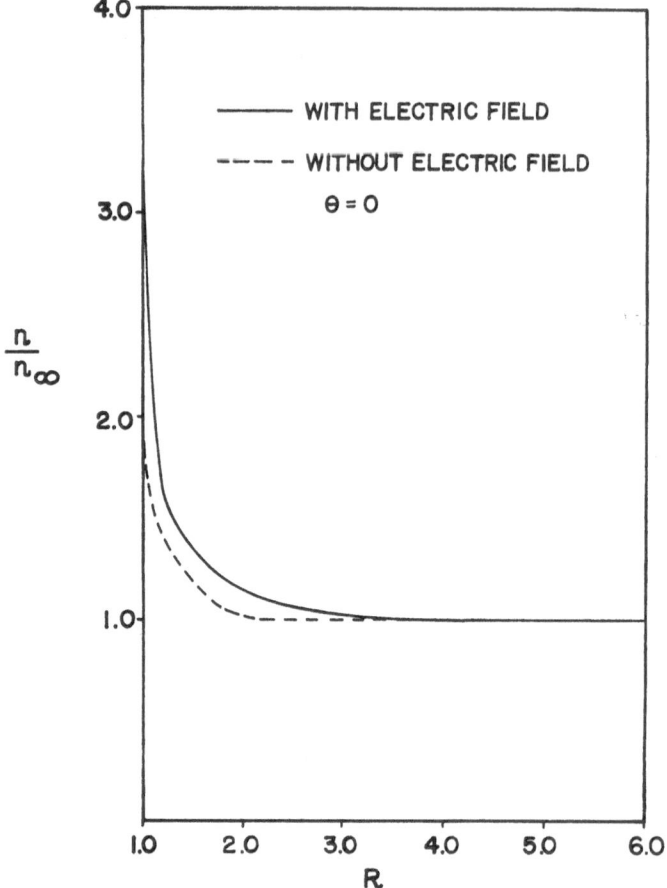

Fig. 6. Proton number density, normalized to undisturbed solar wind density, along the obstacle-Sun
axis $(\theta = 0°)$. $S = U_\infty/(2\,kT_\infty/m)^{1/2} = 9.85$. Results are shown for an angularly-distributed surface
electric potential. The surface repulsive electric field at the subsolar point is taken to be 1 V m^{-1}.
The result for no electric surface potential is essentially the same as that found by Sentman and
Karamcheti (1969) for $S = 10$.

in the wake is clearly seen. Mention has already been made above of the observation
by Neugebauer *et al.* (1973) which may be relevant to the present study. It is seen that
the flow, on the average, is similar to that for continuum flow around a sphere.

  As noted earlier, the interplanetary magnetic field has been explicitly neglected.
Further studies (with the present boundary condition) should, of course, take **B** into
account, but it is possible to make some qualitative observations regarding its effect

on the flow. When **v** is not parallel to **B**, the induced electric field in the object's frame of reference is $-\mathbf{v} \times \mathbf{B}$. Thus, the Lorentz force is proportional to $(\mathbf{v} \times \mathbf{B}) \times \mathbf{B}$. We refer here to the volume current considered in MHD analyses and do not consider charge separation; thus, the $\mathbf{E} \times \mathbf{B}$ drift can be neglected. Assuming, then, that $\mathbf{B} = (B_r, 0, B_\varphi)$ in a spherical coordinate system whose axis is at the center of the Sun, and $\mathbf{v} = (u, 0, 0)$,

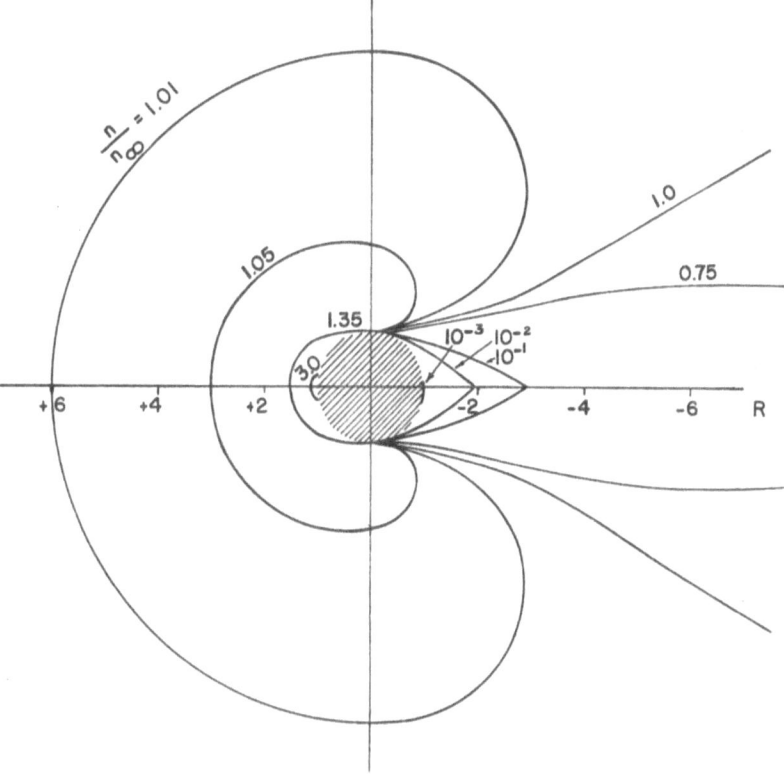

Fig. 7.   Constant density contours with non-central force field potential for $S = 9.85$ in the vicinity of a 'small' spherical obstacle. Surface repulsive electric field at the subsolar point is 1 V m^{-1} and decreases as $\cos^2 \theta$. Note development of a continuum-like 'Mach cone' at the limb, compression near the subsolar point, and expansion immediately beyond the limb.

this force has the following components: $(-u\, B_\varphi^2)\, \hat{r} + (u\, B_r\, B_\varphi)\, \hat{\varphi}$. For the chosen undisturbed velocity components and 'toward the Sun' polarity of the field, thd Lorentz force acts within the ecliptic plane to cause a distortion of the flow fieler The density contours and velocity vector field (Figures 7 through 10) would no longty be perfectly symmetrical. Wang (1971) has shown, for example, that the density distribution shows 'complicated periodic structures' for cases of large gyroradii. The direction of the force, indicated above for the example, will favor a higher density concentration on the sunward, morning quadrant of the flow field. The results

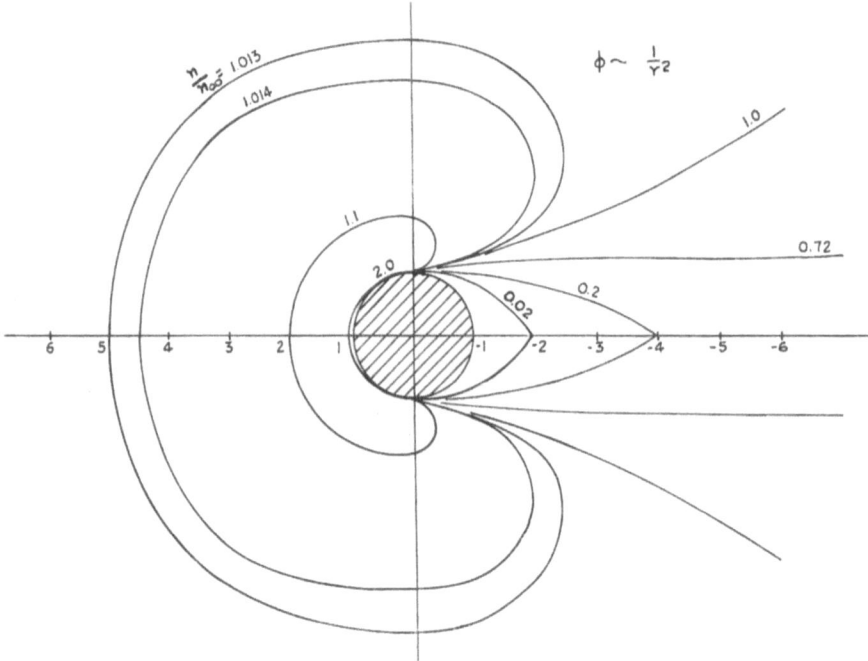

Fig. 8. Constant density contours with central-force-field potential for $n=2$ and $S=9.85$ in the vicinity of a 'small' spherical obstacle. Surface electric field at the subsolar point is 1 V m^{-1}, i.e., the same as for the non-central field in order to make a direct comparison with that case.

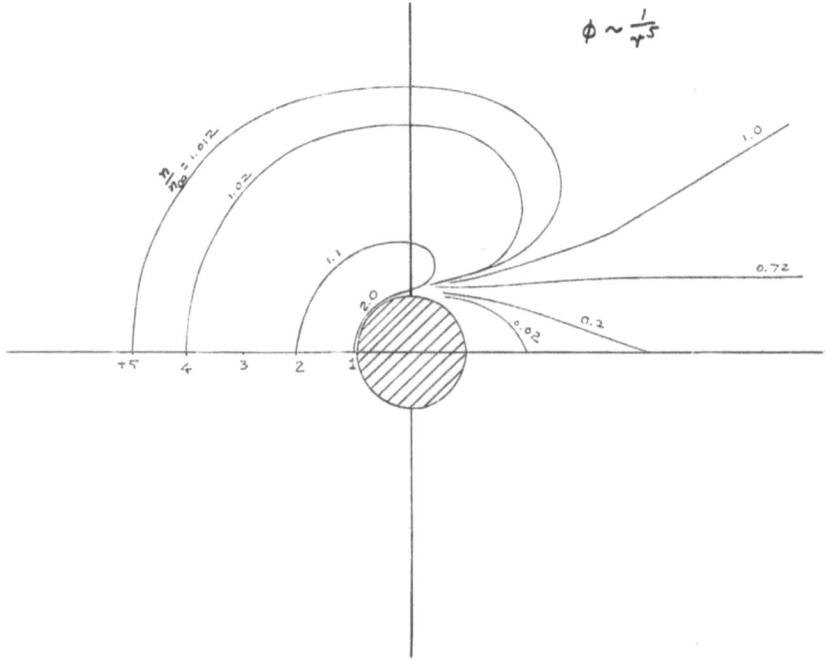

Fig. 9. Constant density contours with central-force-field potential for $n=5$ and $S=9.85$.

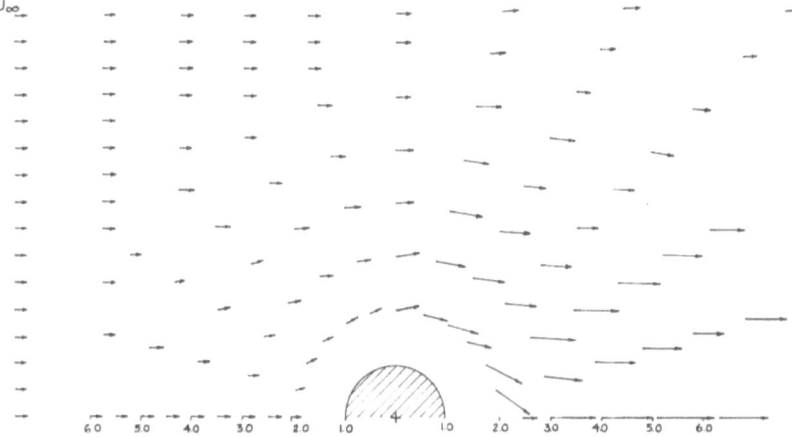

Fig. 10. Velocity vectors (or 'streamline' pattern) with non-central-force field potential for $S = 9.85$ in the vicinity of a 'small' spherical obstacle. Surface electric field at the subsolar point is 1 V m^{-1}. Note turning of the flow on the sunlit hemisphere and acceleration of particles around the limb as in continuum flow.

presented here, then, must be considered to be a qualitative estimate of a self-consistent kinetic calculation which explicitly incorporates the effect of an arbitrarily-oriented, frozen-in, magnetic field. The skewing of the density contours (as well as other parameters) has been explicitly found in the continuum case by Shen (1972) and Hirsh and Reshotko (1971). Some of the latters' work is discussed by Dryer (1970). An additional manifestation of this effect would be the unsymmetrical 'Mach cone' which has been observed by Whang and Ness (1970) in the case of Moon and which is anticipated by the limb density gradient in Figures 7, 8, and 9 for the zeroth order solution.

6. Discussion

Sentman and Karamcheti (1969) have pointed out that the spacing of the constant density contours in front of the sphere is relatively insensitive to variations in the speed ratio, S, once $S > 1$. Behind the sphere, the contours (hence the wake) elongate while there is a decrease of the width of the compression and the 'expansion fan' which extend rearward from the limb. Although results (which include effects of the electrostatic potential) for S different from 10 are not presented here, we believe that they would produce similar effects as noted above. It is seen that the ridge of increased density $(n > n_\infty)$ – due to reflected protons – extends downstream in the direction of what might be called a developing 'Mach cone'. The results suggest, for high Knudsen numbers, that effective collisional encounters would first become important along this ridge of high density gradient. Sentman and Karamcheti (1969) have suggested that "... perhaps it is the high-density ridge that develops into a shock wave as collisions become important." Also, the shock may indeed form in a high Knudsen flow, but it would probably first appear downstream from the body and, as Sentman and Karamcheti point out: "… the stagnation streamline would probably be the last region in the flow to become aware of the presence of the shock!" Flybys of objects (such as Mercury and Pluto which are potential candidates for high Knudsen number flow) should then be scheduled for near-limb passes with a data accumulation rate high enough to detect possible density gradients suggested by Figures 7, 8, and 9. A suggestion by Barnes *et al.* (1971) is possibly related to this discussion. These authors suggest that limb shocks can be caused by localized magnetic anomalies intrinsic to Moon. They further suggest that small length scales (geometric mean of electron and proton gyroradii) are expected to promote instabilities which are, in turn, induced by wave-particle interactions. Pertinent to the present study is the possibility that anomalous transport properties (i.e., 'collisions') would be generated.

Equally important would be detection of streamline patterns suggested by Figure 10. This figure indicates that there could be a sudden deceleration along the subsolar streamlines within a distance of ~1 radius from the surface. Scattering at higher energy densities occurs near the surface in the direction of the on-coming flow, but the protons are quickly turned and deflected around the limb. It is surprising to note that the dark side of the object is impacted by a very low, but finite, density of particles. The observations of Neugebauer *et al.* (1973) and Freeman (1972) noted above may, again, be relevant within the present context.

We also point out that the present kinetic approach to the interaction of the solar wind with 'small' obstacles (i.e., those planetary objects characterized by large Knudsen numbers) is suggested as an alternative to continuum and quasi-kinetic techniques. Indeed the results found in the investigation suggest that, for a given proton thermal gyroradius, a shock would develop first at the limb, gradually building up – as we choose larger obstacles – to a 'conventional' detached bow shock as in Earth's case. Alternatively, for a given 'small' obstacle, the effect would be the same if the solar wind were to get much hotter so that the gyroradius were to decrease gradually by several orders of magnitude.

We state this important point in an alternate way, that is, by discussing an increase in the obstacle's characteristic dimension in a slightly different way. The increase in size could certainly be gained by added material (i.e., choosing a larger planet), but it could also be achieved by invoking the presence of an atmosphere, ionosphere, and/ or a magnetosphere. The physical mechanism for new boundary conditions (ionopause, magnetopause, or even surface penetration) is a condition which could be incorporated in related studies. For example, surface absorption on the sunlit hemisphere, as in the case of Moon, could be incorporated in the expression for the number density of particles which are reflected from the surface (Equation (14)).

Finally, we should note that Wang (1971) has considered the magnetic field effect in his calculation of solar wind-Moon interaction; however, the condition of a completely absorbing surface is used. The reader is reminded that the magnetic field is not explicitly taken into account in the present study; although an estimate of its effect suggests that introduction of the Lorentz force would be confined to a skewing of the density contours and an associated distortion of the flow pattern. Also, provision is not made in the present work for particle absorption (with no subsequent re-emission) in the sunlit hemisphere.

Inasmuch as several statements have been made regarding the development of shock waves, 'Mach cones', and the like, provided the characteristic fluid scale length is small relative to the obstacle's scale length, a final comment is made to relate the present kinetic approach to its extreme 'competitor': continuum theory. It is well known that the 'gain and loss' term in the kinetic description is given by the collision integral. For free particle flow, this term is zero because there are no collisions of any kind. Similarly, in continuum theory – at least at the level of Euler's approximation regarding the absence of transport properties such as viscosity – the collision integral is again found to be zero because the distribution function is found to be Maxwellian. There is, then, no contradiction in the statement that kinetic theory, as discussed herein, preserves the characteristics of the continuum theory.

Acknowledgements

We wish to thank Drs P. Cassen, A. Eviatar, L. Sentman, Y. C. Whang, and E. Whipple for their critical reading of the manuscript and valuable comments and suggestions. We also are grateful to Drs M. Neugebauer and A. Prakash for permission to use their unpublished data from the ALSEP 12 and Explorer 35 instruments and to Dr J. W. Freeman, Jr., for allowing us to modify a figure from one of his publications.

The work done by one of us (S.T.W.) was supported by Marshall Space Flight Center Contract NAS8-25750.

Appendix

We have employed the finite difference technique for obtaining numerical results. We have chosen that 'i' be the label for the r-coordinate and 'j' be the label for the θ-coordinate, starting clockwise with intervals of $15°$ in a range $0 \leqslant \theta \leqslant 180°$. Also we have set up the origin of the coordinate system at the center of the body, therefore the local velocity distribution is expressed as $f_{i,j}(r_i, \theta_j, v_r^{i,j}, v_\theta^{i,j}, v_\varphi^{i,j})$. We have assumed that the local velocity distribution is Maxwellian for $R=10$ (which is effectively $R=\infty$) and $0 \leqslant \theta \leqslant 180°$, then we compute each point successively by using Equation (1), and Equation (2). Newton's equations of motion (i.e., Equation (4)) are used here to track the particle trajectory due to the effects of the non-central force potential field. Because of the non-straight line characteristics, we can only fix our initial points on the boundary; the subsequent points, as said, were determined by Newton's equations of motion. However, our grid points are fixed points, therefore, a Runge-Kutta subroutine is used for the interpolation.

References

Barnes, A., Cassen, P., Mihalov, J. D., and Eviatar, A.: 1971, *Science* **172**, 716.
Dryer, M.: 1970, *Cosmic Electrodyn.* **1**, 115.
Freeman, J. W., Jr.: 1972, *J. Geophys. Res.* **77**, 239.

Greenstadt, E.: 1971, *Icarus* **14**, 374.

Grobman, W. D. and Blank, J. L.: 1969, *J. Geophys. Res.* **74**, 3943.

Hirsh, R. S. and Reshotko, E.: 1971, *AIAA Bull.* **8**, 227.

Karamcheti, K. and Sentman, L. H.: 1965, *SUDAER* No. 236, Dept. of Aero. and Astro., Stanford University.

Lyon, E. F., Bridge, H. S., and Binsack, J. H.: 1967, *J. Geophys. Res.* **72**, 6113.

Michel, F. C.: 1968, *J. Geophys. Res.* **73**, 1533.

Ness, N. F., Behannon, K. W., Scearce, C. S., and Cantarano, S. C.: 1967, *J. Geophys. Res.* **72**, 5769.

Ness, N. F.: 1972, in E. R. Dyer (ed.), *Solar Terrestrial Physics/1970*, D. Reidel Publishing Company, Dordrecht, Holland, p. 159.

Neugebauer, M., Snyder, C. W., Clay, D. R., and Goldstein, B. E.: 1973, *Planetary Space Sci.* **20**, 1577.

Prager, David J. and Rasmussen, Maurice L.: 1967, *SUDAER* No. 299, Stanford University.

Prakash, A.: 1970, *EOS, Trans. Am. Geophys. Union* **51**, 822.

Sentman, L. H. and Karamcheti, K.: 1969, *AIAA J.* **7**, 161.

Shen, W. W.: 1972, *Cosmic Electrodyn.* **2**, 381.

Siscoe, G. L., Lyon, E. F., Binsack, J. H., and Bridge, H. S.: 1969, *J. Geophys. Res.* **74**, 59.

Sonett, C. P. and Mihalov, J. D.: 1972, *J. Geophys. Res.* **77**, 588.

Spreiter, J. R. and Alksne, A. Y.: 1970, in *Ann. Rev. Fluid Mech.* **2**, 313.

Spreiter, J. R., Marsh, M. C., and Summers, A. L.: 1970, *Cosmic Electrodyn.* **1**, 5.

Walbridge, E.: 1968, *EOS, Trans. Am. Geophys. Union* **49**, 708.

Walbridge, E.: 1973, *J. Geophys. Res.*, in press.

Wang, C. P.: 1971, *AIAA J.* **9**, 1148.

Whang, Y. C.: 1968, *Phys. Fluids* **11**, 969.

Whang, Y. C.: 1969, *Phys. Rev.* **186**, 143.

Whang, Y. C.: 1970, *Solar Phys.* **14**, 489.

Whang, Y. C. and Ness, N. F.: 1970, *J. Geophys. Res.* **75**, 6002.

Wolf, R. A.: 1968, *J. Geophys. Res.* **73**, 4281.

DISCUSSION

Criswell: Neuegebauer's observation of higher solar proton densities at the lunar surface than in free space results from compression of the solar wind flow by local remnant magnetic fields on the surface and do not support the potential field model. (Neugebauer *et al.*, 1973). How can a lunar potential field of an electrical nature extend outward 10^5 to 10^6 Debye lengths (i.e. 10^3–10^4 km) into the upstream solar wind?

Wu: The potential effect does not have to be electrical in nature.

Gold: The observed increase of density in front of the Moon surely must be due to compression of the solar wind magnetic field, due to conductivity or permanent magnetic fields in the Moon. It surely cannot be due to any propagation in the upstream direction of any electric field. This, in any physical case, is attenuated by some very large factor, since we are dealing with a very large number of Debye shielding lengths.

Wu: It is true that the density enhancement on the sunlit side of the Moon is due to compression of magnetic field, but the calculation we present here is not necessary for the Moon. The purpose of the present study is to show that in the case of collisionless flow (i.e. free particle flow) the density enhancement in front of an object can be caused by boundary conditions on the surface of the object. The surface potential we used here is hypothetical, which represents the surface properties of the object and is not necessarily an electric potential. Therefore, the Debye shielding length effect can be ignored.

Manka: From your figure it seems that you assumed a positive lunar potential from the subsolar point all the way to the anti-solar point. Could you comment on the effect on your results if you use a negative potential on the dark side?

Wu: Yes, we did use a positive potential all the way (i.e. from subsolar point to anti-solar point). If we use a negative potential on the dark side (i.e. from 90°–180°), I believe that the flow pattern will not change drastically and only slightly modify the numerical values of the density contour on the dark side.

MAGNETIC MEASUREMENTS OF THE SOLAR WIND
INTERACTION WITH THE MOON

B. R. LICHTENSTEIN and P. J. COLEMAN Jr.

Dept. of Planetary and Space Science and Institute of Geophysics and Planetary Physics, University of California, Los Angeles, Calif. 90024, U.S.A.

and

C. T. RUSSELL

Institute of Geophysics and Planetary Physics, University of California, Los Angeles, Calif. 90024, U.S.A.

Abstract. The solar wind is almost completely absorbed or neutralized by the lunar surface leaving a plasma void on the antisolar side of the Moon. The magnetic signature of this interaction, as observed by the Apollo 15 subsatellite, is an enhanced field directly behind the Moon, bounded on either side by two dips in the field strength. On occasion, compressions of the field strength are observed external to either one or sometimes both of these dips. Theories of the interaction postulate either that these compressions are a general feature of the solar wind-Moon interaction modulated by changes in the solar wind parameters or that they are associated with the appearance of specific lunar regions at the limbs. The measurements of the lunar magnetic field with the Apollo 15 and 16 subsatellites, the mapping of projected source positions of limb compressions onto the lunar surface, and the study of the persistence of limb compressions supports the hypothesis that limb compressions are formed when regions of high magnetization are at the lunar limbs.

1. Introduction

The interaction of the solar wind with the Moon was first observed by instruments aboard the lunar satellite Explorer 35 (Ness, 1969). The major features of this interaction are: the absence of an upstream detached bow shock; a plasma void-diamagnetic cavity behind the moon; a rarefaction wave bounding the plasma void; and occasional increases in the magnitude of the magnetic field on the Sun side of the rarefaction wave (Colburn *et al.*, 1967; Ness *et al.*, 1967; Lyon *et al.*, 1967; Michel, 1967). These results led Ness (1972) to conclude that the Moon interacts with the solar wind as a completely absorbing, nonmagnetic and nonconducting body.

Apollo experimental results have significantly modified this view of the Moon's interaction with the solar wind. Lunar surface magnetometers have measured local magnetic fields ranging from 6 γ to over 300 γ (Dyal *et al.*, 1972a, b). Measurements made during the lunar traverses indicate that these fields are microscale features with a scale length of approximately 1 km (Dyal *et al.*, 1971, 1972c). Their assumed source is remanent magnetization in the lunar crust (Strangway, 1970). Simultaneous measurements made by Explorer 35 and the Apollo 12 lunar surface magnetometer have revealed a significant global response by the Moon to time dependent variations of the solar wind magnetic field (Sonett *et al.*, 1971b). This response has been used to construct electrical conductivity profiles of the lunar interior (Sonett *et al.*, 1971a).

In this paper we present some results from the UCLA magnetometer experiment

R. J. L. Grard (ed.), Photon and Particle Interactions with Surfaces in Space, 471–480. All Rights Reserved

aboard the Apollo 15 and 16 subsatellites (Coleman *et al.*, 1972a, 1972d). Figure 1 shows the relative geometry of the orbits of Explorer 35 and the Apollo 15 subsatellite. Explorer 35 has an orbit period of approximately 12 h. The subsatellites were designed

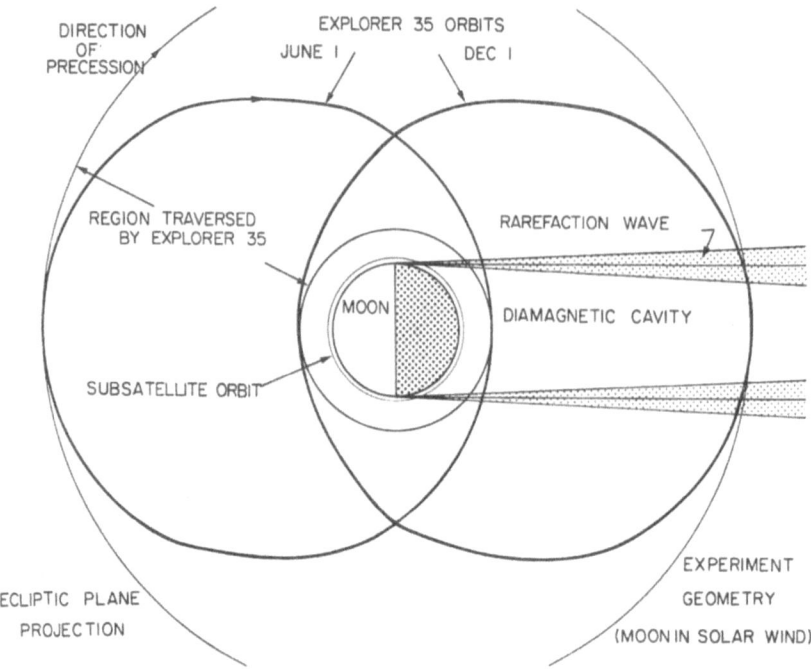

Fig. 1. Orbits of Explorer 35 and Apollo 15 and 16 subsatellites. The Explorer 35 orbits are elliptical with an altitude of perilune of about 850 km and of apolune of 4 lunar radii. The orientation of the orbit ellipse is fixed in inertial space, so that the motion of the Earth-Moon system about the Sun causes apolune to move from the sunward to the anti-sunward side during the course of a year. The Apollo subsatellites are in near circular orbits at an altitude of approximately 100 km. Slight perturbations in the orbital parameters by lunar, solar and terrestrial gravitational forces caused significant changes in the altitude of these orbits with time. As an extreme example, the Apollo 16 subsatellite apolune reached 200 km and perilune 0.

to investigate the plasma environment at altitudes of approximately 100 km above the lunar surface. Both subsatellites had orbital periods of about 2 h and were spin stabilized about an axis perpendicular to the ecliptic plane.

2. The Lunar Magnetic Field

The subsatellite magnetometers were able to measure the intrinsic lunar magnetic field when the Moon was in either lobe of the Earth's geomagnetic tail. These measurements indicate that remanent magnetization exists over much of the lunar surface (Coleman *et al.*, 1972c). Figure 2 shows a contour map of the lunar contribution to

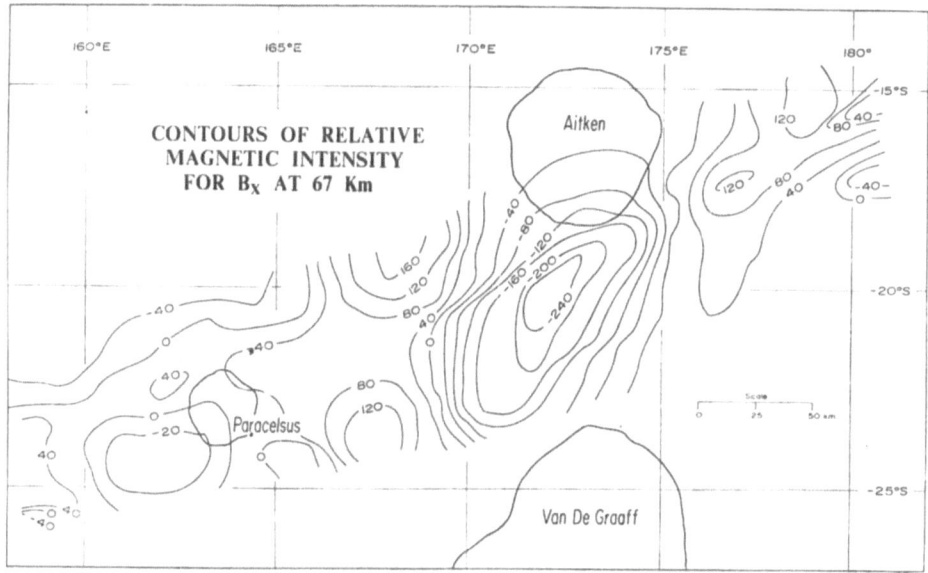

Fig. 2.   Contours of the solar-antisolar component of the lunar contribution to the magnetic field measurement at 67 km altitude while the Apollo 15 subsatellite was in the Earth's geomagnetic tail.

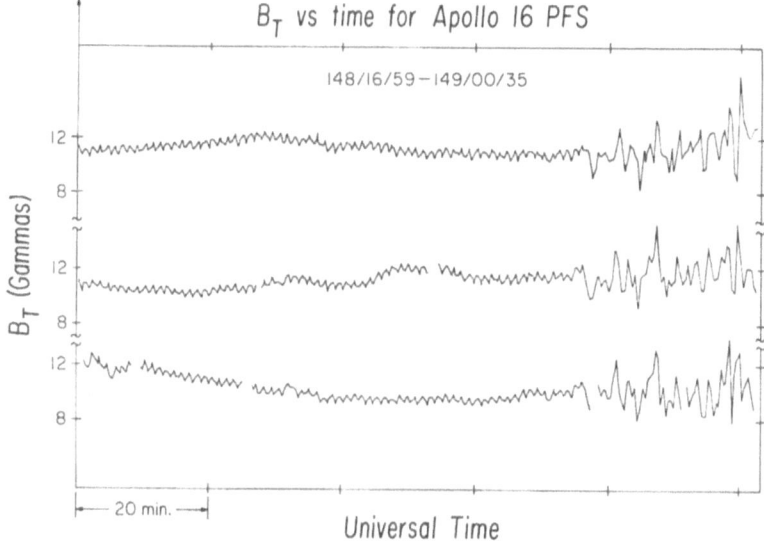

Fig. 3.   The magnitude of the magnetic field in the spin plane of the Apollo 16 subsatellite during three orbits while in the geomagnetic tail. Perilune, which occurred in the center of the region of structured field, was at an altitude of 13 km during this series of orbits. The ripple in the otherwise quiet field is an artifact of the preliminary data processing and is corrected in the final computer processing.

the solar directed component of the magnetic field. The measurements were made at an altitude of 67 km above the Van de Graaff region. Contours are labeled in units of hundredths of a γ and lines drawn every 0.4 γ. Figure 3 shows Apollo 16 subsatellite measurements for three successive orbits while the Moon was in the geomagnetic tail. B_T is the magnitude of the magnetic field transverse to the subsatellite spin axis (i.e., the magnitude in the ecliptic plane). The ripple is due to the subsatellite spin and is removed during computer reduction of the data. The irregular features on the right side occur around perilune, which is 13 km above the lunar surface for these orbits.

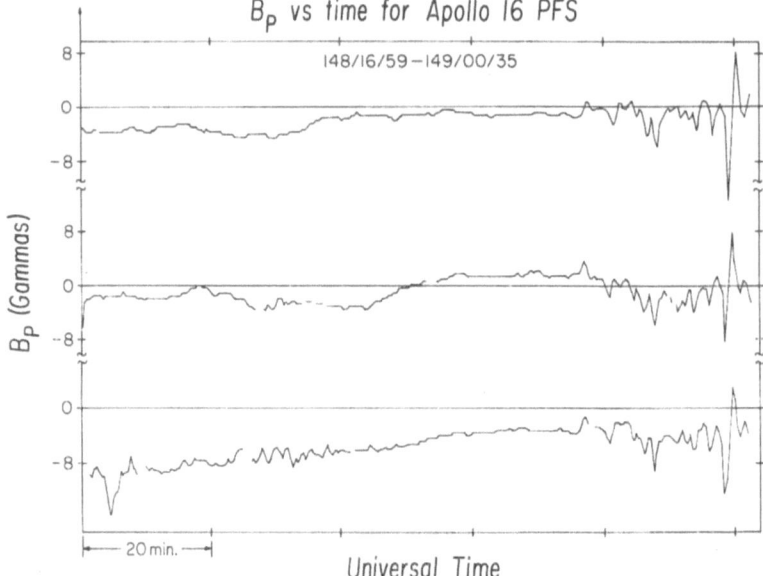

Fig. 4.   The component of the magnetic field parallel to the Apollo 16 subsatellite spin axis for the same three orbits as Figure 3.

Note that the field structure in this region repeats on successive orbits. It seems apparent that the subsatellite magnetometer is measuring mesoscale features of the lunar remanent field, with a scale length of approximately 100 km. Figure 4 shows the component of the magnetic field parallel to the spin axis for the same three orbits. Again, there is repetitive structure near perilune. Figure 5 shows B_T for the last 8 perilune passes of the Apollo 16 subsatellite. At this time, perilune was decreasing at the rate of 0.5 km per revolution. On revolution 424, perilune was approximately 0.4 km above the lunar surface, at longitude 103° E, latitude 10° N. The time scale has been expanded and only 20 min around perilune are shown. The amplitude scale has been compressed from that used on previous figures. The gradual evolution of the field pattern can be seen as perilune drifts across the region. These data were obtained while the Moon was in the magnetosheath and the subsatellite in the plasma void behind the Moon. The largest magnetic field measured was 56 γ.

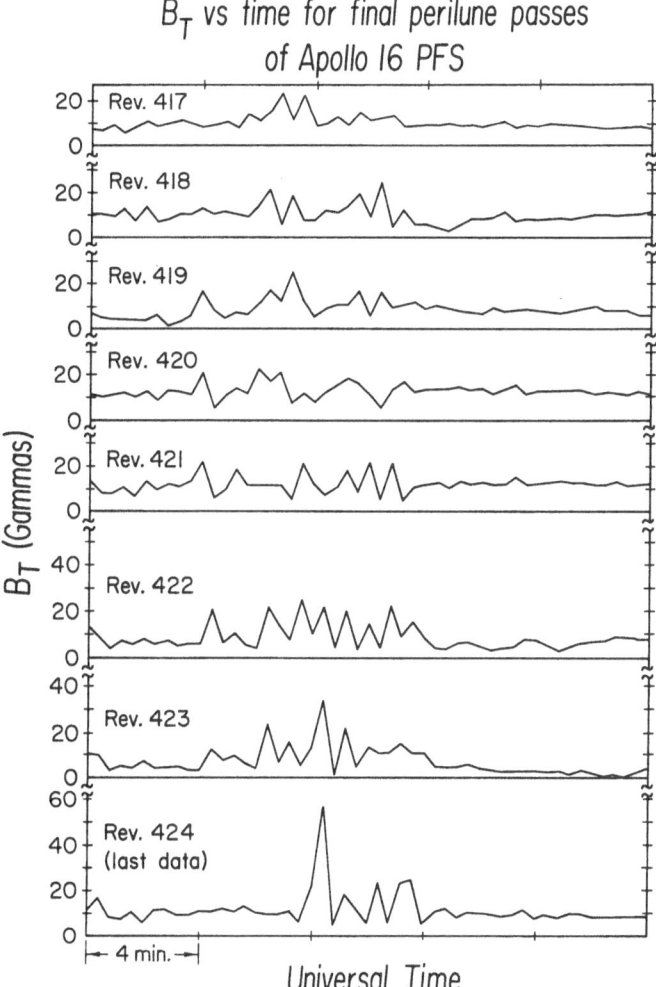

Fig. 5. The magnitude of the magnetic field in the Apollo 16 satellite spin plane for the last 8 orbits, around perilune. Perilune decreased at a rate of $\frac{1}{2}$ km per orbit during this period and reached zero on revolution 425. The time scale has been expanded and the vertical scale compressed from that used in Figures 3 and 4.

3. The Solar Wind-Moon Interaction

The characteristic magnetic signature of the Moon's interaction with the solar wind is best demonstrated by making an average of many orbits. This minimizes the effects of temporal variations. Figure 6 shows the results of averaging B_T for 121 revolutions of the Apollo 15 subsatellite while the Moon was in the solar wind. Sunset and sunrise refer to the entering and leaving of the Moon's optical shadow by the subsatellite. The magnetic enhancement in the plasma void behind the Moon and the presence of dips in the field magnitude associated with the rarefaction wave (Michel, 1968;

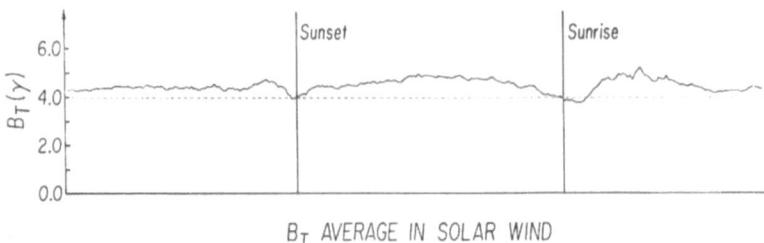

B_T AVERAGE IN SOLAR WIND

Fig. 6.   A 121 orbit average of the magnetic field in the Apollo 15 subsatellite spin plane while in the solar wind.

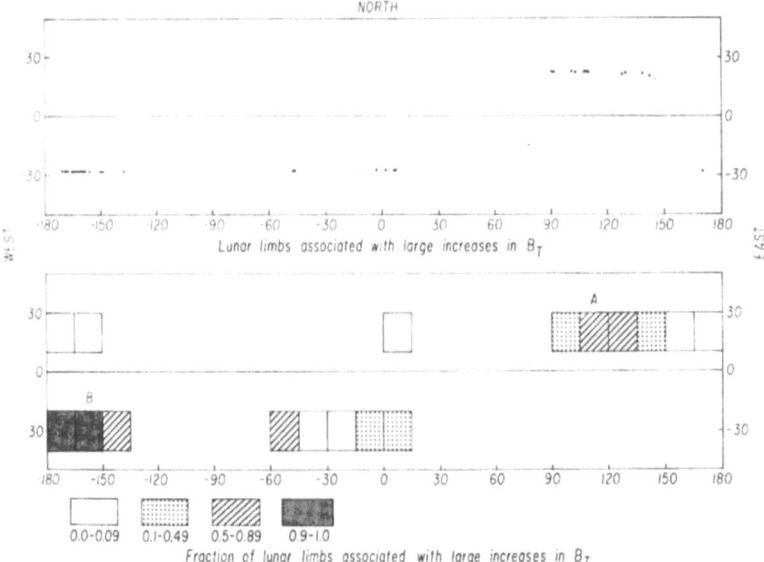

Fig. 7.   The occurrence of limb compressions during the first lunation of Apollo 15. The upper panel shows the projected selenographic latitude and longitude of limb compression sources. The lower panel shows the rate of occurrence.

Spreiter *et al.*, 1970) are qualitatively similar to Explorer 35 magnetic observations made further from the Moon (Colburn *et al.*, 1967, 1971; Ness *et al.*, 1967, 1968; Taylor *et al.*, 1968). This figure also shows an average increase in the field external to the rarefaction dips. This effect in the average is due to the occasional presence of large increases in B_T. Increases in the magnetic field magnitude on the solar wind side of the expansion were originally observed by Explorer 35 magnetometers (Colburn *et al.*, 1967, 1971; Ness *et al.*, 1967, 1968; Taylor *et al.*, 1968). A study of simultaneous Explorer 35 magnetic and plasma measurements by Siscoe *et al.* (1969) has shown that these increases in the field magnitude are associated with increases in the plasma density. There currently is some debate as to whether these increases are due to an interaction with specific regions on the lunar surface or a general property of the solar

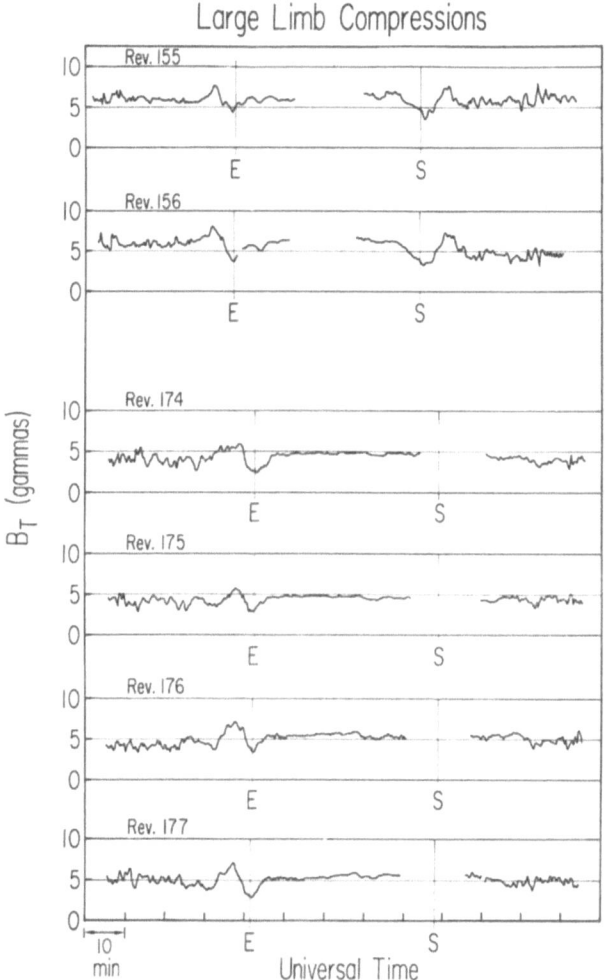

Fig. 8. Limb compressions detected by the Apollo 15 subsatellite. The top two examples are the only cases during the first month when compressions at both limbs were detected. The bottom four orbits illustrate the persistence of limb compressional features. The orbits are 2 h apart.

wind-Moon interaction, governed primarily by solar wind parameters (Sonett and Mihalov, 1972; Whang and Ness, 1972).

In a preliminary study (Coleman *et al.*, 1972b) we assumed that the source of an observed increase in B_T exterior to a rarefaction dip was near the lunar limb. The lunar limbs were defined to be the intersections of the lunar terminator with the subsatellite orbit plane. B_T was scanned for events with an increase of at least 2 γ on the solar side of the expansion dip. We refer to this increase by the general term 'limb compression'. Figure 7 shows the results of mapping the assumed source locations on the lunar surface. Data were taken from the same orbits used to compute the average shown in the previous figure. The upper panel shows the projected source regions for

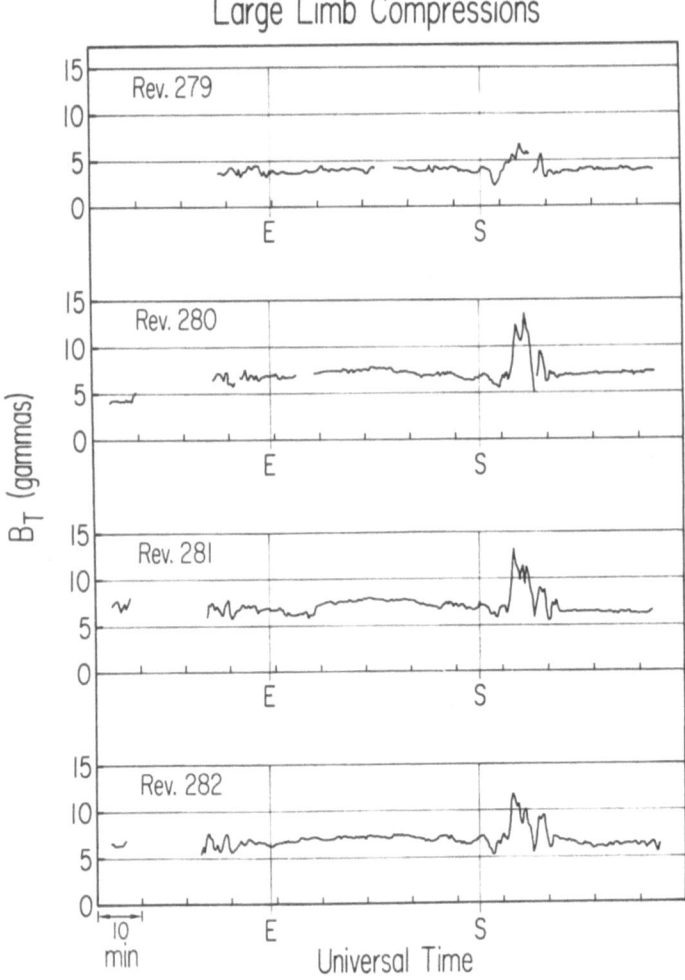

Fig. 9.   Limb compressions detected by the Apollo 15 subsatellite. These orbits illustrate the persistent
occurrence of limb compressions at one terminator and the persistent absence at
the other over a period of 8 h.

each limb compression observed. The lower panel shows the normalized occurrence
rate in boxes of 15° longitudinal width. The 20° latitudinal extent is arbitrary. Each
box represents at least four observed limb crossings. Region B on the figure was the
strongest source region observed. A moderately strong source region was observed
in region A. Examples of limb compressions are shown in Figure 8. Limb compressions in revolutions 174–177 were observed when region A was at one limb. *E* and *S*
refer to eclipse and sunrise as observed by the subsatellite. Note that the eclipse limb
compressions occur within 10 min of the eclipse time. Figure 9 shows large limb
compressions associated with region **B**. Comparison with Figure 8 shows that had
there been limb compressions at the eclipse terminator, they would have been readily

detected in the available data. Clearly, there are limb compressions at only one terminator.

4. Conclusions

The fact that limb compressions are seen only at one of the two limbs on successive orbits indicates that the appearance of limb compressions is not solely dependent on solar wind plasma conditions. It is conceivable that the orientation of the interplanetary magnetic field might control the occurrence of limb compressions, so that they would appear at times at one, or the other, or both limbs as a function of the orientation of the field. We have examined the orientation of the magnetic field in the ecliptic plane for all the limb compressions used in constructing Figure 6, and have found limb compression occurrences at all orientations of the interplanetary field. Thus, the properties of the lunar limb must play the major role in limb compression formation.

We further note that our occurrence map of limb compression sources is in general agreement with the source map of Sonett and Mihalov (1972), constructed using Explorer 35 magnetometer data. The region of strongest magnetization detected at the subsatellite orbit is at southerly latitudes near the 180° meridian on the back side of the Moon. This is precisely the region which, when at the limb, causes the strongest limb compressions observed by Explorer 35 and the Apollo 15 subsatellite. The measurements of magnetic fields of 56 γ at low altitude and 313 γ at the surface indicate that fields comparable to the solar wind dynamic pressure are present. These observations strongly support the hypothesis that lunar limb compressions are due to deflection of the solar wind by regions of strong magnetization near the lunar surface (Barnes *et al.*, 1971).

Acknowledgements

We wish to thank Dr G. L. Siscoe for presenting this paper in our absence. We acknowledge many useful discussions of these data with G. Schubert and the complementary analysis of the lunar magnetic field using these data being undertaken by L. R. Sharp. The research reported here was supported by the National Aeronautics and Space Administration under contract NAS 9-12236.

References

Barnes, A., Cassen, P., Mihalov, J. D., and Eviatar, A.: 1971, *Science* **172**, 716.
Colburn, D. S., Currie, R. G., Mihalov, J. D., and Sonett, C. P.: 1967, *Science* **158**, 1040.
Colburn, D. S., Mihalov, J. D., and Sonett, C. P.: 1971, *J. Geophys. Res.* **76**, 2940.
Coleman, P. J., Schubert, G., Russell, C. T., and Sharp, L. R.: 1972a, in *Apollo 15 Preliminary Science Report*, NASA SP-289, 22-1.
Coleman, P. J., Lichtenstein, B. R., Russell, C. T., Sharp, L. R. and Schubert, G.: 1972b, in D. Criswell (ed.), *Proc. 3rd Lunar Sci. Conf.* **3**, *Geochim. Cosmochim. Acta* Suppl. 3, MIT Press, 2271.
Coleman, P. J., Russell, C. T., Sharp, L. R., and Schubert, G.: 1972c, *Phys. Earth. Planetary Interiors* **6**, 167.
Coleman, P. J., Lichtenstein, B. R., Russell, C. T., Schubert, G., and Sharp, L. R.: 1972d, in *Apollo 16 Preliminary Science Report*, NASA SP-315, 23-1.
Dyal, P., Parkin, C. W., Son, C. P., DuBois, R. L., and Simmons, G.: 1971, in *Apollo 14 Preliminary Science Report*, NASA SP-272, 227.

Dyal, P., Parkin, C. W., and Sonett, C. P.: 1972a, in *Apollo 15 Preliminary Science Report*, NASA SP-289, 9-1.

Dyal, P., Parkin, C. W., Colburn, D. S., and Schubert, G.: 1972b, in *Apollo 16 Preliminary Science Report*, NASA SP-315, 11-1.

Dyal, P., Parkin, C. W., Sonett, C. P., DuBois, R. L., and Simmons, G.: 1972c, *Apollo 16 Preliminary Science Report*, NASA SP-315, 12-1.

Lyon, E. F., Bridge, H. S., and Binsack, J. H.: 1967, *J. Geophys. Res.* **72**, 6113.

Michel, F. C.: 1967, *J. Geophys. Res.* **72**, 5508.

Michel, F. C.: 1968, *J. Geophys. Res.* **73**, 1533.

Ness, N. F.: 1969, *Space Res.* **9**, 678.

Ness, N. F.: 1972, in Dyer (ed.), *Solar Terrestrial Physics/1970*, Part II. D. Reidel Publishing Co., p. 159.

Ness, N. F., Behannon, K. W., Scearce, C. S., and Cantarano, S.C.: 1967, *J. Geophys. Res.* **72**, 5769.

Ness, N. F., Behannon, K. W., Taylor, H. E., and Whang, Y. C.: 1968, *J. Geophys. Res.* **73**, 3421.

Siscoe, G. L., Lyon, E. F., Binsack, J. H., and Bridge, H. S.: 1969, *J. Geophys. Res.* **74**, 59.

Sonett, C. P., Colburn, D. S., Dyal, P., Parkin, C. W., Smith, B. F., Schubert, G., and Schwartz, K.: 1971a, *Nature* **230**, 359.

Sonett, C. P., Dyal, P., Parkin, C. W., Colburn, D. S., Mihalov, J. D., and Smith, B. F.: 1971b, *Science* **172**, 256.

Sonett, C. P. and Mihalov, J. D.: 1972, *J. Geophys. Res.* **77**, 588.

Spreiter, J. R., Marsh, M. C., and Summers, A. L.: 1970, *Cosmic Electrodyn.* **1**, 5.

Strangway, D. W., Larson, E. E., and Pearce, G. W.: 1970, *Science* **167**, 691.

Taylor, H. E., Behannon, K. W., and Ness, N. F.: 1968, *J. Geophys. Res.* **73**, 6723.

Whang, Y. C. and Ness, N. F.: 1972, *J. Geophys. Res.* **77**, 1109.

DISCUSSION

Criswell: The southern region of strong shock production and remnant magnetic fields (165–180° E, 20° S) is also coincident with the roughest portion of the Moon. Photoelectron and remnant field production of shock streamers play complementary roles.

Manka: Can you associate any of the disturbances with momentum transfer from the lunar ionosphere? One would expect this effect to be strongest in a cylinder tangent to the sunlit 'equator' where the 'equator' would be defined by the plane of the interplanetary magnetic field.

Siscoe: That is just the question: what are the effects attributed to? So far the authors have found that magnetic effects give a reasonable explanation, but other contributions might also be possible.

Srnka: Is the scale of the measured surface mangetic fields sufficient to produce a shock in the solar wind?

Siscoe: The scale of the surface fields, at least for the two strong magnetic features that are mapped, is about 100 km. This is comparable with the solar wind ion gyro radii, but is much bigger than electron gyro radii. So some shock-like interaction might be expected.

OBSERVATION OF TM-MODE INDUCTION IN A SIMULATED

SOLAR WIND/MOON INTERACTION

L. J. SRNKA

UKAEA Research Group, Culham Laboratory, Abingdon, Berkshire, and Dept. of Geophysics and Planetary Physics, University of Newcastle-upon-Tyne, England

Abstract. Studies of the interaction of the solar wind with the Moon show that the lunar magnetic field has both steady and transient components. The transient component represents the electrical response of the Moon to electromagnetic action by the solar wind. This response has two modes: a TE mode, driven by interplanetary magnetic field fluctuations and characterized by electrical currents which circulate in the lunar interior; and a TM mode, driven by the solar wind electric field and its variations, and characterized by currents which flow from the Moon and close in the solar wind. The lunar TE mode is observed, but the TM mode has not been detected. The lack of TM mode response is attributed to a cold, resistive lunar crust, and this absence prevents any study of the influence of a surface sheath on the TM mode currents.

A solar wind simulation experiment, VORTEX, has been built to study the plasma aspects of the TM mode response. VORTEX provides a quasi-steady, reproducible, fully ionized argon plasma flow transverse to a magnetic field, by $\mathbf{E} \times \mathbf{B}$ drift around an annular chamber. A preheat discharge provides uniform starting conditions. Typical flow parameters are: density $n_e = 6 \times 10^{20}$ m^{-3}, temperature $T_e = 5$ eV, magnetic field $B_z = 0.1$ T, flow velocity $V_f = 2 \times 10^4$ m s^{-1}, magnetosonic mach number $M_{ms} = 1.4$, and flow time $\tau_f = 0.5$ ms. Cylindrical obstacles of various electrical conductivities are inserted into the flow, along the magnetic field lines.

Framing photographs of the interaction region show well-defined bow waves and downstream wakes. These features change with mach number and the obstacle conductivity. TM mode currents have been detected in a graphite obstacle by a Rogowski coil buried in the obstacle. The induced current varies with plasma density, velocity and magnetic field, but is insensitive to mach number in the steady state. The induced current decreases as the surface condition of the graphite deteriorates. The current is observed to saturate at a level far below the magnetic back pressure limit of the Sonett-Colburn model. The current saturation is consistent with a sheath limit given by the directed ion flux. These observations suggest that a more complex model, which includes surface effects, is needed to describe the solar wind interaction with a Moon-like object if the TM mode response is present.

1. Introduction

The interaction of the solar wind with the Moon has been studied by lunar-orbiting spacecraft and by lunar surface experiments. These studies show that the lunar interaction is characterized by the presence of a diamagnetic plasma wake downstream from the Moon, by the absence of an upstream bow shock wave, and by the impingement of the solar wind directly onto the lunar surface, where the plasma is absorbed and neutralized (see the review by Ness, 1970). These features of the interaction illuminate some of the characteristics and dynamics of the solar wind, and also provide information on the physical and electrical properties of the Moon.

Magnetic field measurements at the lunar surface by the Apollo 12 and Apollo 14 magnetometers show that the lunar magnetic field has both steady and transient components (Dyal and Parkin, 1971). The steady component is attributed to the remanent magnetization of lunar material, and presumably was acquired early in the Moon's development. Measurements of the total lunar magnetic moment (Behannon,

R. J. L. Grard (ed.), Photon and Particle Interactions with Surfaces in Space, 481–497. All Rights Reserved

1968) and data from the portable lunar magnetometer (Dyal *et al.*, 1971) show the steady field to be a local effect. The transient component of the lunar magnetic field represents the electrical response of the Moon to electromagnetic induction by the solar wind (Sonett *et al.*, 1971a), and is a global effect.

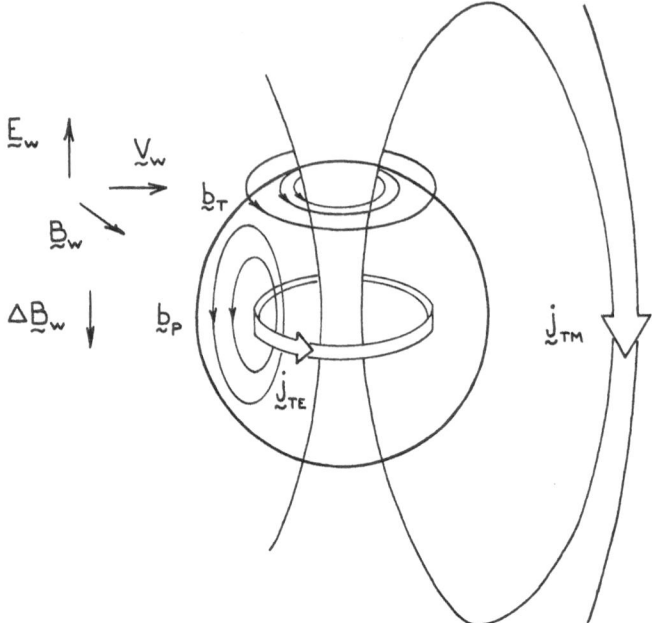

Fig. 1.   Induced currents and fields for a Moon-like solar wind interaction. Fluctuations $\Delta \mathbf{B}_w$ in the interplanetary magnetic field $\mathbf{B}_w$ drive the transverse electric (TE) mode of the response, inducing the currents $\mathbf{j}_{TE}$ and associated poloidal magnetic field $\mathbf{b}_p$. The solar wind electric field $\mathbf{E}_w$ drives the transverse magnetic (TM) mode, inducing the currents $\mathbf{j}_{TM}$ and associated toroidal magnetic field $\mathbf{b}_T$. Asymmetry effects, due to the plasma flow, are not shown.

Theoretical calculations (Blank and Sill, 1969; Schubert and Schwartz, 1969; Schwartz and Schubert, 1969; Sill and Blank, 1970) have shown that the electrical response has two distinct modes, illustrated in Figure 1. Fluctuations $\Delta \mathbf{B}_w$ in the interplanetary magnetic field $\mathbf{B}_w$ drive the transverse electric (TE) mode of the response, which consists of the currents $\mathbf{j}_{TE}$ circulating in the lunar interior, inducing the poloidal magnetic field $\mathbf{b}_p$. This mode is most sensitive to the highest conductivity layer. The solar wind electric field $\mathbf{E}_w$ and its variations drive the transverse magnetic (TM) mode, whose currents $\mathbf{j}_{TM}$ produce the toroidal magnetic field $\mathbf{b}_T$. This mode was first described in its zero-frequency limit by Sonett and Colburn (1967) in terms of a planetary unipolar dynamo model. A special feature of the TM or unipolar mode current is the necessity for the current to cross regions of various electrical conductivity and to close in the solar wind (Sonett and Colburn, 1968), Thus $\mathbf{j}_{TM}$ is most sensitive to the highest impedance part of the equivalent circuit, and can be extinguished by a sufficiently resistive layer. Hollweg (1968) investigated the unipolar dynamo model for a two-layer Moon, and described the combinations of crustal and core electrical conductivities consistent with the absence of a bow shock.

Comparison of the magnetic field fluctuations observed at the lunar surface with fluctuations observed by the Explorer 35 satellite in lunar orbit indicates a strong lunar TE mode response. Lunar electrical conductivity profiles, derived from this data, indicate a cold and resistive crust to be present (Sonett *et al.*, 1971b). This interpretation is consistent with the very weak TM mode response reported. Although a model based on the unipolar mode was proposed by Schwartz *et al.* (1969) to explain the weak limb shock waves sometimes observed, a recent statistical analysis by Colburn *et al.* (1971) together with other supporting evidence suggests that the lunar TM mode response is essentially zero.

The lack of a significant TM response in the lunar interaction prevents any study of the coupling between the induced currents j_{TM}, the lunar surface sheath (Grobman and Blank, 1969), and the electrical structure of the solar wind plasma. This coupling may be important in several ways. Sonett *et al.* (1968) have proposed strong joule heating by TM mode currents, driven by the solar wind from a pre-main sequence T Tauri sun, as a mechanism to chemically differentiate planetesimals and meteorite parent bodies early in the history of the solar system. The problem of the Io-modulated decametric radio emission from Jupiter (Schatten and Ness, 1971) involves the coupling of the Jovian magnetosphere to the satellite Io, and may provide an example of an operating planetary unipolar dynamo (Goldreich and Lynden-Bell, 1969). Sheath effects may have a role in this process (Gurnett, 1972). The TM-mode current may also be important in understanding the interaction of the solar wind with the ionospheres of Venus and Mars. The high electrical conductivity of these ionospheres could effectively couple the collisionless solar wind flow to the planets' atmospheres (Spreiter *et al.*, 1970).

Because of the theoretical and computational difficulties associated with this coupling problem, a solar wind simulation experiment has been built to study the plasma and sheath aspects of the TM mode response.

2. The VORTEX Experiment

Examination of the various approaches to solar wind simulation showed that none of the conventional plasma sources, such as thetatron guns (see Kristoferson, 1969), high-power arcs (Pugh and Patrick, 1967) or shock tubes could provide the reproducibility, long flow time, component of magnetic field transverse to the flow, and supermagnetosonic flow velocity necessary to simulate the TM mode induction process. A new technique was therefore developed, employing a rotating plasma in the homopolar configuration (see Lehnert, 1971 for a review of rotating plasma experiments). This required the production of a supermagnetosonic rotating plasma flow, a unique feature among rotating plasma experiments.

A schematic view of the VORTEX experiment is shown in Figure 2. The dimensions of the discharge chamber were selected on the basis of the Vlasov scaling laws

(Schindler, 1969) for a 1 cm model Moon, and using an MHD theory for the plasma flow. The experiment consists of an annular discharge vessel, or channel, inner radius 0.25 m, outer radius 0.35 m, and height 0.30 m; a magnetic field coil system, providing vertical magnetic fields B_z in the range 0.01 to 0.20 T; two capacitor banks for pre-ionization and acceleration of the plasma; vacuum and gas filling systems, and the usual control and safety circuits.

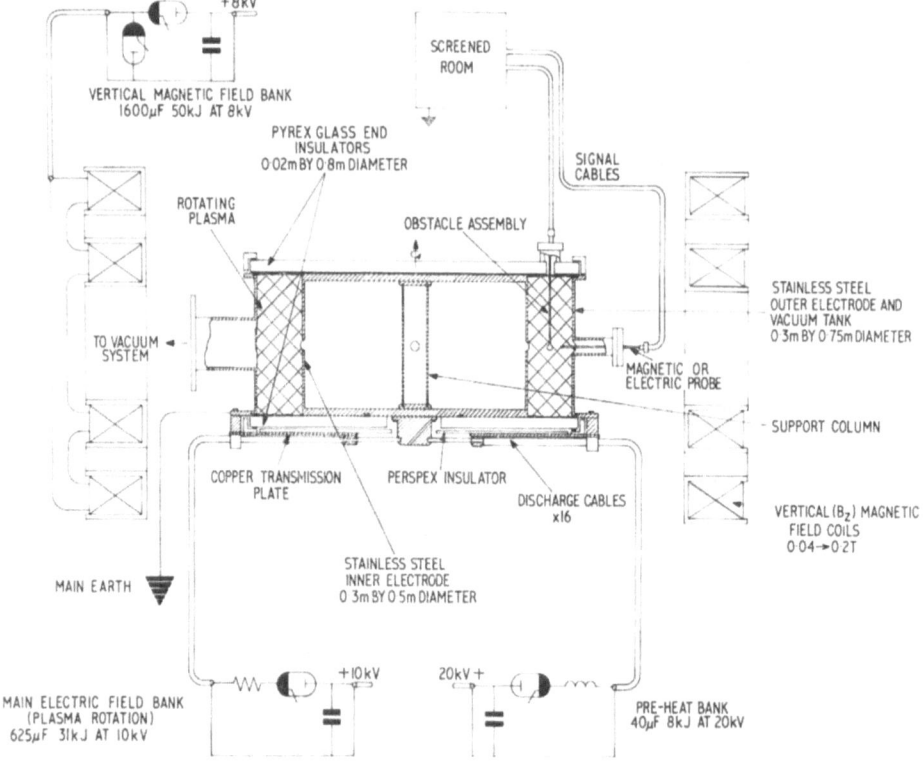

Fig. 2.   Schematic diagram of the VORTEX experiment. Cylindrical obstacles are inserted into the rotating plasma flow with their axes aligned with the magnetic field.

Cylindrical obstacles, shown in Figure 3, are inserted into the discharge channel along the magnetic field lines. Solid obstacles of various electrical conductivities (copper, stainless steel, graphite, glass) were used to obtain framing camera photographs of the interactions. Identical obstacles containing buried Rogowski coils were used to measure the induced obstacle currents I_{OB}. The obstacle is allowed to electrically float at the plasma potential.

The bulk of the experimental work has been devoted to measuring the plasma parameters and optimizing the flow conditions, without an obstacle in the plasma. These results are presented in the following section, as a guide to understanding the obstacle results of Section 4.

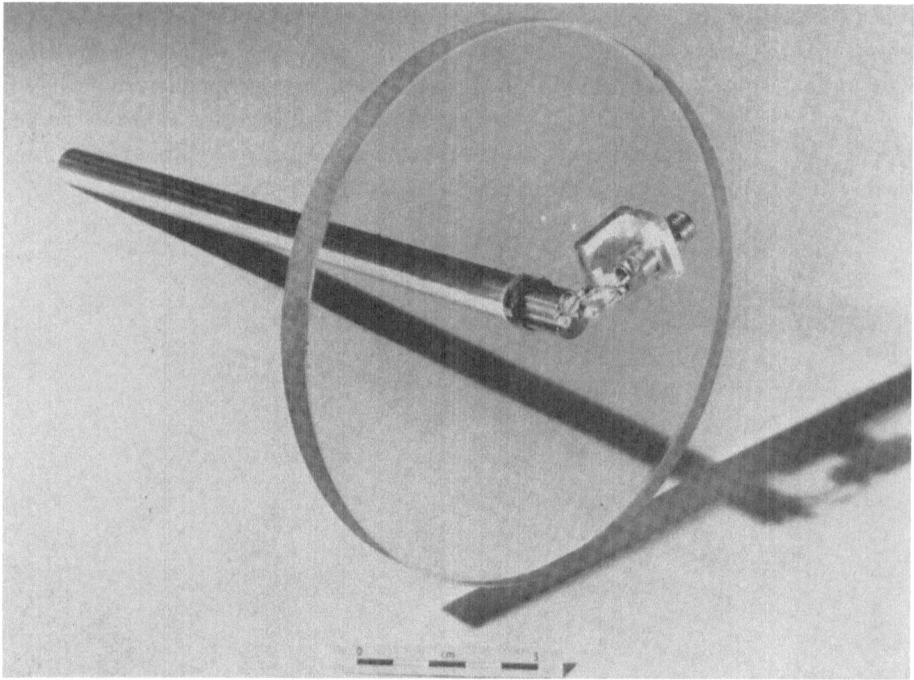

Fig. 3.   Obstacle assembly, showing the Rogowski coil buried within the obstacle. The coil floats at the potential of the obstacle.

3. Characteristics and Dynamics of the VORTEX Plasma

The operating sequence of the experiment is as follows. After evacuation to 10^{-6} torr, the discharge chamber is filled to 20 mtorr with a neutral gas (H, He, Ar) with a constant-flow system. The B_z magnetic field (rise time 25 ms) is then applied, and clamped at its peak value. B_z remains essentially constant throughout the discharge. After an additional 10 ms delay to allow eddy currents in the chamber walls to die away, the preheat discharge (peak volts 15 kV, peak current 200 kA, period 22 μs, duration 110 μs) is applied and ionizes the gas. This establishes a reproducible, uniform afterglow plasma, density $n_e = 3 \times 10^{20}$ m^{-3}, temperature $T_e = 1$ eV, which ensures uniform starting conditions for the main E_R discharge, and helps suppress azimuthal spoking.

An E_R discharge (peak current 30 kA, rise time 100 μs, decay time 200 μs) is then applied to the plasma, producing a radial plasma current j_r. This current produces an azimuthal force $f_\theta = j_r \times B_z$ which accelerates the plasma, driving the rotation. Probe measurements in the plasma confirm a flow velocity $V_f \approx E_r/B_z$, where E_r is the plasma radial electric field, and give no indication of gross instability.

A serious limitation in the use of rotating plasma devices for solar wind simulation is the presence of the Alfvén critical velocity phenomenon (Alfvén, 1954, 1960) in most homopolar configurations. This phenomenon is poorly understood, but it involves the rapid transfer of energy from ions to neutrals when the directed ion energy approaches the ionization energy of the neutrals which collect on the end insulators. This mechanism is present in VORTEX, and prevents supermagnetosonic flow in hydrogen ($M_{ms} \lesssim 0.6$). This limitation can be overcome by using a heavy gas (e.g. argon) and large radial currents, permitting supercritical and supermagnetosonic plasma flow at the sacrifice of the scaling laws for the simulation.

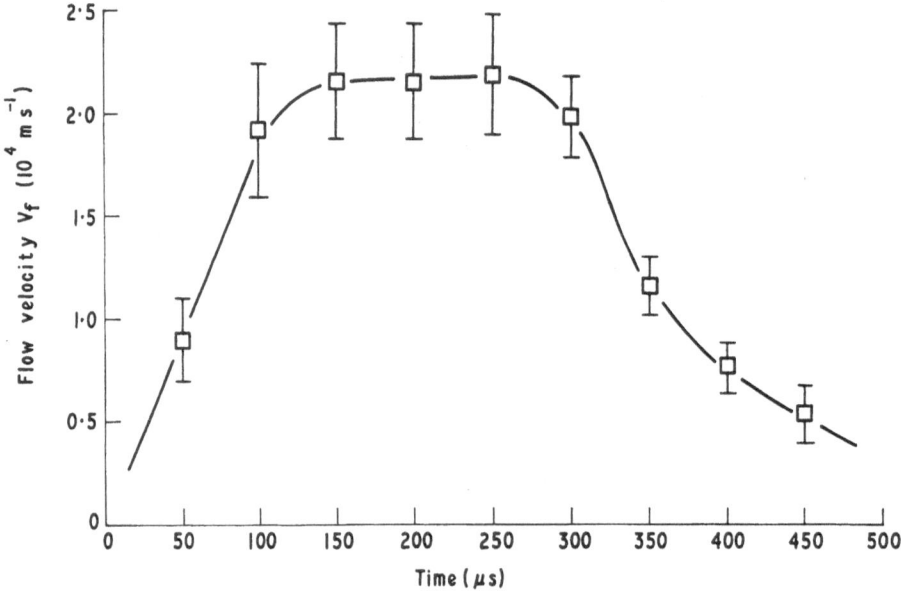

Fig. 4.   Variation of plasma flow velocity with time, measured at the centre of the channel and in the midplane. The plasma density, electron temperature, and magnetosonic mach number vary in a similar way. The discharge current which drives the rotation begins at $t = 0$ and ends at $t \approx 300$ μs.

Typical flow parameters for argon, measured at the centre and in the midplane of the channel, are: density $n_e = 6 \times 10^{20}$ m^{-3}, temperature $T_e = 5$ eV, vertical magnetic field $B_z = 0.1$ T, flow velocity $V_f = 2 \times 10^4$ ms^{-1}, and magnetosonic mach number $M_{ms} = 1.4$. The variation of V_f against time is shown in Figure 4. The early part of the discharge is characterized by a rising V_f, followed by a plateau of 150 μs when V_f is constant. When the j_r plasma current (total current $= I_T$) ceases, the plasma enters a coasting phase, during which time V_f decays slowly with time. The behaviour of the plasma density and temperature is similar to V_f. Note that the plasma continues to rotate after the driving current stops. This is crucial to the detection of the TM mode induced currents in the steady-state obstacle interaction.

During the I_T discharge the plasma undergoes two compressions: a radial compression, due to the centripetal acceleration $m_i V_f^2 / r$, in which the plasma is compressed against the outer electrode; and an axial compression into the midplane, caused by the outward bending of the B_z lines. These dynamic effects produce radial and axial variations in density, velocity, and mach number. The radial variation of magnetosonic mach number in the midplane for times of peak directed plasma flux $\overline{nv}$ is shown in Figure 5. Note that only 5 cm of the 10 cm channel width is super-magnetosonic.

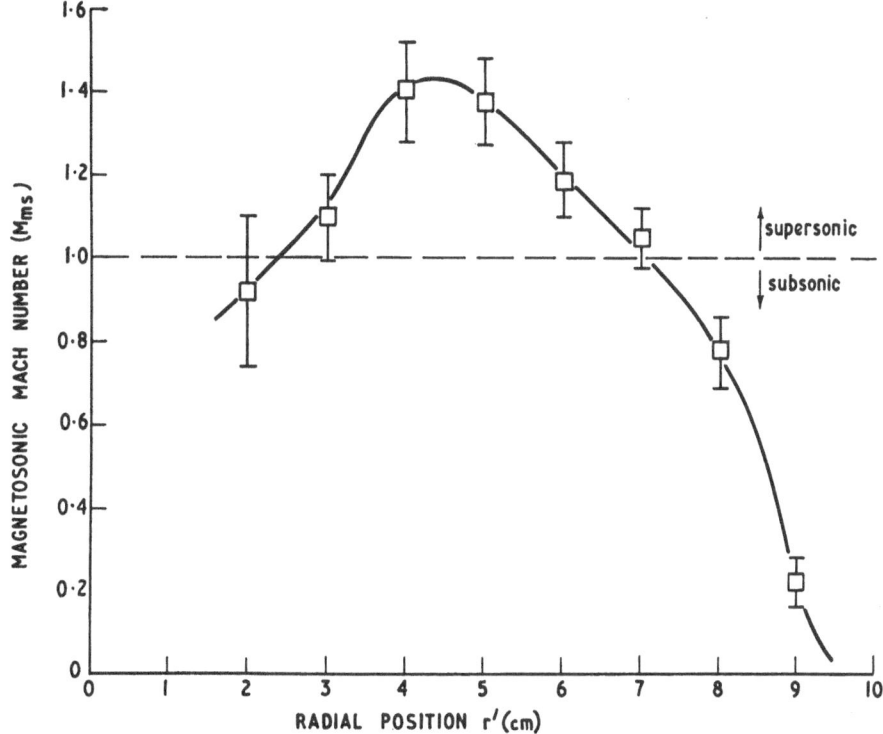

Fig. 5.   Mach number variation with radius at the time of peak directed plasma flux $\overline{nv}$. The regions near the inner electrode ($r' = 0$) and outer electrode ($r' = 10$) remain sub-magnetosonic throughout the discharge.

The plasma conditions were optimized by choosing the value of B_z which yields the largest M_{ms} for the smallest value of ion thermal gyro radius a_i. A table of non-dimensional parameters for the laboratory plasma and solar wind interaction cases is presented in Figure 6. The use of argon plasma means that the ion features of the lunar interaction are not scaled, although the scaling for an asteroidal interaction is reasonable, at least for asteroid radii of 50 km or less. The effects of this scaling error on the interaction are difficult to determine, although the approximations of mono-energetic ions and rectilinear ion motion in the solar wind and laboratory simulation still hold. The worst features of the scaling are the small mean free path, and the large ion gyro radius.

Parameter	Obstacle in Vortex Plasma	Asteroid in Solar Wind	Moon in Solar Wind
L (m)	5×10^{-3}	$\sim 5 \times 10^{4}$	1.74×10^{6}
λ_D/L	1.4×10^{-4}	2×10^{-4}	5.2×10^{-6}
a_e/L	1.8×10^{-2}	6×10^{-2}	1.7×10^{-3}
a_i/L	2.6	1.1	3.5×10^{-2}
a_{di}/L	16.6	13	0.35
λ_{ei}/L	0.94	2×10^{6}	5.2×10^{5}
$c/\omega_{pe}L$	4.4×10^{-2}	3.8×10^{-2}	1.1×10^{-3}
$c/\omega_{pi}L$	6	1.6	4.6×10^{-2}
D/L	10	$\sim 3 \times 10^{8}$	$\sim 10^{7}$
β_{th}	0.15	2	2
β_{dym}	4	130	130
M_s	4	3.5	3.5
M_A	1.45	8	8
M_{ms}	1.38	6	6

Fig. 6.   Scaling considerations for an optimum argon plasma in VORTEX. L is the obstacle radius, λ_D the Debye length, a_e and a_i the electron and ion thermal gyroradii, a_{di} the directed ion gyroradius, λ_{ei} the electron-ion mean free path, c/ω_{pe} and c/ω_{pi} the electron and ion inertial lengths, D the flow dimension, β_{th} and β_{dyn} the plasma thermal and dynamic betas, and M_s, M_A and M_{ms} the sonic, Alfven, and magnetosonic mach numbers, respectively.

4. Observation of TM Mode Currents in the Laboratory Simulation

Unlike the solar wind case, four current systems may be present in a conducting obstacle immersed in a rotating plasma flow, as depicted in Figure 7. The $\mathbf{j}_{TE}$ and $\mathbf{j}_{TM}$ modes in the cylindrical obstacle are equivalent to the current modes shown in Figure 1. The two additional current systems are results of the imperfect nature of the simulation. The $\mathbf{j}_r$ current is produced by the capacitor bank, and provides the plasma rotation. The $\mathbf{j}_\theta$ current is a diamagnetic effect, balancing the centripetal plasma acceleration with a radial gradient in B_z (i.e. the plasma leans radially outwards on the field lines). Thus $\mathbf{j}_r$ ends when the capacitor bank is discharged, but $\mathbf{j}_\theta$ continues as long as the plasma rotates.

Framing camera photographs of the interaction of the argon plasma flow with insulating and conducting obstacles qualitatively confirm the MHD view of the interactions. Figure 8 shows a glass obstacle interaction, taken in total light, exposure time 1 μs. The magnetic field is out of the paper, and the inner electrode is at the bottom in each frame. The obstacle experiences the four plasma regimes during each discharge, with photograph (a) at the beginning and photograph (d) at the end of the discharge.

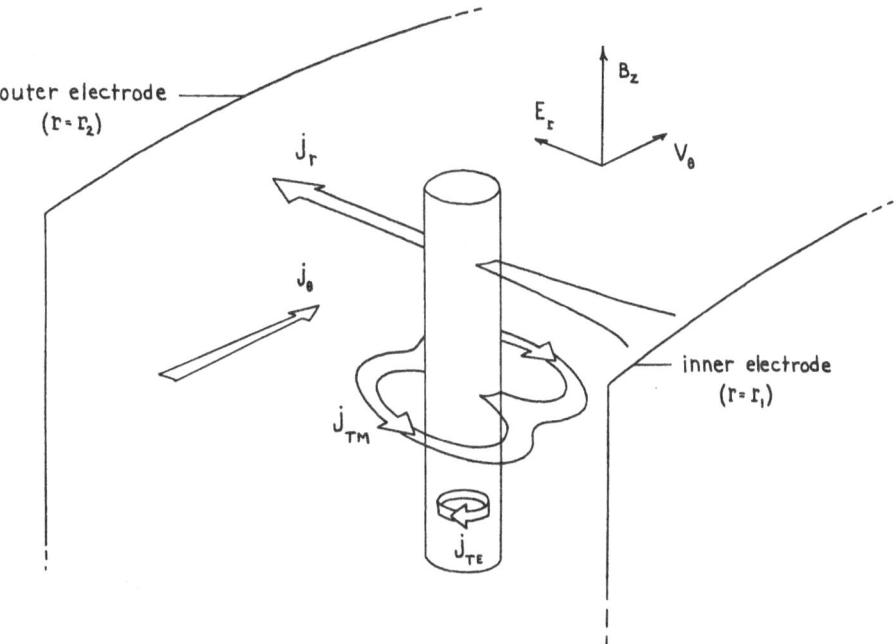

Fig. 7. Conceptual diagram showing the four current systems present in a simulated solar wind/ Moon interaction using a cylindrical obstacle and a rotating plasma. The TE and TM modes shown are analogous to those in Figure 1. The j_r current is supplied by the capacitor bank and provides the plasma rotation. A diamagnetic current j_θ also flows to balance the centripetal acceleration of the plasma.

Note the nearly symmetric wakes and bow waves. Figure 8(c) represents a simulated solar wind/Moon interaction, with a broad plasma wake, straight mach lines, and no appreciable disturbance (i.e. enhanced light) upstream. Figure 9 represents the same condition as Figure 8, but with a graphite obstacle. The magnetic field diffusion time τ_{diff}, where:

$$\tau_{\text{diff}} \approx \mu \dot{\sigma} L^2 \tag{1}$$

is $O(10^{-5})$ s, which is short compared with the steady flow time, so that the interaction is essentially steady-state. The graphite obstacle interaction is characterized by strong radial asymmetry in the bow wave, a curved bow wave front, and considerable light upstream of the obstacle. These features are consistent with the presence of an

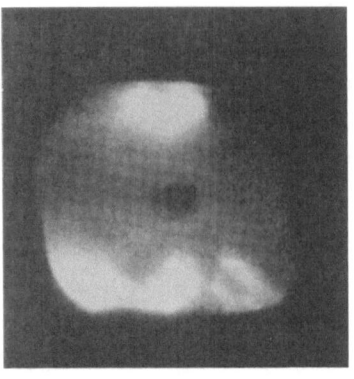

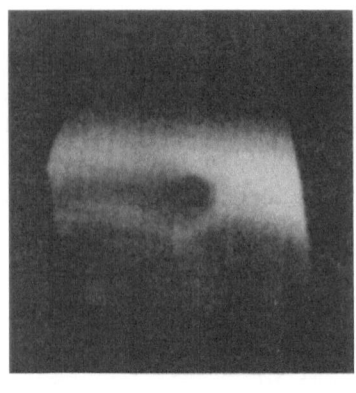

(a) sub-magnetosonic　　　　　　　(b) transonic

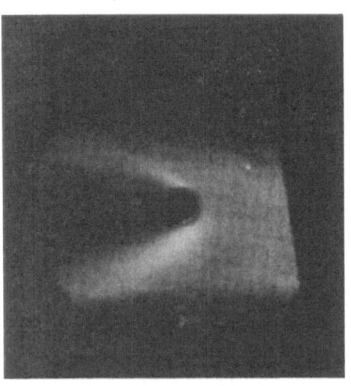

(c) supermagnetosonic
$I_T \neq 0$

(d) supermagnetosonic
$I_T = 0$

Fig. 8.　Framing camera photographs of a glass obstacle interaction, in total light, exposure time 1 μs. The magnetic field is out of the page, and the flow is from the right. The four conditions shown are encountered in the course of each discharge. Note the nearly symmetric bow waves and wakes, and the absence of enhanced light upstream of the obstacle.

added magnetic component in the piston producing the bow wave, as compared to the glass obstacle interaction. This implies additional current systems are present in the graphite obstacle interaction.

　　Instrumented graphite obstacles, inserted into the plasma flow with their Rogowski coils oriented to detect $\mathbf{j}_r$ and $\mathbf{j}_{TM}$ (i.e. with the plane of the coil aligned with the plasma flow), produced the oscillograms shown in Figure 10. The Rogowski coils were

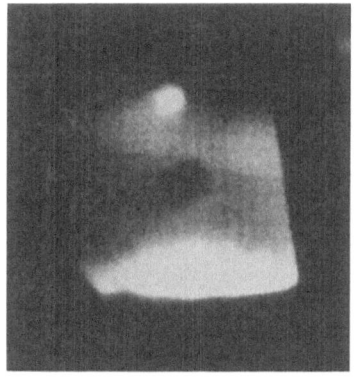

(a) sub-magnetosonic

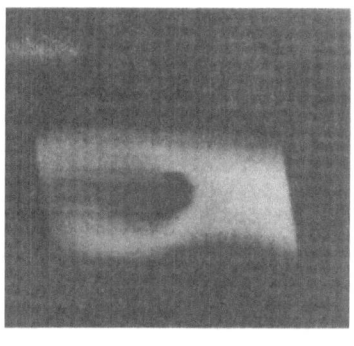

(b) transonic

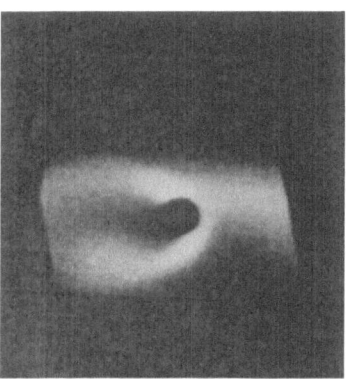

(c) supermagnetosonic
$I_T \neq 0$

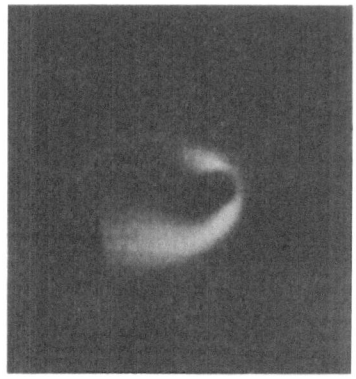

(d) supermagnetosonic
$I_T = 0$

Fig. 9. Framing camera photographs of a graphite obstacle interaction. Plasma conditions and timing are the same as in Figure 8. Note the strong radial asymmetry in the bow wave and the enhanced light upstream of the obstacle in (c) and (d).

calibrated against a known current, and were checked for stray pick-up and frequency distortion with the usual tests. Rotation of the obstacle through 180° reverses the sign of the I_{OB} signal but does not change its magnitude or time dependence. Rotation through 90° produces zero signal ($I_{OB} \leqslant 5$ A). Note that the I_{OB} signal does not follow the I_T history, but continues as long as the plasma rotates. The I_{OB} signal increases with increasing plasma velocity and density, although the exact dependences are not yet determined.

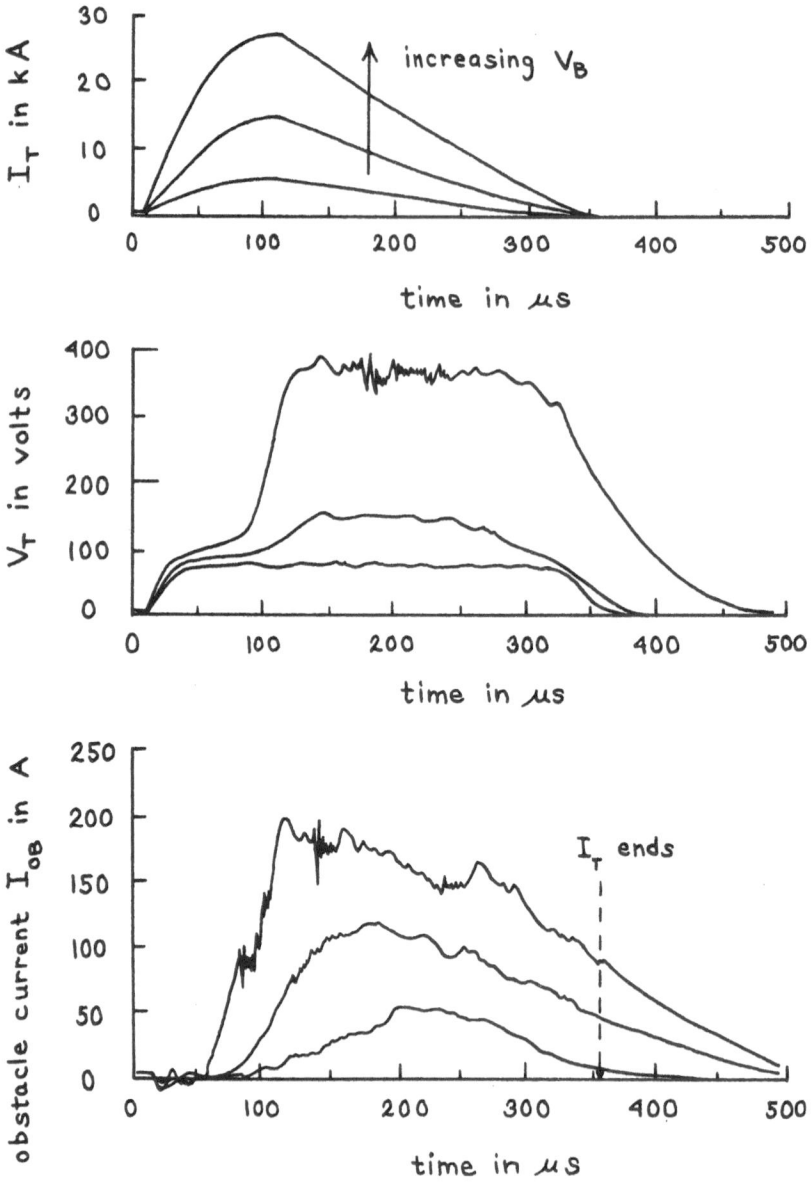

Fig. 10.   Observation of TM-mode induction in a graphite obstacle. Top: total radial current I_T in the plasma; centre: total radial voltage across the plasma, showing increased electric field and flow velocity as I_T increases; bottom: output from Rogowski coil buried in the obstacle, calibrated in amperes. Note that I_{OB} continues after I_T has ended, indicating the presence of the TM mode. The plasma rotation continues to $t \approx 600$ μs in a coasting phase.

The surface condition of the graphite obstacle changes as the number of plasma discharges increases. Surface discolouration, pitting, and other signs of erosion are observed on the upstream-facing surface. Figure 11 shows the decrease in the peak I_{OB} signal, for constant plasma conditions, as the number of discharge increases. This decrease is consistent with a deterioration of the surface electrical properties of the graphite.

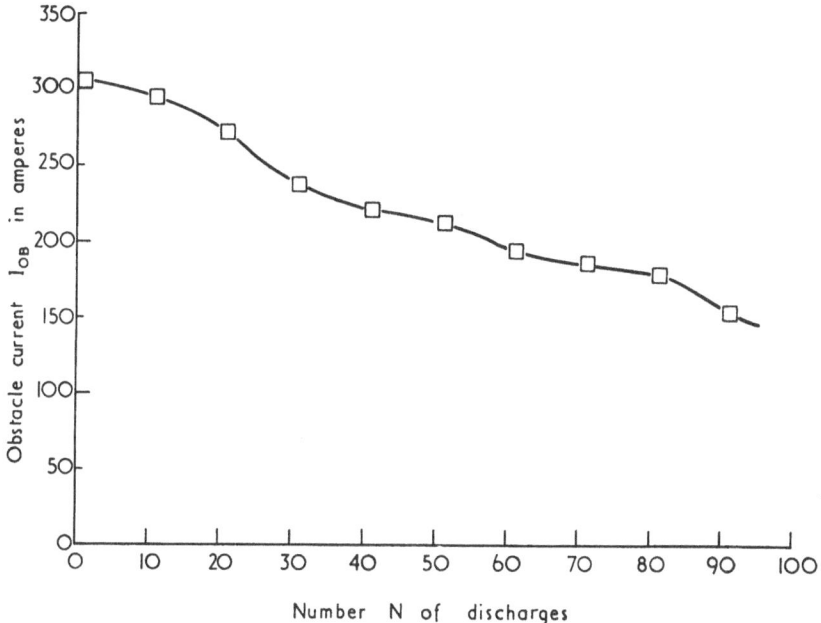

Fig. 11.   Maximum induced current I_{OB} for constant plasma conditions. The surface condition of the graphite obstacle deteriorates as the number of plasma discharges increases. Some sensitivity of the TM mode currents to the surface properties of the obstacle is indicated.

Glass obstacles containing Rogowski coils were also inserted into the flow. In this case, the coils produce a small signal, negative relative to the I_{OB} signal in graphite, directly proportional to I_T, and virtually independent of plasma conditions. This negative signal can also be seen in the graphite interaction. This signal is interpreted as the detection of the local gradient in B_z produced by the interception of $\mathbf{j}_r$ currents by the obstacle wake, some of which must flow in the bow wave.

The above series of observations is interpreted as the detection of TM mode steady induction in an electrically conducting obstacle. The I_{OB} current signal appears to be genuine. A maximum of 440 A flows through the graphite when I_T is peaked. During the I_T discharge the I_{OB} signal likely detects both $\mathbf{j}_r$ and $\mathbf{j}_{TM}$ in the obstacle. However, after I_T has ceased only the $\mathbf{j}_{TM}$ signal should be detected. The Rogowski coils have zero net response to any induced TE mode currents, and $\mathbf{j}_\theta$ should not be collected by the obstacle due to the low-density plasma wake downstream. In addition, $\mathbf{j}_\theta$ would produce I_{OB} signals much smaller than those observed.

5. Discussion and Conclusions

The variation of the obstacle current I_{OB}, measured after the $I_T=0$ point in time, with plasma flow velocity at the obstacle position is shown in Figure 12. The error bars represent plus and minus one standard deviation. Also shown is the magnetic back pressure limit to the TM current system predicted by the Sonett and Colburn (1967)

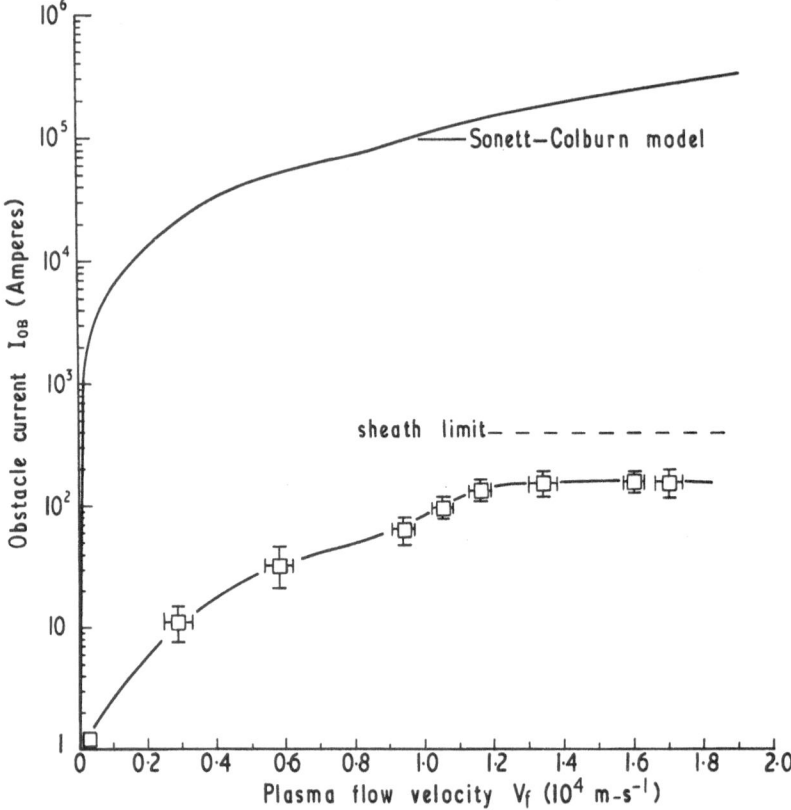

Fig. 12.   Maximum I_{OB} for graphite as a function of plasma velocity. The current predicted by the Sonett-Colburn model is shown, assuming infinite plasma conductivity. The saturated current level of an equivalent floating double Langmuir probe, based on directed ion flux, is also shown.

model, applied to the laboratory case, assuming that the plasma conductivity σ_p is infinite. The total current I_{OB} due to the TM mode is

$$I_{OB} = j_{TM} \times 2hL, \tag{2}$$

where h is the length of the obstacle (0.25m). Using the Sonett and Colburn model, one obtains

$$I_{OB} = 2hL\sigma_{OB}V_f B_z (1 - k), \tag{3}$$

where σ_{OB} is the electrical conductivity of the obstacle and k is the fraction of plasma and field B_z swept to the limbs by the magnetic back pressure $b_T^2/2\,\mu_{OB}$.

At the surface and along the flow stagnation line, provided $h \gg L$,

$$b_T \approx \mu_{OB}\sigma_{OB}V_f LB_z(1-k),\tag{4}$$

where μ_{OB} is the permeability of the obstacle. Equating the magnetic back pressure with the normal component of plasma pressure, the expression for k is then

$$\frac{k}{(1-k)^2} = \frac{\mu_{OB}\sigma_{OB}^2 B_z^2 L^2}{2n_e m_i},\tag{5}$$

For a graphite obstacle in the VORTEX argon plasma, assuming σ_p infinite, $k \approx 1$. This is the strong interaction limit, and the current I_{OB} reduces to

$$I_{OB} \approx V_f \times \left(\frac{8n_e m_i}{\mu_{OB}}\right)^{1/2} \times h.\tag{6}$$

Equation (6) is plotted on Figure 12, using measured values of v_f and n_e. Comparison with the observations shows that the infinite-conductivity k-factor limit, given by (6), is three orders of magnitude larger than the observed I_{OB}.

If the finite plasma conductivity is included in the model, the motional electric field $E_r = V_f \times B_z$ is reduced in the interior of the obstacle from its free stream value to

$$(E_r)_{OB} = \frac{2\lambda}{1+2\lambda} \times V_f \times B_z,\tag{7}$$

where $\lambda \equiv \sigma_p/\sigma_{OB}$ and $(E_r)_{OB}$ is the electric field value in the interior (Sonett and Colburn, 1968). For graphite, and $T_e = 5$ eV, $\lambda = 0.1$ and $(E_r)_{OB} \approx V_f \times B_z/6$. Then $k \approx 0.56$, and the maximum current level remains at $O(10^5)$ A. So the inclusion of finite σ_p still predicts very large I_{OB}.

A final consideration in the Sonnet-Colburn model is the plasma flow power limit. This is an energy limit, set by equating the kinetic power of the flow intercepted by the obstacle's cross-sectional area with the total ohmic dissipation due to the induced currents in the obstacle and the plasma. This is a difficult calculation for the VORTEX case, since the conductivity of the plasma will be much affected by the plasma currents, the wake structure, and so on. Until this computation has been done, little can be said about the plasma flow power limit, except that the power dissipation in the obstacle alone, given these measured current values, is negligible.

A more likely explanation for the observed low values of I_{OB} is to postulate the existence of a stable sheath at the obstacle's surface. Since the obstacle is floating at plasma potential, and is acted upon by the motional electric field E_r, the situation is analogous to a floating double Langmuir probe in a flowing plasma. Provided the difference in voltage across the obstacle is sufficiently large ($\sim 3kT_e/e$), the equivalent double probe would draw saturation current.

The saturation ion current I_{si} for a single probe in a flowing plasma is given in simplified form by Fahleson (1967)

$$I_{si} = A_s n_e e \left(v_f^2 + \frac{8kT_i}{\pi m_i} \right)^{1/2}, \tag{8}$$

where T_i is the ion temperature and A_s is the probe's surface area. In VORTEX, the ion thermal speed is negligible compared with the flow velocity. If we assume a saturated sheath current, so that the analogous double probe is drawing maximum ion current, then I_{OB} becomes

$$I_{OB} \approx I_{si} = A_s e n_e V_f. \tag{9}$$

Using the known radial and axial variations of plasma density and velocity in VOR-TEX, (9) can be used to predict an approximate saturated sheath value for the obstacle current. Substituting the measured values and summing over the area of the obstacle which is exposed to the flow, one obtains an upper limit

$$(I_{OB})_{\text{saturation}} \lesssim 420 \text{ A}.$$

This sheath limit value is indicated in Figure 12. Note that the measured values of I_{OB} are within a factor of two of the sheath limit. A more detailed calculation of the sheath current limit is clearly needed for a proper comparison of theory and experiment.

Observations of TM mode currents in a laboratory simulation of the solar wind/Moon interaction suggest that a more complex model is needed to describe the interaction if the TM mode response is present. It would be desirable to include both surface properties and sheath characteristics into the model. Within the validity of the laboratory simulation this experiment indicates that an electrostatic surface sheath may control the TM response. An important question to be ans wered is whether this sheath effect can be extended to solar wind interactions, where surface photoemission is the dominant particle process.

References

Alfvén, H.: 1954, *On the Origin of the Solar System*, Oxford University Press, Oxford.
Alfvén, H.: 1960, *Rev. Mod. Phys.* **32**, 710.
Behannon, K. W.: 1968, *J. Geophys. Res.* **73**, 7257.
Blank, J. L. and Sill, W. R.: 1969, *J. Geophys. Res.* **74**, 736.
Colburn, D. S., Mihalov, J. D., and Sonett, C. P.: 1971, *J. Geophys. Res.* **76**, 2940.
Dyal, P. and Parkin, C. W.: 1971, *J. Geophys. Res.* **76**, 5947.
Dyal, P., Parkin, C. W., Dubois, R. L., and Simmons, G.: 1971, *NASA SP-72*, 227.
Fahleson, U.: 1967, *Space Sci. Rev.* **7**, 238.
Goldreich, P. and Lynden-Bell, D.: 1969, *Astrophys. J.* **156**, 59.
Grobman, W. D. and Blank, J. L. 1969, *J. Geophys. Res.* **74**, 3943.
Gurnett, D. A.: 1972, *Astrophys. J.* **175**, 525.
Hollweg, J. V.: 1968, *J. Geophys. Res.* **73**, 7269.
Kristoferson, L.: 1969, *J. Geophys. Res.* **74**, 906.
Lehnert, B.: 1971, *Nucl. Fusion* **11**, 485.
Ness, N. F.: 1970, *NASA-GSFC*, Rep. X-692-70-141, Goddard Space Flight Centre, Greenbelt, Md.

Pugh, E. and Patrick, R.: 1967, *Phys. Fluids* **10**, 2579.
Schatten, K. H. and Ness, N. F.: 1971, *Astrophys. J.* **165**, 621.
Schindler, K.: 1969, *Rev. Geophys.* **7**, 51.
Schubert, G. and Schwartz, K.: 1969, *The Moon* **1**, 106.
Schwartz, K. and Schubert, G.: 1969, *J. Geophys. Res.* **74**, 4777.
Schwartz, K., Sonett, C. P., and Colburn, D. S.: 1969, *The Moon* **1**, 7.
Sill, W. R. and Blank, J. L.: 1970, *J. Geophys. Res.* **75**, 201.
Sonett, C. P. and Colburn, D. S.: 1967, *Nature* **216**, 340.
Sonett, C. P. and Colburn, D. S.: 1968, *Phys. Earth Planet. Interiors* **1**, 326.
Sonett, C. P., Colburn, D. S., and Schwartz, K.: 1968, *Nature* **219**, 926.
Sonett, C. P., Dyal, P., Parkin, C. W., Colburn, D. S., Mihalov, J. D., and Smith, B. F.: 1971a, *Science* **172**, 256.
Sonett, C. P., Colburn, D. S., Dyal, P., Parkin, C. W., Smith, B. F., Schubert, G., and Schwartz, K.: 1971b, *Nature* **230**, 359.
Spreiter, J. R., Summers, A. J., and Rizzi, A. W.: 1970, *Planetary Space Sci.* **18**, 1281.

DISCUSSION

Shawhan: Can you do diagnostics to determine the properties of the sheath around the object?

Srnka: The problem is that the object is so small that it is not really possible to use present probes. One needs a larger chamber and object.

Willis: Could you clarify what you mean by the effect of 'surface conditions', i.e. are you referring to surface topography affecting the plasma flow boundary conditions, or to electrical properties of the surface?

Srnka: The properties of the graphite surface I refer to are the colour, texture, and so on, observed when the obstacle is removed from the plasma after a number of plasma discharges. The speculation is that the interaction of hot argon plasma with the graphite surface alters the surface electrical conductivity of the graphite. Because of the dependence of the TM mode currents on surface conductivity, one would expect the current to decrease.

3.4. SOLAR X-RAY INTERACTION WITH THE LUNAR SURFACE

LUNAR COMPOSITION FROM APOLLO ORBITAL
MEASUREMENTS

I. ADLER, JACOB I. TROMBKA, and LO I. YIN

NASA/Goddard Space Flight Center, Greenbelt, Md. 20771, U.S.A.

Abstract. An X-ray spectrometer carried in the Service Module of the Apollo 15 and 16 spacecraft were employed for compositional mapping of the lunar surface. The surface measurements involved observations of the intensity and characteristic energy distribution of the fluorescent X-rays produced by the interaction of the solar X-rays with the lunar surface as well as naturally occurring gamma-rays and cosmic ray produced gamma-rays. A large scale compositional map of approximately twenty percent of the lunar surface was obtained for the first time. It was possible to demonstrate differences between the highlands and the mare and to learn something about the composition of the Moon's hidden side. Results obtained from the X-ray experiment and the gamma-ray experiment are consistent with those obtained from lunar sample analysis.

1. Introduction

With Apollo 15 began a unique series of orbital experiments, aimed at measuring a number of important lunar parameters, remotely from orbit. Among the experiments were a number of spectrometers, which were part of an integrated geochemical package, designed to remotely determine chemical information about a substantial portion of the lunar surface (some of it quite inaccessible to manned landings). The experiments were based on the utilization of interaction phenomena of the lunar surface with such radiation as solar X-rays, solar protons and cosmic rays.

Figure 1 summarizes our view of the lunar surface radiation environment around

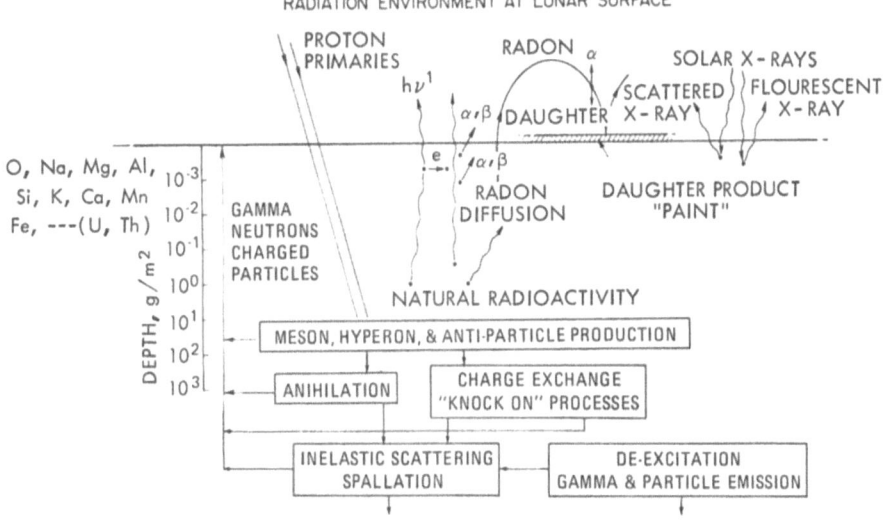

Fig. 1. The lunar surface radiation environment.

R. J. L. Grard (ed.), Photon and Particle Interactions with Surfaces in Space, 501–513. All Rights Reserved
Copyright © 1973 by D. Reidel Publishing Company, Dordrecht-Holland

which the various components of the combined geochemistry experiment were designed. As shown in the figure both natural and induced sources of radiation are
present. Gamma-ray emission from the lunar surface comes from three major sources:
natural radioactivity, activity induced by high energy solar particles and activity induced by primary cosmic rays. The natural radioactivity is due to the isotopes of K,
U and Th. The mechanism for the production of gamma-rays of discrete energy by
high energy particles include primary nuclear reactions, neutron capture, neutron
inelastic scattering and activation. The inelastic scattering induced activity is
primarily a fast neutron reaction.

Solar X-rays absorbed on the lunar surface produce fluorescent X-rays characteristic of some of the elements making up the lunar surface. The relative yields of these
X-rays depends on the intensity and spectral character of the solar X-ray flux as well
as the relative abundance of the elements.

2. Lunar X-Ray Fluorescence Experiment

Numerous calculations had shown that a typical 'quiet' solar X-ray spectrum was
energetic enough to produce measurable amounts of characteristic X-rays from all the
abundant elements of atomic number 14(Si) or smaller. During periods of more intense solar activity it was expected that characteristic X-rays from higher atomic
number elements would also be produced.

An examination of the quiet Sun solar X-ray flux with low resolution instruments
such as proportional counters, reveals a spectral distribution that decreases very
sharply with increasing X-ray energies. If a strictly thermal mechanism of production
is assumed, coronal temperatures are found to range between 10^6 and 10^7 K. Such
variations in temperature produce changes in both flux and spectral composition.
From our knowledge of the processes of fluorescent X-ray production we must therefore expect from the lunar surface, not only a variation in X-ray fluorescent intensities
but in the relative intensities from the various elements observed. For example, if the
solar spectrum hardens (increased fluxes of higher energies), an enhancement of the
intensities from the heavier elements relative to the lighter ones would be observed.

An X-ray monitor was employed during the mission to follow the possible variation
in solar X-ray intensity and spectral shape. In addition, detailed simultaneous measurements of the solar X-ray spectrum were obtained during the mission from the various
Explorer satellites that continuously monitor the solar radiation.

A simplified estimate of the emitted solar X-ray flux based on Solrad (Kreplin and
Horan, personal communication) observations is shown in Figure 2. This follows
from combining a coronal temperature of about 1.5×10^6 K with a hot spot temperature of about 3×10^6 K in proportions determined by the Solrad data and using the
model developed by Tucker and Koren (1971). Superimposed on this curve along
the energy axis are the K shell absorption edge energies for various elements from Na
through K. Calculations of expected characteristic X-ray yields has been made by
Gorenstein (1970) and Eller (personal communication) for the 1–8 Å region based

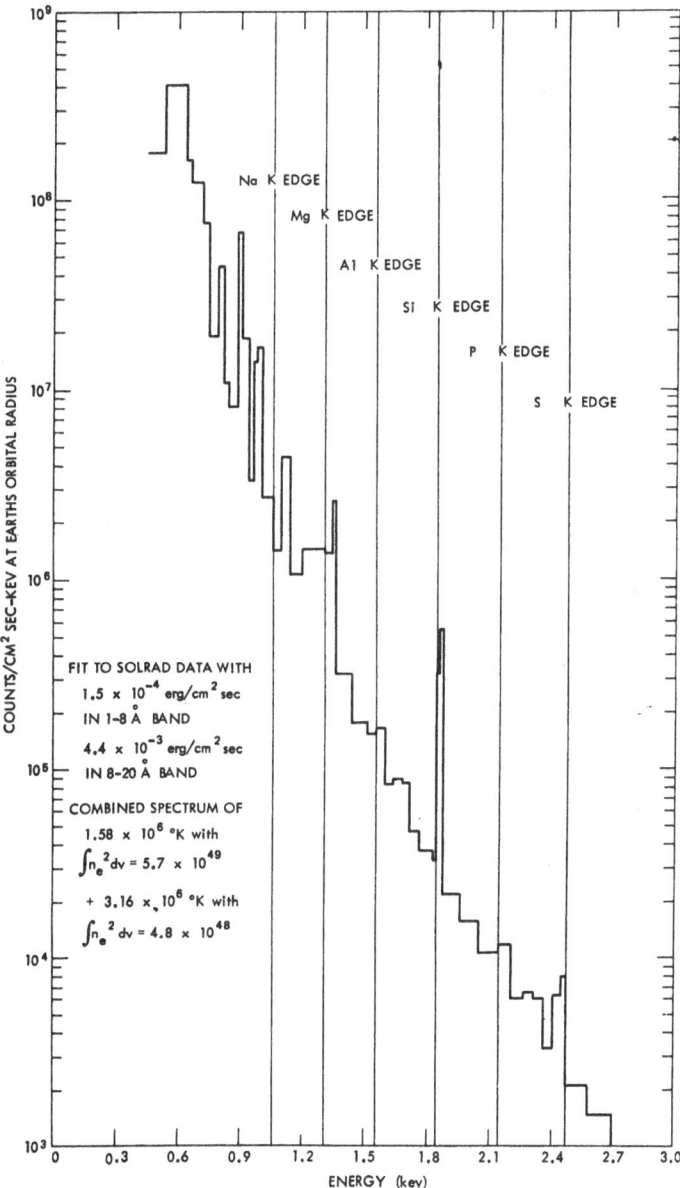

Fig. 2. The solar X-ray spectrum calculated for quiet Sun conditions. Superimposed are the K shell absorption energies for a number of elements from Na to K.

on an assumed coronal temperature of 4×10^6 K and a solar free-free continuum From these calculations it was obvious that the lunar X-ray brightness was relatively low and thus high resolution devices such as crystal spectrometers were not feasible. In order to obtain adequate counting statistics in flight it was necessary to use large area proportional counters with highly transmitting windows.

Details of the X-ray fluorescence Experiment are shown in Figure 3. The X-ray sensing assembly consists of three gas filled proportional counters (P–10), mechanical collimators, calibration sources for in-flight-calibration and the associated electronics. The individual detectors each had beryllium windows approximately 30 cm^2 in area and approximately 4×10^{-2} mm thick.

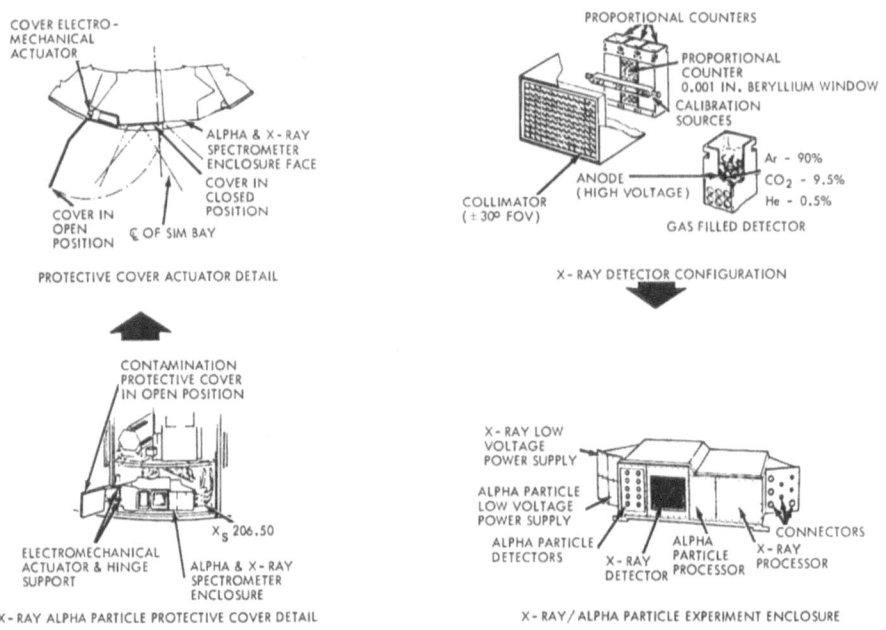

Fig. 3.   Details of the X-ray Fluorescence Experiment.

Two methods of energy resolution were employed. The output of the detectors was energy analyzed by seven discriminator channels covering in equal intervals either 0.5–2.75 keV in a high gain mode or 1–5.5 keV in the low gain mode. As an additional method of energy discrimination, one detector was operated bare; two had selected X-ray filters. A magnesium foil filter covered one detector window and the other had an aluminum filter. The magnesium filter preferentially filters the aluminum and silicon radiation; the aluminum filter is most selective for the silicon radiation. The collimator assembly was such as to provide a 60 deg field of view At the orbital altitude of about 110 km this provided a projected instantaneous surface area of about 110 × 110 km.

3. Gamma-Ray Experiment

From the very beginning of the planning for lunar exploration, it was assumed that K, U and Th would be among the key elements to an understanding of lunar evolution. In terrestrial processes the radioactive decay of these elements is considered to be the source of energy leading to volcanism and magnetic differentiation; and in turn, the concentration of these elements becomes a measure of the extent of differentiation (the K, U and Th tend to concentrate in the late stage crystalline rocks such as the granites). By the Apollo 15 mission, excellent determinations of K, U and Th had been made on the various samples returned during the Apollo 11, 12 and 14 missions. Because K, Th and U are such important indicators of geochemical processes, the attempt to map these elements globally about the Moon is obviously of great significance.

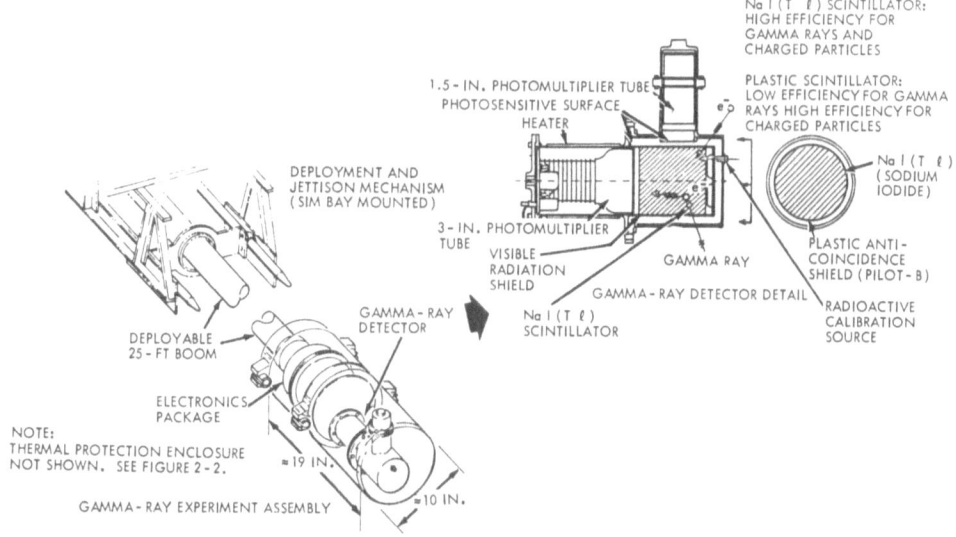

Fig. 4. Exploded view of the Gamma-ray Experiment.

The gamma-ray assembly used for the orbital mapping (Figure 4) consisted of three major subassemblies: a gamma-ray scintillation detector, the electronics and a thermal shield. The detector was a right-cylindrical NaI (thallium activated) crystal, about 7×7 cm. The crystal had a thin mantle of a scintillating plastic crystal which was optically isolated from the primary NaI crystal. Both detectors were used in anti-coincidence. The primary detector has an approximate resolution of 8% for the 0.661 MeV line of ^{137}Cs. The plastic scintillator was used to eliminate the effects of charged particle cosmic ray flux within the field of view of the detector. When the experiment was used in gathering data, it was deployed on a boom to a distance of about 7.5 m from the spacecraft to remove the detector as far as possible from the naturally occurring radiation and the activity induced in the spacecraft by the cosmic-ray flux. Background measurements were determined by making measurements with the experiment stowed as well as at intermediate distances of 2.5 and 4.5 m. The electronics consisted of a 512 channel analog-to-digital converter built around a 2-MHz oscillator.

Figure 5 shows the Science Instrument Module (SIM) and its relationship to the Command-Service Module (CSM) of the Apollo vehicle. Some of the functions concerning the X-ray and gamma-ray experiment involved astronaut participation; as for example the gamma-ray boom deployment and the opening and closing of the X-ray experiment's protective covers during the maneuvers involving the firing of the control jets. Not shown in Figure 5 is the X-ray solar monitor, the small proportional counter with a pin hole aperture, on the opposite side of the spacecraft which was used to monitor the Sun's X-rays simultaneously with the measurements of the lunar surface X-rays.

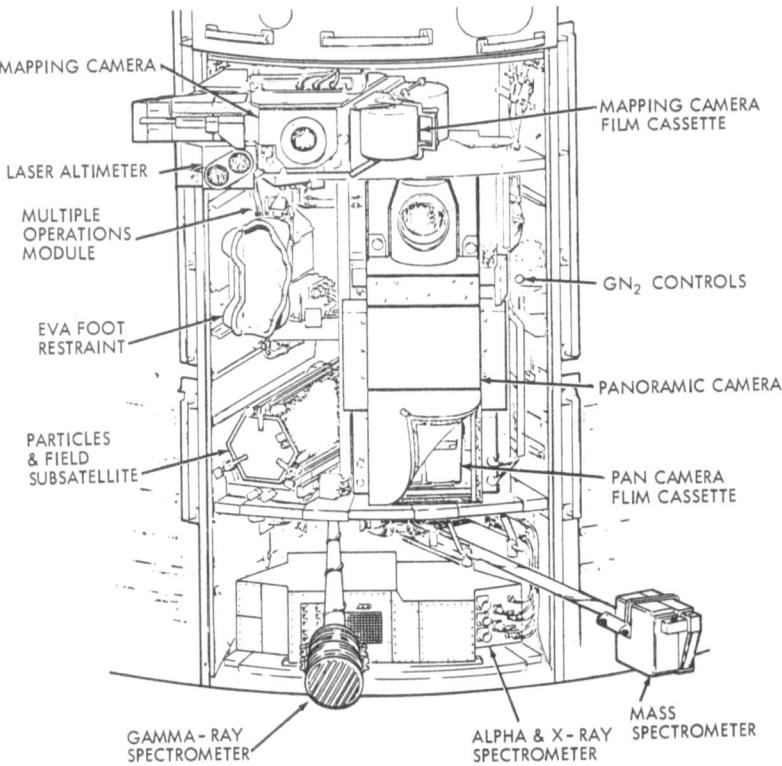

Fig. 5.   The Science Instrument Module (SIM). The X-ray Experiment and Gamma-ray Experiment are shown at the base of the SIM.

4. Observations During Apollo 15 and 16

The Apollo 15 mission was to the Hadley Rille area while the objective of the Apollo 16 flight was the Descartes region. The Apollo 15 flight was at a lunar orbital inclination of 26 deg as compared to the 9 deg inclination of the Apollo 16 flight, thus the projected ground track of the 15 flight covered a larger projected surface area than the 16 flight. There was some overlap of coverage between the 15 and 16 flights which made it possible to compare results of both. The agreement for the overlap areas proved to be sufficiently close (within 10%) so that no normalization was required. The total coverage of the lunar surface for both missions was about 20% of the Moon's surface area.

As stated in the experiment description, the flux and energy distribution of the solar X-rays were expected to have a significant effect on the intensity of the fluorescent X-rays measured. This is clearly demonstrated in Figures 6 and 7. Figure 6 shows the integrated intensities (1–3 keV) as registered during the Apollo 15 flight by the solar monitor for the approximate period corresponding to the surface measure-

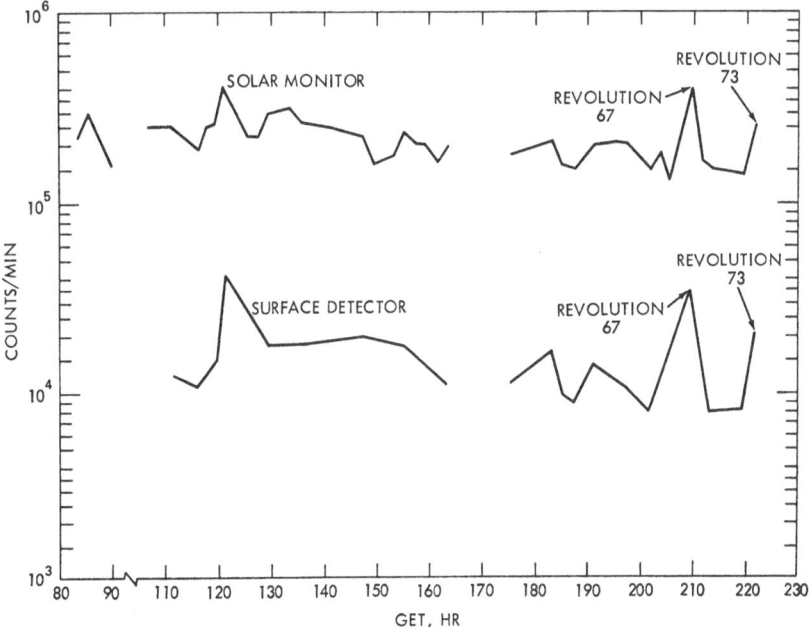

Fig. 6.   Comparison of solar and surface X-ray fluxes.

ments (all values were computed at the subsolar point). Also shown are the corresponding surface measurements. With the exception of such orbits as 67 and 73, the solar flux was fairly stable, varying by less than ±30% of the mean value. In Figure 7 a comparison is drawn between the Al/Si and Mg/Si intensity ratios for orbits 67 and 68 for the same longitudes. It can be seen that for the more active period (orbit 67) the Al/Si and Mg/Si ratios were depressed relative to the more normal orbit 68. This is probably due to the hardened solar X-ray spectrum which more efficiently excites the silicon relative to the aluminum and magnesium. Thus the moon itself provides an excellent indicator of changes in the solar spectrum.

We have observed a regular variation in solar X-ray illumination as well as fluorescence as a function of spacecraft position. The maximum occurs at the subsolar point and then falls nearly symmetrically towards both terminators. On this basis it is clear that the illuminated lunar surface has an ambient soft X-ray flux which peaks, not

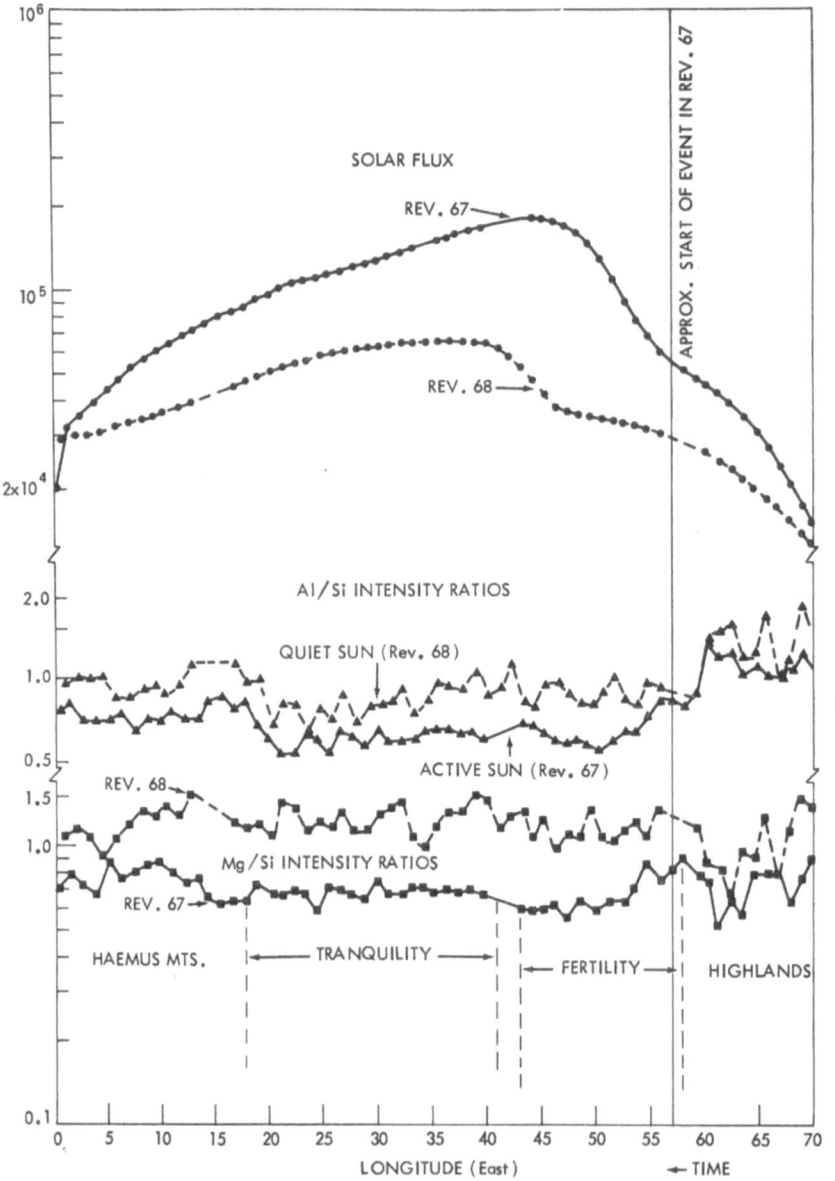

Fig. 7. Al/Si and Mg/Si intensity ratios for orbits 67 and 68. Orbit 67 represents and more active period. The intensity ratios are depressed during orbit 67 because of enhanced excitation of the Si radiation.

surprisingly, at the subsolar point and decreases for geometric reasons as one moves towards the terminators. A calculation of the surface flux based on the observed X-ray intensities during the orbital phase of the Apollo 15 flight has yielded some intriguing numbers. For example, the counting rate observed in the unfiltered X-ray detector is conservatively taken as 400 counts s^{-1} at the subsolar point. On this basis, if corrections are made for geometry and detector efficiency, the flux at the lunar surface for the integrated X-ray energies between 1–3 keV is of the order of 1000 counts cm^{-2} s^{-1}. One interesting implication of this is that one can hope to do X-ray chemical analysis at the surface by using the Sun as the only excitation source and an energy sensitive detector.

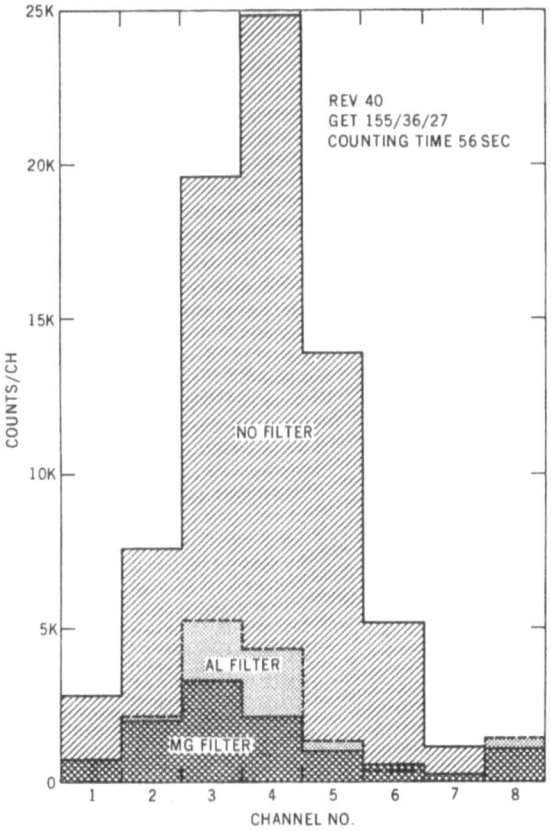

TYPICAL X-RAY SPECTRUM (DET 1-3)-SUBSOLAR POINT

Fig. 8.   Raw X-ray data shown in histogram form.

A brief discussion of the results of the geochemical measurements are presented below. More detailed descriptions have been published elsewhere (Adler *et al.*, 1972a, b, c). A typical example of the appearance of the raw data is shown in histogram form in Figure 8. These data were initially reduced to Al/Si and Mg/Si intensity ratios and ultimately to chemical ratios. Figure 9 shows the variation of Al/Si and Mg/Si concentration ratios along the projected ground tracks for Apollo 15 (upper envelope) and Apollo 16 (lower envelope). The upper values correspond to Al/Si and the lower numbers to Mg/Si concentration ratios. These tracks have been divided into areas

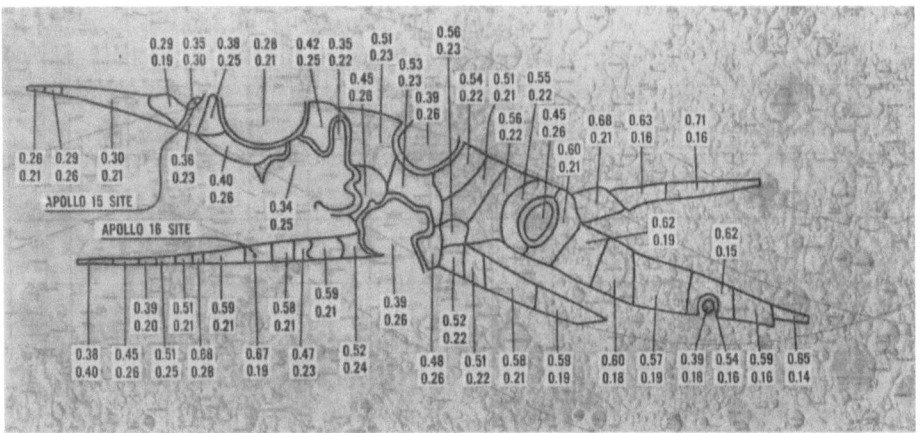

Fig. 9.   Al/Si and Mg/Si chemical ratios along the projected Apollo 15 and 16 ground tracks. The upper envelope is Apollo 15 and the lower envelope Apollo 16. The upper values in each square are the Al/Si figures and the lower values the Mg/Si.

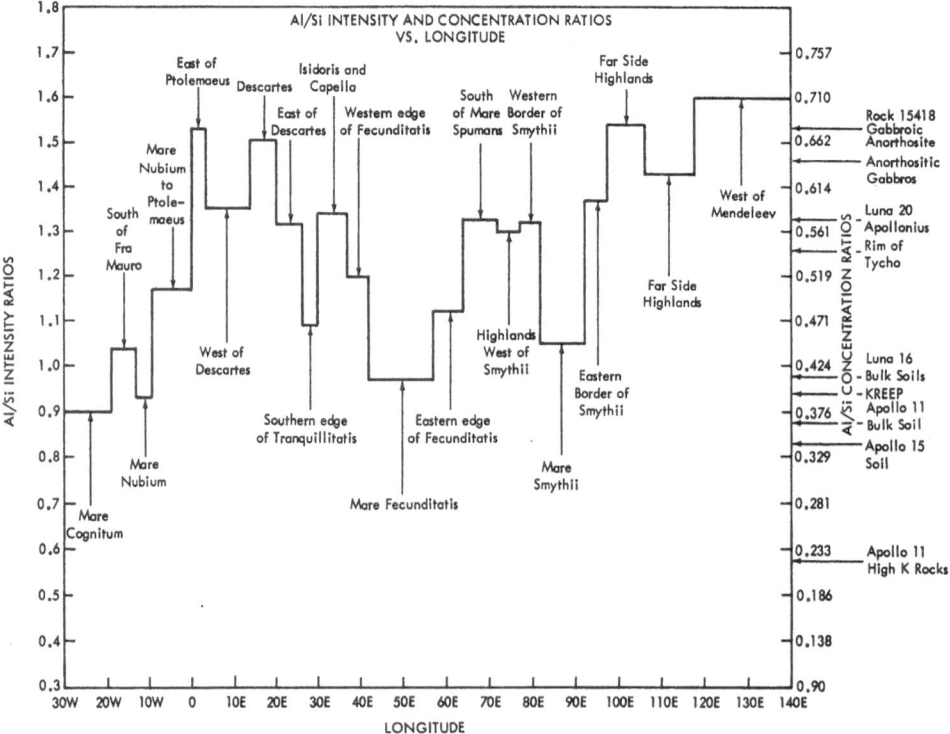

Fig. 10.   Longitudinal plot of Al/Si values along the Apollo 15 ground track. The values on the right hand axis come from the analysis of various lunar materials.

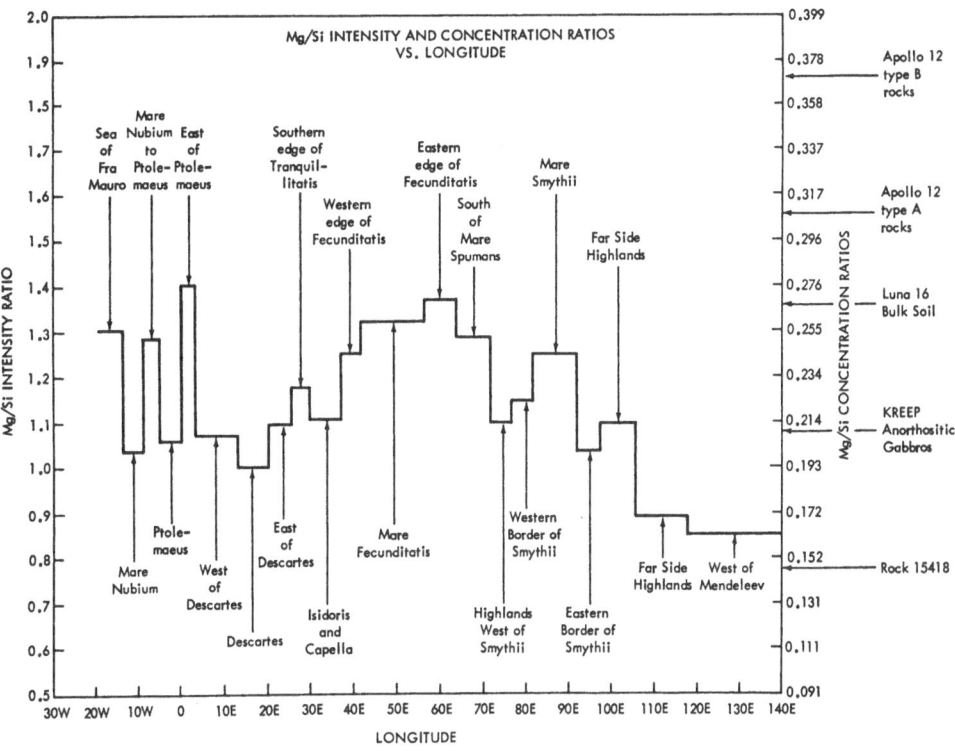

Fig. 11.   Longitudinal plot of Mg/Si ratios along the Apollo 15 ground track.

based in part on obvious geologic features and in part on concentration contours. These relationships for Al/Si are shown in Figures 10 and 11 in greater detail for the various features overflown. The values for various analyzed materials are shown along the right hand axis for reference. A number of observations can be drawn from a study of these figures:

(1) The Al/Si ratios are highest in the highlands and considerably lower in the mare areas. The extreme variation is about a factor of two. The Mg/Si concentration ratios show the opposite relationship.

(2) There is a general tendency for the Al/Si values to increase from the western mare areas to the eastern limb highlands.

(3) There are distinct chemical contrasts between such features as the small mare basins and the highland rims. The rim areas as one would predict are intermediate between the mare areas and the surrounding highlands.

Another interesting study involves a comparison of Al/Si intensity ratios versus the optical albedo values. These observations are particularly significant in view of the long standing discussion as to whether these albedo differences are solely representative of topographic differences or also a reflection of compositional differences among surface materials. The data from both Apollo flights showed an excellent correspon-

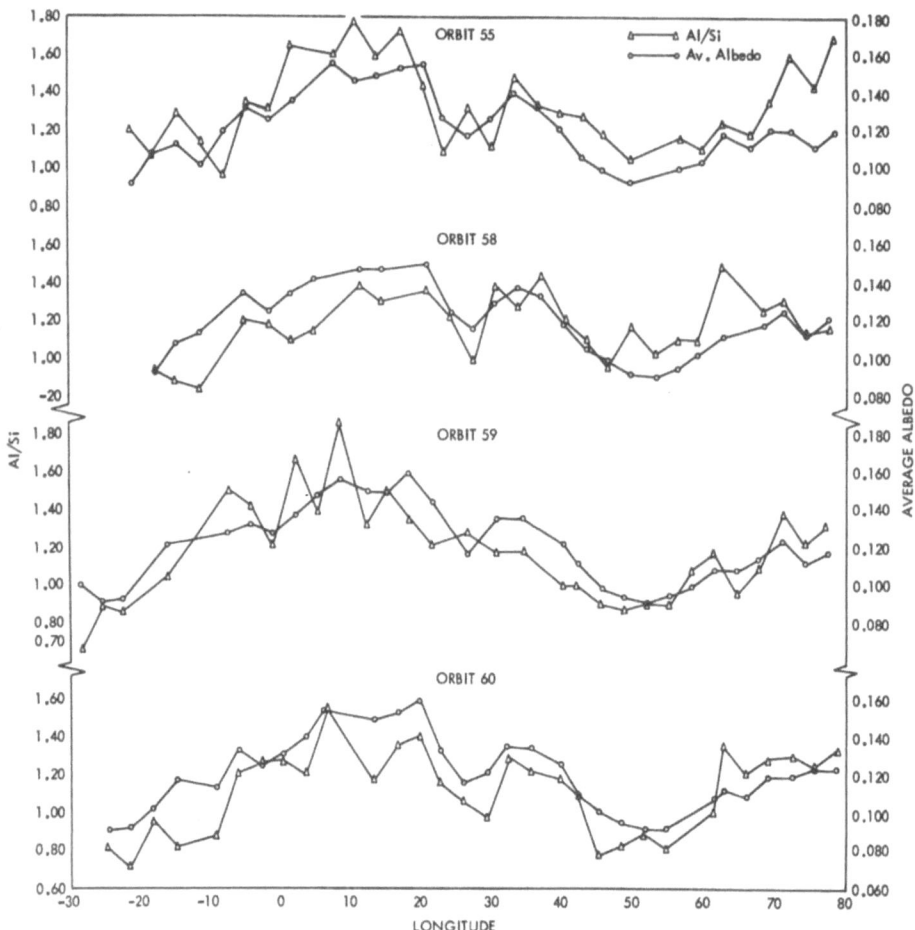

Fig. 12.   Al/Si intensity ratios vs optical albedo values along a series of Apollo 16 ground tracks.

dence between Al/Si values and the optical albedos. An excellent example from the Apollo 16 flight is shown in Figure 12.

Finally one can now also summarize the gamma ray observations obtained during the Apollo 15 and 16 missions. 85% of the gamma-ray emission from the lunar surface was continuum. 90% of the discrete line emission was due to the natural radioactive emissions in the energy range up to about 3 MeV. The induced activity was dominated by what appears to be gamma-rays produced by inelastic neutron scatter. All other effects inducing discrete line emission seem to be small or even negligible. Thus fast neutron effects appear to dominate, a fact consistent with the evidence of extremely low hydrogen content in the lunar surface. The solar particle flux seems to have little effect; the light and dark side activity was not significantly different. In fact, if the solar particle flux did make a significant contribution, one would look for such an effect in the energy region above 3 MeV; however this was not observed.

5. Conclusions

The lunar radiation environment has been exploited in order to study the chemical composition of the lunar surface remotely. The solar produced X-ray fluorescence, the natural gamma-rays and the cosmic-ray induced gamma-rays have been measured from orbit. These measurements have also provided us with new information about the ambient radiation at the Moon's surface.

References

Adler, I., *et al.*: 1972a, *Science* **175**, 436.
Adler, I., *et al.*: 1972b, *Science* **177**, 256.
Adler, I., *et al.*: 1972c NASA SP-289, 17-1.
Eller, E.: personal communication.
Gorenstein, P., Gursky, H., Adler, I., and Trombka, J. I.: 1970, *Adv. X-Ray Anal.* **13**, 330.
Kreplin, R. W. and Horan, D. M.: personal communication.
Tucker, W. H. and Koren, M.: 1971, *Astrophys. J.* **168**, 283.

3.5. EROSION PROCESSES ON THE LUNAR SURFACE

SPUTTERING AND DARKENING OF THE GRAINS ON THE
LUNAR SURFACE

T. GOLD

Center for Radiophysics and Space Research, Cornell University, Ithaca, N. Y. 14850, U.S.A.

Abstract. Sputtering experiments have been carried out in the lunar laboratory at Cornell (first by B. Hapke) since 1964. These have shown that solar wind exposure will lead to the deposition of a dark layer on grains of most rocks. The nature of this layer is not yet known with certainty, but it is thought to be chiefly due to reduced metals. This confirms the supposition, first put forward in 1955, that the albedo of any part of the lunar surface is dependent on the length of time for which it has been exposed. This albedo effect is likely to dominate over effects due to regional chemical differences.

The surface of the Moon is very dark despite the fact that most fine rock powders of the consistency of the lunar soil are very light. Indeed we now know that ground-up lunar rocks would also be much lighter than the lunar soil. A special explanation is needed for this.

Already in 1964 my colleague Hapke and I discovered that the entire range of observed optical properties of the Moon could be matched by rock powders subjected to bombardment similar to the solar wind (Hapke, 1966). At that time few lunar scientists considered a fine powder as being a possible material for the lunar surface and in fact no untreated rock powder showed the required optical properties.

The optical observations of the moon had given brightness and polarization as a function of lunar phase angle, and of course there was the observation of the generally low albedo, around 7%, with the characteristic difference that rough ground was generally a few percent lighter than flat ground.

We started from the supposition that the Moon's surface consisted of powder, since we could see many signs of erosion and transportation and we could not understand this for more substantial objects than very small particles. We then observed that while we could not match the lunar reflection and polarization law with any but very special powdered rocks, the effects of proton and α-particle bombardment in the laboratory was such as to make many rock powder surfaces extremely similar to the Moon's surface. The dosages required were, when referred to a solar wind of 300 km s^{-1} and 5 particles per cm^3, approximately 10000 yr of bombardment, during which time a layer of about 1mm would have been treated. Although the grains are for the most part much smaller than 1 mm, the rough microtopography allows access to a layer of about that depth.

We ascertained that the darkening observed under laboratory bombardment with kilovolt protons and α-particles was not due to some contamination in the vacuum system, by subjecting in each experiment also certain powders to the same bombardment that had the property of not darkening at all. It even became clear that the darkening process resulted from the attachment of material from one grain on another since we could find powders which by themselves did not darken, but which darkened in a mixture of two types. Thus, no hypothesis of contamination of the bombarding beam could account for this result.

We do not know yet what the precise composition and structure is of the darkened layers caused by the bombardment, but certain points have been experimentally verified.

(1) The deposit is caused by a sputtering process, and the amount of darkening caused depends on the nature of both the originating and the receiving surface.

(2) The darkening occurs almost entirely on surfaces that are not exposed to the direct beam of incoming particles, but that are exposed to secondary particles ejected by the beam. Thus, smooth surfaces do not darken significantly, and in the case of the complex structure of surfaces made up of grains, it is the undersides and protected areas that darken. Small rock grains are usually initially translucent and therefore even the surface appearance is changed.

(3) Minerals containing much iron seem to darken more quickly and reach a darker saturation than those that are poor in iron.

We are therefore inclined to think that the sputtering process tends to deposit some reduced metal which, if it is not oxidized on the receiving surface, or otherwise chemically combined, will in general look dark. Probably iron is the most effective among the metals in minerals.

The top surfaces that are exposed to the incoming beam have more material removed from them than added, and they do not darken therefore. Areas accessible only to secondaries suffer addition of sputtered deposits, and those darken.

The metallic appearance of particles in the lunar soils that are large enough to be studied under optical microscopy is probably due to sputtered deposits. These particles are not electrical conductors, and the metallic appearance is not removed with common acids except by the addition of hydrofluoric acid, suggesting that the coating is a mixture of reduced metal and silicates.

This coating may be the one observed by Borg et al. (1971) under high voltage electron microscopy as an amorphous layer on most grains. It is also possible, however that these observations are more sensitive to the effects of implantation of solar wind ions and the crystallographic destruction resulting, than to the deposit of small amounts of reduced metal. Since both effects are dependent on surface exposure they may well appear highly correlated without, however, being identical.

The result of the investigations in our laboratory in the years from 1961 to 1965 was to persuade us that the surface of the Moon was indeed composed of a fine powder of rocks that had darkened by an amount depending on the time of exposure to the solar wind. This then agreed with the earlier suggestions (Gold, 1955), firstly that the material was a powder so that surface transportation could have taken place; and secondly that darkening is an effect of surface exposure so that denuding ground is generally less darkened than depositing ground. Thus highlands would in general be lighter than flat areas of lowlands. We believe now that this is the correct explanation of the major albedo differences on the Moon, and that chemical differences play only a minor role, since it is clear that exposure ages of soil vary, and that the sputtering darkening effects are very strong.

This explanation then makes clear also why the higher albedo rays are seen only around the youngest craters, since they will be darkened in the course of time like the surrounding ground. Since micrometeorites undoubtedly mix up the top few millimeters fairly frequently, a layer has to be darkened that is much thicker than would be needed in the static case, before the ray has been effectively obliterated.

If the Moon possesses deep deposits of material similar to the top soil seen, (as I suggest it does (Gold, 1955, 1971)). then the present rate of solar wind bombardment would not be sufficient to have darkened it all. One millimeter in 10000 yr would mean that not more than an average depth over the whole Moon of 400 m could have been darkened unless the solar wind was much more intense in the past. In actual fact the figure would have to be much less since one cannot suppose optimal exposure everywhere and all the time. Much area will be exposed to saturation and beyond, thus making the available solar wind less effective for the total darkening effect caused. It will therefore be interesting in future investigations to observe whether perhaps the deeper craters have only a thin layer of dark material underlaid by similar but lighter material. Perhaps the rays are merely a manifestation of such a structure. Surface transportation will of course have caused great regional differences in the thickness of the dark layer.

Acknowledgement

This work was supported in part under NASA Grant NGL-33-010-005 and NASA Grant NGR-33-010-137.

References

Borg, J., Maurette, M., Durrieu, L., and Jouret, C.: 1971, *Proc. 2nd Lunar Sci. Conf.* **3**, 2027, MIT Press.
Gold, T.: 1955, *Monthly Notices Roy. Astron Soc.* **115**, 585, 1955.
Gold, T.: 1971, *The Nature of the Surface of the Moon, Space Research XI*, Akademie-Verlag, Berlin, 1971.
Gold, T.: 1971, *Proc. 2nd Lunar Sci. Conf.* **3**, 2675, MIT Press.
Gold, T.: 1971, *Proc. Am. Philos. Soc.* **115**, 2.
Hapke, B.: 1966, in W. Hess, D. Menzel, and J. O'Keefe (eds.), *The Nature of the Lunar Surface*, Johns Hopkins Univ. Press, Baltimore.

DISCUSSION

Willis: Adler *et al.* (1972, to be published) have also conducted Ar⁺ sputtering experiments on a variety of metal oxide films and shown exposure darkening effects due to reduction and sputtering of the metal. This would substantiate your own experiments on rock powders.

Criswell: Do the lunar grains greater than 25 μ in radius, which have radiation track damage on only one end, display dark coatings on the opposite end? This is a severe test of the darkening model proposed.

Gold: This has not been looked for.

THE LUNAR ELECTRONOSPHERE AND IMPLICATIONS FOR EROSION ON THE MOON

EVAN HARRIS WALKER

U.S. Army Ballistic Research Laboratories, Aberdeen Proving Ground, Md. 21005, U.S.A.

Abstract. Bodies in space that are subject to the solar ultraviolet emit photoelectrons. A portion of these escape, forming a current from the body balanced at equilibrium potential by accretion from the surrounding plasma. For a positively charged body a much larger flux of less energetic photo-electrons will be released from the surface, but fail to escape from the body. Their effect is to produce an inner screening of the body's electric charge. The characteristics of this photoelectron sheath and the equilibrium surface potential are treated for spherical and flat bodies subject to the solar wind electron and ion flux.

The enhanced electric field resulting from the electronosphere gives rise on the Moon to electro-static dust transport. Evidence for the presence of electrostatic erosion as a contributory process in shaping lunar features on the 1- to 10-m scale is found.

This transport mechanism, together with small and large-scale meteoric processes and radiation darkening of the exposed surface material, is adequate to account for the principal lunar features.

1. Introduction

It is sometimes not appreciated that the acquisition of data is the basis upon which we build understanding and knowledge of the universe and not the end-product. The Apollo missions to the lunar surface have made available a very great amount of information. It is appropriate at this point that we should collate this information into a meaningful conception of the processes operating on the moon to shape its features.

Erosion processes are a major factor in shaping the surface of planeraty bodies. In the case of the Moon, the major processes of erosion arise from meteoric impact, solar wind ablation, and electrostatic transport. In speaking of erosion, large-scale meteoric impact, so dominant in the creation of the positive relief of the Moon's surface, should be separated from the small-scale impact events that slowly erode away these features, creating an ever finer matrix of craters. Such a saturation of craters (Walker, 1967) for wide ranges of sizes is quite evident in many areas of the Moon, particularly for the large-scale impact craters of about 1-km diam or more and for the quite small scale, 1 cm or less. There are areas, the maria, that appear relatively free of craters. While there are craters there, these ancient structures, some 3.7×10^9 yr old, should be saturated, even in the presence of endogenous processes (the existence of which is yet to be established). In spite of this, there appears to be a general smooth-ing of the surface. For these large-scale features, this is to be expected from the general nature of the event forming the maria (Walker, 1969). The impact of very large meteoric bodies or planetesimals obliterates previously existing structures but, since the vast size of the crater exceeds the range of the ejecta, this material and its kinetic energy is returned to the crater, forming almost immediately the general smooth mare regions.

R. J. L. Grard (ed.), Photon and Particle Interactions with Surfaces in Space, 521–544. All Rights Reserved
Copyright © 1973 by D. Reidel Publishing Company, Dordrecht-Holland

On the smaller scale, we observe from the Apollo photographs of the lunar surface that rocks generally are half buried in fine granular material that appears to have slowly risen about the rock. Where the slope of the rock is gradual, there are signs of a 'gentle lapping' of the material up the sides of the rock. The scale of this 'lapping' is about a decimeter.

Solar wind erosion, clearly evident in certain effects in the quite fine structure of long-exposed rocks, does not suffice to explain the magnitude or character of this surface smoothing.

The velocity of the ejecta from impacts has a lower (general) cutoff of about 10 m s^{-1}, due to the vacuum contact welds between grains on the Moon. Such material would coat rocks in debris to an average height of 15 m, while the impact of larger bodies would fill in around boulders and rocks in the irregular manner expected of saturation cratering, not smoothly and cleanly about a sharp level. From the manner in which these objects are filled in at the base, it would appear that the mean vertical velocity of the erosion material is quite small, less than 1 m s^{-1}. Without a supporting transport mechanism, the range of such material could not be more than about one meter, entirely inadequate to account for the surface smoothness around these objects. It is for this reason that it appears that electrostatic transport of micro-meteoric ejecta participates in the erosion of the Moon and principally accounts for surface smoothing on the 1-m scale. The purpose here is to elucidate this erosion mechanism.

2. Review of Electrostatic Erosion Problems

Work of Öpik and Singer (1960) indicated that the lunar surface would carry a positive electric charge on the daylight side of the Moon as a result of the solar photon flux. Their work indicated a charge of 20 to 40 V. It was shown by Singer and Walker (1962a) that the photoelectrons ejected with too small an energy to escape from the Moon would nevertheless form a photoelectron layer, *electronosphere*, near the moon. A value of $E = 67$ V m^{-1} for the electric field at the surface was calculated. Manka (1973) calculates values of 4.4 V and 3.4 V m^{-1} at the subsolar point. Under certain circumstances, the potential and electric field becomes rather higher; Reasoner and Burke (1973) find values of 200 V and 20 V m^{-1} obtaining as the Moon passes through the geomagnetic tail of the Earth.

The absence of an atmosphere capable of shielding the lunar surface from meteoric impact made it clear that the surface of the Moon should likely have a layer of dust. The appearance of certain telescopically visible features indicated that large-scale erosion by or fluidity of this dust might be present on the Moon. However, the fact that in a vacuum most materials form fairly strong contact welds has made it clear that any significant fluidity of the dust would have to derive from an outside mechanism and could not be due to Brownian motion (Berg, 1964).

Gold (1955) first suggested that electrostatic effects may be adequate to give the dust on the Moon sufficient fluidity to explain the flatness of the maria. Singer and

Walker (1962a, b) calculated the magnitude of the effect, using the data available at that time. The physical picture envisioned that incoming micrometeorites disrupt a lunar surface layer that is made up of loosely bound pulverized material. Following the impact, much of the material is ejected with a certain mass and velocity spectrum. Following this ejection, electrostatic forces can cause transport of some of the material. They concluded that the electrostatic transport mechanism would not be adequate to explain large-scale erosion processes. They calculated that over 5×10^9 yr an average of 0.4 to 2.0 g cm^{-2} would have been transported and deposited in shadow areas. Gold (1962), on the other hand, treating the problem qualitatively, suggested that the electrostatic field might be sufficient to tear loose dust particles from the surface, leading to more extensive erosion. However, Singer and Walker (1962a) showed that the electrostatic field at the surface would not be adequate for this.

Two crucial points enter here; one is that a dust particle on the surface of the Moon, while at a lunar potential of as much as 200 V, cannot have the large charge which it would possess if it were freely moving in space (Singer, 1956). Essentially, it forms part of the lunar surface and therefore has a proportionate charge. Thus a dust particle with a radius of 10 μ would end up with a small fraction of an electronic charge rather than 1000 charges, as Gold assumed. Secondly, no potential differences of any appreciable magnitude can develop on the lunar surface at neighboring points. In spite of the fact that the electrical conductivity of the lunar surface is very small, enough photoelectrons are produced by solar UV radiation to allow the shorting out of any potential differences. The trajectories of the photoelectrons are deviated by any potential differences in just a way as to minimize them.

Grannis (1961) has estimated the rate of mass transport by a mechanism that depends on the buildup of a potential on individual dust grains as a result of a random accretion process. It was assumed that the number of ions and electrons hitting the grain and photoelectrons leaving the grain fluctuates from the mean, giving rise to a charge on the grain that deviates from the equilibrium value. It was suggested that two neighboring grains, as a result of their fluctuating charge, might experience a sufficiently large force to bring about the ejection of one of the grains from the surface giving rise to a mass transport. Walker (1962) has calculated the probability for the occurrence of large charge fluctuations. His statistical treatment shows that this mechanism will not lead to a large transport across the lunar surface. It appears, therefore, that the lunar electric field can only affect dust particles that are somehow injected into it and that this kinetic energy of injection is derived from the impact of extralunar material.

Gold had also assumed that the dust levitated by the electrostatic processes would be *preferentially* deposited in the low-lying maria. Singer and Walker (1962b) showed this mechanism to lead to deposition in local shadow areas. Walker (1966) has shown, however, that meteoroidal impact erosion does lead to an exchange of ejecta material between the maria and the highland regions and that this exchange process is adequat to explain granular layer thickness and albedo differences that exist between the maria and the highlands.

Gehrels (1964) proposed large-scale electrostatic dust clouds to explain the optical properties of the lunar surface. Hapke (1965a) has shown that levitation of a large cloud of electrostatically suspended particles is not feasible. It might be noted here that these optical properties were more adequately accounted for by Hapke's concept of a 'fairy castle' structure supposed to exist in the fine-scale lunar surface structure. That such a process might occur as a result of electrostatic processes was anticipated by Salisbury's experiments, and the magnitude of the effect is consistent with the present calculations. Such structures, in fact, are observed on the lunar surface.

In the present treatment, the arguments are based on the following steps in the physical picture. First, in the disruption and ejection, the very finest of the smoke particles are accelerated by the electric field and escape, while the larger particles cannot travel far and therefore land in the immediate vicinity of the impact. Only a fraction of the ejected material has the proper mass that is compatible with electrostatic transport.

However, the dynamics of the charging process of this selected group of particles must be considered in detail. We calculate the time constant for charging and show that the particle will acquire its full charge only after it has surmounted the inner screening layer of the lunar surface.

Next the motion and lifetime of these charged dust particles must be discussed with some care. It is shown that they hop around above the lunar surface, being elastically reflected by the lunar surface electric field. In order to be deposited, such a particle must penetrate the electric field and attach itself to the surface. This may happen when it loses its positive charge as it enters shadow areas. We come to the conclusion that this effect will lead to a preferred deposition of dust in shadow areas, such as in crevices and about the base of rocks, boulders, and other surface irregularities.

We will consider the process in as quantitative a manner as possible based on the present information concerning the nature of the lunar surface and the environment to which it is exposed. Four steps are involved: (1) the disruption and ejection of lunar surface material by meteoric impacts; (2) the charging of the dust particles; (3) force and energy considerations of the flotation process; (4) motion and deposition of the dust particles.

3. Photoelectric Screening and Charging Effects

It has been realized for some time that bodies in space may acquire an electric charge; for example, interstellar grains presumably carry a *negative* charge (Spitzer and Savedoff, 1950). The problem has been considered in more detail for interplanetary dust particles by taking into account also the photoelectric effect of the solar ultraviolet radiation (Singer, 1956); the equilibrium potential of a body in space was shown to depend on the competition between the release of photoelectrons through the incoming ultraviolet radiation and the accretion of electrons from the surrounding plasma. Since the ultraviolet spectrum extends to very high energies, there will always be some photoelectrons produced that have a high enough energy to escape from the presumably positively charged dust particle. With various assumptions concerning the solar ultraviolet spectrum and the density of interplanetary plasma, a positive potential of the order of 50 to 100 V was derived for dust particles. (However, in a medium of high electron density, as has been pointed out by Whipple (1960), the potential will be nagative.) Using the solar ultraviolet spectrum (Hinteregger, 1961), Öpik and Singer (1960) derived a potential for the surface of the Moon of about 20 to 40 V positive.

Calculations by Grobman and Blank (1969) detailed for the solar minimum, including the effect of an isotropic emission of electrons, photoelectron energy spectrum effects and angle of incidence θ of the photons and solar wind, lead to values in the range $+0.6$ to 10.2 V. Setting $\beta = n_0 v_t / \gamma_0$, where n_0 is the solar wind density, v_t their thermal speed, and γ_0 the (photon) quantum yield of the lunar surface, and taking the ion temperature of the solar wind of density 5 cm^{-3} to be 10^5 K streaming with a velocity of 400 km s^{-1}, they obtain the potential distribution on the sunlit lunar surface as given in Figure 1, where W is the photoelectric work function. More recent determinations have been made by Willis et al. (1973), Reasoner and Burke (1973), and Manka (1972). Their values will be used in the calculations below.

Let us now turn to the electrostatic screening effects due to the photoelectrons. It is apparent that, while the steady-state potential of a positively charged body is determined by the number of photoelectrons which have sufficient energy to escape, there are produced a large number of low-energy photoelectrons that do not escape; they are simply liberated at the surface by ultraviolet photons, ascent to a certain distance from the body, and then return to the surface. The effect of the electrons in 'ballistic' rather than escape orbits will be to produce an 'inner screening' of the electric charge of the body.

Let us find the variation of potential and space charge as a function of distance from the center of a spherical body. An approximate solution to the problem can be obtained as follows. Assume the electron density to be constant and equal to n_0, where $\frac{1}{2}n_0$ is the density of photoelectrons per cm^3 at the surface moving away from the surface; the factor of 2 is due to the return of these electrons to the surface. The surface density of positive charges σ is

$$\sigma = n_0 x_c, \tag{1}$$

where x_c is the distance that the electron distribution extends above the surface. From the Poisson equation we have

$$\frac{\partial^2 \phi}{\partial x^2} = -\frac{e^2 n_0}{\varepsilon_0}, \tag{2}$$

where ϕ is the potential energy. Substituting

$$z = (\phi_0 - \phi)/U, \tag{3}$$

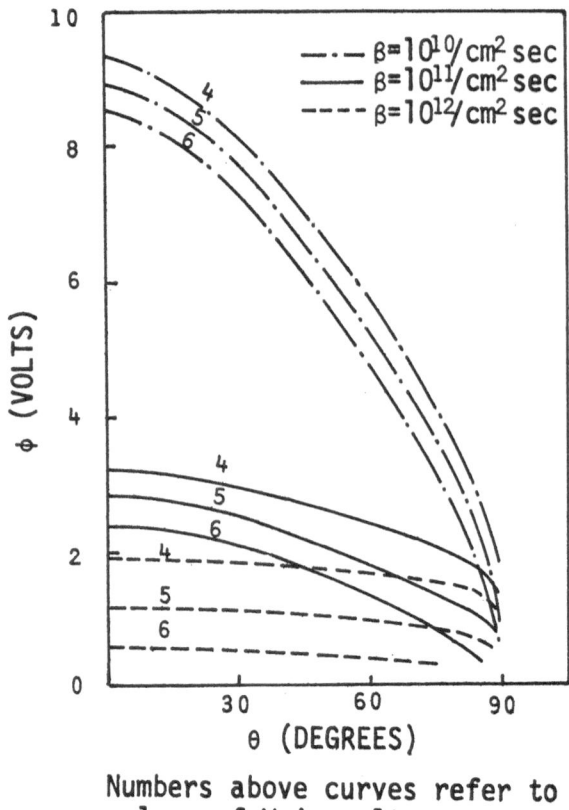

Fig. 1. Electrostatic potential distribution of the sunlit lunar hemisphere during solar minimum. The parametric dependence upon W and $\beta = n_0 v_t / \gamma_0$ is shown for the fixed value of u corresponding to $V = 400$ km s^{-1}, $T = 10^5$ K. For a solar wind density of 5 particles cm^{-3}, $\beta = 10^{11}$ cm^{-2} s^{-1} corresponds to a quantum yield $\gamma_0 = 8.7 \times 10^{-3}$.

where ϕ_0 is the potential energy at the surface and U is the initial energy of the photo-electrons, and integrating, we obtain

$$z' = z'_0 + \frac{e^2 n_0}{\varepsilon_0 U} x, \tag{4}$$

where $x = 0$ at the surface and z'_0 is proportional to the surface value of the electric field. The second integration yields

$$z = z'_0 x + \frac{e^2 n_0}{\varepsilon_0 U} x^2. \tag{5}$$

The surface electric field can be obtained from the value of σ in Equation (1); thus

$$z'_0 = \frac{e^2 n_0}{\varepsilon_0 U} x_c. \tag{6}$$

Since $z=1$ at $x=x_c$, we obtain for x_c

$$x_0 = \sqrt{\frac{\varepsilon_0 U}{3e^2 n_0}}. \tag{7}$$

If, instead of a constant density, one assumes an exponential variation $n_0 \exp(-x/|x_c|)$ of n with distance, the result is

$$x_c = \sqrt{\frac{\varepsilon_0 U}{e^2 n_0}} \tag{8}$$

which is seen to differ only by a small factor from x_c.

4. Derivation of the Density Equation Around a Spherical Body

If u is the velocity of the photoelectrons at the surface and θ_0 is the particle's angle to the normal to the surface, then the conservation of angular momentum requires that (see Figure 2)

$$uR \sin \theta_0 = rv \sin \theta, \tag{9}$$

where R is the radius of the body, r is the distance from the center of the body to an

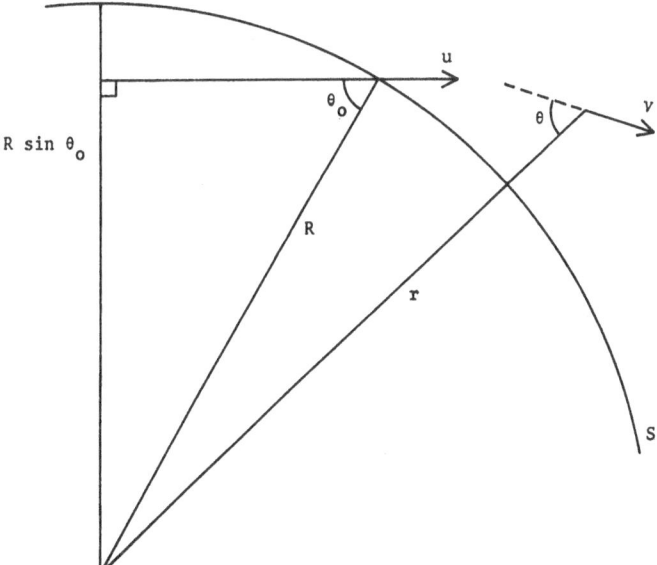

Fig. 2. The parameters used to describe the motion of a photoelectron in the electric field of a spherical body. The initial position R and velocity u at the surface S of the spherical body and the position r and velocity v along the trajectory are shown. The angles θ_0 and θ are the angles to the radius vector on the surface and at a point on the trajectory, respectively.

arbitrary point above the surface, and θ is the angle to the normal at r. For the conservation of the energy U, we write

$$U = \tfrac{1}{2} m u^2 = \tfrac{1}{2} m v^2 + \phi_0 - \phi . \tag{10}$$

Solving Equation (9) for v and substituting in Equation (10), we get

$$\sin^2 \theta = \frac{R^2 \sin^2 \theta_0}{r^2 \{1 + [(\phi - \phi_0)/U]\}} . \tag{11}$$

The maximum angle at which the electron can just manage to reach a position r is $\theta_0 = \theta_{\max}$, where

$$\theta_{\max} = \arcsin \frac{r}{R} \sqrt{1 + [(\phi - \phi_0)/U]} . \tag{12}$$

If we have a given energy injection spectrum, $F(U)$, we can find the density contribution to the screening due to the photoelectrons. The procedure is analogous to finding the density of atoms in a planetary exosphere (Öpik and Singer, 1959, 1961). Assuming the injection to be isotropic and writing $F(U) \, dU$ for the flux of electrons injected with energy U in the range dU, the contribution to the density at a position of potential ϕ by electrons in the interval dU is

$$dn = 2F(U) \, dU \int_{\theta_{\max}}^{0} \frac{R^2}{r^2} \frac{\cos \theta_0 \sin \theta_0 \, d\theta}{v_r} , \tag{13}$$

where v_r is the velocity component normal to the surface for a particle with initial angle θ_0. This is given by

$$v_r = v \cos \theta = \sqrt{\frac{2U}{m} \left(1 + \frac{\phi - \phi_0}{U}\right) \left(1 - \frac{R^2 \sin^2 \theta_0}{r^2 \{1 + [(\phi - \phi_0)/U]\}}\right)} . \tag{14}$$

Thus,

$$n = 2 \int_{0}^{U_0} \sqrt{\frac{m}{2U}} \left\{ \int_{\theta_{\max}}^{0} \frac{r^2}{R^2} \left[1 + \frac{\phi - \phi_0}{U} - \frac{R^2 \sin^2 \theta_0}{r^2}\right]^{-1/2} \times \right.$$
$$\left. \times \cos \theta_0 \sin \theta_0 \, d\theta_0 \right\} F(U) \, dU . \tag{15}$$

If the photoelectric emission is not isotropic but varies as $\cos \theta_0$, we have as the general solution for the electron density

$$n = 2 \int_{0}^{U_0} \sqrt{\frac{m}{2U}} \left\{ \int_{\theta_{\max}}^{0} \frac{r^2}{R^2} \left[1 + \frac{\phi - \phi_0}{U} - \frac{R^2 \sin^2 \theta_0}{r^2}\right]^{-1/2} \times \right.$$
$$\left. \times \cos^2 \theta_0 \sin \theta_0 \, d\theta_0 \right\} F(U) \, dU . \tag{16}$$

The factor 2 in Equations (15) and (16) takes account of the contribution of the returning electrons. In addition, we have the contribution of the escaping electrons, given by a similar integral but with limits from U_0 to U_{max}. This contribution is usually negligible in the immediate vicinity of the body and will be neglected. We also neglect the contribution of external plasma electrons.

If $R \approx r$ and the flux is monoenergetic and isotropic, then we have for the electron density

$$n = n_0 \int_{\theta_{max}}^{0} \frac{\cos\theta_0 \, d(\cos\theta_0)}{\{[(\phi - \phi_0)/U] + \cos^2\theta_0\}^{1/2}} \, . \tag{17}$$

The integration of Equation (17) with the value of θ_{max} given in Equation (12) gives

$$n = n_0\sqrt{1 - z}, \tag{18}$$

where z is defined in Equation (3). Poisson's equation becomes

$$\frac{d^2 z}{dx^2} = -\frac{e^2 n_0}{\varepsilon_0 U}\sqrt{1 - z} \tag{19}$$

(see Walbridge, 1969), which has a solution

$$\frac{dz}{dx} = 2\sqrt{2k/3}\,(1 - z)^{3/4} \tag{20}$$

$$z = 1 - [1 - (k/6)^{1/2}\, x]^4, \tag{21}$$

where

$$k = e^2 n_0/2\varepsilon_0 U \tag{22}$$

and including a factor of 4 to take account of the average vertical velocity and the return factor

$$n_0 = 2N_v\gamma/v_0, \tag{23}$$

where N_v is the photon flux, γ the quantum yield, and v_0 the electron velocity. These expressions can be used together with experimental values for N_v and γ to establish the electric field and charge density near the surface of bodies in space.

Solution of the Poisson Equation using Equation (17) has been obtained by Grard and Tunaley (1971).

Let us now consider the problem of the photoelectric screening of a very small body. For the interplanetary dust particles, the photoelectric screening is unimportant except for those particles that are very close to the sun. This is due to the small cross section of the particles and the subsequent small ejection of photoelectrons. Since the cross section of a particle is πa^2 where a is the radius of the particle, the rate of ejection of photoelectrons is $\pi a^2 \gamma N_v$. Hence the average time interval between the ejection of a photoelectron and its reabsorption is about 7.2×10^{-8} s. The density of these electrons about the dust particle will be approximately

$$n = \frac{\pi a^2 \gamma N_v \times 7.2 \times 10^{-8}}{\frac{4}{3}\pi x_1^3}, \tag{24}$$

where x_1 is the value for the height of the screened field as obtained for a flat surface. It is seen that the density varies as the square of the radius for small particles. Taking the previous values of $\gamma = 0.1$, N_v, and x_1 we obtain

$$n = 0.153\,a^2. \tag{25}$$

Thus, in order that n be 1 cm^{-3}, the radius a must be 2.56 cm. The problem of the photoelectric screening of a dust particle is thus of little consequence. The effect is important when the body is of the order of or larger than the screening length.

5. Disruption and Ejection of Lunar Surface Material

The relation between the mass of material M upon the impact of a meteoroid of mass M_i and velocity v_i and the velocity v of the ejecta is given by (Walker, 1967)

$$dM = Bv^{-\beta}\,dv, \tag{26}$$

where $\beta = 2.55$ and B is evaluated in terms of the total mass ejected from the crater by the impact

$$M = \frac{M_i v_i^2}{2K} = B \int_{v_a}^{v_i} v^{-\beta}\,dv, \tag{27}$$

where $\frac{1}{2}M_i v_i^2$ is the kinetic energy delivered to the surface. The coefficient K has a value of 5.5×10^8 ergs g^{-1} for impacts in basalt or granite. For impacts in the granular lunar surface having a yield strength of $\sigma = 4.45 \times 10^4$ dynes cm^{-2} obtained for the Surveyor spacecraft and a density of the surface material $\varrho = 1.6$ g cm^{-3} (Costes et al., 1970),

$$v_a = \sqrt{\sigma/\varrho} = 167 \text{ cm s}^{-1}. \tag{28}$$

This gives 2.13×10^6 ergs gm^{-1} for K; the resulting value for B is $1.012 \times 10^{-3}M_i v_i^2$. If we take $v_i = 5 \times 10^5$ cm s^{-1}, the ratio of mass ejected to incident mass μ is

$$\mu = M/M_i = 5.87 \times 10^4. \tag{29}$$

This is somewhat of an overestimation, since craters in granular surfaces are shallower than those in rock. However, only the low-angle, low-velocity ejecta will be retained in Equation (31), thus compensating for this fact. We will find below that only a portion of this material can be floated, since most of the ejected particles will not fall into the allowed velocity or mass range.

The median size of the lunar regolith particles falls in the 42 to 98 μm range (Apollo Preliminary Examination Team, 1972; see also Walker, 1965), as shown in Figure 3.

The maximum number of particles that will be floated per impact is limited by two factors. First, the mass of the particle must be sufficiently small to be floated but not so small that it will be completely ejected from the Moon. Thus, the amount of material floated will be limited by a factor J which expresses the relative abundance of particles of suitable mass to be floated present on the lunar surface. A value for J will be derived later.

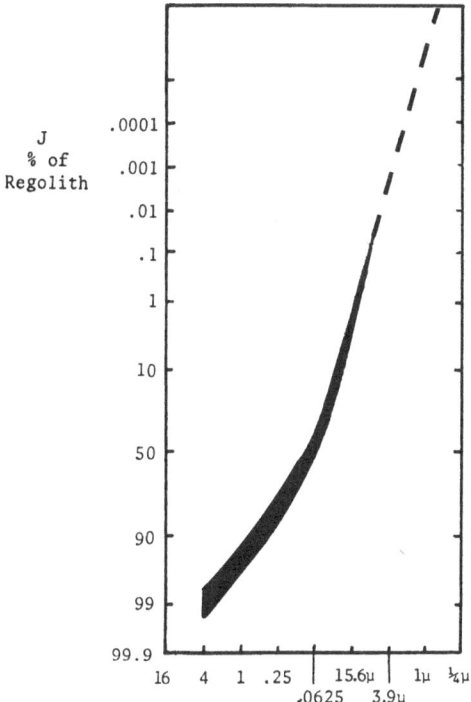

Fig. 3.   Cumulative particle size distribution for the lunar regolith.

Seond, in order to be floated, the *vertical* component of the ejection velocity must be less than a certain limiting value v_0; its value is given in Equation (42). For particles ejected isotropically with velocity v, this limiting velocity determines the maximum angle of ejection θ_m:

$$\theta_m = \arcsin\left(v_0/v\right). \tag{30}$$

This yields $(1-\cos\theta_m)$ as the fraction that can be floated at a velocity v, where $v \geqslant v_0$. The total amount of mass μ_J that will be floated per impact, in terms of M_i, the mass of the incident particle, is given by the product $(J/M_i)\,dM(v)$, the mass with velocity in the range v to $v+dv$, which is of proper size to be floated, times $(1-\cos$ $-\theta_m) = 1 - \sqrt{1-v_0^2/v^2}$ for $V_i \geqslant v \geqslant v_0$ or times 1 for $v < v_0$, integrated over all possible v.

It will be shown that v_0 is maller than the value v_a found in Equation (28) [see Equation (42)]. Thus the limits are from v_c to V_i,

$$\mu_J = (J/M_i) \int_{v_a}^{V_i} \left(1 - \sqrt{1 - v_0^2/v^2}\right) \mathrm{d}M\,(v), \tag{31}$$

where $\mathrm{d}M$ is given by Equation (26). In terms of $w = v/V_i$, $w_c = v_c/V_i$ and $w_0 = v_0/V_i$, we have

$$\mu_J = \frac{JB}{M_i V_i^{\beta - 1}} \int_{w_a}^{1} \left(1 - \sqrt{1 - w_0^2/w^2}\right) w^{-\beta}\,\mathrm{d}w. \tag{32}$$

Equation (32) can be integrated approximately, since $w \gg w_0$ in the range $w_a \leqslant w \leqslant 1$; the result is (including* the factor 2)

$$\mu_J = \frac{JB w_0^2}{M_i V_i^{\beta - 1}(1 + \beta)} \left[w_a^{-(1+\beta)} - 1\right]. \tag{33}$$

6. The Charging of the Dust Particles

The particles that are ejected from the lunar surface by the impacts of meteoric bodies can now become charged by the same kind of process that brings about the charging of interplanetary dust (Singer, 1956): the escaping flux of high-energy photoelectrons, which is itself a function of the potential of the surface, must balance the flux of electrons from the ambient plasma to the surface. Since the dust particles are located in the lunar electronosphere, there will be an enhancement of the flux of electrons to the surface of a charged dust particle over the flux received by the lunar surface. This enhancement may be expressed by the factor f where

$$f = \frac{v_e N_0'(1 + \phi/E_e)}{4u_e N_0}, \tag{34}$$

where N_0' is the density of the photoelectrons in the neighborhood of the dust particle, E_e is their average energy, and v_e is their average velocity; N_0 is the density of the ambient electrons and u_e is their average velocity. The potential at the surface of the dust particle is ϕ. Since the integrated flux spectrum of solar photons (Hinteregger, 1961; Hall and Hinteregger, 1970), varies so rapidly, the factor f can be easily accomodated by taking the potential of the dust particle to be only slightly smaller

---

* We should note that Equation (32) is not complete in that it does not include the effects of secondary particles. That is to say, particles ejected from the lunar surface by micrometeoric impacts and not floated will return to the surface to produce further particle ejections. Since half of the momentum of an incoming particle appears in the form of ejected particles and, in turn, half of their momentum will go into ejecting other particles, the total effect is the sum of $1 + \frac{1}{2} + \frac{1}{4}$ etc., i.e. 2. The factor in Equation (32) should therefore be doubled.

than the potential of the lunar surface. However, since most of the dust particles are ejected from the surface at a velocity of from one to a few hundred meters per second, they will be ejected above the region of highest electron density and there the potential which they attain will be higher. Fluctuations in the potential of the dust particle will not occur as the particle experiences successive rebounds. Since the rate of emission of photoelectrons from the surface of a dust particle is $\gamma N_v \pi a^2$, where γ is the quantum efficiency, N_v is the flux of photons with sufficient energy to produce escaping photo-electrons, and a is the particle's radius, and since the number of electrons on the particle at a potential ϕ is given by

$$Z = (4\pi\varepsilon_0 a\phi)/e, \tag{35}$$

the time constant t_0 is

$$t_0 = \frac{4\varepsilon_0\phi}{\gamma N_v ae}, \tag{36}$$

where e is the electron charge and ε_0 is the permittivity of space. As an example, if we take $\gamma = 0.01$, $a = 1.5$ μm, and use the solar flux data to obtain ϕ vs N_v, then the time required for the potential to change from zero to 10 V is 0.5 s. For a dust particle ejected with the lower limit velocity $v_a = 1.67$ m s^{-1}, the time required to pass through the photoelectron sheath will be 0.24 s, while the total flight time of the particle to return to the surface is 2.00 s. Thus, the particle will pass unchanged through the electric field, becoming fully charged before returning to the surface. It gains no energy from the electric field. Such extra energy would have prevented the rebound of the particle.

7. Force and Energy Considerations of the Flotation Process

For those particles that are initially ejected with a velocity of 167 cm s^{-1} in the up-ward direction, their in-flight travel time to return to the surface is 2.00 s. Thus, during this time a particle becomes fully charged. However, although the particle may be small enough to be reflected by the electric field, the kinetic energy of the particle must also be overcome. Let us first find what size particle can be supported and then find the limitation imposed by its kinetic energy. If we set the gravitational force equal to the electric force, we get an *upper limit* as follows:

$$mg = ZeE, \tag{37}$$

where m is the particle mass, g is the acceleration of gravity on the Moon's surface, and E is the electric field intensity. Substitution for the mass in terms of the density ϱ and the radius of the particle a gives the expression for the maximum radius a_{max}:

$$a_{max} = [(3\varepsilon_0\phi E)/(\varrho g)]^{1/2}. \tag{38}$$

Equation (38) is used below to obtain the maximum size particles entering into the electrostatic transport of the regolith.

In addition to the upper limit on the particle size, there is a *lower limit* produced by the escape of particles from the Moon. The region of the inner screening which is due to the large density of photoelectrons falls off rapidly within a few centimeters. There is also a region that is screened primarily by the plasma surrounding the Moon. This electric field has a screening lengths of about 15 m (Öpik and Singer, 1960). If we take this potential to be $\bar{\phi}$, the amount of energy that will be imparted to a particle that moves through this region is given by $Ze\bar{\phi}$, where Z was given in Equation (35). In order for the particle to exceed the escape velocity by this process

$$Ze\bar{\phi} > mgR, \tag{39}$$

where R is the radius of the Moon. This gives for the minimum radius a_{min} of particles that can be retained by the Moon

$$a_{min} = \left(\frac{3\varepsilon_0\phi\bar{\phi}}{\varrho gR}\right)^{1/2}, \tag{40}$$

where ϕ is the potential on the particle. If we assume that ϕ and $\bar{\phi}$ are both equal to 10 V, then we obtain $a_{min} = 7.56$ Å. Thus, only atomic or molecular particles will be ejected; indeed, this is an important mechanism preventing the buildup of a lunar atmosphere.

The maximum allowable vertical velocity v_0 is obtained from the condition that the particle has sufficient kinetic energy to penetrate the lunar field; i.e.,

$$\tfrac{1}{2}mv_0^2 = Ze\phi_0, \tag{41}$$

where ϕ_0 is the potential of the lunar surface. Equation (41) gives

$$v_0 = (2\phi_0 g/E)^{1/2}, \tag{42}$$

which is independent of the density ϱ. For example, using a value of $\phi_0 = 4.4$ V and $E = 3.4$ V m^{-1} (see Manka, 1972), we obtain $v_0 = 2.1$ m s^{-1}. Thus, the vertical velocities range up to ~ 1 m s^{-1} for the largest size particles; a wider range is allowed for particles with a smaller radius. On the other hand, the horizontal velocities range from about one to several hundred meters per second.

The total mass of material injected per impact, in terms of the mass of the incident particle, is obtained by using the results of Equation (42) in Equation (33).

The time-averaged value $\bar{\mu}$ of this quantity is obtained by multiplying μ_J by the fraction of the time $\Delta t/t$ that the surface is maintained at any given potential,

$$\bar{\mu} = \mu_J \Delta t/t. \tag{43}$$

Table I shows the results of these relations for the potential and electric field values for the lunar surface under various conditions, i.e., for the Moon in the solar wind at the subsolar point on the Moon's surface, for the Moon in the Earth's magnetospheric

TABLE 1

Tabulation of data for electrostatic injection and transport

Electric field data	ϕ (V)	E(V m^{-1})	$\Delta t/t$	v_0 (m s^{-1})	a_{max} (m)	J	$\bar{\mu}$
Solar wind at subsolar point (Manka)[c]	4.4	3.4	0.45	2.1	0.39	(10^{-10})	1.7×10^{-6}
High latitude tail at subsolar point (Manka)[c]	17	42	0.05	1.2	2.66	(10^{-5})	5.8×10^{-3}
Plasma Sheath at Terminator (Manka)[c]	1.5	0.33	0.05	3.9	0.07	(10^{-12})	6.5×10^{-9}
Solar wind at subsolar point (Willis et al.)[c]	2	2	0.45	1.8	0.20	(10^{-11})	1.2×10^{-7}
Moon in tail of Earth's magnetosphere (Reasoner and Burke)[c]	200	20	0.10	5.8	6.31[a]	5×10^{-4}	9.2
Solar wind at subsolar point (Singer and Walker, 1962a, b)[c]	20	60	0.45	1.0[b]	3.45	0.1 to 0.5	5×10^{-3} to 5×10^{-2}

( ) Extrapolated value
[a] Agreement with cutoff value for J, Figure 3.
[b] $v_a = 4.5$ m s^{-1} (1.7 m s^{-1} for all other cases)
[c] Referenced values to be found in this text.

tail, etc. The fraction of time $\Delta t/t$ that these conditions exist is given. Equation (42) is used to calculate v_0, the maximum ejecta velocity allowing electrostatic floatation, along with a_{max} obtained from Equation (38). The value of J is obtained from Figure 3. The last column lists the value of $\bar{\mu}$ obtained from Equation (43).

From these values, we see that the electrostatic transport is unimportant except during the passage of the Moon through the Earth's magnetosphere. But during that passage, electrostatic transport becomes dominant. Even after taking account of the small time spent by the Moon in the magnetospheric tail of the Earth, the material electrostatically transported will amount to 9.2 times the flux of meteoric material to the Moon's surface. For a meteoric flux of 5×10^{-16} g cm^{-2} s^{-1} to the lunar surface, this gives 650 g cm^{-2} transported over the 4.5×10^9 yr age of the Moon, an average of 4 m of material over the Moon. This material, deposited preferentially in shadow areas, is adequate to account for the smooth appearance of much of the maria; and, in fact, this electrostatic transport may account for the apparent lifetime of 1-m boulders on the lunar maria, given as 3×10^8 yr (Shoemaker *et al.*, 1970). During that time, $\frac{1}{4}$m of material would be transported and deposited preferentially about boulders. Since the major part of the transport occurs only during the passage of the Moon through the magnetospheric tail of the Earth, preferred deposition orientations should be expected to give rise to the streaked surface appearance of the Moon (see cover photograph, *Science* **175**, No. 4020). This fact, together with the orbital dynamics responsible for the orientation of the major lunar anomalies facing the Earth, is the principal cause of the front-back asymmetry of the Moon's surface.

8. Origin of the Principal Lunar Surface Features

While it is generally acknowledged that meteoric bodies, cometary nuclei, and planetesimals have produced the craters of the continentes of the Moon and played a role in the initial stages in the formation of the maria, many consider that endogenous processes giving rise to vast lava flows to have been the principal factor in shaping the present flat and dark appearance of the maria.

Gold (1955) has argued strongly against the lava flow interpretation of the maria, but his theory of the maria in which impact craters have been filled with dust transported from the continentes does not appear tenable in view of the arguments given by Öpik (1962) or the present consideration of the electrostatic transport process, though smoothing of features to a scale of a few meters is to be expected of the electrostatic processes.

The existence of any substantial endogenous processes on the Moon still lacks any significant substantiation from the Apollo finds. The material found is best understood as resulting from repeated impact crushing, mixing, shocking, and occasionally melting of the surface material. Extensive lava flows are not found.

The question to be asked is, "Can the gross features of the Moon's surface be understood in terms of impact (small and large scale), electrostatic transport, and radiation darkening of the regolith?"

9. Structure of Large-Scale Craters

Hess and Nordyke (1961) in a report on the scaling laws governing explosion craters noted that the proportion of fallback material in large craters was greater than for small craters, and he suggested that the maria might be large impact craters with a very high proportion of fallback material. The difficulty with his evidence was threefold. First, the significant difference between explosion craters and impact craters is in their ejecta velocity distributions, there being essentially no high velocity ejecta issuing from explosion craters. Secondly, there was actually no data that would allow one to scale up to the large lunar craters and certainly not up to craters the size of the maria. Finally, air resistance was probably a very important factor in producing the ejecta fallback in the case of the explosion craters.

The suggestion, nevertheless, offers an interesting possibility as to the origin of the lunar maria if a satisfactory means can be found to scale up to the large size craters necessary. It is primarily to this scaling problem that we will address ourselves.

The velocity distribution of the ejecta from impact craters is given by Equation (26). This can be reexpressed in terms of the radius by assuming the crater at any time during its formation approximates to half an oblate ellipsoid having a semi-major axis 2.5 (a factor obtained from experimental craters formed in rock) times its semi-minor axis. If the semi-major axis is r, the volume will be

$$V = \frac{4\pi}{15} r^3.$$

(44)

If we multiply Equation (44) by the density of the target ϱ, we can equate this to the integral of Equation (26), giving us, on solving for r:

$$r = \left[\frac{15B}{4\pi\varrho(\beta - 1)} (v^{1-\beta} - v_b^{1-\beta}) \right]^{1/3},$$

(45)

where v_b is the upper limit on the ejecta velocity.

The material that is ejected at r will land at r' where

$$r' = r + l,$$

(46)

l being the range of the ejecta. The range of the ejecta is given with sufficient accuracy, assuming the acceleration of gravity g is constant and the surface is flat. We therefore have

$$r' = r + \frac{2v^2}{g} \sin\alpha \cos\alpha.$$

(47)

By substituting Equation (45) into Equation (47), we obtain the distance at which the ejecta lands as a function of the ejecta velocity. Let us assume that $v_b \gg v$ so that this term can be neglected. For impacts in basalt targets, the density ϱ is about 2.7 g cm^{-3},

B is a function of the mass m, and velocity v_i of the projectile; if we assume a cratering efficiency of 5.5×10^8 ergs g^{-1},

$$B = 6.43 \times 10^{-2} \, mv_i^2 . \tag{48}$$

If we set $T = Mv_i^2 \times 10^{-18}$, T representing the equivalent number of metric ions traveling 10 km s^{-1}, and substitute into Equation (45), we obtain

$$r = 2.43 \times 10^5 \, T^{1/3} v^{-2/3} . \tag{49}$$

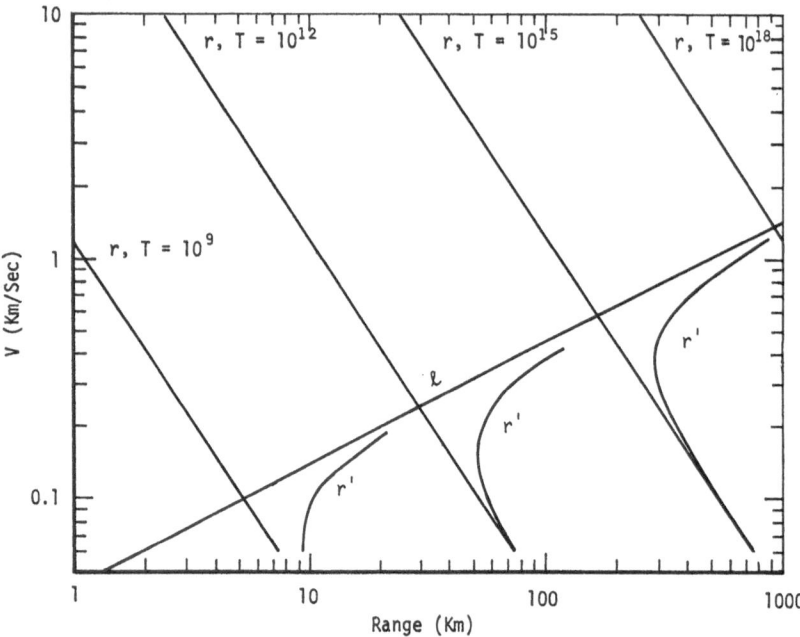

Fig. 4.   A plot of the velocity v with which material is ejected during the cratering process as a function of the distance from the point of impact. In addition, the range l of ejecta as a function of the velocity of ejection (assuming $\alpha = 60$ deg) is shown. The sum of r and l gives the distance at which the ejecta will land as a function of v; this quantity is labeled r'.

This equation is plotted in Figure 4 for various values of the parameter T. The value of l is plotted in Figure 4 as a function of v where $g = 1.62$ cm s^{-2} and $\alpha = 60$ deg, a value that is sufficiently close to what is observed experimentally to serve our purposes. To obtain r', the curves for r and l are added. The resulting curves for r' are shown in Figure 4.

The radius R of the crater produced by a given impact is determined by the lower limit on the ejecta velocity. For basalt targets, this lower limit is about 60 m s^{-1}. This cutoff point in the $r(v)$ curves is also indicated in Figure 4. It will be observed that, for the large craters, much of the ejecta does not escape. If we assume that this fall-back material becomes evenly distributed over the crater floor, as it obviously must, since the ejecta fallback material will return such a large fraction of the original kinetic energy, then we can calculate the crater depth. For the crater geometry we have assumed here (see Equation (44)), the equation for the depth d becomes

$$\frac{d}{R} - \frac{25}{12}\left(\frac{d}{R}\right)^3 - \frac{4}{15}\left(\frac{r_0}{R}\right)^3 = 0, \tag{50}$$

where R is the radius of the completely formed crater and r_0 is the radius within which all the ejecta escape from the crater. We find, using Figure 4, that for $T = 10^9$, a crater 7.4 km in diameter is formed with a depth of 1.13 km, all the ejecta escaping. For $T = 10^{10}$ we obtain a crater 74.0 km in diameter with a depth of 2.7 km, and for $T = 10^{15}$ the diameter and depth are 740 km and 4.7 km, respectively. In Figure 5, we have plotted the resulting theoretical diameter depth curve, together with the diameter vs depth data for 320 lunar craters and 8 maria (Baldwin, 1963). The lower portion of the theoretical curve is consistent with the experimental data for experimentally produced craters.

The above consideration satisfactorily explains the circularity and general dimensions of the maria. Most of the kinetic energy of the incident planetesimals forming the maria would fail to escape from the crater. This energy would be adequate to level the floor of the maria, though not to the extent of its present smoothness. Small scale meteoric and micrometeoric impacts, together with the electrostatic transport mechanism, give rise to the present smooth surface.

The occurrence of ghost craters is easily explained by the present theory. Toward the edge of the maria the ejecta velocity would be so small that the ejecta would travel no more than a few kilometers. This distance is not sufficient to obliterate completely large distinct craters existing before the formation of the maria. The walls of the ancient crater would be broken down, but not completely dispersed. The low albedo of the floor of these craters requires further consideration. The low albedo of terrestrial lava provides a ready solution to this problem for the advocates of the lava flow theory of the lunar maria. However, the fact that the lunar surface is covered by granular material (Walker, 1965) requires the proponents of the lava flow theory of the maria to acknowledge that the low albedo is derived from the lava lying below the surface. The difficulty with this is that new craters characteristically have a high albedo. One can no longer argue that these craters have a higher albedo because the lava has been pulverized. The entire mare surface consists of the same pulverized material. This fact clearly shows that the low albedo results from exposure to ion bombardment by the solar wind. Extensive studies have been carried out by Wehner et al. (1963, 1964), and Hapke (1965b). The question that must be considered here is 'Why should there exist a difference in albedo between the continentes and the maria?'

Walker (1966) has shown that micrometeoric erosion is a long-range erosion process in which the mass transport is dominated by the higher velocity ejecta. The significance is that eroded material will not simply migrate to the foor of the nearest slope, but rather there will be an exchange of material involving larger areas of the lunar surface. Thus, material will migrate from the highland areas to the low-lying maria. This fact suggests that the dust or granular material covering the highland areas

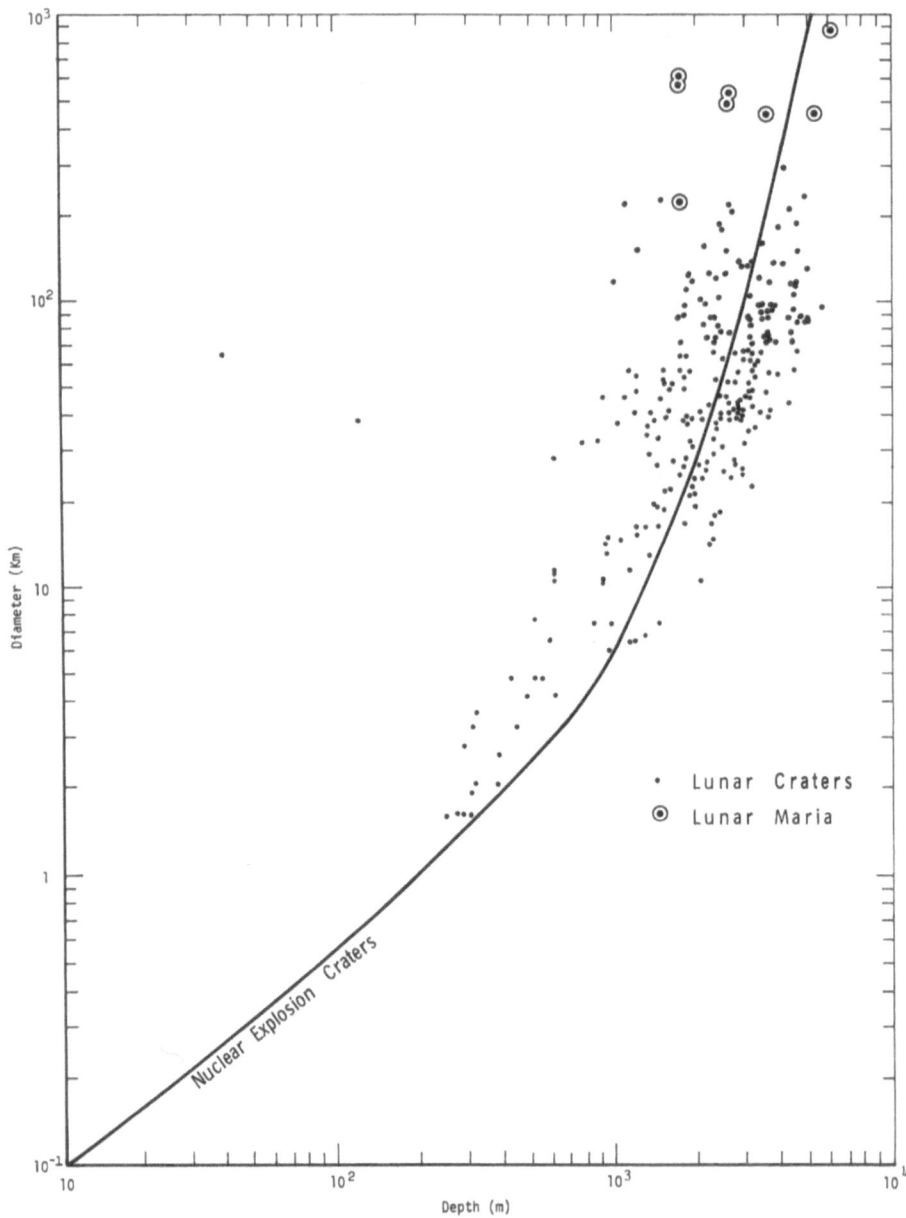

Fig. 5. A plot of the diameter D as a function of the depth d for 320 lunar craters (see Table 7, Baldwin (1963)) and 8 Maria. The theoretical curve, including the effects of the fallback material, is also shown; the lower portion of this curve is from data for nuclear explosion craters.

(Öpik, 1962) will be newer material than that in the maria, the older material gradually accumulating in the maria, becoming darker with increased exposure to the solar wind and radiation.

Let us determine the average age of exposure of the surface particles in the highland regions, neglecting for the moment the material coming to the highlands from the maria. If t_0 is the time required for the average particle placed on the lunar surface to be covered by debris, the average age t of a particle on the surface will be

$$t = t_0 (1 + P + P^2 + \cdots) = t_0/(1 - P), \tag{51}$$

where P is the probability that the particle has previously been exposed to the surface. Assuming the surface to be old so that every point on the surface has been disrupted by an impact (primary, secondary, etc.), we can write

$$P = 1 - P_n, \tag{52}$$

where P_n is the probability that a given grain represents newly eroded material; that is to say, the probability that a grain was formed by the pulverization of subsurface rock immediately prior to its deposition on the surface. The value of P_n is simply the ratio of new material in ejecta compared to the total amount of ejecta produced in a given time interval.

Let us obtain an expression for P_n. We assume for simplicity an impact forming a crater with a depth d less than the depth of the overlay D produces no new granular material, and assume all the ejecta to be new granular material when $d > D$. The generating function $G(r)$ for craters having a radius r is related to the rate of formation of craters in the size range r to $r + dr$ by

$$d^2N/dt = G(r)\,dr; \tag{53}$$

and if we include secondary craters, we write

$$d^2N'/dt = H(r)\,dr. \tag{54}$$

If r_{min} and r_{max} represent the smallest and largest craters formed on the surface, P_n is given by

$$P_n = \int_{2.5D}^{r_{max}} VH(r)\,dr \Big/ \int_{r_{min}}^{r_{max}} VH(r)\,dr, \tag{55}$$

where V is the crater volume in terms of r as given by Equation (44).

Walker (1967) gives an expression for $H(r)$ based on a theoretical calculation of the effects of secondary cratering and Brown's (1960) estimate of the functions $G(r)$ for the Moon,

$$H(r) = ar^{-3.41}(1 + 1.087 \times 10^4 r^{-0.93}). \tag{56}$$

We will use Equation (56) for $H(r)$ in Equation (55). If we assume r_{max} to be 5×10^6

cm (a larger value would be inconsistent with the probable age of the surface layers under consideration), r_{min} to be 10^{-2} cm as observed by the Surveyor vehicles, and assume $D = 10$ cm in the highland regions, we obtain

$$P_n = 0.153, \tag{57}$$

so that the average exposure age of the surface grains in the highlands t_n will be, by Equation (51),

$$t_h = 6.54 \, t_0 . \tag{58}$$

Taking $D = 10^3$ cm in the maria, we obtain

$$P_n = 0.102, \tag{59}$$

which gives, for the average exposure of age the surface grains in the maria,

$$t_m = 9.84 \, t_0 . \tag{60}$$

Data given by Wehner *et al.* (1964) indicate that the albedo A of a mineral generally varies as

$$A = C_1/(t + C_2) \tag{61}$$

(for example, for Tuff their data would yield $C_1 = 1.8 \times 10^3$, $C_2 = 6.5 \times 10^3$, where t is in years of exposure). If we assume $t_0 \gg C_2$ in the case of the Moon, we would have for the ratio of the maria albedo to the highland albedo approximately

$$A_m/A_h = t_h/t_m = 0.664 . \tag{62}$$

The fact that the observed ratio of the albedo of the maria to that of the highlands is 0.61 (Öpik, 1962) indicates that the mechanism described here is a quite satisfactory mechanism for producing the observed albedo differentiation between the highlands, including peaks located in the maria, and the maria.

If we include the effects of ejecta exchange between the maria and the highlands, Equation (51) becomes

$$t = t_0 + Q't' + Q(t_0 P + t_0 P^2 + \cdots) = t_0 [1 + QP/P_n] + Q't' , \tag{63}$$

where Q is the portion of material originating 'locally' (in the highlands when calculating t for the highlands and in the maria when calculating for the maria); the quantity Q' is the portion of material originating in the opposite region (in the maria when calculating t for the highlands, etc.), and t' is the average exposure time for the opposite region.

Since the surface material is eroded from the highlands to the maria, we can largely ignore the small amount of material transported from the maria into the highlands so that Equation (63) reduces to Equation (51) for the highlands, yielding $6.54 \, t_0$ as given in Equation (58).

The value of t_m will depend on the transport of material from the highlands to the maria. To determine Q' for the maria, we calculate the rate at which new surface material is being produced in the highlands and compare this with the total rate of its production for both regions, since this will equal Q' as the depth in the highlands becomes constant. This gives

$$Q'_m = \frac{S_h \int\limits_{2.5 D_h}^{r_{max}} VH(r)\,dr}{S_h \int\limits_{2.5 D_h}^{r_{max}} VH(r)\,dr + S_m \int\limits_{2.5 D_m}^{r_{max}} VH(r)\,dr}, \tag{64}$$

where S_h and S_m are the areas of the highland regions and mare regions, respectively; D_h and D_m are the depths of the surface material in the highland and mare regions. Using Equation (56) for $H(r)$ and setting $S_h = S_m$, we obtain for Q'_m a value of 0.601, so that t_m is 8.44 t_0. The ratio t_h/t_m is thus 0.773. This is also a satisfactory result, considering the limitations of the present discussion.

10. Conclusions

The present considerations indicate electrostatic transport of lunar surface material is important in shaping the surface features on a scale of 1 to 10 m. This process taken together with the impact process, both small and large scale, and the radiation darkening of the exposed surface material serves to account for the principal surface features of the Moon.

References

Apollo 15 Preliminary Examination Team: 1972, *Science* **175**, 363.
Baldwin, R. B.: 1963, *The Measure of the Moon*, University of Chicago Press, Chicago.
Berg, C. A.: 1964, *Nature* **204**, 461.
Brown, H.: 1960, *J. Geophys. Res.* **65**, 1679.
Costes, N. C., Carrier, W. D., Mitchell, J. K., and Scott, R. F.: 1970, *Science* **167**, 739.
Gehrels, T.: 1964, *Icarus* **3**, 491.
Gold, T.: 1962, in Z. Kopal and Z. Michailova (eds.), *Proc. IAU Pulkovo Meeting*, December 1960.
Gold, T.: 1955, *Monthly Notices Roy. Astron. Soc.* **115**, 585.
Grannis, P. D.: 1961, *J. Geophys. Res.* **66**, 4293.
Grard, R. J. L. and Tunaley, J. K. E.: 1971, *J. Geophys. Res.* **76**, 2498.
Grobman, W. D. and Blank, J. L.: 1969, *J. Geophys. Res.* **74**, 3943.
Hall, L. A. and Hinteregger, H. E.: 1970, *J. Geophys. Res.* **75**, 6959.
Hapke, B.: 1965a, *Icarus* **5**, 154.
Hapke, B.: 1965b, *Ann. New York Acad. Sci.* **123**, 711.
Hess, W. N. and Nordyke, M. D.: 1961, *J. Geophys. Res.* **66**, 3405.
Hinteregger, H. E.: 1961, *J. Geophys. Res.* **66**, 2367.
Manka, R. H.: 1973, this volume, p. 347.
Öpik, E. J.: 1963, in S. F. Singer (ed.), *Progres in the Astronautical Sciences*, North-Holland Publishing Co., Amsterdam, Chap. 5.
Öpik, E. J. and Singer, S. F.: 1959, *Phys. Fluids* **2**, 653.
Öpik, E. J. and Singer, S. F.: 1960, *J. Geophys. Res.* **65**, 3065.

Öpik, E. J. and Singer S. F.: 1961, *Phys. Fluids* **4**, 221.

Reasoner, D. L. and Burke, W. J.: 1973, this volume, p. 369.

Shoemaker, E. M., Hait, M. H., Swann, G. A., Schleicher, D. L., Dahlem, D. H., Schaber, G. G., and Sutton, R. L.: 1970, *Science* **167**, 452.

Singer, S. F. and Walker, E. H.: 1962a, *Icarus* **1**, 7.

Singer, S. F. and Walker, E. H.: 1962b, *Icarus* **1**, 112.

Singer, S. F.: 1956, in J. A. van Allen (ed.), *Scientific Uses of Earth Satellites*, University of Michigan Press, Ann Arbor.

Spitzer, L., Jr. and Savedoff, M. P.: 1950, *Astrophys. J.* **111**, 593.

Walbridge, E. W.: 1969, *Icarus* **10**, 342.

Walker, E. H.: 1969, *Am. Geophys. Union. Trans.* **50**, 221.

Walker, E. H.: 1967, *Icarus* **7**, 233.

Walker, E. H.: 1966, *J. Geophys. Res.* **71**, 5007.

Walker, E. H.: 1965, *Trans. Am. Geophys. Union* **46**, 136.

Walker, E. H.: 1962, *J. Geophys. Res.* **67**, 2586.

Wehner, G. K., KenKnight, C. E., and Rosenberg, D. L.: 1964, Applied Science Division, Litton Systems Inc., Minneapolis, Minn., *6th Quarterly Status Report* (Contract NASW-75), Report No. 2669.

Wehner, G. K., KenKnight, C. E. and Rosenberg, D. L.: 1963, *Planetary Space Sci.* **11**, 1257.

Whipple, F. L.: 1960, in O. O. Benson Jr. and H. Strughold (eds.), *Physics and Medicine of the Atmosphere and Space*, John Wiley, New York, p. 48.

Willis, R. F., Anderegg, M., Feuerbacher, B., and Fitton, B.: 1973, this volume, p. 389.

DISCUSSION

Manka: One comment: I think that in light of secondary coefficients greater than unity, measured by the ESTEC group, it may be best to check the small positive dark side potential for the case of the Moon in the plasmasheet.

Walker: The high potential is contingent on a value of the emissivity being less than unity for the lunar surface. If the value given by the ESTEC group is accurate, and there is no reason at the moment to question this, then the small value given by Manka should be used. This still leaves us with high potential for the sunlit lunar surface in the magneto tail, as noted in the calculations. Since there was general agreement that the secondary coefficients would be greater than unity, the dark high side potential has not been included in the present study.

HORIZON-GLOW AND THE MOTION OF LUNAR DUST

DAVID R. CRISWELL

Lunar Science Institute, 3303 Nasa Road 1, Houston, Tex. 77058, U.S.A.

Abstract. Surveyor-7 photographed a bright glow along the western lunar horizon one hour after local sunset. This horizon-glow (HG) must result from the forward scattering of sunlight by electrically charged dust grains [a (grain radius) $= 5 \times 10^{-4}$ cm] which are electrostatically levitated 3 to 30 cm above rocks or surface irregularities located in the lunar terminator zone. The levitation condition is $E(\text{V cm}^{-1}) \geqslant 270 \times [a(\mu)]^{1/2}$. Column densities the order of 50 grains cm^{-2} are produced. The electrostatic field is generated in the following manner. High energy (500–1500 eV) photoelectrons are ejected from directly illuminated surfaces in the terminator zone. A positive monopole charge is produced which forces the return of subsequently ejected photoelectrons to the vicinity of the illuminated surface. Computer modeling indicates that 1% to 5% of the returned flux will accrete on adjacent completely dark areas. A stable, multipole charge distribution is generated between the light and dark areas. The associated intense multipole electric field ($E \gtrsim 10^3$ V cm^{-1}) can levitate micron size soil grains located in the charged regions. Approximately 10^{-2} gr cm^{-2} yr^{-1} of surface material are 'churned' by this process. The photoelectron work function of the lunar material, flux level of solar X-rays ($\lambda \sim 25$ Å), and the attenuation of solar ultraviolet photons by multiple scattering (as controlled by local surface geometry and reflectivity) are the physically significant factors.

1. Surveyor Observations of Horizon-Glow

Figure 1 is a composite of two Surveyor-7 photographs of the local western horizon (200 m distance) from the landing site. The top portion was taken approximately one-hour after local sunset while the bottom was taken during the preceding afternoon. The bright patchy region is the horizon-glow (HG). HG typically extends 2°–3° on each side of the sunset line, persists for 90 min after local sunset (Gault *et al.*, 1968a, b), is closely associated (<3 cm) with the silhouette of the horizon (Allen, 1968) and displays no polarization in the analog photography (Shoemaker *et al.*, 1968). Examination on the digital representation of Figure 1 reveals the HG to be 3 to 30 cm in verticle extent which rules out scattering by surface grains, a residual atmosphere, or secondary meteorite ejecta as sources (Criswell, 1972). Surveyor-7 HG luminance decreased monotonically from an initial value $B_7 \simeq 0.3$ cd/cm^2 (Rennilson, personal communication).

Figure 2 is an example of HG obtained during the Surveyor–6 mission (Rennilson, 1968). The apparent luminance $B_6' \simeq 5 \times 10^{-4}$ c cm^{-2} was much lower than the Surveyor-7 HG and was due to the greater horizon-distance ($\simeq 2$ km) and the image smear resulting from operating the television system in the highest sensitivity mode. The linear extent of the HG clearly distinguished it from the elliptically shaped solar corona. The actual intrinsic luminance of the Surveyor-6 HG is comparable to the HG of Surveyor-7. Refer to Criswell and Rennilson (1973) for an extensive analysis of the Surveyor-HG data.

2. Levitation Mode

Figure 3 illustrates the situation which is proposed to explain the HG. The rock,

R. J. L. Grard (ed.), Photon and Particle Interactions with Surfaces in Space, 545–556. All Rights Reserved
Copyright © 1973 by D. Reidel Publishing Company, Dordrecht-Holland

which is assumed to be an HG source in Figure 1, is partially shadowed by a small western ridge. An electric field is generated about this rock. The field levitates a dust grain which became electrically charged when the field was originally produced. The charged dust grain rises into the sunlight and scatters a portion of the sunlight to the completely shadowed Surveyor spacecraft. Light scattering is accomplished by large-sphere diffraction which produces the lobed intensity pattern $(I = I_0(a^2/r^2)x^2 \times$ $\times |J_1(x \sin\theta)/x \sin\theta|^2$, $J_1 \sim$ 1st order Bessel function and $x = 2\pi a/\lambda)$. The angular width

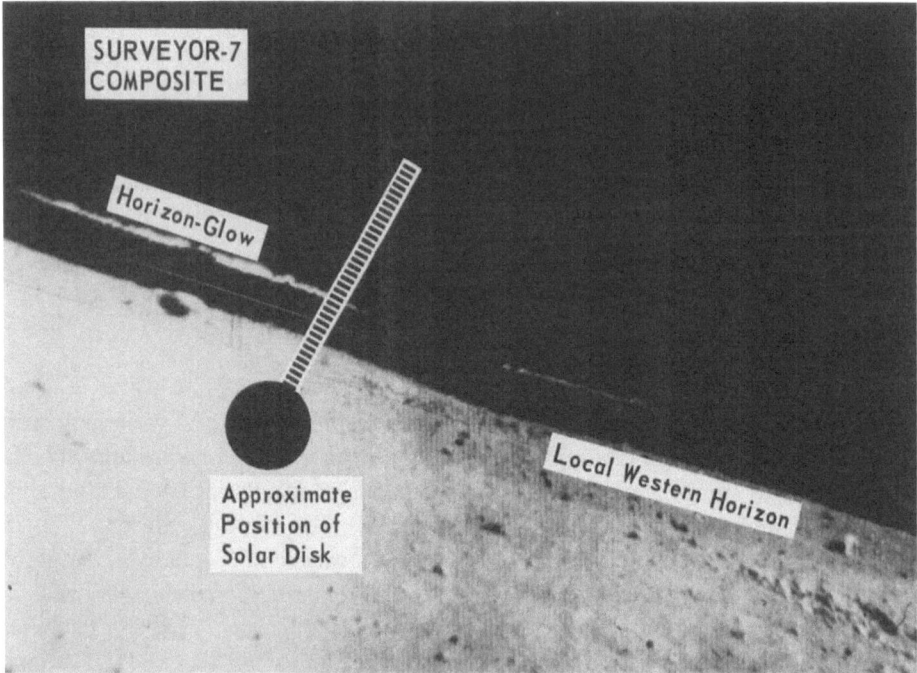

Fig. 1.   Composite of a morning and evening photograph of the local western horizon from the Surveyor-7 landing site on the northwestern rim of Tycho. Note the horizon rock just to the left of the sunset line.

of the diffraction pattern is $\theta_0 = \arcsin(3.832\lambda/2\pi a)$ where for the horizon-glow $\theta_0 \simeq 3°$, $a \sim$ radius of dust grain, and $\lambda \sim$ wavelength of the scattered light (Van de Hulst, 1957). Dust grains of $a \simeq 5$ to $6\,\mu$ in radius will produce this diffraction pattern for $\lambda \simeq 0.4$ to $0.5\,\mu$ light to which the Surveyor television systems were sensitive. Levitation can also occur across the ridge.

The levitating electric field is generated in a two-step process illustrated in Figure 4. Photoelectrons of 500 to 1500 eV are emitted from this sunlit surface due to the absorption of soft solar X-rays with wavelengths less than 25 Å. Escape of these energetic electrons will continue until sufficient positive charge is left on the rock to force the photoelectrons to return to the vicinity of the rock. A portion of the returning photoelectrons will accrete on the completely dark area. A potential difference of

500 to 1500 V will be generated across the cm-scale boundary between the sunlit and completely dark area. Computer modeling of this situation reveals the fractional accretion rate to be $f(\Delta U = U_\infty) \simeq 0.05$, $f(\Delta U = 1.1\ U_\infty) \simeq 0.01$, and $f(\Delta U \simeq \simeq 1.2\ U_\infty) \simeq 0$, where eU_∞ is the energy of the X-ray photons.

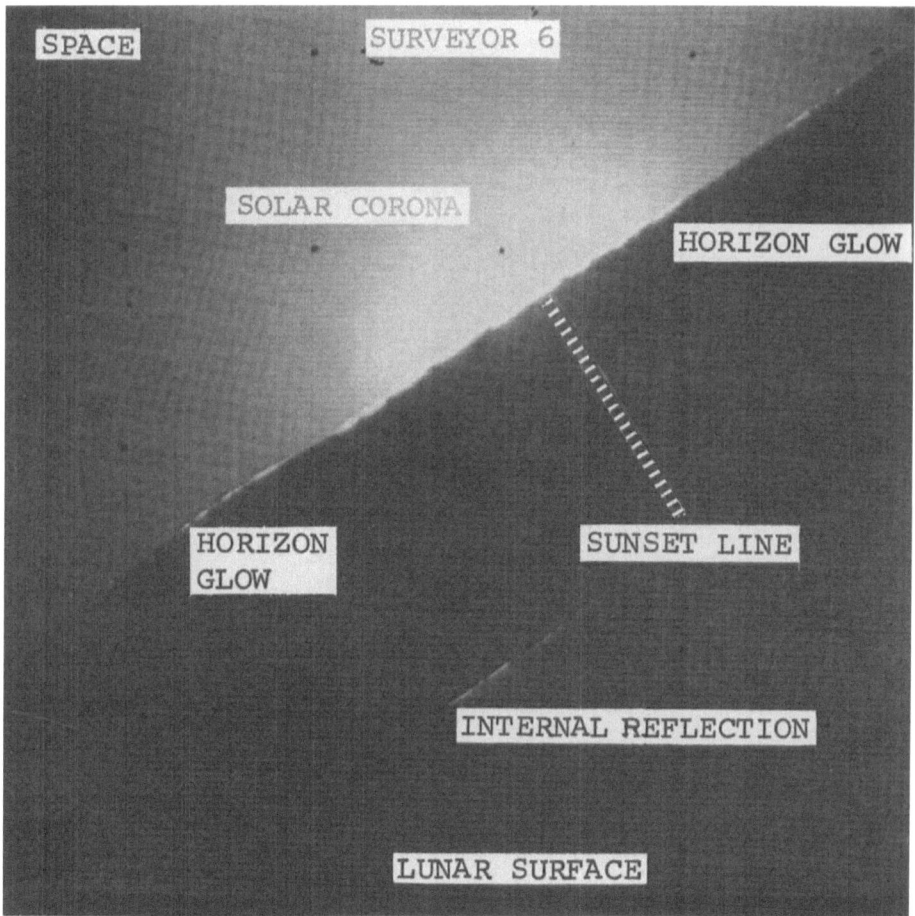

Fig. 2. This Surveyor-6 photograph, taken approximately 1 h after local sunset, clearly demonstrates that HG is not an early manifestation of the solar corona. HG is intrinsically much brighter than the corona in the early stages and is linear rather than elliptically shaped as is the corona.

Figure 5 illustrates the complex pattern of multipolar electric fields which exist about the rock. A two component electric field surrounds the rock. The highly positively and negatively charged areas produce the intense multipole electric field ($E \gtrsim 1000$ V cm^{-1}). Dust grains residing in these areas are electrically charged. The multipole field levitates the charged dust particles. The net positive charge on the rock produces a weak, pervasive electric field. This monopole field extends out a meter or so, that is one debye length, and terminates on remote portions of the lunar soil or in the solar wind. The meter-sized volume can be electrically neutral overall.

Figure 6 illustrates the two means by which the electric field may be prevented from developing or discharging. Rapid ($\lesssim$ hours) multiple discharge is unlikely because of the extremely low electrical conductivity of lunar dust and rocks ($\eta \lesssim 10^{-17} \, \Omega^{-1} \, cm^{-1}$)

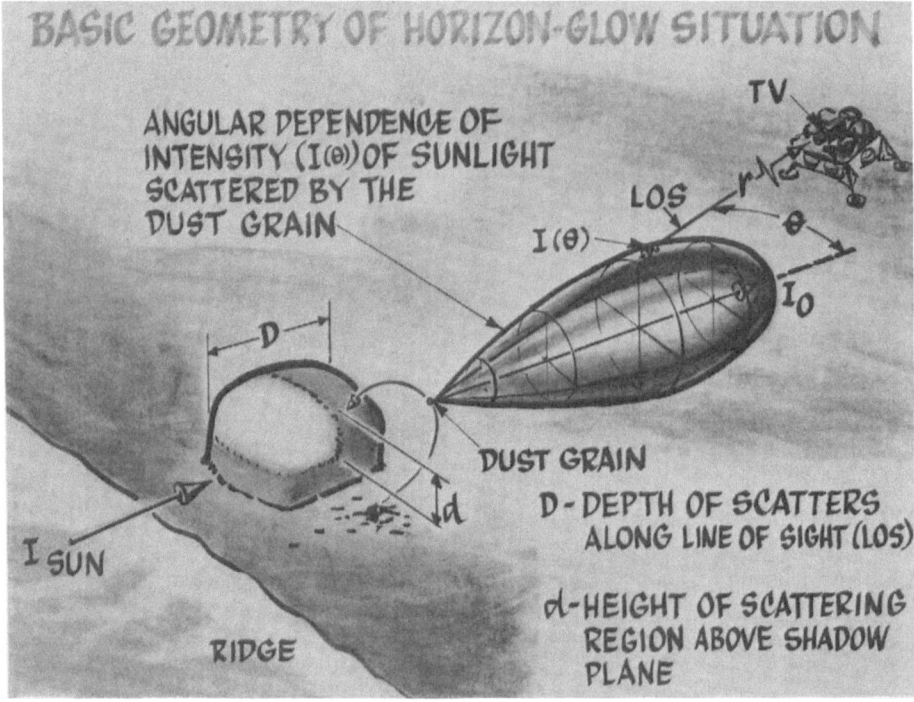

Fig. 3.   This schematic depicts the relative position of the Surveyor-7 spacecraft and the western source region of HG associated with the large rock left of center in Figure 1. The estimates of column densities of the levitated grains assumes $D \simeq d$.

required for this process. This low conductivity will result from the extremely dry state of the rocks and the existence of numerous interfaces and boundaries between grains of different chemical composition (Parkhomenko, 1967). There has been a steady decrease in the experimentally measured conductivity of lunar material as more care has been used in eliminating terrestrial contamination. Olhoeft *et al.* (1972) have obtained conductivity values of dry (nitrogen gas storage) lunar dust which decrease exponentially with temperature and are $\eta \simeq 10^{-17}$ to $10^{-16} \, \Omega^{-1} \, cm^{-1}$ at 0 °C. In this model the dark regions about the rock and the shadowed rock surface will have temperatures less than -100 °C which insures the required low conductivity (Shorthill, 1973).

The monopole field (left side of Figure 6) must be generated before the multipole electric field will develop. The monopole field would not occur if low energy photoelectrons could flow in from the dark region (Figure 7) just below the directly illuminated area. Such an electron flow will replace escaping 500 to 1500 eV photoelectrons. An electron influx to the overall region from the directly illuminated rock, westward

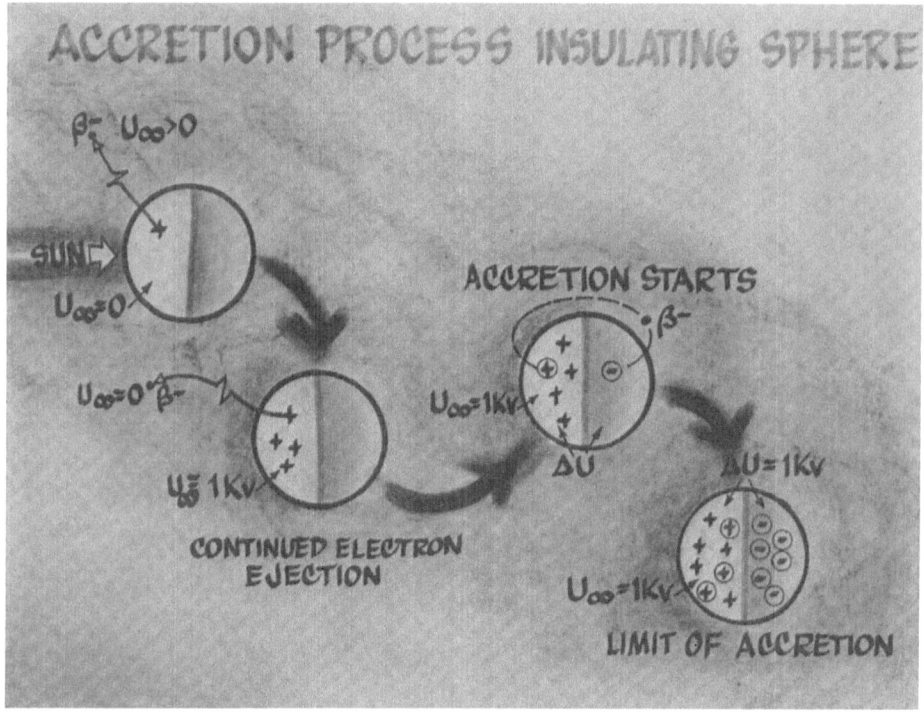

Fig. 4. Creation of a multipole electric field is a two stage process. This illustration assumes that only 1 keV photons are emitted from the Sun. Thus, monopole development and accretion are separable processes. In fact the continuous energy distribution of photons and photoelectrons means the two processes will occur simultaneously with $U_\infty \simeq \Delta U$ photons dominating the accretion current at any given instant.

sunlit areas, or the solar wind would complete the overall current loop in this neutralizing case.

Neutralizing photoelectrons are evoked in greatest number by solar ultraviolet photons with energies somewhat greater than the photoelectric work function (w) of the lunar material. These photons are scattered from the sunlit surface (Figure 8) to the dark surface in front of the rock, and then back to the sub-illuminated region. Less than 5×10^{-7} of the directly incident photons reach the dark region of the rock The attenuation is calculated assuming lambert scattering at both surfaces, the top quarter of a cylindrical rock being illuminated and conservatively using albedos of 0.05 for the dust and rock (Lebedinsky *et al.*, 1968). Using the data of Feuerbacher *et al.* (1972) we get a maximum value of the ratio for discharge to escape current for the Surveyor-7 case of

$$R = (\text{Attenuation}) \left[\frac{Y(12\,\text{eV})}{Y(500\,\text{eV})} \frac{\Phi(>12\,\text{eV})}{\Phi(>500\,\text{eV})} \right] \lesssim 3 \times 10^{-3},$$

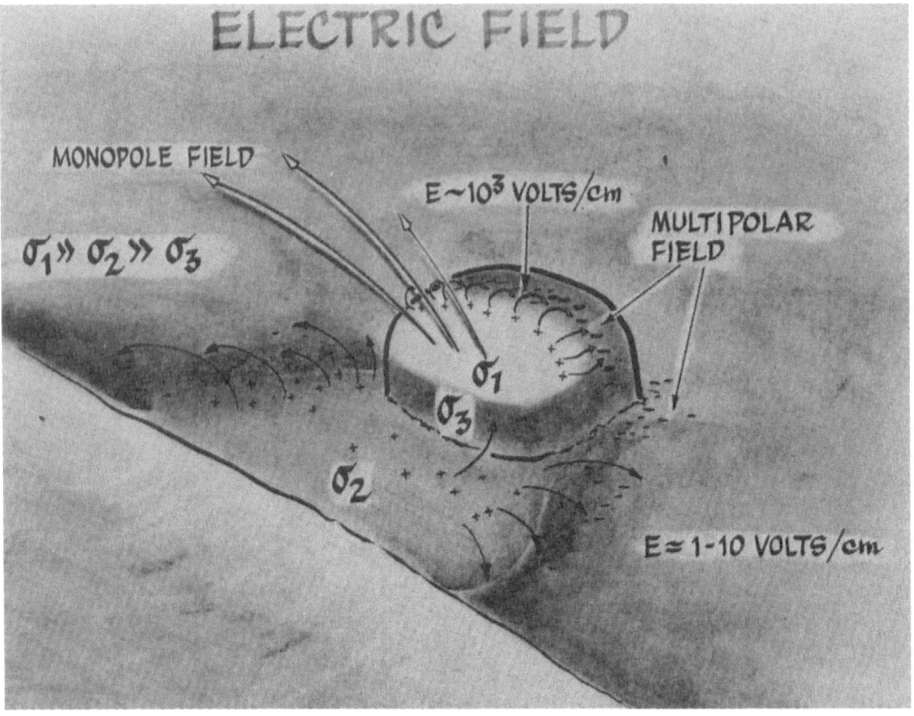

Fig. 5. Isolated, and partially illuminated rocks will be the center of a weak monopole electric field. Much stronger multipole fields will extend across adjacent light/dark boundaries and will drive the dust motion. Similar fields extend from the bright to dark sides of the foreground ridge. The sigmas indicate relative surface charge densities.

where $Y \sim$ photoelectron yield and $\Phi(>\varepsilon) \sim$ photon flux for energies greater than $\varepsilon(\text{eV})$ and $Y(12\,\text{eV}) = 5 \times 10^{-2}$, $Y(500\,\text{eV}) = 10^{-2}$, $\Phi(>12\,\text{eV}) = 5.6 \times 10^{10}$ cm^{-2} s^{-1}, and $\Phi(>500\,\text{eV}) = 5 \times 10^{7}$ cm^{-2} s^{-1}. Monopole discharge was not a problem for the Surveyor-7 situation. Thus, for normal solar X-ray conditions monopole discharge should not be a problem.

3. X-Ray Levels and HG Occurrence

There is sufficient uncertainty in our knowledge of photoelectron work functions (w), yields, and lunar albedo in the hard ultraviolet and soft X-ray to preclude a precise determination of R. However, these quantities should be insensitive to the absolute values of the photon flux (i.e. $\Phi(\simeq w)$ and (soft X-rays)). It is reasonable to see if the detection, or lack of detection, of HG by the Surveyors 1, 5, 6, and 7 corresponds in some way to the solar X-ray output.

As previously noted, the intrinsic brightness of the Surveyor 7 and 6 HG's were nearly equal. The contrast arose because the Surveyor-6 horizon was 10 times further

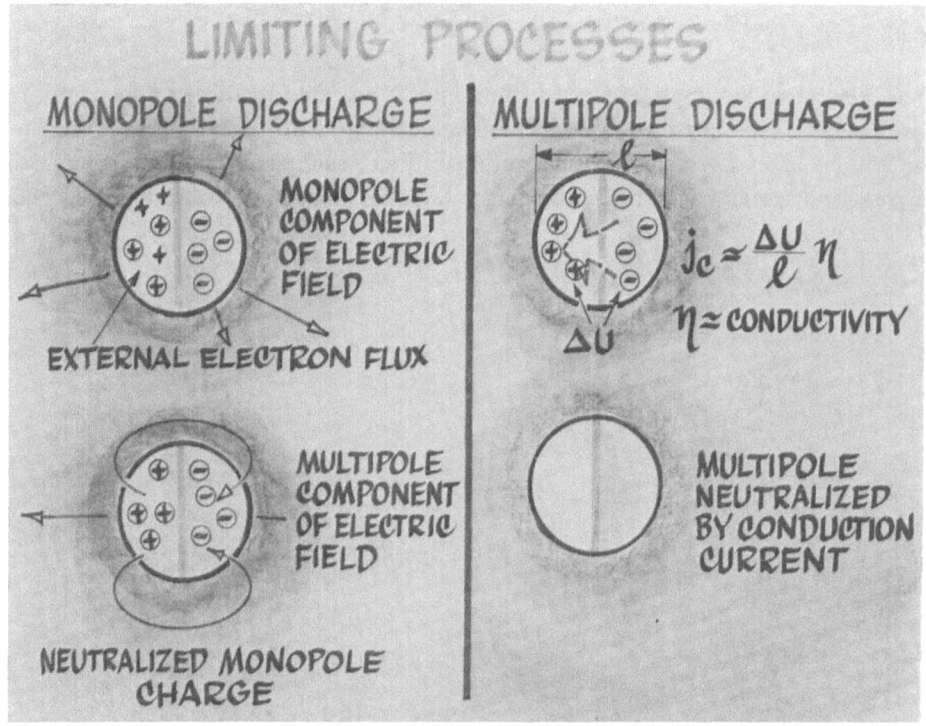

Fig. 6. Two processes limit the creation and intensity of the multipole field. A source of external electrons may replace the charging current and prevent the growth of the monopole field or eliminate the monopole field (unbalance positive charge is uncircled). Secondly, conduction current, or a source of positive ions on the sunward side, can discharge the multipole field. The extremely granular and diverse mineralogical nature and complete lack of water in the lunar rocks and dust indicates that conduction currents are negligible. However, strong internal polarization fields can be expected.

away than that of Surveyor-7 (Table I). However, Surveyors 6, 5, and 1 all landed in maria regions in which the local horizon was $1\frac{1}{2}$–2 km distance (approximately the spherical Moon horizon). One must examine the second generation negatives from Surveyor-5 in order to discern the HG (National Space Science Data Center). It can not be resolved in the photos presented in the mission reports. The Surveyor-5 HG is present but less contrasty than the Surveyor-6 example. Surveyor-1 did not detect a linear HG but may have detected a very short-lived spot of local HG immediately at sunset (Norton et al., 1966, 1967; Rennilson, 1968).

TABLE I

8–25 Å X-ray intensities ($\times 10^7$ photons cm^{-2} s^{-1})

Surveyor	1	5	6	7
Glow [a]	none	same	same	very bright
Full disk output	1.5	3	3	5
East limb output	<.1	0.2	0.5	1.5

[a]Relative to the solar corona.

It is necessary to consider the distribution of X-ray sources over the solar disk for two reasons. First, there was considerable reluctance on the part of Surveyor mission operations to expose the television optics to direct sunlight for fear of ruining the system and certainly increasing the noise level for post-sunset observations. Thus,

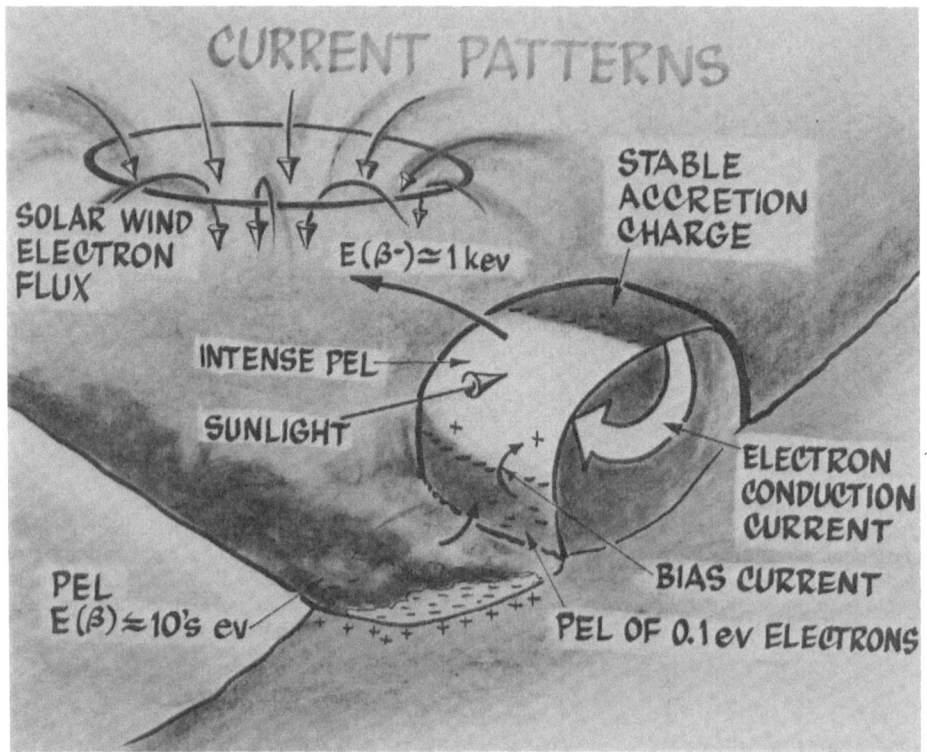

Fig. 7.   Light scattered from the directly illuminated surface (Figure 8) to the foreground of the rock and then back to the rock will evoke a photoelectron layer between the soil line and the illuminated surface. The resultant bias current of 0.1 eV photoelectrons must be less than the escape current in order for a monopole charge to develop. The 10 eV electrons in the foreground will have only temporary direct access to the sunlit surface. Eventually the foreground electric field will increase in strength and retain them in the foreground region. The rock and multipole field are smaller than a debye length so collective plasma effects should not discharge the multipole. The HG region is shielded from the direct solar wind at sundown due to the aberration of the solar wind velocity by the orbital motion of the Moon about the Sun.

only the later stages of HG activity would have been seen which would correspond to a partially obscured solar disk in the HG source region. Secondly, the solar disk displays strong spacial contrast in the soft X-rays. Figure 9 (Vaiana *et al.*, 1968) is an X-ray photograph of the Sun in the 8–25 Å region. The bright spots are calcium plage regions. Thus ,the plage regions on the unobscured disk which are actually illumina-

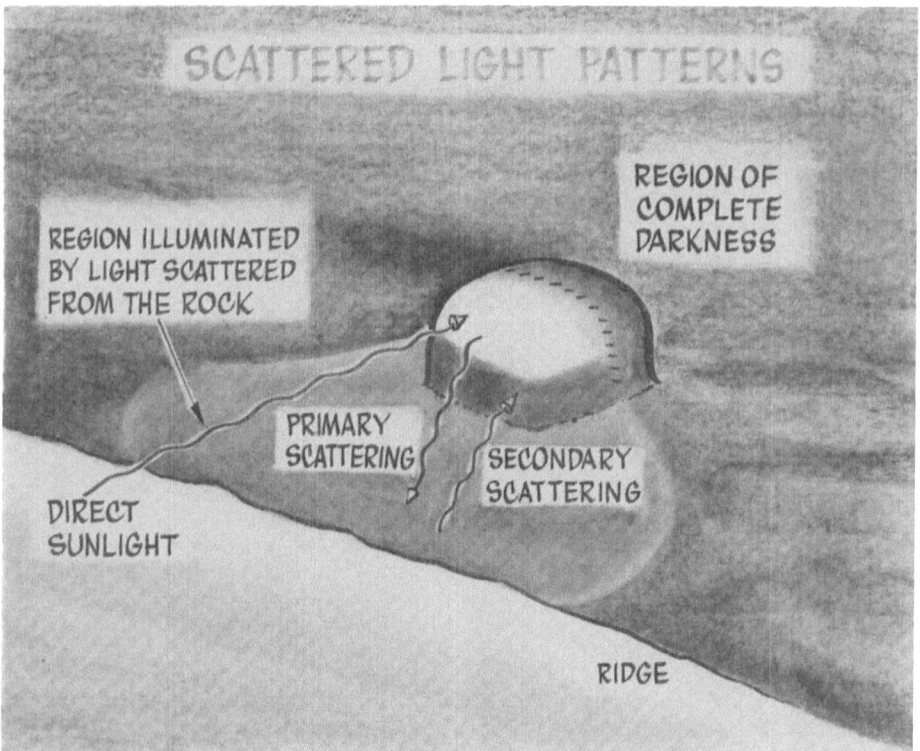

Fig. 8. These are the scatter paths of light that control the photoelectron leakage currents. It is assumed that X-rays are attenuated sufficiently that only negligible high energy photoelectrons are evoked in the foreground region.

ting the HG-source region during the 1 h period following Surveyor sunset will determine the X-ray flux level.

Estimates were made of the X-ray output of the individual plage areas on the eastern and western halves of the solar disk for the sunset period for each Surveyor mission. Plage areas and 10-cm flux levels for each were obtained from the *Solar-Geophysical Data* series (U.S. Dept. of Commerce) and the McMath records. X-ray flux levels were inferred from available satellite data (Teske, 1970; Drake *et al.*, 1969; Wende, 1972). The X-ray output of each plage was proportioned to the product of 10-cm flux level and plage area to the sum of this product for all the plage areas on the disk (R. G. Teske, personal communication). The last line in Table I is the total X-ray emission from the eastern half of the solar disk. The Surveyor-1 case in which

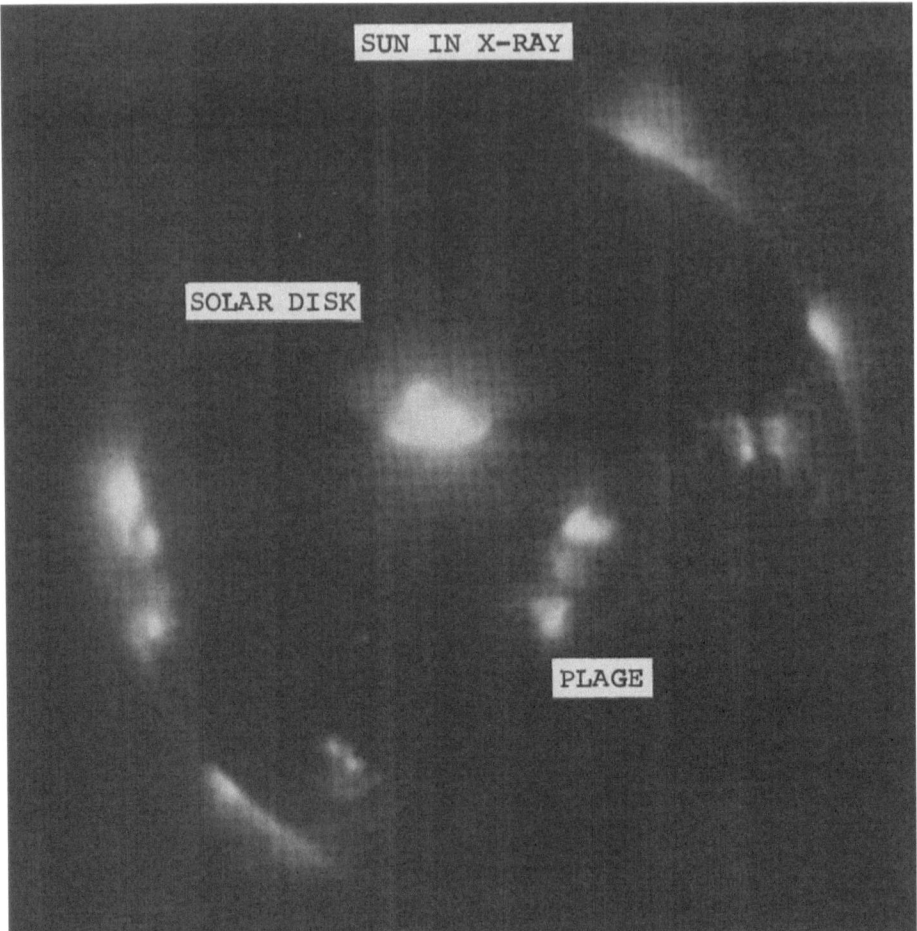

Fig. 9. This is a photograph of the sun obtained by a glancing incidence X-ray telescope in the 8–25 Å band (496–1550 eV). The bright sources are calcium plage regions and their associated solar flares. Notice that the flares can extend the order of 0.1 *R*(Sun) from the solar disk. At sunset, a lunar region could be illuminated by an extremely strong limb flare but not receive any hard or mid-ultra-violet radiation. Potential differences in excess of 10000 V might be temporarily generated.

HG was not clearly detected corresponds to both the minimum of whole disk output and a very low level of eastern limb output of the Sun. A word of caution is however necessary. Slight misfocusing of the Surveyor-1 camera system could have diffused the line image of a HG signal and prevented its detection. Surveyor-7 was the first flight in which a pre-mission decision was made to search for HG on the basis of the very clear Surveyor-6 observation.

4. Discussion

Detection of horizon-glow by three of the five Surveyor spacecraft constituted one of the major puzzles of lunar science which came to light in the early stages of direct lunar exploration. Electrostatic mixing and transport of lunar fines must be a very significant factor in properties of the soil such as radiation exposure, gas retention, rock burial and exposure, crater evolution, surface texture, and layering. A column density of the order of 50 grains cm^{-2} of 5 μ grains implies a rate of mass motion the order of 10^{-2} g cm^{-2} yr^{-1} which is nearly 10^{+7} times faster a turnover rate than that produced by micrometeorites (Shoemaker *et al.*, 1970).

All previous work dealing with electrostatic transport on the lunar surface have dealt with the fully illuminated regions (Gold, 1955, 1973; Walker, 1962; Singer and Walker, 1962; Heffner, 1965). Electric fields generated under such a condition will be very weak due to neutralizing effects of low energy photoelectrons. Significant physical processes which are driven by photoelectrically created electrostatic fields are very likely limited to the terminator band on the dark side and depend on the creation of electrical multipole fields in order to attain significant field strengths over macroscopic distances ($\gtrsim 1$ cm).

As a final point, the extremely low electrical conductivity of the lunar surface material produces a situation in which very small electrical current sources can generate large potential differences. Thus, the intensity of the hierarchy of multipole electric fields will generally be governed by the very hard ultraviolet and soft X-ray output of the sun. Mid-ultraviolet photons will generate the leakage currents which limit local potential differences.

Acknowledgment

This work was conducted at the Lunar Science Institute, which is operated by the Universities Space Research Association under Contract No. NSR 09-051-001 with the National Aeronautics and Space Administration. This paper constitutes the Lunar Science Institute Contribution No. 113.

References

Allen, L. Harold: 1968, *Surveyor Project Final Report Part 2. Science Results*, JPL Tech. Rep. 32-1265, p. 459.
Criswell, David R.: 1972, *Proc. 3 Lunar Sci. Conf., Suppl. 3, Geochim. Cosmochim. Acta*, **3**, 2671.
Drake, Jerry F., Sr., Gibson, O. S. B. Jean, and VanAllen, J. A.: 1969, *Solar Phys.* **10**, 433.
Feuerbacher, B., Anderegg, M., Fitton, B., Laude, L. D., Willis, R. F., and Grard, R. J. L.: 1972, *Revised Abstracts of Papers Presented at the 3 Lunar Sci. Conf.*, p. 254.
Gault, D. E., Adams, J. B., Collins, R. J., Kuiper, G. P., O'Keefe, J. A., Phinney, R. A., and Shoemaker, E. M.: 1968a, *Surveyor Project Final Report Part 2. Science Results*, JPL Tech. Rep. 32-1265, p. 401.
Gault, D. E., Adams, J. B., Collins, R. J., Kuiper, G. P., Masursky, H., O'Keefe, J. A., Phinney, R. A., and Shoemaker, E. M.: 1968b, *E. Surveyor 7 Mission Report, Part 2. Science Results*, JPL Tech. Rep. 32-1264, p. 308.
Gold, T.: 1955, *Monthly Notices Roy. Astron. Soc.* **115**, 585.
Gold, T.: 1973, this volume, p. 517.
Heffner, H.: 1965, N66-16171, Minn. Univ. Report of August 1965, 'TYCHO' meeting.
Lebedinsky, A. I., Krasnopolsky, V. A., and Aganina, M. V.: 1968, in A. Dollfus (ed.), *Moon and Planets II*, North-Holland Publ. Co., Amsterdam, p. 47.
Norton, R. H., Gunn, J. E., Livingston, W. C., Newkirk, G. A., and Zirin, H.: 1966, *Surveyor 1 Mission Report, Part 2: Scientific Data and Results*, JPL Tech. Rep. 32-1023, p. 87.

Norton, R. H., Gunn, J. E., Livingston, W. C., Newkirk, G. A., and Zirin, H.: 1967, *Surveyor V Mission Report, Part 2: Science Results*, JPL Tech. Rep. 32-1246, p. 115.

Olhoeft, G. R., Frisillo, A. L., and Strangway, D. W.: 1972, *The Apollo 15 Lunar Samples*, Lunar Science Institute, p. 575.

Parkhomenko, E. I.: 1967, *Electrical Properties of Rocks*, Plenum Press, New York.

Rennilson, J. J.: 1968, *Surveyor Project Final Report, Part 2: Science Results*, JPL Tech. Rep. 32-1265, p. 119.

Remilson, J. J. and Criswell, D. R.: 1973, 'Surveyor Observations of Lunar Horizon-Glow', submitted to *The Moon*.

Shoemaker, E. M., Batson, R. M., Holt, H. E., Morris, E. C., Rennilson, J. J., and Whitaker, E. A.: 1968, *Surveyor 7 Mission Report, Part 2: Science Results*, JPL Tech. Rep. 32-1264, p. 66.

Shoemaker, E. M., Hait, M. H., Swann, G. A., Schleicher, D. L., Schaber, G. G., Sutton, R. L., Pahlem, D. H., Goddard, E. N., and Waters, A. C.: 1970, *Proc. of Apollo 11 Lunar Sci. Conf.* 3. 2399.

Shorthill, Richard W.: 1973, *The Moon* 7, 22.

Singer, S. F. and Walker, E. H.: 1962, *Icarus* 1, 112.

Teske, Richard G.: 1970, NASA Report N70-41817.

Vaiana, G. S., Reidy, W. P., Zehnpfennig, T., Van Speybroeck, L., and Giaconi, R.: 1968, *Science* 161, 564.

Van de Hulst, H. C.: 1957, *Light Scattering by Small Particles*, John Wiley and Sons.

Walker, E. H.: 1962, *J. Geophys. Res.* 67(6), 2586.

Wende, Charles D.: 1972, *Solar Phys.* 22, 492.

DISCUSSION

Feuerbacher: The process you just described is limited to a very special geometry, so each sunset will deposit dust on certain places. Don't you expect the build-up of some kind of dunes from this process?

Criswell: This process will tend to smooth the lunar surface by moving particles to geometrically low areas which become negatively charged at very low sun angles. However, there may be a minimum roughness scale length (r) which will be maintained due to the finite angular extent of the Sun (θ). Interestingly, the Apollo 15 bistatic radar reveals that $r \simeq 30\mathrm{m} \simeq D/\tan\theta$, where $D \simeq$ horizon glow height (20–30 cm) and $\theta \simeq \frac{1}{2}°$ over maria regions. Highland regions where major topography controls secondary light scattering have $r < 30$ m at the few cm wavelengths of the bistatic radar (T. Howard, private communication).

Jones: What limits this levitation process to particle sizes in the region of that required to give the $3°$ diffraction angle that was mentioned?

Criswell: All particle sizes are levitated, but the larger particles are not levitated very far; whereas smaller particles are levitated to greater heights, but do not scatter sufficient light for their effect to be seen (scattering αr^4) by the Surveyor television system which responds only in the optical band $\simeq 4000$–6000 Å.

ELECTROSTATIC TRANSPORTATION OF DUST
ON THE MOON

T. GOLD and G. J. WILLIAMS

Center for Radiophysics and Space Research, Cornell University, Ithaca, N. Y. 14850, U.S.A.

Abstract. The study of detailed photography of the lunar surface makes clear that some surface transportation process has been active. Theory and laboratory experiments indicate that electrostatic effects resulting from secondary electron emission are the dominant cause of movement of small grains on the surface. The various electrostatic actions are discussed, and a host of unexpected phenomena are described that have turned up in the course of the laboratory experiments.

There is much evidence that the surface transportation of lunar soil has been a major process in the shaping of the lunar surface. Some such transportation has taken place as a result of meteoritic impacts, but the evidence is that another process has been active and indeed more important. This evidence is:

(1) Rocks on the lunar surface that have clearly been tossed by impact events into their present location are generally found now neatly partially embedded in the soil and with the protruding part virtually free from soil. There is no evidence in the surrounds of each rock of the large disturbance that must once have existed when the rock fell and came to rest in the soft, powdery soil. If micrometeoritic impacts had been responsible for the major movements of soil since that date it seems unavoidable to conclude that either the evidence of the impacts of the rocks would still show in the surrounding soil, or there would be a general blanketing, a 'snowed over' appearance of the rocks. Neither is the case.

(2) Core samples returned in the Apollo program show occasionally extremely sharp divisions into layers. In one case (Laul *et al.*, 1971) one layer contained 10^5 times as much of a trace element as layers above or below. Other stratification effects of particle size, albedo, etc. have also been noted. If micrometeorite and meteorite 'plowing' had been the main process these layers would have been mixed up. It is necessary to think that in addition to the meteorite plowing, which must of course have taken place, a deposition process had also occurred which had added layer after layer gently enough not to mix up the material and at a mean rate fast enough to make the meteorite impact mixing only a minor effect by comparison.

(3) Many features seen on the lunar surface, especially in Lunar Orbiter pictures, require some process of downhill transportation to have been active that settles the material in quite characteristic patterns. One is the 'elephant hide' pattern seen on all inclined slopes. This is a pattern of lines that are approximately at a constant angle to the line of steepest descent, criss-crossing and making up an unevenness on any one slope of a well-defined scale. Another effect, even more striking is the phenomenon of 'shoulders'. Frequently where a steep hillside meets the plain the junction line is characterized by a convex deposit of material not the concave, gentle changeover that

is common on the Earth. There are thousands of kilometers of such junction lines on the Moon, and the amounts of material involved are very great. Some downhill movement seems to take place on the slopes, that is usually impeded near the flat plains, for reasons that are not known, and a very steep junction line results. There are hundreds of Lunar Orbiter pictures showing this effect.

We have studied experimentally and theoretically a considerable variety of processes that might be responsible for these effects. We believe that electrostatic effects are the most likely cause of movement, and we have demonstrated in the laboratory a range of electrostatic transportation phenomena that appear to be very effective in lunar surface conditions.

Such processes can indeed transport the particles in the size range that makes up most of the lunar soil, and the rare large particles that are found may have been tossed by impacts to their present positions.

Photoelectron charging effects may play a role, but we have not succeeded in the laboratory to demonstrate any movement even with intense ultraviolet light. We do understand, however, and can demonstrate in the laboratory, quite a range of transportation effects that are caused by electron bombardment in the energy range of a few hundred volts. Such electrons are very much more effective in causing movement than any other bombardment that the Moon is subjected to, and we think it likely, but by no means certain, that they are the major cause of the observed effects. We are directed to this also by another consideration, namely the very great difference in the appearance between the front and the back of the Moon.

Any gravitational effect exerted by the Earth would show the symmetry that the back and the front are treated alike, while the limb would be treated differently; the effect would have mainly a quadrupole symmetry, not the symmetry of a displaced monopole. Yet it is the latter symmetry that is seen. Thermal or other radiation striking the Moon from the Earth is quite negligible, and it is difficult therefore to find an influence the Earth could have exerted to so change the two halves. Yet it is only the Earth that treats the two halves differently; from the position of the Sun or any other object the back and the front are of course treated alike.

The difference is too great to be attributed to chance. Also it is now known from a laser altimeter flown in an Apollo mission, that the back has indeed large depressions like the front, but that these have apparently not suffered a similar filling. It is only in the filling process that there appears to have been a great difference between the two halves. One influence of the Earth that could affect the surface transportation effects is the particle bombardment occurring when the Moon is in the geomagnetic tail. Characteristically energetic electrons are present there while they are not present in the undisturbed solar wind. We do not know yet whether these strike the front preferentially, but this information will come from observations of the electron flux on low orbit lunar satellites.

Charging of grains causes movement when there is a tendency for different grains to adopt very different charges, and when therefore very strong electric fields are set up on a very small scale. Two reasons exist for such differential charging.

Any insulator when bombarded with electrons of an energy near the first crossover point of the secondary emission curve, develops a great instability in the potential to which it is driven. A point initially above the cross-over point will start by emitting more electrons than come in and will therefore become more positive and therefore the incoming electrons will be even more energetic. Thus it will be even more above the crossover point, and the process continues until there is no point that is more positive that the secondary electrons can reach, so that they have to fall back on the emitting region. A point that starts out slightly below crossover will become more and more negative until the incoming electrons cannot reach it any more. In our laboratory case this means that particles are driven to adopt either anode or cathode potential, and that they are therefore separated in potential by the full beam voltage. For a typical value of a cross-over point this is in the neighborhood of 300 V.

Another reason for differential charging is the local topography. Grains that are on high spots have a larger angle to lose secondaries, while grains in hollows will catch the secondaries from exposed neighbors. Thus even if the secondary emission coefficients for smooth surfaces were exactly the same, grains would still be driven to different potentials.

The electric forces resulting make the grains move and expose a different set of surfaces and a different microtopography. A new situation is set up and the movement continues.

If two very different powders are mixed up then the movement is often seen to be impeded. We think this is due to the fact that one type of particle may then have the tendency to charge always in one sense and the other type in the other sense, due to a large difference in the secondary emission coefficient. Under these conditions chains of oppositely charged particles are set up and these have permanent stability; no movement changes the sense of charging of any one particle.

Many other effects are observed in laboratory studies. On the whole, however, powders are caused to migrate downhill by electron bombardment in the energy range 200–800 V. Actual lunar soil has been tested also and behaves similarly. The rate observed in terms of the electron fluxes used, would seem quite adequate to account for all the migration needed for filling of deep basins, i.e. several kilometers, in the age of the lunar material.

We believe that the seismic evidence greatly favors the explanation of a filling of basins with powder rather than with lava. The general existence of bedrock at a shallow depth is not compatible with the combined evidence of the acoustic and long-wave radar observations. Transportation phenomena must therefore be discussed, in order to account in any case for the smaller scale transportation evidence mentioned, but probably also for the large scale filling of basins. The electrostatic mechanism described here is the one we now consider the most effective.

Acknowledgements

This work was supported under NASA Grant NGL-33-010-005.

References

Gold, T.: 1971, *Proc. 2 Lunar Sci. Conf.* **3**, 2675, Mit Press.
Laul, J. C., Morgan, J. W., Ganapathy, R., and Anders, E.: 1971, *Proc. 2 Lunar Sci. Conf.* **3**, 1139, Mit Press.

DISCUSSION

Shawhan: Are the time scales for the changes in lunar features short enough to expect to see changes?

    Gold: No. Transport of material is only about 1 μ deposit per year.

    Knott: Do you have any evidence for an anisotropic particle flux in the magnetosheath which could cause the difference in structure between front and rear lunar surface?

    Gold: No, but we expect the movement of dust to be caused by both secondary emission and photo-effect, and the secondary emission effect is the only one we know that could affect the front and back differently.

    Willis: Do you think that your secondary electron emission charging is responsible for transport over distances comparable to mare regions or is it a much more localised phenomenon? Since the moon enters the plasmasheet for only a few days per month, it would suggest that the charged dust particles have a long life time (for long-range horizontal transport). Is this true?

    Gold: I do suppose that the transportation occurs over distances comparable with mare regions, but of course only very slowly. Any one grain is just occasionally lifted up and transported a small distance. Nothing needs to have a lifetime from one month to the next, but the suggestion is merely that each time the Moon is in the plasmasheet, some movement takes place. The mean rate required to account for all erosion and transportation shown now on the surface is only a deposition of 1 cm in 10 000 yr.

    Feuerbacher: I cannot see how this process works in the presence of a dense photoelectron sheath. To me it seems that this is a current-balance problem. As the in- and outgoing photocurrent is about 10 times the particle flux, all potential differences should easily be smoothed out.

    Gold: It is of course very critical whether the photoelectron current is ten times the electron flux or perhaps only twice that. Also the topography and the microtopography will cast shadows, and there will be substantial variations in the photoemission coefficient of different grains. All this results in potential differences between grains that can be of the same order as the energy of the majority of the photoelectrons.

The guiding of the return flux of photoelectrons towards the positively charged grains and away from the negative ones is a very ineffective process for the case that the space charge layer is very high compared with the size of the grains. A returning photoelectron approaches the surface to within a distance of the order of the size of the grains (a few microns) before there is any deflection. At that stage it has already the full velocity corresponding to its emission, having come back from a distance of the order of one meter. The amount of deflection that can be caused by the different charges of the grains will be proportional to the square of the height at which deflection occurs, and therefore of the grain size, while the deflection necessary to direct electrons to particular grains would be proportional to grain size. Thus large grains can be discharged readily while small ones cannot.

Potential differences of a few volts are sufficient to move grains, and those may occur despite the equalizing effects of the photoelectrons.

4. LECTURES IN CONCLUSION

PLASMA-VEHICLE INTERACTIONS IN SPACE– SOME ASPECTS ON PRESENT KNOWLEDGE AND FUTURE DEVELOPMENT

ULF FAHLESON

Dept. of Plasma Physics, Royal Institute of Technology, Stockholm, Sweden

Although plasma-vehicle interactions are of considerable interest in themselves the present emphasis on this complicated subject is largely motivated by practical necessity. Several contributions to this symposium have dealt with undesired or unexpected effects encountered on space vehicles. In some cases these effects have led to misinterpretation of scientific measurements, in other cases to complete failure of experiments. In a few cases there is even a suspicion that effects of this kind may have been the cause of major spacecraft malfunctions.

Scientists from what traditionally has been widely different fields have contributed their specialized knowledge and a general feeling after this very fruitful meeting is that we are close to a reasonable *qualitative* understanding of the most outstanding problems. On the other hand we are probably still very far from a detailed *quantitative* description.

Progress in this field is achieved by various kinds of efforts, theoretical as well as experimental.*

1. Theoretical Approach

The properties of plasma and radiation in space are fairly well known (Siscoe, p. 23) as well as the fundamental surface processes of interest (Lucas, p. 3). Thus it is in principle possible to attack the problems of plasma-vehicle interactions with purely theoretical methods. However, the complexity and multitude of processes involved offer tremendous difficulties. This may best be understood by a comparison with some other field of research, e.g. aerodynamics. The ordinary aerodynamic equations contain essentially four independent dimensionless numbers: the Reynolds' number, the Mach number, the Prandtl number and the specific heat ratio. In general, that field can certainly be described as being well understood, but the progress has taken decades of intense research and would certainly not have been possible without huge investments in laboratory facilities and experimental investigations. Still there is nothing like (and will never be) a unified theory describing the complete flow field around a body. Instead the development has led to a number of highly specialized theories, each one applicable only to a certain part or a certain phenomenon in a flow field, like shock theory, boundary layer flow, transsonic expansion, turbulent wake

---

* The papers referred to in these concluding remarks can be found in this volume at the page indicated after the author's name.

R. J. L. Grard (ed.), Photon and Particle Interactions with Surfaces in Space, 563–569. All Rights Reserved
Copyright © 1973 by D. Reidel Publishing Company, Dordrecht-Holland

flow etc. Further, the analytical or analytic-numerical approach is in general limited to two-dimensional (plane or axi-symmetric) problems.

Plasma-vehicle interactions in space are certainly not simpler than the corresponding aerodynamic problems. Table I gives a list of non-dimensional numbers used in different kinds of plasma-vehicle interaction problems. Although not all of these numbers are independent and not all are simultaneously important, it is obvious that a multitude of interesting combinations are possible leading to a variety of different solutions, each one valid for some very special situation. What is still more serious is that even for a vehicle of simplest possible shape almost all situations of practical importance involve a 3-dimensional problem. The cause for this is the flow velocity in combination with the magnetic field (important at low altitudes) or the anisotropic photo emission (important at high altitudes). Theoretical models (Schröder, p. 51; Walker, p. 73, Tunaley and Jones, p. 59; Weil *et al.*, p. 101) may be produced for a large number of special cases – and should be so. Although few of them will be directly applicable to typical real situations, they are essential for our understanding, also of more complicated situations. No single model will ever emerge as the universal solution to the plasma-vehicle interaction problem. In the best case we will eventually, like in aerodynamics, get a number of different approximations, each one able to describe some special feature of a special situation. The support of extensive laboratory and space experiments will be indispensable for such a development. Although comparison of observations and theoretical models (Samir, p. 193; Cauffman, p. 153) can not be expected to yield detailed agreement, the importance of such comparison can hardly be stressed too much.

A method which offers great possibilities for the future is numerical simulation (Soop, p. 127). The method is simple and straight-forward, but requires huge computer facilities to keep track of thousands of particles during thousands of computation steps. Present computers suffice only for two-dimensional problems but new generations may be faster and have larger memory capacities. An advantage with numerical simulation is that it may be applied to quite complicated, fairly realistic flow situations and that also unstable flow situations can be treated.

A vivid memory from this symposium is the demonstration of the distortion of an AC electric field inside a cavity (Kaiser and Kendall, p. 91). The modification due to the Debye sphere around a vehicle should be much less serious, but close to the vehicle serious disturbances should be expected from wave reflexions from the body and from the unsymmetric time-variable photoelectron cloud. Further investigations of these effects are urgently needed and a likely result may be that short dipole measurements in the vicinity of a space vehicle are very seriously distorted.

2. Experimental Evidence

A number of reports are available of anomalous behaviour of satellite experiments (Wrenn and Heikkila, p. 221; Norman and Freeman, p. 231; Montgomery *et al.*, p. 247; DeForest, p. 263; Fredricks and Scarf, p. 277), but it is often difficult

TABLE I

Non-dimensional numbers that often are of fundamental importance:

$$L/\lambda_D,\ L/r_{ge},\ v/v_i,\ i_{ph}/i_{e0},\ i_{Hee}/i_{e0},\ i_{sec}/i_{e0},\ e\Phi/kT_{e0},\ \frac{kT_{e0}\sigma_s}{e i_{Hee}L^2},\ T_{Hee}/T_{e0},\ T_{ph}/T_{e0},\ T_{sec}/T_{e0},$$

angle $(\bar{v}, \bar{S})$, angle $(\bar{v}, \bar{B})$, angle $(\bar{S}, \bar{B})$.

Non-dimensional numbers that usually are of less importance:

$$L/\lambda_{mtp},\ L/r_{gi},\ v/v_e,\ m_i/m_e,\ T_i/T_{e0},\ \frac{eEL}{kT_{e0}},\ \gamma,\ M_A,\ R_M$$

where:

L	= body dimension
λ_D	= Debye length in plasma
r_{ge}, r_{gi}	= electron and ion gyro radii
$\bar{v}$	= flow velocity
v_e, v_i	= electron and ion thermal velocities
i_{ph}	= unlimited photo emission current density
i_{e0}	= random electron current density in plasma
i_{Hee}	= current density due to high energy electrons
i_{sec}	= current density due to secondary electrons
Φ	= body potential relative to plasma
σ_s	= body surface conductivity
T_{e0}	= thermal electron temperature
T_i	= ion temperature
T_{Hee}	= characteristic energy of energetic electrons
T_{sec}	= characteristic energy of secondary electrons
T_{ph}	= characteristic energy of photo electrons
k	= Boltzmann constant
e	= electronic charge
$\bar{S}$	= direction to sun
$\bar{B}$	= direction of magnetic field
λ_{mtp}	= mean free path
m_e, m_i	= electron and ion masses
E	= large-scale electric field in plasma
γ	= specific heat ratio
M_A	= Alfvén number
R_M	= Magnetic Reynolds' number

to find out afterwards what was the cause of a specific event. The reason for this is usually insufficient information about the actual conditions. Although we have detailed knowledge about spacecraft design, which materials were used etc., we usually do not know how they were treated prior to launch or how they may have changed their properties in the space environment. We know also fairly well *typical* plasma parameters in space, but usually we do not know these parameters in the *special* situation when some unexpected effect appeared.

Evidence is accumulating now, however, that the most serious difficulties are connected with electrical charging of space vehicles to high potentials. This has been found to happen especially often on satellites in the outer magnetosphere when the photo emission vanishes in the Earth's shadow.

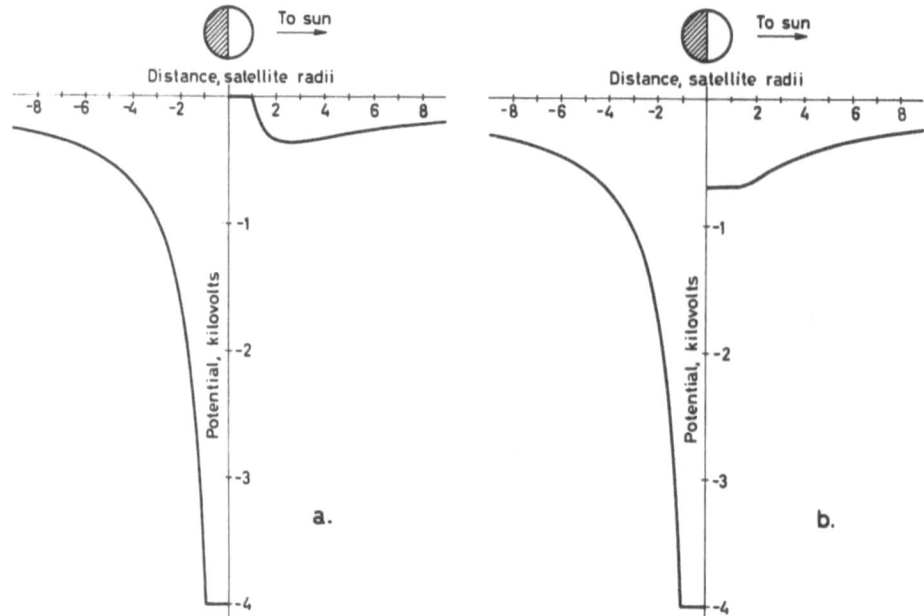

Fig. 1. Potential along Sun-spacecraft line due to charges on spacecraft surface. Case a: dark side at −4 kV, sunlit side at zero potential. Case b: dark side at −4 kV, no retarding potential outside center of sunlit side (so photoelectrons can escape).

Many details of the vehicle charge-up are still little known and it may be appropriate to add some personal thoughts on this subject. In the presence of a high energy electron flux in the outer magnetosphere the dark side of a satellite with highly insulating surface will usually go to high negative potentials. If the high energy electron flux is small compared to the photo emission current of the satellite material (which is usually the case) it might be expected that the illuminated side should still assume a positive potential of a few volts or at least stay close to zero potential. However, as shown in Figure 1 this will not be the case. Assuming a long plasma Debye length and accepting the result of Soop (p. 127) that the charges of the electron cloud usually influence the potential outside a vehicle little compared to the charges on the vehicle itself, we get a first approximation of the potential around the satellite by solving Laplace's equation. Tne result is that, due to the charges on the dark backside surface, a potential barrier will exist all around the satellite. Assuming − 4 kV potential on the backside, the barrier on the front side will be − 360 V (Figure 1a), thus effectively preventing photoelecton escape. As a result the front side will also go negative until it can get rid of a sufficient number of photoelectrons. When going negative the front side will push the barrier further negative and in the specific example the front side will not come sufficiently close to the barrier potential until at − 680 V (Figure 1b). This means that not only the backside but the entire satellite will be driven highly negative due to an insulating surface. Even if only 10% of the surface were insulating we might easily get a situation where every part of the satellite were

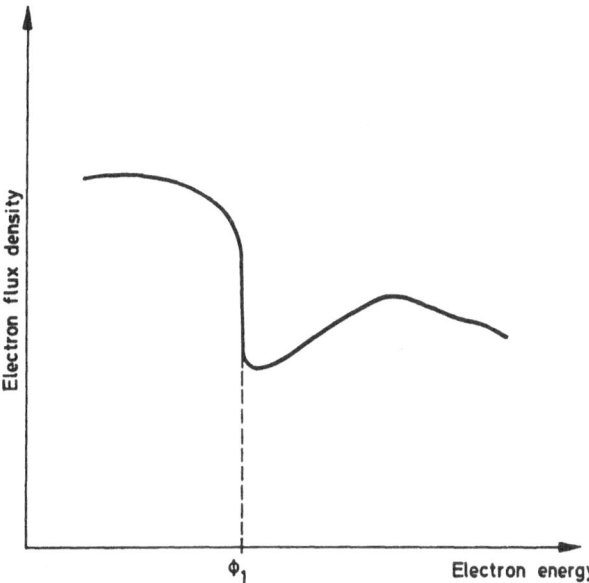

Fig. 2. Imaginable shape of electron spectrum recorded on a satellite in the Earth's shadow. Electrons below Φ_1 are mainly reflected secondaries, electrons above Φ_1 originate from outside but may be retarded by an unknown potential. Observe that the secondary electron cutoff at Φ_1 does not prove that the spacecraft is positive.

driven to at least minus 100 V. This would be sufficient to ruin all attempts to measure low energy plasma or DC electric fields from the satellite. If the same satellite were also rotating such that a varying fraction of the insulated surface were exposed to sunlight, violent potential fluctuations would result, ruining also AC electric field measurements and at least certain kinds of particle anisotropy measurements.

Next let us look at a vehicle in the Earth's shadow. According to Grard *et al.* (p. 163) the most common satellite materials have secondary emission yields below unity, so large parts of the surface may charge up to some kilovolts of negative potential. Even if this happened to only a few percent of the surface, we should expect essentially the same situation as that described above with a potential barrier and a generally negative satellite. An electron detector would see a retarded electron spectrum, but if situated close to a part with secondary yield exceeding one it would also see secondary electrons reflected from the local potential barrier that has to cover that part of the surface. The recorded spectrum can be imagined to look like Figure 2. For this reason it might be dangerous to refer to the observation of returning secondaries as a proof of a positive satellite potential. Furthermore, the electrons recorded at energies just above the secondary electron cutoff may not necessarily be thermal ambient electrons, since they may have been retarded by the unknown potential of the barrier, which may be anything from a few volts to a few kilovolts. Simultaneous proton measurements may help to clarify the situation, but if the barrier is only a few tens of volts it may be hard to detect.

3. Electrical Cleanliness

It is evident that many present satellites offer an akward environment for many types of experiments, in extreme cases so akward that it may cause physical damage to sensitive equipment or even failure of entire satellite systems by spark generation (Fredricks and Scarf, p. 277).

A general agreement at this symposium has been that we know fairly well how these malfunctions can be avoided (Rosenbauer, p. 139; De Forest, p. 263), and that we have to require that every possible measure should be taken to eliminate them from future satellites. A starting point for such measures may be the definition of one or several degrees of *electrical cleanliness* involving requirements on a minimum satellite surface conductivity (Grard *et al.*, p. 163) or a maximum allowed fraction of the area that may be insulated. Other requirements may be that the emission properties of the satellite should be independent of spin angle and that no parts at differing potentials should be exposed to the plasma. For electric field and plasma experiments, an active command activated satellite potential regulating system (e.g. along the lines demonstrated here by Polychronopulos and Goodall, p. 309) would be highly desirable as well as arrangements to keep photo emission at some suitable level. A minimum requirement for such experiments could be some facilities for measurements of the satellite potential. Probably some of these requirements will have to be solved by the experimenters themselves, but it should be made understood that most of these measures are strictly necessary to guarantee successful operation.
Very important is also the production of more homogeneous and stable surfaces with prescribed resistivity and emission properties, as well as improved knowledge of how surfaces change in the space environment.

We have learned that considerable progress is possible in this field (Bujor, p. 323, Anderegg *et al.*, p. 331; Köstlin and Atzei, p. 333) and that new and improved materials may become available as soon as the necessity of having them has been generally recognized.

4. Boom-Mounted Experiments

Whichever measures are taken a space vehicle will never offer a *perfect* environment. Certain especially sensitive sensors like electric field probes and low energy plasma detectors will have to be mounted on booms sufficiently far from the vehicle. Such booms offer special technical difficulties.

First of all, the boom surface can not be a good insulator since this would involve the risk of high negative potentials on its dark side. Secondly, the boom can neither be a good conductor all the way since it would then propagate potentials and disturbances to the sensors. The best solution may be to bias the boom to the sensor potential or some steady potential. A short insulating section that may be necessary should be as far away as possible from the sensor. Another solution may be a boom coating sufficiently resistive to insulate the sensor from the vehicle, but sufficiently conductive to prevent charge accumulation at its shadow side. A bias system for the sensors, special guard rings or deflexion electrodes may be used to prevent pickup of photo electrons originating from the vehicle.

In this way it should be possible to fly even extremely sensitive experiments on booms of limited length. In general it should be sufficient to have a boom length long compared to the vehicle dimensions, but not necessarily long compared to the plasma Debye length. On the other hand observations and simulations have shown (Cauffman, p. 153; Soop, p. 127) that the photoelectrons do not screen off potential disturbances from the satellite even if their Debye length is quite short. Instead they are in general one of the most serious sources of disturbance. Interesting possibilities would be to replace the photo emission by the emission of an electron beam from the spacecraft or to guide the photo electrons be special electrodes or grids so they leave in suitable directions with a minimum of interference.

5. Summary

We are in the beginning of a phase of development of various methods of electrical control of the spacecraft environment. Undoubtedly very much can be gained by further progress in this field and action is urgently needed to utilize this knowledge as early as possible on forthcoming space probes and satellites.

DISCUSSION

Polychronopulos: The picture of potential distribution around an illuminated spacecraft with insulated surface appears to be inconsistent with the picture presented by previous experimental papers.

Fahleson: It is documented that in certain situations parts of satellites have charged up to negative potentials of several kilovolts. In a situation where this occurs, the potential barrier around the satellite and the charge-up of the entire vehicle as described by me seems to follow as an inevitable consequence of the fundamental equations. As far as I know no presently reported observations rule out the possibility of space-craft charging at least qualitatively along the lines indicated here. Quantitatively, I may here have overestimated the charging, however, since certain limiting factors like ion focusing, secondary emission and capacitive effects for a spinning vehicle have been neglected in this first investigation.

Kaiser: It would be especially helpful at this time to focus the discussion around the GEOS satellite. I would therefore like to hear from ESTEC colleagues concerning the remedies that it will be realistic to apply to GEOS so that problems such as charging of the vehicle etc. may be minimised.

Knott: The only experiment I know of where proton and electrons have been measured simultaneously has been reported by DeForest and from this, strong electrostatic charging became evident.

Fredricks: Is the resistivity of a thin conducting film the same in directions parallel to the film and across its thickness? Do the Philips-ESTEC measurements measure resistance between points separated on the film surface, such as between dark and sunlit sides of the satellites?

Knott: The resistances measured are resistances between two opposite sides of a square of any size and a given thickness.

Gold: Is it known that there really is no significant photoemission from areas out of the sunlight? In particular the Lα scattered back in the solar system may be significant and perhaps needs to be measured.

Fahleson: Photo-emission due to back-scattered Lα radiation is certainly small compared to that due to direct radiation. In many cases it is certainly small, also compared to the plasma electron current in the magnetosheath. A careful investigation is necessary, however, before it can be stated whether this is generally true or not.

PARTICLE INTERACTIONS WITH CELESTIAL OBJECTS –
CONCLUDING REMARKS

T. GOLD

Center for Radiophysics and Space Research, Cornell University, Ithaca, N. Y. 15, U.S.A.

I intend this summary just to be a reminder of some of the points that came up in our meetings that may deserve further thought and discussion. When one hears a particular item for the first time one's response is not the same as after a little more time and perhaps a little more thought. For that reason it is useful to be reminded again now and to open another round of discussion, and that is the purpose of my present summary.

The meeting started out very nicely I thought, in the same way in which a good school starts out when a new class of pupils comes in, by essentially teaching them all that they are supposed to know very quickly as so to be sure there is no huge gap in their knowledge. We had two introductory talks that taught this new class the subject matter in pretty short order. My feeling is that, on listening to the conversations, my own included, we are still not very strong on certain aspects of theoretical understanding. We still have constantly trouble, all of us, to understand what is the situation above a surface in the photoelectron plasma; how far can one treat this like a Debye length situation and how far not? It seemed to me, in listening to all this, that we would all do well now to study the details, which have been known for many years, of the ordinary engineering discussion of the surroundings of a cathode. One understands the height of the charge layer very well for that case. There is a lot in that discussion that would have been useful to us and I think we'll have to look into that a little. The general feeling I had is that there could be still very large errors here and there, in discussions of charging of surfaces and in discussions of the interaction of photoelectrons, incoming and secondary electrons and plasma currents of one kind or another. One is still treading on very uncertain ground there, and quite large errors could still be made. A deepening of understanding seems to be needed in that regard. It was a very good thing at this conference, I thought, to combine the technology and the science: the discussion by engineers concerned with understanding the satellite situation in detail combined with that of people who are concerned with interstellar grains or the surface of the Moon or the surface of Io or whatever. A new outlook that has to be brought into areas of theoretical astronomy or space physics is that of the surface electrical property domain – that is a very important matter. In the case of the Moon discussion, for example, it is something that has been very sadly lacking. You can go to lunar conferences and find hardly any mention of the various physical processes that might occur in a vacuum, in a plasma, in an electron bombardment or in photon bombardment on the Moon, while you might hear a great deal of all the discussion of Earth-like processes that might have been going on on the Moon.

R. J. L. Grard (ed.), Photon and Particle Interactions with Surfaces in Space, 571–576. All Rights Reserved
Copyright © 1973 by D. Reidel Publishing Company, Dordrecht-Holland

Those that have no counterpart on the Earth are very rarely discussed, and it is refreshing to me to come to a conference here where there is a mixture of people who are really principally interested in this kind of surface physics.

We heard then – let me turn to the more detailed points – discussions of the potential of various natural bodies, the surface of the Moon chiefly, and the way in which this potential is affected. The circumstances to which the Moon is exposed were discussed, in the different regions of the plasma through which it goes, as were the trajectories of particles released from the Moon, the lunar ionosphere as it was called this very fast moving temporary atmosphere that the Moon will generate around itself. This is in detail of course very similar to the discussion of the space vehicle charging effects. We heard the story also of the electrical properties that some other bodies in the solar system might exhibit, in particular Io, and its electrical effects which have been observed in the strong magnetic field of Jupiter. I was sorry that other solar system bodies were not discussed, just the Moon and Io. Lots of other bodies in the solar system may have interesting and strange electrical effects taking place on their surfaces. There are some satellites of the major planets which are quite remarkable in certain respects: there is one that is half black and half white. There are enormous differences in all kinds of observed surface properties between the different satellites. Therefore it would be quite interesting to discuss the electrical environment that some of these satellites must be living in. However, we did have Io, but I think we should have had some of the others.

Then we had a discussion of the grains in space and the electrical properties surrounding them and that seemed to me to be the beginning or a large field of study. What happens electrically to grains in space is probably a terribly important thing for the understanding of a whole lot of strange effects noted in connection with grains; among those we have the fine detail of their distribution in space; of their alignment causing polarization; of the details of their optical scattering and absorption laws which are known, which are in some cases very unusual, and that may have to be understood in terms of surface forces that cause the growth of the grains, that shape them, that cause their accumulation, their sticking together. Perhaps all this area is dominated by electrical effects, or at any rate one would feel very insecure discussing the agglomeration of grains without knowing what the electrical environment of the grains is. Equally we think that the bodies of the solar system formed by an initial agglomeration of grains. It is virtually certain now that the bodies of the solar system cannot be formed except from the gradual accumulation of small solid particles condensed out of a vapor into solid form. So then we have the question of large amounts of small grains milling around in the original solar system's space, and adhering together, to grow eventually to a size where gravitation is important to collect them further; and in that circumstance it is again quite clear that the electrical surface effects will be very important, if not dominant. We must understand how particles would condense into a variety of different kinds perhaps, due to their crystallographic growth patterns, and how as a result of that they would charge differently; and then how as a result of that, they would tend to stick together with different charges and maybe snowball in that way. This might become the discussion of the initial growth of the planets. The stiction of the grains has always been a great problem. It has been known for a long time that it will be necessary to invent a force that holds grains together, and I personally think that that is in the realm of electrostatics.

I must say that in the discussion of potentials around bodies and on surfaces of the Moon, etc. this conference seems to have served a useful purpose in bringing together the members of Rice University to talk with each other. I hope that they have made acquaintance with each other here, and when they go back, will speak to each other occasionally and discuss some of these items there. At any rate they certainly, together, made an impressive contribution to this meeting in all these areas.

I was personally very interested in the simulation of the Moon in the flow of plasma, that was reported this morning. Simulation experiments of this kind may not be a way of understanding what really goes on, because of the difficulty of reproducing all the non-dimensional quantities correctly, and it is quite clear that one cannot make a real simulation of the Moon in the solar wind plasma. But on the other hand what such experiments do is something much more than giving us merely the answer to a particular problem. They show us the diversity of nature in complex circumstances and remind us that we must not be too naive in offering any particular explanation. They show us that complex phenomena occur, even under very well controlled conditions, that are beyond he reach of simple-minded explanations, and that is a good education for anybody trying to explain natural phenomena. In other words we must not be too surprised if the Moon does a few tricks. Even under the laboratory circumstances, which are different in detail of course, we see a variety of effects we would not have predicted. So one loses some of one's naivete by looking at the laboratory simulations and that is an important part of one's education.

Then we heard the discussion of the 'bright' horizon phenomenon on the moon and I want to comment a little on that. I was at JPL at the time when this was discovered, and I was very excited about it. A number of false explanations were put forward at the time that put people's minds at rest, that there wasn't anything of significance that had taken place, that one had merely overlooked the diffraction to be expected around the edge of the sunlit part of the Moon and that was all there was to it. This glibness interrupted what otherwise would have been, I think, an intensive effort, while the Surveyor was still working, to find out more about it. Then under careful further thought it became quite clear that indeed it was a very strange phenomenon; there was no reasonable explanation in terms of mere diffraction. It is true that if you have the edge of a roof of a church, say, with the Sun having set just behind the church, then you will see a bright line on the tip of that edge, which is in practice not even the diffraction over the edge (although it itself would make a bright line) but is the dust or a small scale roughness on the top of that church roof edge. That indeed will make a bright line. But there are arguments against that being the explanation. One of those was presented, namely it seems that the line had a thickness that was resolved by the camera; although some investigators have claimed that the focus of the camera might be in doubt, and therefore that the apparent thickness of the line is no certain indication. But the other point is of course that the probability that the terminator, being the line to which the last rays of sunlight strike on the Moon, is coincident with the horizon for $2\frac{1}{2}$h – that probability is fantastically small. That is a probability that would be large for a church roof with a lot of empty space behind it, through which the Sun would sink down and still illuminate the curch roof edge, but it is a very small probability when one is dealing with a big body and all the time the terminator is moving at a certain rate back away from you. Then it is very unlikely that the movement of the terminator has been arrested just exactly at your horizon line for a long period of time, and that it should be almost a continuous line 3° to each side, and remain so arrested for the whole of that time is fantastically unlikely. Yet the same effect was seen 3 times. It is just a totally unreasonable story. So it is clear that the bright horizon does have a story to tell. I have actually suggested to the Russians running the Lunokhod that this was an interesting enough story to try to investigate with such means. The last version of the Lunokhod could not be so used, for mechanical reasons, but I am told that there is interest in this and perhaps a later Lunokhod will be equipped to look at the bright horizon line. This would be an observation which from a mobile vehicle would be immensely superior, because it means that any chance effects due to local topography would be gotten rid of when in the next lunar night the observation is from a different site. Then one gets a new observation each lunar night, and as the last Lunokhod lived for nearly a year one would expect to get 12 different bright horizon observations accumulated. So perhaps that will be an important observation.* The bright horizon is so important because, of course, if, for any reason whatever (be it the reasons that Dr Criswell mentioned or some others), any dust is levitated on the Moon, that is the circumstance in which it will be most directly observable, and a very small quantity that is lifted above the surface will show itself. One would see a far smaller quantity than could be seen by any other direct observation on the Moon; a quantity which is so small that one would need some years of exposure of a test surface before one would see it contaminated significantly. It is an extremely sensitive thing to look at the Sun below the horizon to see whether the atmosphere is clear, of course. The one test that we have from the Surveyor days was that the atmosphere is not clear, that it does scatter.

One other general remark I want to make. It seems to me to become quite clear that many points we discussed here now are points one could have discussed already years ago, points that did not depend on any new information. These are points where one is driven to think about them by the observations, but where the knowledge existed already before to make the same theoretical deductions. Why couldn't we have discussed the details of charging of different surfaces under different plasma and bombardment circumstances before; then we would have thought a lot more about it before the experiments went up. How nice it would have been if somebody had really thought about the electron sheath on the Moon in some detail and the charging

---

* It has now been reported that Lunokhod 2 has indeed performed such an observation. The result reported is a remarkably large amount of scattered light above the horizon, interpreted as indicating that dust grains are raised above the surface by an unknown process.

circumstances that might arise here or there, and then taken up the experiments on the Apollo missions. That would have given us a lot of answers that we now don't have. Now we do the conversation a little bit too late. Now we can easily think of a lot of experiments we would have loved to have done on the Moon to discover the different charging effects in different circumstances. It would have been, after all, quite easy to carry up experiments that would measure the electrical situation in the shadow of a rock and the electrical situation in the sunlight, and it would have been quite easy to take two probes and stick them into the ground and measure the electrical conductivity *in situ* on the Moon between two sites a few yards apart connected by cable, with far higher accuracy than can be done on the returned material. One could have easily done the experiment of having exposed surfaces and measured the current that can be transported horizontally as a result of an applied potential difference between two probes, a current due to the photoelectron layer on the lunar surface. Then from that one would have been able to get a very good figure of the photo-emissivity of the ground, because just a measurement of the conductivity and a little bit of theory will allow one to know how many photoelectrons there are. That would have all been a lot better than to do the observations on the returned material, and I feel quite sure that if there has been much more interest and many more scientists who focussed their attention on these problems before the missions, that one would have had such experiments, because they are quite simple. We see that we are doing the job too late. But let us at least be sure that the kind of uncertainties we now have in our minds still, be it about theory or about circumstances on the natural bodies, that those are cleared up by the next set of missions. Let us be sure to press for the correct type of observations to be made in future. We are not going to have any Apollo missions after the one that is going very soon, but no doubt there will be other lunar programs. Let us define the experiments that will be needed to clear up the many uncertainties we still have and press for those to become part of the next round of exploration. I personally think that these physical measurements in the vicinity of the Moon will throw far more light on the important questions about the formation of the lunar surface and the circumstances in the past in the solar system than very much of the geologists' discussions of the Moon, based far too much on similarities with the Earth that for the most part don't exist, and ignoring far too much the circumstances we have discussed here, that are totally different from any that can occur on the Earth.

We have had the very excellent results here from the ESTEC group on the determination of the lunar material for its photoelectric properties. They have come out with a figure a great deal lower than most people had guessed for the photo-emissivity. I have every reason to think that these measurements were correct, but on the other hand one must still have some doubts as to the surface properties of the returned lunar material, which has been subjected to various contaminations, different from what occurred on the Moon. It is not yet quite certain one can trust the laboratory measurements for this reason; still they are the best figures we have to date, and they are certainly very valuable, but in future I think we must measure photoemissivity directly on the Moon to see whether there is agreement.

These were the main topics that I recall as having been discussed here and I, for my own part, just wish that a large part of this kind of discussion could be taken back and replayed, as it were, at the next Lunar Science Conference, where this subject is otherwise treated with much less respect. It is refreshing for me to come to Europe where the discussion of the Moon is still much more free, where many more independent points of view can still be aired. I was very glad to see such a healthy outlook and a skeptical and searching attitude at this conference.

DISCUSSION

Knott: Have you tried the effect of bombardment with electrons of energies much higher than the first crossover? Under this condition I would expect the dust to be lifted from the surface.

Gold: We did not extend our experiment to such high energies.

Willis: Regarding your earlier talk (Gold and Williams, p. 557) you mentioned that experiments with lunar dust were rather intriguing in that the dust tended to agregate into large (~ 200 μm) clumps, which themselves move under electrostatic charging. Would you please describe this phenomenon in more detail and comment on its relevance, if any, in grain size segregation effects? Also, is the mechanism of compaction of the thick lunar dust surface purely electrostatic?

Gold: Lunar dust did show tendencies to clump, more so than the other powders. Large clumps of the order of 200 μm were formed and observed to move. We found this surprising since we thought that electrostatic charging and transport would tend to segregate out the smaller particles. We believe that the lunar dust transport involves mainly the small particles, the observed larger ones. in excess of 100 μm in the samples being due to meteorite interactions with the lunar surface. Core-section sampling of the lunar soil shows striations due to transport and compaction.

Criswell: Intergrain binding of surface grains should not be significant because the material turn-over rate in horizon glow regions is 10^2–10^3 times greater than micro meteorite turn-over rates.

Gold: Perhaps the HG regions are special.

Jones: It was mentioned that on average the erosion would take place downhill. Statistically however, one would expect cases of uphill movement caused by electrostatic effects. Is there any evidence of this?

Gold: Rocks have been observed which had lunar dust 'deposited' on them up to a certain distinct level. This could be evidence for uphill migration on a small scale. On a larger scale, at present it is difficult to know what evidence to look for.

Polychronopulos: In the experiment simulating electrostatic charging of the Moon dust, what would be the effect of using an electron beam with a large energy spread in it?

Gold: It would certainly reduce the amount of the observed effect, but not completely obliterate it.

Reasoner: The secondary electron instability can still occur in the presence of a distributed incident spectrum for the reason that the secondary electron yield function is not a strongly varying function of energy. Therefore, the instability at the crossover point is not 'washed out'.

Feuerbacher: The electrostatic movement process depends on the slope, being higher for large slopes. How do you explain the appearance of shoulders? The transport process should let the slopes fade out smoothly.

Gold: It is clear that an usual process must have been active on a large scale on the Moon which does not result in slopes fading out as is the case for all common erosion processes. The electrostatic process may explain this phenomenon, and we have certainly seen it on a small scale in the laboratory. What happens there is that a mixture of two different powders is much less mobile than either singly, and junction lines therefore tend to stay fixed until one slope becomes so steep that material falls down over the other type. If on the Moon the small impacts mix material over distances of a few meters then, if the mare and the mountain material are different, a steep slope will develop in the mixed region until it is so steep that transportation occurs despite the presence of the mixture. I do not claim that this must be the right explanation, but the appearance certainly suggests strongly that something impedes movement near the junction line.

INDEX OF AUTHORS

ASTROPHYSICS AND SPACE SCIENCE LIBRARY

Edited by

J. E. Blamont, R. L. F. Boyd, L. Goldberg, C. de Jager, Z. Kopal, G. H. Ludwig, R. Lüst,
B. M. McCormac, H. E. Newell, L. I. Sedov, Z. Švestka, and W. de Graaff

22. L. N. Mavridis (ed.), *Structure and Evolution of the Galaxy. Proceedings of the NATO Advanced Study Institute, held in Athens, September 8–19, 1969.* 1971, VII + 312 pp.

23. A. Muller (ed.), *The Magellanic Clouds. A European Southern Observatory Presentation: Principal Prospects, Current Observational and Theoretical Approaches, and Prospects for Future Research. Based on the Symposium on the Magellanic Clouds, held in Santiago de Chile, March 1969, on the Occasion of the Dedication of the European Southern Observatory.* 1971, XII + 189 pp.

24. B. M. McCormac (ed.), *The Radiating Atmosphere. Proceedings of a Symposium Organized by the Summer Advanced Study Institute, held at Queen's University, Kingston, Ontario, August 3–14, 1970.* 1971, XI + 455 pp.

25. G. Fiocco (ed.), *Mesospheric Models and Related Experiments. Proceedings of the 4th ESRIN-ESLAB Symposium, held at Frascati, Italy, July 6–10, 1970.* 1971, VIII + 298 pp.

26. I. Atanasijević, *Selected Exercises in Galactic Astronomy.* 1971, XII + 144 pp.

27. C. J. Macris (ed.), *Physics of the Solar Corona. Proceedings of NATO Advanced Study Institute on Physics of the Solar Corona, held at Cavouri-Vouliagmeni, Athens, Greece, 6–17 September 1970.* 1971, XII + 345 pp.

28. F. Delobeau, *The Enivronment of the Earth.* 1971, IX + 113 pp.

29. E. R. Dyer (general ed.), *Solar-Terrestrial Physics/1970. Proceedings of the International Symposium on Solar-Terrestrial Physics, held in Leningrad, U.S.S.R., 12–19 May 1970.* 1972, VIII + 938 pp.

30. V. Manno and J. Ring (eds.), *Infrared Detection Techniques for Space Research, Proceedings of the Fifth ESLAB-ESRIN Symposium, held in Noordwijk, The Netherlands, June 8–11, 1971.* 1972. XII + 344 pp.

31. M. Lecar (ed.), *Gravitational N-Body Problem, Proceedings of IAU Colloquium No. 10, held in Cambridge, England, August 12–15, 1970.* 1972, XI + 441 pp.

32. B. M. McCormac (ed.), *Earth's Magnetospheric Processes. Proceedings of a Symposium Organized by the Summer Advanced Study Institute and Ninth ESRO Summer School, held in Cortina, Italy August 30–September 10, 1971.* 1972, VIII + 417 pp.

33. Antonin Rükl, *Maps of Lunar Hemispheres.* 1972, V + 24 pp.

34. V. Kourganoff, *Introduction to the Physics of Stellar Interiors.* 1973, VII + 115 pp.

35. B. M. McCormac (ed.), *Physics and Chemistry of Upper Atmospheres, Proceedings of a Symposium Organized by the Summer Advanced Study Institute, held at the University of Orléans, France, July 31–August 11, 1972.* 1973, VIII + 389 pp.

36. J. D. Fernie (ed.), *Variable Stars in Globular Clusters and in Related Systems, Proceedings of the IAU Colloquium No. 21, held at the University of Toronto, Toronto, Canada, August 29–31, 1972.* 1973, IX + 234 pp.

SOLE DISTRIBUTORS FOR U.S.A. AND CANADA:

Vols. 2–6, and 8: Gordon & Breach Inc., 150 Fifth Ave., New York, N.Y. 10011
Vols. 7 and 9–28: Springer Verlag New York, Inc., 175 Fifth Ave., New York, N.Y. 10011